Environmental Science and Engineering

Environmental Engineering

Series Editors

Ulrich Förstner
Robert J. Murphy
W. H. Rulkens

For further volumes:
http://www.springer.com/series/3172

Jaime Klapp · Abraham Medina
Anne Cros · Carlos A. Vargas
Editors

Fluid Dynamics in Physics, Engineering and Environmental Applications

 Springer

Editors

Jaime Klapp
Instituto Nacional de Investigaciones
  Nucleares, La Marquesa, Ocoyoacac
Mexico
Mexico

Abraham Medina
ESIME Azcapotzalco, Instituto
  Politécnico Nacional
Mexico
Mexico

Anne Cros
Instituto de Astronomia y Meteorologia
Guadalajara
Mexico

Carlos A. Vargas
Departamento de Ciencias Básicas
Universidad Autónoma
  Metropolitana-Azcapotzalco
Mexico
Mexico

ISSN 1431-2492
ISBN 978-3-642-27722-1   ISBN 978-3-642-27723-8   (eBook)
DOI 10.1007/978-3-642-27723-8
Springer Heidelberg New York Dordrecht London

Library of Congress Control Number: 2012950374

Printed on acid-free paper

Springer is part of Springer Science+Business Media (www.springer.com)

Preface

This book presents recent experimental and theoretical advances in fluid dynamics in physics and engineering. It begins with invited lectures given during the International Enzo Levi Spring School 2011 held at *San Nicolas de Hidalgo Michoacán University* in Morelia, May 5–6, 2011, and invited seminars presented in the XVII National Congress of the Fluid Dynamics Division of the Mexican Physical Society, held in Guadalajara, Jalisco, Mexico, November 8–11, 2011.

The Spring School is organized every year in honour of Prof. Enzo Levi, a well-known Mexican scientist that dedicated his research to the study of fluids. He was one of the founders of the Instituto de Ingeniería (Engineering Institute) of the Universidad Nacional Autónoma de México (UNAM), and of the Instituto Mexicano de Tecnología del Agua (Mexican Institute for Water Technology) of the National Water Commission. He was the mentor of several generations of Mexican Engineers.

During the two day school, lectures were given by well-known national and international scientists. In 2011, many people attended the meeting with 50 researchers and more than a 100 graduate and undergraduate students. A wide variety of topics were presented by young national researchers that included geological structures, astrophysics, oceanography, SPH, multiphase flow, and complex fluids. Moreover, two American and one European very well-known researchers presented important aspects of fluid dynamics: instabilities that develop in rotating flows, presented by Patrice Le Gal, from Aix-Marseille University, France, a kind of solitary waves, presented by Patrick Weidman, from the University of Colorado, and spatiotemporal complexity by Michael Schatz, from Georgia Institute of Technology.

The Annual Fluid Dynamics Congress has a different format. In 2011 it lasted 4 days and was mainly constituted by short oral presentations of students and researchers. There is also a Gallery of Fluid Motion where posters exposing fluid motion photographs of special beauty are presented. The first three papers in the Gallery of Fluids correspond to the first three prices given during the Congress. Moreover, eight invited speakers presented works related to different themes belonging to fluid dynamics. As for international well-known invited speakers,

Herman J. H. Clercx, from Eindhoven University of Technology, presented a seminar about the formation of coherent structures in 3D laminar flows; Christophe Eloy, from Aix-Marseille University, France, spoke about the hydrodynamical constraints on the shape of fishes; José Rafael Castrejon Pita, from the University of Cambridge, England talked about drop formation and the consequent problems in the printing industry; Leonardo Trujillo, from the Physics Center, IVIC, Venezuela, showed how granular media can be described from a theoretical hydrodynamical point of view, and Jorge Arreola came from San Luis Potosí, Mexico, to speak about ions transport in biological membranes. Three local speakers presented the main research areas developed in Guadalajara University: astrophysics with Silvana Navarro, who showed how symbiotic stars can eject matter via jets, oceanography with Guillermo Gutiérrez, and rheology with Armando Soltero.

The short oral presentations are organized by themes: Geophysics, Porous Media, Microfluidics, Astrophysics, Multiphase Flow, Heat Transfer, Rheology, Magnetohydrodynamics, Hydraulics, Fluid-Structure Interaction, Granular Flow, and Viscous Flow.

The book is aimed to fourth-year undergraduate and graduate students, and to scientists in the field of physics, engineering, and chemistry that have interest in fluid dynamics from the experimental and theoretical point of view. The material includes recent advances in experimental and theoretical fluid dynamics and is adequate for both teaching and research. The invited lectures are introductory and avoid the use of complicated mathematics. The other selected contributions are also adequate to fourth-year undergraduate and graduate students.

The editors are very grateful to the Institutions who made possible the realization of the International Enzo Levi Spring School and the XVII National Congress of the Fluid Dynamics Division of the Mexican Physical Society, especially the Consejo Nacional de Ciencia y Tecnología (CONACYT), the Consejo Mexiquense de Ciencia y Tecnología (COMECYT), the *Sociedad Mexicana de Física, the San Nicolas de Hidalgo Michoacan University, the Mexican Secretaria de Relaciones Exteriores*, Guadalajara University, Metropolitan Autonomus University-Azcapotzalco, UNAM, IPN, Cinvestav-Abacus, DEFSA and Cambridge University, England.

We acknowledge the help of the Edition Committee: Salvador Galindo Uribarri, Estela Mayoral Villa, Elizabeth Teresita Romero Gúzman, Eduardo de la Cruz Sánchez, Roberto Zenit, Marcos José Solache Ríos, Edmundo del Valle Gallegos, Abel López Villa, René Vargas, Galileo Domínguez and in particular Cynthia Centeno Reyes and Roberto González Galan for their important and valuable contribution to the final manuscript.

Mexico City, Mexico, May 2012   Jaime Klapp
                                                                 Abraham Medina
                                                                     Anne Cros
                                                                Carlos A. Vargas

Acknowledgments

The production of this book has been sponsored by the Consejo Mexiquense de Ciencia y Tecnología (COMECYT), Consejo Nacional de Ciencia y Tecnología (CONACYT), Universidad de Guadalajara, Universidad Autónoma Metropolitana-Azcapotzalco, Instituto Nacional de Investigaciones Nucleares (ININ), Centro de Investigación y de Estudios Avanzados del Instituto Politécnico Nacional (CINVESTAV-Abacus), and Sociedad Mexicana de Física (SMF).

Contents

Part IV   Meteorology and Pollution

Contributors

F. Aragón SEPI ESIME Azcapotzalco, Instituto Politécnico Nacional, Av. de las Granjas 682, Col. Santa Catarina, 02250 Mexico, DF, Mexico

C. E. Alvarado-Rodríguez División de Ciencias Naturales y Exactas, Universidad de Guanajuato, Campus Guanajuato, Noria Alta S/N, Guanajuato, Mexico

V. S. Álvarez Salazar SEPI ESIME Azcapotzalco, Instituto Politécnico Nacional, Av. de las Granjas 682, Col. Santa Catarina, 02250 Mexico, DF, Mexico

Gabriel Ascanio Circuito Exterior, Cd. Universitaria, Instituto de Investigaciones en Materiales, C.P. 04510 Coyoacan, D.F., Mexico; Circuito Exterior S/N, Cd. Universitaria, Centro de Ciencias Aplicadas y Desarrollo Tecnológico, C.P. 04510 Coyoacan, Mexico

Héctor Barrios-Piña Explotación de Campos en Aguas Profundas, Instituto Mexicano del Petróleo, Eje Central Lázaro Cárdenas 154, Gustavo A. Madero, C.P. 07730 Mexico, D.F., Mexico, e-mail: hbarrios@imp.mx

Eduardo Basurto-Uribe Departamento de Ciencias Básicas, Universidad Autónoma Metropolitana-Azcapotzalco, Mexico City, Mexico, e-mail: ebasurto@correo.azc.uam.mx

J. Becerra Facultad de Ciencias, UNAM, Mexico, D.F., Mexico, e-mail: cmi@fciencias.unam.mx

O. Begovich Centro de Investigación y de Estudios Avanzados der Instituto Politécnico Nacional, Av. Del Bosque 1145, C.P. 45019 Guadalajara, Jalisco, Mexico, e-mail: epadilla@ gdl.cinvestav.mx

O. C. Benítez-Centeno Instituto Tecnológico de Zacatepec. Prolongación Av. Palmira esq. Apatzingan, Col. Palmira, 62490 Cuernavaca, Morelos, Mexico; Calzada del Tecnológico No. 27, 62780 Zacatepec, Morelos, Mexico, e-mail: omarcbc@gmail.com

J. C. Cajas Facultad de Ciencias, Universidad Nacional Autónoma de México, Mexico, D.F., Mexico, e-mail: jc.cajas@gmail.com

I. Carvajal Mariscal Av. Instituto Politécnico Nacional s/n, Unidad Profesional "Adolfo López Mateos", Col. Lindavista, Del. Gustavo A. Madero, C.P. 07738 Mexico, Mexico

Rafael Castrejón-García Centro de Investigación de Energía, UNAM, Priv. Xochicalco s/n, 62580 Temixco, Morelos, Mexico, e-mail: rcg@cie.unam.mx

José Rafael Castrejón-Pita Department of Engineering, University of Cambridge, CB3 0FS, 17 Charles Babbage Road, Cambridge, UK, e-mail: jrc64@cam.ac.uk

O. Cazarez-Candia Eje Central Lázaro Cárdenas, Instituto Mexicano del Petróleo, Norte 152 San Bartolo Atepehuacan, Gustavo A. Madero, C.P. 07730 Mexico, D.F., Mexico; Calzada del Tecnológico No. 27, Instituto Tecnológico de Zacatepec, 62780 Zacatepec, Morelos, Mexico, e-mail: ocazarez@imp.mx

C. Centeno-Reyes Eje Central Lázaro Cárdenas, Instituto Mexicano del Petróleo, Norte 152 San Bartolo Atepehuacan, Gustavo A. Madero, C.P. 07730 Mexico, D.F., Mexico, e-mail: ccenteno@imp.mx

A. Cervantes Daniel Instituto de Matemáticas, UNAM, Circuito Exterior CU, 04510 Mexico D.F., Mexico

Sergio Chiva-Vicent Departamento de Ingeniería Mecánica y Construcción, Universidad Jaime I. Castellón de la Plana, Valencia, Spain, e-mail: schiva@emc.uji.es

Satyan Chowdary Department of Chemical Engineering, Indian Institute of Science, 560 012 Bangalore, India

J. H. Clercx Herman Department of Applied Physics, Fluid Dynamics Laboratory and J.M. Burgers Center for Fluid Dynamics, Eindhoven, The Netherlands, e-mail: h.j.h.clercx@tue.nl

J. Corral Luis Universidad de Guadalajara, Instituto de Astronomía y Meteorología, Guadalajara, Mexico, e-mail: lcorral@astro.iam.udg.mx

A. Cros Departamento de Física, Universidad de Guadalajara, Av. Revolución 1500, Col. Olímpica, 44430 Guadalajara, Jalisco, Mexico, e-mail: anne@astro.iam.udg.mx

C. Cruz-Gómez Raúl Departamento de Física, Instituto de Astronomía y Meteorología, Universidad de Guadalajara, Guadalajara, Mexico , e-mail: rcruz@astro.iam.udg.mx

Juan Pablo Cruz-Pérez Instituto de Física y Matemáticas, Universidad Michoacana de San Nicolás de Hidalgo. Edificio C-3, Cd. Universitaria, 58040 Morelia, Michoacán, Mexico

Juan Carlos Degollado Departmento de Astrofísica Teórica, Instituto de Astronomía, Universidad Nacional Autónoma de México, Circuito Exterior Ciudad Universitaria, Apdo 70-264, CU, 04510 Mexico, D.F., Mexico; Departamento de Matemáticas y Departamento de Física, Centro Universitario de Ciencias Exactas e Ingeniería, Universidad de Guadalajara, Revolución 1500, Colonia Olímpica, 44430 Guadalajara, Jalisco, Mexico, e-mail: jcdegollado@ciencias.unam.mx

E. De La Cruz-Sánchez Instituto Nacional de Investigaciones Nucleares, Carretera México-Toluca s/n, La Marquesa, 52750 Ocoyoacac, Estado de México, Mexico, e-mail: eduardo.delacruz@inin.gob.mx

Zambrano Héctor Manuel De La Rosa Departamento de Física, Universidad de Guadalajara, Av. Revolución 1500, Col. Olímpica, 44430 Guadalajara, Jalisco, Mexico

G. Domínguez-Zacarías Programa de Investigación de Recuperación Hidrocarburos, Eje Central Lázaro Cárdenas Norte 152, C.P. 07730 Mexico, D. F., Mexico, e-mail: gdzacari@imp.mx

R. Duarte-Pérez Instituto Nacional de Investigaciones Nucleares, Carretera México-Toluca s/n, La Marquesa, 52750 Ocoyoacac, Estado de México, Mexico

C. Echeverría Colegio de Ciencias Humanidades, UNAM, Cataratas y Llanura S/N, Jardines del Pedregal, C.P. 04500 Coyoacán, D.F., Mexico

C. Echeverría-Arjonilla Facultad de Ciencias, Ciudad Universitaria 3000, Col. Copilco Universidad, C.P. 04360 Del. Coyoacán, D.F., Mexico

Tiburcio Fernández-Roque Escuela Superior de Ingeniería Mecánica y Eléctrica, Unidad Ticoman, Instituto Politécnico Nacional, 07340 Mexico, D.F., Mexico

Jorge Flores Dirección General Académica, Universidad Autónoma Chapingo, km 38.5 carretera México Texcoco, C.P. 56230 Chapingo, Estado de México, Mexico

David Flores-García Escuela Superior de Ingeniería Mecánica y Eléctrica, Unidad Ticoman, Instituto Politécnico Nacional, 07340 Mexico, D.F., Mexico, e-mail: jdbp09_david72@hotmail.com

J. Omar Flores-León Facultad de Ingeniería, Universidad Nacional Autónoma de México, C.P. 04510 Mexico, D.F., Mexico, e-mail: joomar35@hotmail.com

Marcelo Funes-Gallanzi Ardita Aeronautica SA de CV San Jose 3016, Guadalajara, Jalisco, Mexico, e-mail: mfg@ardita-aeronautica.com

S. Galindo-Uribarri Instituto Nacional de Investigaciones Nucleares, Carretera México-Toluca s/n, 52750 Ocoyoacac, Estado de México, Mexico, e-mail: salvador.galindo@inin.gob.mx

M. E. García-Guadalupe Astronomy and Meteorology Institute—University of Guadalajara, Av. Vallarta 2802. Col. Arcos Vallarta, C.P. 44130 Guadalajara, Jalisco, Mexico

Z. I. González-Acevedo Instituto Nacional de Investigaciones Nucleares, Carretera México-Toluca s/n, 52750 Ocoyoacac, Estado de México, Mexico

Pedro González-Casanova Instituto de Matemáticas, UNAM, Circuito Exterior CU, 04510 Mexico D.F., Mexico, e-mail: casanovapg@gmail.com

Arturo González-Casillas Instituto Mexicano de Tecnología del Agua, IMTA, Paseo Cuauhnáhuac 8532 Colonia Progreso, C.P. 62550 Jiutepec, Morelos, Mexico, e-mail: rmercado@tlaloc.imta.mx

José Antonio González-Cervera Instituto de Física y Matemáticas, Universidad Michoacana de San Nicolás de Hidalgo. Edificio C-3, Cd. Universitaria, 58040 Morelia, Michoacán, Mexico, e-mail: gonzalez@ifm.umich.mx

R. González-Galán Facultad de Ciencias, Universidad Autónoma del Estado de México, El cerrillo Piedras Blancas, 50200 Toluca, Estado de México, Mexico, e-mail: rgonzalez470@yahoo.com.mx

R. González-López Instituto Mexicano del Petróleo, Explotación de Campos en Aguas Profundas, Eje Central Lázaro Cárdenas 154, Gustavo A. Madero, C.P. 07730 Mexico, D.F, Mexico

Christian Gout INSA Rouen, LMI, Av. de l'Université, BP 08, 76801 Saint Etienne du Rouvray Cedex, France, e-mail: rafael.resendiz@gmail.com

P. Guido-Aldana Instituto Mexicano de Tecnología del Agua, IMTA, Paseo Cuauhnáhuac 8532 Colonia Progreso, C.P. 62550 Jiutepec, Morelos, Mexico

Fidel Gutiérrez AVNTK, S.C. Av. Chapalita 1143, Guadalajara, Jalisco, Mexico, e-mail: fidel@avntk.com

G. Juliana Gutiérrez-Paredes SEPI ESIME Azcapotzalco, Instituto Politécnico Nacional, Av. de las Granjas 682, Col. Santa Catarina, 02250 Mexico, DF, Mexico

Sergio Valente Gutiérrez-Quijada Facultad de Ciencias, Ciudad Universitaria 3000, Col. Copilco Universidad, C.P. 04360 Del. Coyoacán, D.F., Mexico

Catalina Haro-Pérez Departamento de Ciencias Básicas, Universidad Autónoma Metropolitana-Azcapotzalco, Mexico City, Mexico, e-mail: cehp@correo.azc.uam.mx

Andrzej Herczynski Department of Physics, Boston College, Chestnut Hill, MA 02467-3811, USA

Jorge Hernandez-Tamayo Escuela Superior de Ingeniería Mecánica y Eléctrica, Unidad Ticoman, Instituto Politécnico Nacional, 07340 Mexico, D.F., Mexico

R. Herrera-Solís Sede, Exploración y Producción, Eje Central Lázaro Cárdenas Norte 152, C.P. 07730 Mexico, D.F., Mexico

F. J. Higuera División de Ciencias Naturales y Exactas, Universidad de Guanajuato, Campus Guanajuato, Noria Alta S/N, Guanajuato, Mexico

Ian Hutchings Department of Engineering, University of Cambridge, CB3 0FS, 17 Charles Babbage Road, Cambridge, UK, e-mail: imh2@cam.ac.uk

Mauro Íñiguez-Covarrubias Instituto Mexicano de Tecnología del Agua, IMTA, Paseo Cuauhnáhuac 8532 Colonia Progreso, C.P. 62550 Jiutepec, Morelos, Mexico, e-mail: rmercado@tlaloc.imta.mx

R. Islas-Juárez Sede, Exploración y Producción, Eje Central Lázaro Cárdenas Norte 152, C.P. 07730 Mexico, D.F., Mexico

L. Héctor Juárez Departamento de Matemáticas, UAM-I, Av. San Rafael Atlixco 186, Col. Vicentina, 09340 Mexico, D.F., Mexico

Guadalupe Juliana-Gutiérrez SEPI ESIME Azcapotzalco, Instituto Politécnico Nacional, Av. de las Granjas 682, Col. Santa Catarina, 02250 Mexico, DF, Mexico

K. Kesava-Rao Department of Chemical Engineering, Indian Institute of Science, 560 012 Bangalore, India

J. Klapp Instituto Nacional de Investigaciones Nucleares, ININ, Carretera México-Toluca Km 36.5, 52750 La Marquesa, Estado de México, Mexico; Departamento de Matemáticas, Cinvestav del I.P.N., 07360 Mexico, D.F., Mexico, e-mail: jaime.klapp@inin.gob.mx

Anoop Kumar Department of Chemical Engineering, Indian Institute of Science, 560 012 Bangalore, India

Patrice Le Gal Institut de Recherche sur les Phénomènes Hors Équilibre, UMR 6594 CNRS, Universités d'Aix-Marseille I & II, Technopôle Château-Gombert, 49 rue F. Joliot-Curie, BP 146, 13384 Marseille Cedex 13, France, e-mail: legal@irphe.univ-mrs.fr

Marcos Ley Koo Facultad de Ciencias, Ciudad Universitaria 3000, Col. Copilco Universidad, C.P. 04360 Del. Coyoacán, D.F., Mexico

Irineo López Dirección General Académica, Universidad Autónoma Chapingo, km 38.5 carretera México Texcoco, C.P. 56230 Chapingo, Estado de México, Mexico

A. López-Villa SEPI ESIME Azcapotzalco, Instituto Politécnico Nacional, Av. de las Granjas 682, Col. Santa Catarina, 02250 Mexico, DF, Mexico, e-mail: abelvilla77@hotmail.com

C. Málaga Facultad de Ciencias, UNAM, Mexico, D.F., Mexico, e-mail: cmi@fciencias.unam.mx

Ernesto Mancilla Circuito Exterior, Cd. Universitaria, Instituto de Investigaciones en Materiales, C.P. 04510 Coyoacan, D.F., Mexico; Circuito Exterior S/N, Cd. Universitaria, Centro de Ciencias Aplicadas y Desarrollo Tecnológico, C.P. 04510 Coyoacan, Mexico

F. Mandujano Facultad de Ciencias, UNAM, Mexico, D.F., Mexico, e-mail: frmas@ciencias.unam.mx

José Manuel Martínez-Magadán Instituto de Ciencias Nucleares, Universidad Nacional Autónoma de México (UNAM), Apartado Postal 70-543, 04510 Mexico, D.F., Mexico

E. Mayoral-Villa Instituto Nacional de Investigaciones Nucleares, Carretera México-Toluca s/n, La Marquesa, 52750 Ocoyoacac, Estado de México, Mexico, e-mail: estela.mayoral@inin.gob.mx

Abraham Medina-Ovando SEPI ESIME Azcapotzalco, Instituto Politécnico Nacional, Av. de las Granjas 682, Col. Santa Catarina, 02250 Mexico, DF, Mexico, e-mail: abraham_medina_ovando@yahoo.com

A. Méndez-Ancona Exploración y Producción Región Sur, Instituto Mexicano del Petróleo, Periférico Carlos Pellicer Cámara, No. 1502, C.P. 86209 Villahermosa-Tabasco, Mexico

Santos Mendez-Diaz Universidad Autónoma de Nuevo León, Av. Pedro de Alba s/n, C.P. 66451 San Nicolás de los Garza, NL, Mexico, e-mail: santos.mendezdz@uanl.edu.mx

P. Mendoza-Maya Centro Nacional de Investigación y Desarrollo Tecnológico, Prolongación Av. Palmira esq. Apatzingan, Col Palmira, 62490 Cuernavaca, Morelos, Mexico, e-mail: pedro_m_maya@yahoo.com.mx

J. R. Mercado-Escalante Instituto Mexicano de Tecnología del Agua, IMTA, Paseo Cuauhnáhuac 8532 Colonia Progreso, C.P. 62550 Jiutepec, Morelos, Mexico, e-mail: rmercado@tlaloc.imta.mx

Arturo Morales-Fuentes Universidad Autónoma de Nuevo León, Av. Pedro de Alba s/n, C.P. 66451 San Nicolás de los Garza, NL, Mexico, e-mail: arturo.moralesfs@uanl.edu.mx

Claudia Moreno Departmento de Astrofísica Teórica, Instituto de Astronomía, Universidad Nacional Autónoma de México, Circuito Exterior Ciudad Universitaria, Apdo 70-264, CU, 04510 Mexico, D.F., Mexico; Departamento de Matemáticas y Departamento de Física, Centro Universitario de Ciencias Exactas e Ingeniería, Universidad de Guadalajara, Revolución 1500, Colonia Olímpica, 44430 Guadalajara, Jalisco, Mexico, e-mail: claudia.moreno@cucei.udg.mx

S. L. Moya-Acosta Centro Nacional de Investigación y Desarrollo Tecnológico, Prolongación Av. Palmira esq. Apatzingan, Col Palmira, 62490 Cuernavaca, Morelos, Mexico

José Luis Muñoz-Cobo Departamento de Ingeniería Química y Nuclear, Universidad Politécnica de Valencia, Camino de Vera s/n, C.P. 46022 Valencia, Spain, e-mail: jlcobos @ iqn.upv.es

Eduardo Nahmad-Achar Instituto Mexicano del Petróleo (IMP), Eje Central Lázaro Cárdenas S/N, Mexico D.F., Mexico

Y. Nahmad-Molinari Instituto de Física, Universidad Autónoma San Luis Potosí, Av. Manuel Nava 6, Zona Universitaria San Luis Potosí, 78210 San Luis Potosí, Mexico

G. Navarro-Silvana Universidad de Guadalajara, Instituto de Astronomía y Meteorología, Guadalajara, Mexico, e-mail: silvana@astro.iam.udg.mx

Rubén Nicolás-López Instituto Mexicano del Petróleo, Eje Central Lázaro Cárdenas No. 152, Col. San Bartolo Atepehuacan, C.P. 07730 Mexico D.F., Mexico, e-mail: rnlopez@imp.mx

Gualberto Ojeda-Mendoza Departamento de Física, CINVESTAV-IPN, Av. Instituto Politécnico Nacional 2508, 07360 Mexico, D.F, Mexico, e-mail: gojeda@fis.cinvestav.mx

Alejandro Ortega Instituto de Ciencias Nucleares, Universidad Nacional Autónoma de México (UNAM), Apartado Postal 70-543, 04510 Mexico, D.F., Mexico

Miguel Ortega SEPI ESIME Azcapotzalco, Instituto Politécnico Nacional, Av. de las Granjas 682, Col. Santa Catarina, 02250 Mexico, DF, Mexico

M. Ortega-Rocha Exploración y Producción Región Sur, Instituto Mexicano del Petróleo, Periférico Carlos Pellicer Cámara, No. 1502, C.P. 86209 Villahermosa-Tabasco, Mexico

José A. Ortiz-Martínez Depto. de Metal Mecánica y Mecatrónica, Instituto Tecnológico de Veracruz, Calzada Miguel A. de Quevedo 2779, Col. Formando Hogar, 91860 Veracruz, Mexico

Guillermo E. Ovando-Chacon Depto. de Metal Mecánica y Mecatrónica, Instituto Tecnológico de Veracruz, Calzada Miguel A. de Quevedo 2779, Col. Formando Hogar, 91860 Veracruz, Mexico, e-mail: geoc@itver.edu.mx

Sandy L. Ovando-Chacon Depto. de Química y Bioquímica, Instituto Tecnológico de Tuxtla Gutiérrez, Carretera Panamericana Km. 1080, 29000 Tuxtla Gutierrez, Chiapas, Mexico, e-mail: ovansandy@hotmail.com

E. A. Padilla Centro de Investigación y de Estudios Avanzados der Instituto Politécnico Nacional, Av. Del Bosque 1145, C.P. 45019 Guadalajara, Jalisco, Mexico, e-mail: epadilla@gdl.cinvestav.mx

D. Pastrana Facultad de Ciencias, Universidad Nacional Autónoma de México, Mexico, D.F., Mexico, e-mail: dpastrana@ciencias.unam.mx

F. Peña-Polo Centro de Física, Instituto de Venezolano de Investigaciones Científicas (IVIC), Apartado Postal 20632, Caracas 1020, Venezuela, e-mail: franklin.pena@gmail.com

S. Peralta Politécnico Nacional Escuela Superior de Ingeniería Mecánica y Eléctrica, Sección de Estudios de Posgrado e Investigación, Av. de las granjas, Col. Santa Catarina, Delegación Azcapotzalco, Mexico

G. Pérez-Ángel CINVESTAV-Mérida, Ap. Postal 73, Cordemex, Mérida, Yucatán, Mexico

S. Pérez-Morales Exploración y Producción Región Sur, Instituto Mexicano del Petróleo, Periférico Carlos Pellicer Cámara, No. 1502, C.P. 86209 Villahermosa-Tabasco, Mexico

Nora Pérez-Quezadas Instituto de Geofísica, Universidad Nacional Autónoma de México, 04510 Mexico, D.F., Mexico; Instituto Nacional de Investigaciones Nucleares, La Marquesa, 52750 Ocoyoacac, Estado de México, Mexico

A. Pérez Terrazo SEPI ESIME Azcapotzalco, Instituto Politécnico Nacional, Av. de las Granjas 682, Col. Santa Catarina, 02250 Mexico, DF, Mexico

A. Pizano-Moreno Centro de Investigación y de Estudios Avanzados der Instituto Politécnico Nacional, Av. Del Bosque 1145, C.P. 45019 Guadalajara, Jalisco, Mexico, e-mail: epadilla@gdl.cinvestav.mx

D. Porta-Zepeda Facultad de Ciencias, Ciudad Universitaria 3000, Col. Copilco Universidad, C.P. 04360 Del. Coyoacán, D.F., Mexico

Juan C. Prince-Avelino Depto. de Metal Mecánica y Mecatrónica, Instituto Tecnológico de Veracruz, Calzada Miguel A. de Quevedo 2779, Col. Formando Hogar, 91860 Veracruz, Mexico, e-mail: jcpa@itv.edu.mx

H. Ramírez-León Explotación de Campos en Aguas Profundas, Instituto Mexicano del Petróleo, Eje Central Lázaro Cárdenas 154, Gustavo A. Madero, C.P. 07730 Mexico, D.F., Mexico, e-mail: hrleon@imp.mx

H. U. Ramírez-Sánchez Astronomy and Meteorology Institute—University of Guadalajara, Av. Vallarta 2802. Col. Arcos Vallarta, C.P. 44130 Guadalajara, Jalisco, Mexico, e-mail: ramirez@astro.iam.udg.mx

José Alfredo Ramos-Leal Instituto Potosino de Investigación Científica y Tecnológica A.C., IPICyT, Camino a la Presa San José 2055, C.P. 78216 San Luis Potosí, Mexico, e-mail: jalfredo@ipicyt.edu.mx

D. Danny Reible Department of Civil, Architectural, Environmental Engineering, Bettie Margaret Smith Chair of Environmental Health Engineering, The

University of Texas at Austin, 1 University Station C1786, Austin, TX 78712-0273, USA, e-mail: reible@mail.utexas.edu

Rafael Reséndiz Departamento de Matemáticas, UAM-I, Av. San Rafael Atlixco 186, Col. Vicentina, 09340 Mexico, D.F., Mexico, e-mail: rafael.resendiz@gmail.com

Claude Rey Laboratoire M2P2, UMR 6181 CNRS—Universités d'Aix-Marseille, Technopôle Château-Gombert, 38 rue F. Joliot-Curie, 13451 Marseille, France

L. R. Reyes-Gutiérrez Instituto Nacional de Investigaciones Nucleares, ININ, Carretera México-Toluca Km 36.5, 52750 La Marquesa, Estado de México, Mexico, e-mail: raymundo.reyes@inin.gob.mx

J. F. Reyes-Tendilla Instituto de Física, Universidad Autónoma San Luis Potosí, Av. Manuel Nava 6, Zona Universitaria San Luis Potosí, 78210 San Luis Potosí, Mexico

R. Rodríguez-Castillo Instituto de Geofísica, Universidad Nacional Autónoma de México, 04510 Mexico, D.F., Mexico, e-mail: ramiro@geofisica.unam.mx

C. Rodríguez-Cuevas Facultad de Ingeniería, Universidad Autónoma de San Luis Potosí, Dr. Manuel Nava # 8, Zona Universitaria Poniente, C.P. 78290 San Luis Potosí S.L.P., Mexico

G. M. Rodríguez-Liñán Instituto de Física, Universidad Autónoma San Luis Potosí, Av. Manuel Nava 6, Zona Universitaria San Luis Potosí, 78210 San Luis Potosí, Mexico

Abraham Rojano Dirección General Académica, Universidad Autónoma Chapingo, km 38.5 carretera México Texcoco, C.P. 56230 Chapingo, Estado de México, Mexico, e-mail: abrojano@hotmail.com

Luis F. Rojas-Ochoa Departamento de Física, CINVESTAV-IPN, Av. Instituto Politécnico Nacional 2508, 07360 Mexico, D.F, Mexico, e-mail: lrojas@fis.cinvestav.mx

E. T. Romero-Guzmán Instituto Nacional de Investigaciones Nucleares, ININ, Carretera México-Toluca S/N, C.P. 52045 La Marquesa, Mexico

Oscar Rubio Mechatronics Student, ITESM Campus Guadalajara, Guadalajara, Mexico, e-mail: O.Rubio.Rivera@gmail.com

Raquel Salazar Dirección General Académica, Universidad Autónoma Chapingo, km 38.5 carretera México Texcoco, C.P. 56230 Chapingo, Estado de México, Mexico

Martha Yadira Salazar-Romero Facultad de Ciencias, Ciudad Universitaria 3000, Col. Copilco Universidad, C.P. 04360 Del. Coyoacán, D.F., Mexico

E. Salinas-Rodríguez Universidad Autónoma Metropolitana-Iztapalapa, San Rafael Atlixco 186, Col. Vicentina, D. F. 09340, Mexico, e-mail: asor@xanum. uam.mx

J. Sánchez-Sesma Instituto Mexicano de Tecnología del Agua, IMTA, Paseo Cuauhnáhuac 8532 Colonia Progreso, C.P. 62550 Jiutepec, Morelos, Mexico

F. Sánchez-Silva Av. Instituto Politécnico Nacional s/n, Unidad Profesional "Adolfo López Mateos", Col. Lindavista, Del. Gustavo A. Madero, C.P. 07738 Mexico, Mexico

Uwe Schmidt Division Biosystems Engineering, Humboldt Universität zu Berlin, Albrecht-Thaer-Weg 14195, Berlin

L. Di G. Sigalotti Centro de Física, Instituto de Venezolano de Investigaciones Científicas (IVIC), Apartado Postal 20632, Caracas 1020, Venezuela, e-mail: leonardo.sigalotti@gmail.com

C. Stern-Forgach Facultad de Ciencias, Universidad Nacional Autónoma de México, Ciudad Universitaria, C. Exterior s/n, Coyoacán, 04510 Mexico, Mexico, e-mail: catalina@ciencias.unam.mx

A. Soria Departamento de IPH, Universidad Autónoma Metropolitana-Iztapalapa, Av. San Rafael Atlixco No. 186, Col. Vicentina, 09430 Mexico, D.F., Mexico, e-mail: asor@xanum.uam.mx

Enrique Soto Circuito Exterior, Cd. Universitaria, Instituto de Investigaciones en Materiales, C.P. 04510 Coyoacan, D.F., Mexico

Ismael Soto Instituto de Ciencias Nucleares, Universidad Nacional Autónoma de México (UNAM), Apartado Postal 70-543, 04510 Mexico, D.F., Mexico

F. M. Speetjens-Michel Department of Mechanical Engineering, Energy Technology and J.M. Burgers Center for Fluid Dynamics, Eindhoven, The Netherlands, e-mail: h.j.h.clercx@tue.nl

A. Torres SEPI ESIME Azcapotzalco, Instituto Politécnico Nacional, Av. de las Granjas 682, Col. Santa Catarina, 02250 Mexico, DF, Mexico

C. Treviño Facultad de Ciencias y UMDI Sisal, Universidad Nacional Autónoma de México, Yucatán, Mexico, e-mail: ctrev@gmail.com

L. Trujillo Centro de Física, Instituto de Venezolano de Investigaciones Científicas (IVIC), Apartado Postal 20632, Caracas 1020, Venezuela; The Abdus Salam, International Centre for Theoretical Physics, ICTP, Trieste, Italy, e-mail: leonardo.trujillo@gmail.com

Carlos A. Vargas Departamento de Ciencias Básicas, Universidad Autónoma Metropolitana-Azcapotzalco, Av. San Pablo 180, Azcapotzalco, 02200 Mexico, DF, Mexico, e-mail: cvargas@correo.azc.uam.mx

Eslí Vázquez-Nava Depto. de Ingeniería Química, Universidad Veracruzana, Adolfo Ruiz Cortines s/n, Fracc. Costa Verde, Boca del Rio, 94294 Veracruz, Mexico, e-mail: evazquezn@gmail.com

Enrique Vázquez-Semadeni Centro de Radioastronomía y Astrofísica, UNAM Campus Morelia, 58089 Michoacán, Mexico, e-mail: e.vazquez@crya.unam.mx

Oscar Velasco-Fuentes Departamento de Oceanografía Física, CICESE, Ensenada, Mexico, e-mail: ovelasco@cicese.mx

Stéphane Viazzo Laboratoire M2P2, UMR 6181 CNRS—Universités d'Aix-Marseille, Technopôle Château-Gombert, 38 rue F. Joliot-Curie, 13451 Marseille, France

Patrick D. Weidman Department of Mechanical Engineering, University of Colorado Boulder, Boulder, CO 80309-0427, USA

Fei Yan Department of Civil, Architectural, Environmental Engineering, The University of Texas at Austin, 1 University Station C1786, Austin, TX 78712-0273, USA, e-mail: feiyan@austin.utexas.edu

Roberto Zenit Instituto de Investigaciones en Materiales, Universidad Nacional Autónoma de México, Ciudad Universitaria, C. Exterior s/n, México, 04510 Mexico, D.F., Mexico, e-mail: zenit@unam.mx

Jan Carlos Zuñiga Department of Mathematics, Universidad de Guadalajara, Av. Revolución 150, Guadalajara, Jalisco, Mexico, e-mail: 3juan.zuniga@red.cucei.udg.mx

Part I
Invited Lectures

Applied Fluid Mechanics in the Environment, Technology and Health

J. Klapp, L. Di G. Sigalotti, L. Trujillo and C. Stern

Abstract The objective of this chapter is to review the importance of fluid dynamics research and its impact on science and technology. Here we consider four particular areas of study, namely environmental fluid mechanics, turbulence, nano- and microfluids, and biofluid dynamics, with deeper emphasis on environmental flows. Each of these topics is illustrative of how improved scientific knowledge of fluid dynamics can have a major impact on important national needs and worldwide economies, as well as help developed nations to maintain their leadership in the production of novel technologies.

J. Klapp (✉)
Instituto Nacional de Investigaciones Nucleares, ININ, Km. 36.5,
Carretera México-Toluca, 52750 La Marquesa, Estado de México, Mexico
e-mail: jaime.klapp@hotmail.com

J. Klapp
Departamento de Matemáticas, Cinvestav del I.P.N., 07360 México, D.F., Mexico

L. D. G. Sigalotti · L. Trujillo
Centro de Física, Instituto Venezolano de Investigaciones Científicas, IVIC,
Apartado Postal 20632, 1020 Caracas, Venezuela
e-mail: leonardo.sigalotti@gmail.com

L. Trujillo
e-mail: leonardo.trujillo@gmail.com

L. Trujillo
The Abdus Salam, International Centre for Theoretical Physics, ICTP, Trieste, Italy

C. Stern
Facultad de Ciencias, Universidad Nacional Autónoma de México, Ciudad Universitaria,
C. Exterior s/n, 04510 Coyoacán, D.F., Mexico
e-mail: catalina@ciencias.unam.mx

J. Klapp et al. (eds.), *Fluid Dynamics in Physics, Engineering and Environmental Applications*, Environmental Science and Engineering,
DOI: 10.1007/978-3-642-27723-8_1, © Springer-Verlag Berlin Heidelberg 2013

1 Introduction

The science of fluid mechanics describes the motion of liquids and gases and their interactions with solid boundaries. It is a broad, interdisciplinary field that touches almost every aspect of our daily lives, and it is central to much of science and engineering. It is one of the most challenging and exciting fields of scientific activity mainly because of the complexity of the subject and the breadth of the applications. Fluid dynamics impacts astrophysics, biology, chemistry, medicine, electronics, defense, homeland security, transportation, manufacturing, energy, and the environment. The quest for deeper understanding has inspired numerous advances in applied mathematics, computational physics, and experimental techniques.

A central problem is that the governing differential equations, the Navier-Stokes equations, have no general analytical solution, and so much effort has been concentrated in developing mathematical models from which computer-aided numerical solutions can be obtained for specific applications. The dawn of the twentieth century marked the beginning of the numerical solutions of differential equations in mathematical physics and engineering. The development of modern *computational fluid dynamics* (CFD) began with the advent of the digital computer in the early 1950s. Historically, finite difference and finite element methods for solving numerically the Navier-Stokes equations dominated the CFD community. Simplicity in the formulations and computations contributed to this trend. However, in recent years meshfree particle methods have started to become increasingly popular. One of the methods that has been developed extensively in fluid dynamics, heat transfer and solid mechanics, as a promising alternative to grid-based schemes, is the method of Smoothed Particle Hydrodynamics (SPH) (Lucy 1977; Gingold and Monaghan 1977) because of its intrinsic simplicity to code, to include new physics, and to handle irregular and deformable boundaries. The ultimate goal of CFD practitioners is to be aware of the advantages and disadvantages of all available methods so that, if and when, supercomputers grow manyfold in speed and memory storage, this knowledge will be an asset in determining the computational scheme capable of rendering the most accurate results.

The advent of parallel supercomputers for numerical simulations and analysis and of newly available diagnostic methods for experiments, have made fruitful research in fluid dynamics to rely mostly on computational and experimental work. Computational research in astrophysics and gravitational physics, ranging from planetesimal formation (Benz 2000; Boss and Durisen 2005) to star (Bodenheimer et al. 1980; Boss 1981, 1991; Bonnell and Bate 1994; Bate 1998; Sigalotti and Klapp 2001; Arreaga-García et al. 2007; Klessen et al. 2011) and galaxy formation (Steinmetz 1996; Berczik and Kolesnik 1998; Springel et al. 2005) to large-scale cosmological flows (Centrella and Melott 1983; Abel et al. 1998; Bryan 1999; Springel et al. 2001), has strongly contributed to shape CFD as a new discipline. Cutting-edge research in astrophysics today depends heavily on the use of numerical simulations to act as a bridge between the limited data gathered from ground- and space-based observatories and the predictions of simplified analytical models. The studies may range from the exploration of simple models on desktop computers to large-scale numerical simulations requiring the petaflop supercomputers available at high-performance computing centers. Over the past 20 years CFD has extended to a broad spectrum of fluid-dynamical systems arising in the real world. For instance, applications to environmental flows, micro- and nano-fluidics, biological and biomedical flows, turbulence, aerodynamics, magnetohydrodynamics, hydraulics, fluidized granular media, multiphase and compositional flows, non-Newtonian fluids of engineering significance, and flow control are some important examples.

The solution of most complex fluid mechanics problems involves some sort of experimental investigation. CFD is being used for many practical problems today, but usually in combination with an experimental verification to test the quality of the computational results. An important achievement of modern experimental fluid mechanics lies on the invention of sophisticated devices and development of techniques for the measurement and analysis of whole, instantaneous fields of scalars and vectors. These techniques have allowed experimentalists to analyze complicated flow patterns in the laboratory, providing quantitative information on the interacting processes involved in such flows. In experimental research, it is the local flow velocity that is of interest rather than the total flow rate as in most common industrial flow measurement requirements. *Velocimetry* methods fall into two broad categories: those which measure the flow velocity at a single point, and those which offer velocity data over a plane or within a volume. In the former category, the humblest of devices for measuring flow velocity directly is the Pitot-static tube (Pitot 1732; Darcy 1858). Modern methods include the use of hot-wire/hot-film anemometry in liquid and gas domains by introducing a probe into the flow (Lekakis 1996) and non-invasively using laser Doppler velocimetry (LDV) (Yea and Cummings 1964). A detailed account of the advantages and limitations of both techniques can be found in Refs. (Fingerson and Freymuth 1983; Bruun 1995). Since its inception in 1966, LDV has progressed, and aided by developments in other technologies, notably in lasers, fibre-optics, and computers, it has become a most useful tool under a wide range of conditions. Velocity field measurements in a plane, or even in a volume, can be obtained today using the *particle image velocimetry* (PIV) (Westerweel 1993; Raffel et al. 2007; Adrian and Westerweel 2011). It is an optical method that uses tracer particles for flow visualization. When the particle concentration is low, it is possible to follow individual particles in three-dimensional space using the *particle tracking velocimetry* technique, while *laser speckle velocimetry* is employed in cases where the particle concentration is high enough that it becomes difficult to observe individual particles in an image. PIV has also been used to measure the velocity field of the free surface and basal boundary in granular flows such as those in shaken containers, tumblers, and avalanches (Lueptow et al. 2000; Jain et al. 2002; Pudasaini et al. 2005). The setup for granular PIV is also well-suited for other non-transparent media such as sand, gravel, quartz, and other materials that are common in geophysics. State-of-the-art techniques for measurements of scalar fields include *tomographic interferometry* and *planar laser-induced fluorescence* (Seitzman and Hanson 1993). In addition, accurate droplet and particle sizes in multiphase flows can be measured using the *phase Doppler particle analysis* (PDPA), which is today recognized by the spray industry as the current "gold standard" in droplet size measurement.

Improved scientific knowledge of fluid dynamics can significantly enhance the future of nations, and its application to real world problems is expected to have major impact on important national needs. These include improvements in transportation and energy efficiency, prediction and mitigation of environmental problems, development of novel technologies based on microfluidics, improvements to security and defense, and major contributions to health. For instance, predicting the flow of blood in the human body, the behavior of microfluidic devices, the aerodynamic performance of airplanes, cars, and ships, the cooling of electronic components, or the hazards of weather and climate, are all aspects of modern life that require a detailed understanding of fluid dynamics, and therefore substantial research. The outcomes of this applied research may lead, for example, to improved predictions of hurricane landfall and strength by investigating the mechanisms that govern their formation, growth, and interaction with the global weather system; to more efficient vehicles by reducing the friction between the vehicle surface and the surrounding air and by understanding how to control vortex formation; and to a new generation of micro-scale devices that will include combustors to replace batteries, advanced flow control devices to cool electronic systems, and labs-on-a-chip to manipulate and interrogate DNA. Already, the number of channels in microfluidic devices is growing at a rate faster than the exponential growth in electronic data storage density. Moreover, arsenic contamination in natural water is a worldwide problem with particular relevance to the developing world (Choong et al. 2007). Many other dangerous heavy metals, like cadmiun and nickel, which exist as ions in solution or small nanometer-size particles, are also found in groundwater, posing similar threats to human health. In this case, any remediation or separation strategies must allow for a continuous flow, as occurs in standard filters or water-cleaning processes, and must recognize the unique chemical and transport dynamics relevant to these nanoscale particulate and ion-surface interactions.

Intensive research in fluid dynamics has often been motivated (and sponsored) by industrial applications of home and personal care, oil-field services, fiber coating, float-glass manufacturing, screen-printing for solar panels, and medical/ clinical applications among many others. Despite of their importance, some of these applications may have lacked any detailed mathematical analysis, with current approaches relying heavily on experimental observations and empirical hypotheses. The difficulty lies in that a number of these applications involve non-Newtonian fluids. During the past 50–60 years, there has been a growing recognition of the fact that many substances of industrial significance, especially of multiphase nature (foams, emulsions, dispersions and suspensions, slurries, for instance) and polymeric melts and solutions (both natural and man made) do not conform to the Newtonian postulate of the linear relationship between the shear stress σ and the rate of strain $\dot{\gamma}$. Accordingly, these fluids are variously known as non-Newtonian, non-linear, complex, or rheologically complex fluids (Chhabra and Richardson 2008). Indeed so widespread is the non-Newtonian fluid behavior in nature and in technology, that it would be no exaggeration to say that Newtonian fluids are the exception rather than the rule! In principle, one can always set up the equation of continuity and Cauchy's momentum equation (written in their compact form for an incompressible fluid) for a non-Newtonian fluid in as much as the same way as we do for a Newtonian one, that is

$$\nabla \cdot \mathbf{v} = 0, \tag{1}$$

$$\rho \frac{d\mathbf{v}}{dt} = -\nabla p + \rho g + \nabla \cdot \sigma, \tag{2}$$

where for Newtonian flow characteristics, i.e., at constant temperature and pressure, in simple shear, the deviatoric stress tensor σ is proportional to the rate of deformation tensor $\dot{\gamma}$, and the constant of proportionality is the familiar dynamic viscosity η. A significant research effort has been expended in seeking a similar relation for σ for non-Newtonian fluids which should be able not only to predict shear-dependent viscosity, yield stress, visco-elastic effects in shear and extensional flows, rheopexy, and thixotropy, but should also satisfy the requirements of frame invariance, material objectivity, etc. (Bird et al. 1983; 1987). Given the diversity of the materials, this is indeed a tall order to expect that a single constitutive equation will perform satisfactorily under all circumstances for all types of materials. Notwithstanding the significant advances made in this field, the choice of an appropriate constitutive relation is guided by intuition and by experience. Critical appraisals of the current state of the art and useful guidelines for the selection of an appropriate constitutive expression for σ are available in the literature (Tanner 2000; Morrison 2001; Graessley 2004; Kroger 2004). Even when the non-linear inertial terms are neglected altogether in the momentum equation (corresponding to a zero Reynolds number flow), the resulting equations are still highly non-linear due to the constitutive equation. Therefore, one frequently resorts to numerical solutions which themselves pose enormous challenges in terms of being highly resource intensive and in terms of acute convergence difficulties (Owens and Phillips 2002). On the other hand, experimentalists also confront similar challenges both in terms of material characterization (rheometry) as well as in terms of the interpretation and representation of data using dimensionless groups (Macosko 1994; Coussot 2005).

2 Fluids in the Environment

There is no doubt that all living creatures are immersed in a fluid, be it the air of the atmosphere or the water of a river, lake, or ocean; even, soils are permeated with moisture, without which life would be impossible. It is precisely the mobility of fluids that makes them so useful to the maintenance of life. Therefore, natural fluid motions in the environment are vital and there has been a strong incentive to study them, particularly those of air in the atmosphere and of water in all its streams, from underground aquifers to surface flows in rivers, lakes, estuaries, and oceans. The study of these flows has received considerable attention over the years, and several distinct disciplines have emerged, like for instance, meteorology, climatology, hydrology, hydraulics, limnology, and oceanography. Environmental concerns compel experts in those disciplines to consider problems that are essentially similar: the flow around a body of complex geometry, the dispersion of a dilute contaminant, and the transfer of a substance and/or heat between the fluid and a boundary. This overlap between the various disciplines concerned with the environmental aspects of natural fluid flows has given rise to a body of knowledge that is known as *environmental fluid mechanics* (EFM). As a matter of fact, the scope of EFM is the study of motions and transport processes in Earth's hydrosphere and atmosphere on a local or regional scale, from millimeters to kilometers, and from seconds to years. At larger scales, the Coriolis force due to Earth's rotation must be considered, and this is the topic of geophysical fluid dynamics. Sticking purely to EFM, one will be concerned with the interaction of flow, mass, and heat with man-made facilities and with the local environment. Accordingly, EFM does not extend to fluid flows inside living organisms, such as air flow in lungs and blood flow in the vascular system, although these can be classified as natural. In the same way, EFM differs from classical fluid dynamics in that the latter is chiefly concerned with artificial (engineered) fluid motions such as flows in pipes and around airfoils, in pumps, turbines, and other machineries that utilize fluids.

The objective of EFM also differs from that of hydraulics, which deals exclusively with free-surface water flows (Chow 1959; Sturm 2001). Traditionally, problems in hydraulics have addressed prediction and control of water levels and flow rates, but the realm of hydraulics has recently been shifting considerably toward environmental concerns (Singh and Hager 1996; Chanson 2004), mainly because it has become important to estimate the effects of bottom turbulence on erosion and sedimentation, as it has been to calculate water levels and pressures against structures. On the other hand, geophysical fluid dynamics, which studies the physics of atmospheric and oceanic motions on the planetary scale (Cushman-Roisin and Beckers 2011), is another branch of fluid mechanics that overlaps with EFM. The two main ingredients of geophysical fluid dynamics are stratification and rotation, while those of EFM are stratification and turbulence. Other cousin disciplines are limnology, also called *freshwater science*, which studies all inland waters, including lakes and ponds, rivers, springs, streams, and wetlands (Imberger 1998); and hydrology, which deals with the study of the movement, distribution, and quality of water on Earth and other planets, including the hydrological cycle, water resources, and environmental watershed sustainability (Bras 1990; Ward and Trimble 2004). Hydrology has been a subject of investigation and engineering for millennia, predating Ancient Greeks and Romans.

Plumes and *thermals* are common features in environmental fluids. For instance, plumes occur whenever a persistent source of buoyancy creates a rising motion of the buoyant fluid, away from the source. Good examples are that of hydrothermal vents at the bottom of the ocean and the rising of freshwater from the bottom of the sea at submarine springs in karstic regions. Another occurrence is the common urban smokestack plume, which consists of warm gas rising not only under its own buoyancy but also under the propulsion of momentum (inertia). Such plumes are more properly categorized as buoyant jets or forced plumes. In any case, what drives a plume is its heat flux, defined as the amount of heat being discharged through the exit hole per unit time. The dynamics of plumes is strongly affected by stratification. In a stratified environment, they encounter temperatures becoming closer to their own so that they progressively lose buoyancy. At some level, they lose all of it and begin to spread horizontally. Such is the case of smokestack plumes in a calm (no wind) and stratified atmosphere typical of early mornings. Early mathematical developments explored the dynamics of plumes in terms of the buoyancy flux defined as $F = \alpha g / \rho_0 C_p$, where α is the fluid's thermal expansion coefficient, g the Earth's gravitational acceleration, and ρ_0 and C_p the fluid's reference density and heat capacity at constant pressure (Rouse et al. 1952; Turner 1973). Because of the heterogeneous structure of plumes, with entrainment and dilution taking place along their sides, most of these studies were based on cross-plume averages of the vertical velocity and the buoyancy $g' = \alpha g T'$, where T' is the temperature anomaly, or temperature inside the plume. The vertical velocity and the temperature anomaly are both functions of the height within the plume (Morton et al. 1956). More recent studies based on numerical simulations aided by laboratory experiments and field observations, have made good predictions of the level at which the vertical velocity vanishes and of the evolution of the horizontal spreading radius with time (List 1982; Cho and Chung 1992; Venkatakrishnan et al. 1999; Bhat and Krothapalli 2000; Suzuki and Koyaguchi 2007). In contrast to plumes, a thermal is a finite parcel of fluid consisting of the same fluid as its surroundings but at a different temperature. Because of its buoyancy, a cold thermal sinks (negative buoyancy), while a warm thermal rises (positive buoyancy). Convection in the atmosphere does indeed proceed by means of rising thermals (Priestley 1959). Experiments conducted in the laboratory have shown that all thermals roughly behave in similar ways, i.e., as they rise (or sink), they entrain surrounding fluid and become more dilute, thereby slowing down in their ascent (or descent) and their actual shape can vary considerably from one set of observations to another (Scorer 1997).

In many flows of environmental interest there are situations where a wave is capable of extracting energy from the system. It does so by drawing either kinetic energy from pre-existing motion or potential energy from background stratification. In either case, the wave amplitude grows with time, and the wave is said to be unstable. A trivial example of an extreme type of stratification is just a two-layer system in which a lighter layer of fluid floats over a heavier layer such as in a lake in summer. It is well-known from physical principles that gravity waves can propagate on the interface separating these two layers, not unlike waves propagating on the surface of a pond (Thorpe 1971). If the layers flow at different velocities, i.e., when shear is present, these waves may grow with time and lead to overturning in the vicinity of the interface. These breaking waves, called *billows*, generate mixing over a height a little shorter than their wavelength and the phenomenon associated with them is known as the Kelvin–Helmholtz instability (Lord Kelvin 1871; von Helmholtz 1868). In actual environmental systems, there exists a zone of gradual changes from one layer to the other, and the mathematical analysis becomes much more complicated than for the simple two-layer model (Lawrence et al. 1991) and so we must resort to numerical calculations (Bader and Deiterding 1999; Orazzo et al. 2011). Other types of instability may exist, as for instance, the barotropic and baroclinic instabilities which are the processes through which momentum and heat are redistributed within a symmetric vortex by Rossby waves. The barotropic instability, which propagates horizontally, chiefly transports momentum, while the baroclinic instability propagates vertically and transports heat. The study of these instabilities has been one of the main objectives of dynamic meteorology over the last three decades as the baroclinic instability is the principle source for midlatitude synoptic-scale disturbances. Computer models of these instabilities have been developed and successfully applied to a number of meteorological and oceanographical problems (Zabusky et al. 1979; Dritschel 1989; Smith 1991; Pullin 1992; Blumen et al. 2001).

Almost all environmental fluid flows are turbulent. However, the highly intermittent and irregular character of turbulence defies analysis, and there does not yet exist a unifying theory of turbulence, not even one for its statistical properties. Most common manifestations of turbulence in natural fluid flows involve homogeneous turbulence, shear turbulence, and turbulence in stratified fluids. If the flow volume of interest is far removed from bounding surfaces, then statistical characteristics can be assumed to be independent of spatial position. Such (real or ideal) turbulence is termed *homogeneous*, and is often also *isotropic*, unless there is some mechanism imposing a preferred direction on the system, such as a constant magnetic field in a magnetofluid, in which case it may be anisotropic. At a basic level, a turbulent flow can be interpreted as a population of many eddies, or vortices, of different sizes and strengths, embedded in one another and forever changing, giving a random appearance to the flow (Kolmogorov 1941). Most environmental fluid systems are much shallower than they are wide, as is the case of the atmosphere, oceans, lakes, and rivers. In these systems friction acts to reduce the horizontal velocity from some finite value in the interior of the flow to zero at the bottom boundary, thus creating a vertical shear. The turbulent nature of these flows greatly complicates the search for a velocity profile, and much of what we know is derived from observations of actual flows, either in the laboratory or in nature. However, the turbulent nature of the shear flow along a flat or rough surface includes variability at short time and length scales, and the best observational techniques for the detailed measurements of these flows have been developed for laboratory experiments rather than for outdoor situations (Kline et al. 1967; Wei and Willmarth 1991; Garratt 1992; Pope 2000). In order to solve more general problems in turbulence, an attempt has been made to assimilate the mixing caused by turbulence to an enhanced viscosity. This amounts to a search for a turbulent viscosity, or eddy viscosity, that would replace in turbulent flows the molecular viscosity of laminar flows (Smagorinsky 1963; Baldwin and Lomax 1978). Moreover, since no complete theory of turbulence exists, there has been a need to distill somehow the results of observations into some empirical rules. A computer simulation model that incorporates one or several of these rules is said to include a closure scheme. A large number of closure schemes has been proposed over the years, with varying degrees of success. A couple of them, which have each been tested extensively in the context of environmental flows are described in Mellor and Herring (1973) and Mellor and Yamada (1982). A known mathematical model for turbulence used in CFD is the so-called *large-eddy simulation*, initially proposed to simulate atmospheric air currents (Smagorinsky 1963). This model is currently applied in a wide variety of engineering applications, including combustion (Pitsch 2006), acoustics (Wagner et al. 2007), and simulations of the atmospheric boundary layer (Stoll and Porté-Agel 2008). A few recent applications of the technique to geophysical flows include breaking waves and tidal bores (Lubin et al. 2010) and turbulence in the planetary boundary layer (Moeng and Sullivan 2002).

Mathematical models describing the intrusion of a fluid into another are essential for studying air quality in the lower atmosphere, which is the portion of space where we live and breathe, and where all of our gaseous emissions are discharged. The physical processes that affect this region, form an important element of EFM and entire textbooks exist in the literature that are devoted to the topics of the Atmospheric Boundary Layers and Air Pollution Meteorology. In particular, the lowest atmosphere is far from being a simple system. Its physics include convection during the day and stratification at night, complications due to complex terrain (i.e., surface elements such as buildings, forests, hills, and mountains) as well as large weather events, including replacement of air masses by prevailing winds, clouds, and precipitation. Whereas the atmosphere is more than 100 km thick, weather systems, like cyclones, anticyclones, storms, and hurricanes, hardly occupy more than the bottom 10 km, a layer called the troposphere. This layer is in permanent state of turmoil and its ubiquitous instabilities, while challenging the fluid dynamicists, are highly beneficial to us, on two levels. First, the instabilities create weather patterns, which cause precipitation, thus providing a freshwater supply to us on land. Second, turbulent mixing generates dispersion and dilution of our pollutants. A well-known effect that causes climatic implications is the *greenhouse* effect due to heat retention because of absorption of solar radiation by pollutants, like carbon dioxide and other gases. Pollution has a root cause, *a source*, from which contaminants are released in either predictable or accidental ways. These travel through one or several media, such as the atmosphere, a river, or a food chain (the *pathway*). Along the way, they are diluted and modified. Eventually, they encounter objects, animate or inanimate, on which they have adverse effects (the *receptor*). Environmental management is brought to bear on one, two, or all three components of the system. Protection of a beach from an off-shore oil pill or relocation of residents away from a contaminated area are examples of management at the receptor level. In particular, pathway management must be preceded by a careful analysis of both the transporting mechanisms and the transformations that can occur along the way. Such analysis is the object of *Environmental Transport and Fate* (Hemond and Fechner 1994). This discipline has two distinct aspects: it implicates both the physics of transport (*How do pollutants travel? Where do they go?*) and the biochemistry of fate (*What do pollutants become? What other secondary pollutants do they generate?*). In this context, EFM is the chapter that addresses the physical aspects of fluid pathways. Common sources of pollution are wastewater discharges from pipes into rivers, or lakes, and forced plumes exiting from industrial smokestacks. The various types of intrusion are categorized according to whether they inject momentum, buoyancy, or both in the ambient fluid, and whether they persist in time. For instance, turbulent jets are good examples of a continuous injection of momentum such as occurs in nature when a river empties in a lake, or estuary, or occasionally when wind blows through an orographic gap. But, perhaps the most clearly defined jets are those produced when a fluid is discharged in the environment through a relatively narrow conduit, such as an industrial discharge released through a pipe on the bank of a river, lake, or coastal ocean. Intermittent injections of momentum, or *puffs*, may also occur in many situations. For instance, buildings in urban areas are exposed to pollutants daily, and could possibly be exposed to hazardous toxic gases released into the environment. Heating, ventilation, and air conditioning systems also affect gas dispersion in urban environments, and numerical simulations have been developed to estimate how much pollutants penetrate into a specific building, the concentrations of contaminants at pedestrian level due to heavy traffic, and the effects of gas dispersion from stationary pollutant sources (Hanna et al. 2002; Britter and Hanna 2003; Coirier et al. 2005; Hanna et al. 2006; Fernando et al. 2010).

Rivers and streams are types of open-channel flows, i.e., conduits of water with a free surface. In contrast to canals, ditches, aqueducts, and other structures designed and built by humans, rivers and streams are the products of natural geological processes and, as a consequence, are quite irregular. They have the ability to scour their beds as well as carry and deposit sediments. Through their watershed, they also gather, convey, and disperse almost any substance that enters water on land. In this way, streams and rivers are actors of environmental transport and fate. Such materials can be contaminated, and therefore one pollution transport mechanism in a river is by successive erosion, as the solid particles are entrained into the moving water, and sedimentation, as they are deposited on the river bed. Experimental observations have shown that the entrainment of a solid particle lying on the bed into the flow depends primarily on the size of the particle and the stress of the moving water onto the bed (Chanson 2004; Ward and Trimble 2004). Such natural flows are also important because they distribute nutrients to species, habitats, and ecosystems, thus maintaining biodiversity. Although the physical principles governing this type of flows are well understood, traditional problems in hydraulics have evolved into grand challenges. These include sustainable water supply and drainage, control of sediment transport, sediment-water-pollution exchange, subsurface seepage, river restoration, wetlands and plant canopies, bank stability and exchanges between rivers and floodplains, and fish ladders and habitats. All of these have substantial economic implications. More than 200 methods are used today worldwide to prescribe river flows. However, very few of them are comprehensive and holistic, accounting for seasonal and inter-annual flow variations as needed to support the whole range of ecosystem services that healthy rivers provide (Tharme RE 2003; King et al. 2003; Richter et al. 2006; Poff NL et al. 2010). To facilitate environmental flow prescriptions, a number of computer models and tools have been developed by groups such as the US Army Corps of Engineers (USACE) at the Hydrologic Engineering Center to capture flow requirements defined in a workshop setting or to evaluate the implications of environmental flow implementations.

Lakes and reservoirs are other types of flows with a free surface. The former are natural bodies of water, where flow from one or several rivers is impounded by a natural obstacle, and the latter are artificial lakes created by a dam blocking a river. They differ from rivers by their greater depths and their weaker velocities. Generation of hydro-electric power, flood control, freshwater supply for households and irrigation, and control of water quality are some of the reasons why people build reservoirs. The average time spent by a water parcel from its time of inflow into the lake to that of outflow from it, i.e., the *residency* time, is an important characteristic of a lake because the longer this time is, the more likely the water parcel is affected by local processes such as heating or cooling, sedimentation, and biological or chemical transformations. Therefore, primary scientific research was addressed to study the seasonal variations of the water level and temperature, amount of dissolved oxygen as a percentage of saturation, vertical stratification, level of pollution, acidification due to atmospheric precipitation, and wind-driven circulation in some important lakes around the world (Fisher et al. 1979; Graf and Mortimer 1979; Henriksen and Kemp 1988; Schnoor 1996). A huge number of papers on CFD models applied to lakes, reservoirs, and associated rivers which study erosion and sediment deposition in rivers, wind-induced currents and radioactive tracer movements in lakes, fluvial geomorphology and bed changes, impacts of river regulations on fish habitat, open-channel flow with submerged vegetation, spatio-temporal distribution of phyto-plankton in small productive lakes, etc., exists in the literature.

Evidently, the study of fluid mechanics allows many environmental flows to be understood, modeled, and predicted, with important consequences for engineering, planning, and policy-making. The social contribution of EFM has been rather pervasive. Only 20 years ago, great leaps in the accuracy of hurricane track forecasts were made. Recently there has been attained a 50 % reduction in the prediction error. Similarly, fluid dynamics provides the basis for computer fore-casting models, some of which have made major strides in recent years, including improved predictions related to El Niño, the ozone hole and tsunami propagation; transport of pollutants; hydrological forecasting; jet stream tracking for air traffic routing; and airborne laser defense. The looming specter of biochemical terrorism has heightened the need for emergency evacuation planning, which relies heavily on transport models for toxic substances based on fluid-dynamics principles.

Perhaps no other issue is more important in environmental forecasting than the so-called sub-grid scale parameterizations, i.e., the mathematical representation of processes occurring within the computational grid boxes of predictive models. These processes, which involve turbulence, mixing, eddies, waves, convection, and diffusion, are understood qualitatively, but not quantitatively, due to the non-linearity of natural phenomena. Motions at different scales, from the milli-meter-sized whirling motions that dissipate energy and help suspend nano- and microscale particles (e.g., aerosols, pollutants, and biota) to turbulent wind gusts and oceanic waves, as well as interactions between the different scales, continue to be fertile ground for cutting-edge research. Opportunities are plentiful for studying real life-size flow configurations. The new area of sensor-model fusion, where modeling, data gathering and processing, and its incorporation into CFD models occur simultaneously, offers great potential for improved environmental predictions at lower cost. Fundamental advances in sensor placement, data acquisition, high-performance computing, and automated recognition of physical processes are in continuous progress, and major social benefits can be anticipated from research-based knowledge of environmental flows.

3 Nano and Micro-Scale Flows

Many recent advances in science and technology are aimed at making smaller devices. The electronics industry provides the best example of the gains in productivity, efficiency, scale, and even in new culture-changing products that result from designing and controlling small devices. Similar advances and applications in fluid dynamics are occurring at a rapid pace. The resulting technology is called *microfluidics* when the typical sizes of the fluid-carrying channels are smaller than 1 mm and *nanofluidics* when the typical sizes involved are smaller than 1 μ. The ability to control fluids in channels of such small dimensions is leading to advances in basic research and technological innovations in biology, chemistry, engineering, and physics. For instance, micron-sized accelerometers are being used to deploy air bag systems in automobiles. Novel bioassays consisting of microfluidic networks are designed for patterned drug delivery. The advances are more significant in research focusing on new materials, new fabrication methods, cooling of electronic devices, multiphase flows in labs-on-a-chip, and efforts to understand basic processes in individual biological cells.

Microdevices behave differently from the objects we are used to handle in our daily life (Gad-El-Hak 1999). The inertial forces, for example, are quite small, and surface effects dominate the behavior of these small systems. Friction, electrostatic forces, and viscous effects due to the surrounding air or liquid become increasingly important as the devices become smaller. At these scales, surface tension forces are dominant, and so micropumps and microvalves have been fabricated taking advantage of this fact (Evans et al. 1997). Early applications can be found in the micro- and nanoscale design of computer components such as the Winchester-type hard disk drive mechanism, where the read/write head floats 50 nm above the surface of the spinning platter (Tagawa 1993). Turning to micro-electro-mechanical systems (MEMS), one of the first microfabricated products is a polysilicon, surface micromachined side-driven motor whose operation and performance have been studied extensively (Mehregany et al. 1990; Trimmer 1997). Microdevices that involve *particulate flows* for sorting, analysis, and removal of particles or cells from a sample, with both liquid and gas microflows have been also designed (Telleman et al. 1998; Ho and Tai 1998). Microparticles, from 20 nm to about 3 μm, can also be used to fabricate microdevices, such as pumps and valves, which in turn can be used for microfluidic control (Hayes et al. 2001; Doyle et al. 2002). Colloidal micropumps and colloidal microvalves of the size of a human red blood cell are already in existence, and have been used for active microfluidic control (Terray et al. 2002).

Detailed modeling of microflows necessitates either using atomistic (particle-based) simulations or adding correction terms to the macroscopic full simulation methods. For example, in the slip flow regime it is reasonable to employ the Navier-Stokes equations modified at the wall surface with appropriate velocity slip and temperature jump conditions (Karniadakis et al. 2005). There exists a dynamic similarity between microflows and the rarefied flows encountered in a low-pressure environment. This similarity is used as a fundamental assumption for modeling microflows. In gas microflows we encounter four important effects, namely rarefaction, compressibility, viscous heating, and thermal creep. The competing effects of compressibility and rarefaction result in a non-linear pressure distribution in microchannels in the slip flow regime. Compressibility produces a curvature in the pressure distribution, whereas rarefaction reduces it. Viscous heating, due to dissipation, is important for microflows, especially in creating temperature gradients within the domain, while thermal creep causes a variation of pressure along microchannels in the presence of tangential temperature gradients (Fukui and Kaneko 1988). This mechanism is significant for transport through porous media in atmospheric conditions (Vargo and Muntz 1996). On the other hand, modeling of liquids in microdomains requires a different approach. At mesoscopic scales a continuum description suffices, whereas in submicron dimensions atomistic models are required. Besides the slip condition, other phenomena may be present such as wetting, adsorption, and electrokinetics. In particular, the wetting of the solid surfaces by liquids can be exploited in microfluidics to determine precisely defined routes based on surface tension gradients and different types of surfaces, i.e., hydrophobic or hydrophilic. Moreover, adsorption is important in interactions of liquids with nanoporous materials such as glasses. Liquids confined in microgeometries exhibit supercooling of the liquid-solid phase transition, which can be quite substantial, and depends strongly on the geometry of the pores (Tell and Maris 1983). Finally, electrokinetic effects provide a means of maintaining a certain level of flowrate with practically uniform profiles (Karniadakis et al. 2005).

Fluid flows in nanometer scale channels and pores play an important role in determining the functional characteristics of many biological and engineering devices and systems. Ionic channels are naturally occurring in nanotubes found in the membranes of all biological cells (Hille 2001). It is therefore important to understand the flow of water and electrolytes in naturally existing nanoscopic pores in the presence of a strong permanent charge. Another application occurring in nanotubes in which nanoflows are gaining considerable attention is the translocation of DNA through a nanopore (Nakane et al. 2003). Molecular gates are other emerging application of nanoflows (Kuo et al. 2003; Jin and Aluru 2011). For instance, nanoporous membranes can be used to interface vertically separated microfluidic channels to create a truly three-dimensional fluidic architecture. Integration of these circuits into a single chip can thus enable very complicated fluidic and chemical manipulations. A high-density microfluidic device containing 2056 integrated channels has already been developed, which can be used as a memory storage device whose behavior resembles random-access memory (Thorsen et al. 2002). There are a number of other applications such as fuel cell devices, drug delivery systems, chemical and biological sensing, and energy conversion devices. With advances in nanofabrication techniques, it is now possible to fabricate devices with diameters ranging from a few angstroms to few hundred nanometers. Such advances make it possible to understand fundamental physical mechanisms in nanoflows through a detailed comparison between experimental and theoretical studies.

4 Turbulent Flows in Engineering and Technology

Turbulent flows are ubiquitous in our transportation systems, process industries, and natural environment. In contrast to laminar flows, exemplified by honey pouring from a jar, turbulent flows are chaotic, three-dimensional, and unsteady over a large range of scales. To have a rough idea of the vastly different sizes in which non-linear interactions take place, consider a bumpy aircraft flight which may be caused by eddies of atmospheric turbulence. These eddies are in general much larger than the aircraft, whereas the drag on the wings and fuselage are caused by turbulent eddies smaller than a millimeter. From a scientific viewpoint, turbulence is a notoriously difficult subject, and research in the field today relies on the use of modern technologies such as laser diagnostics in the laboratory and high-performance computing for simulations. Techniques as digital PIV can be used in experimental investigations to determine the three-dimensional fluctuating velocity fields in turbulent motion.

Understanding turbulence may have direct impact on the improvement of several technologies, for example, in the mixing of fuel and air in all types of engines. Turbulent flow is essential in these processes, and enhancement of the mixing rates would yield significant benefits in fuel economy and reduced pollutant emissions. On the other hand, the drag on heavy trucks and other vehicles is largely due to the effectiveness of turbulent flows in transporting momentum, which is unwanted in this context. Thus research on the complicated flow patterns behind trucks can lead to improved designs with reduced drag. Similarly, the cost of pumping oil and gas through pipelines is directly proportional to the frictional losses due to turbulence. In other cases, the goal is to suppress deleterious instabilities and the ensuing turbulence. Inertial-confinement fusion is an example whose success relies on the control of the Rayleigh-Taylor instability. In hypersonic flight, delaying the transition to turbulence can make the difference between successful reentry from space and the loss of a mission.

An unwanted by-product of turbulent flow is acoustic noise. Aircraft noise is responsible for much of the community reluctance to increase airport capacity, which impacts operations at many airports with strong restrictions. As a consequence, turbulence-generated noise is a growing concern in the aviation industry for both fixed-wing aircrafts and rotorcrafts. The key challenge to the aeroacoustic engineering community is the development of noise abatement technologies without sacrifying system performance. Unsteady aerodynamics and turbulence generated by rotorcraft blades, aircraft engine turbomachinery, and jet exhaust are sources of noise. Understanding and predicting the impact of abatement technologies is a critical capability required to bring quieter products to the market.

5 Biological Fluid Dynamics

Biological fluid flows are present in many aspects of life. Basic functions like reproduction, growth, feeding, metabolism, and locomotion, are all sustained by the motion of internal fluids. The proper function of many organs such as the brain, eyes, lungs, heart, liver, and kidneys depends critically on fluid transport processes that provide rapid exchange of molecules between tissues and blood. Any disruption or deficiency in biofluid transport at the nano (sub-cellular), micro (cellular), or macro (vessel/organ) scale can result in major vascular diseases such as atherosclerosis, aneurysms, heart failure, strokes, hydrocephalus and glaucoma, each with devastating effects on health. Thus, a comprehensive understanding of biological flows may lead to the development of better diagnostics and cost effective medical treatments.

One of the most common problems related to biological fluid flow in the human body are cardiovascular diseases, which are the leading cause of mortality and morbidity in the western world. The observation that atherosclerotic plaques typically occur at arterial bifurcations and bends has led to the now almost universal acceptance that local hemodynamic factors, in particular, wall shear stresses, play a role in the disease's initiation and, perhaps more importantly, its progression (Malek et al. 1999). For example, the annual cost of cardiovascular diseases in the United States economy amounts to 300 billion dollars and this figure increases when other diseases related to biological flows are included. Projections suggest that these costs will increase dramatically as the aging "baby boom" generation becomes more susceptible to this pathology. Biofluid mechanics poses some of the most difficult basic science and engineering problems to experimental and computational fluid dynamicists, as well as to the medical device industry. These difficulties arise from the wide range of size scales involved (from cell to organ), the compliant nature of boundaries in the fluid (moving vessel walls and deformable cells), and the inherent complexity of biological fluids (e.g., blood). A classical example of these complexities can be found in the role of blood flow in atherosclerosis and vascular remodeling (Mulvany et al. 1996; Korshunov and Berk 2004; Chatzizisis et al. 2007). Where progress has been made, it has been stimulated by the availability of novel biological and medical imaging techniques (Udupa and Herman 2000; Makowski et al. 2011) as well as powerful computational and modeling tools (Steinman 2002; Löhner et al. 2003; Wittek et al. 2011). In particular, impressive progress has been made in the field of medical imaging, which has allowed cardiologists to visualize plaque build-up in patients' arteries in three-dimensions, highlighting potential problems before symptoms develop. Such data combined with powerful CFD tools has also made it possible to develop patient-based flow modeling to predict the outcome of surgical cardiovascular interventions. On the other hand, CFD simulations and laboratory fluid dynamics measurement techniques are taking cancer research to a real new dimension (Conolly et al. 2003; Leyton-Mange et al. 2006; Liang et al. 2008). A challenge today is to understand why and how the migration of cancer cells is affected by the fluid dynamics of the system (Hoskins et al. 2009). Since both cancer and white blood cells are flexible, improvements to early simulation models today point to the inclusion of cell deformation effects.

The field of biofluid mechanics also extends beyond medical applications. For instance, biological models can be used to design fluid transport systems that are vital to national defense. Novel designs of unmanned aerial and underwater vehicles can be traced to basic science research on locomotion of flying and swimming animals (Liu 2005). For example, a unifying theory of winged locomotion could explain the magical mid-air maneuvers of birds and insects, and guide the design of flying robots. Though scientists understand the principles underlying many flight-enhancing physiologies, from birds' hollow bones to dragonflies' flexible wings, the biomechanics of turning was in many ways a mystery (Hedrick et al. 2009; Tobalske 2009). These applications amongst others, such as aerial surveillance, underwater mine detection, and the inspection of hazardous sites, along with the likely medical advances that can be anticipated, provide a rich variety of potential benefits from research in biofluid dynamics.

Acknowledgments L. T. acknowledges the organizers of the XVII Annual Meeting of the Fluid Dynamics Division (XVII-DDF) of the Mexican Physical Society, with special mention to Anne Cros. This work has been partially supported by CONACyT-EDOMEX-2011-C01-165873 project.

References

Abel T, Bryan GL, Norman ML (1998) Numerical simulations of first structure formation. Mem Soc Astronom Ital 69:377–384
Adrian RJ, Westerweel J (2011) Particle image velocimetry. Cambridge University Press, Cambridge
Arreaga-García G, Klapp J, Sigalotti LDiG, Gabbasov R (2007) Gravitational collapse and fragmentation of molecular cloud cores with GADGET-2. Astrophys J 666:290–308
Bader G, Deiterding R (1999) A distributed memory adaptive mesh refinement package for inviscid flow simulations. In: Jonas P, Uruba V (eds) Proceedings of colloquium on fluid dynamics. Institute of Thermodynamics (Academy of Science of Czech Republic), Prague, pp 9–14

Baldwin BS, Lomax H (1978) Thin-layer approximation and algebraic model for separated turbulent flows. AIAA Paper, pp 78–257

Bate MR (1998) Collapse of a molecular cloud core to stellar densities: the first three-dimensional calculations. Astrophys J 508:L95–L98

Benz W (2000) Low velocity collisions and the growth of planetesimals. Space Sci Rev 92:279–294

Berczik P, Kolesnik IG (1998) Gasodynamical model of the triaxial protogalaxy collapse. Astron Astrophys Trans 16(3):163–185

Bhat GS, Krothapalli A (2000) Simulation of a round jet and a plume in a regional atmospheric model. Mon Weather Rev 128:4108–4117

Bird RB, Dai GC, Yarusso BJ (1983) The rheology and flow of viscoplastic materials. Rev Chem Eng 1:1–83

Bird RB, Armstrong RC, Hassager O (1987) Dynamics of polymeric liquids, vol I and II. Wiley, New York

Blumen W, Banta R, Burns SP, Fritts DC, Newsom R, Poulos GS, Sun J (2001) Turbulence statistics of a Kelvin-Helmholtz billow event observed in the night-time boundary layer during Cooperative Atmosphere-Surface Exchange Study field program. Dyn Atmos Oceans 34:189–204

Bodenheimer P, Tohline JE, Black DC (1980) Fragmentation in rotating isothermal protostellar clouds. Space Sci Rev 27:247–252

Bonnell IA, Bate MR (1994) The formation of close binary systems. Mon Not R Astronom Soc 271:999–1004

Boss AP, Durisen RH (2005) Sources of shock waves in the protoplanetary disk. In: Krot AN, Scott ERD, Reipurth B (eds) Chondrites and the Protoplanetary Disk. ASP conference series, vol 341, San Francisco, pp 821–838

Boss AP (1981) Collapse and fragmentation of rotating, adiabatic clouds. Astrophys J 250:636–644

Boss AP (1991) Formation of hierarchical multiple protostellar cores. Nature 351:298–300

Bras RL (1990) Hydrology: an introduction to hydrologic science. Addison-Wesley, New York

Britter RE, Hanna SR (2003) Flow and dispersion in urban areas. Annu Rev Fluid Mech 35:469–496

Bruun HH (1995) Hot-wire anemometry. Oxford University Press, Oxford

Bryan GL (1999) Fluids in the universe: adaptive mesh refinement in cosmology. Comput Sci Eng 1(2):46–53

Centrella J, Melott AL (1983) Three-dimensional simulation of large-scale structure in the universe. Nature 305:196–198

Chanson H (2004) Environmental hydraulics of open channel flows. Elsevier Butterworth-Heinemann, Oxford

Chatzizisis YS, Coskun AU, Jonas M, Edelman ER, Feldman CL, Stone PH (2007) Role of endothelial shear stress in the natural history of coronary atherosclerosis and vascular remodeling: molecular, cellular, and vascular behavior. J Am Coll Cardiol 49:2379–2393

Chhabra RP, Richardson JF (2008) Non-Newtonian flow and applied rheology. Butterworth-Heinemann, Oxford

Cho JR, Chung MK (1992) A k-$\epsilon-\gamma$ equation turbulence model. J Fluid Mech 237:301–322

Choong TSY, Chuah TG, Robiah Y, Greogory-Koay FL, Azni I (2007) Arsenic toxicity, health hazards and removal techniques from water: an overview. Desalination 217:139–166

Chow VT (1959) Open-channel hydraulics. McGraw-Hill College, New York

Coirier WJ, Fricker DM, Furmanczyk M, Kim S (2005) A computational fluid dynamics approach for urban area transport and dispersion modeling. Environ Fluid Mech 15(5):443–479

Conolly RB, Kimbell JS, Janszen D, Schlosser PM, Kalisak D, Preston J, Miller FJ (2003) Biologically motivated computational modeling of formaldehyde carcinogenicity in the F344 rat. Toxicol Sci 75(2):432–447

Coussot P (2005) Rheometry of pastes, suspensions and granular materials. Wiley, New York

Cushman-Roisin B, Beckers J-M (2011) Introduction to geophysical fluid dynamics: physical and numerical aspects. Elsevier Inc, Amsterdam

Darcy M (1858) Note relative à quelques modifications à introduire dans le tube de Pitot. Annales des Ponts et Chaussées N° 204:351–359

Doyle PS, Bibette J, Bancaud A, Viory J-L (2002) Self-assembled magnetic matrices for DNA separation chips. Science 295:2237–2237

Dritschel DG (1989) Contour dynamics and contour surgery: numerical algorithms for extended high-resolution modelling of vortex dynamics in two-dimensional, incompressible flows. Comput Phys Rep 10:79–146

Evans JD, Lipemann D, Pisano, AP (1997) Planar laminar mixer. In: MEMS-97, The tenth annual international workshop on MEMS (Jan 26–30, 1997)

Fernando HJS, Zajic D, Di Sabatino S, Dimitrova R, Hedquist B, Dallman A (2010) Flow, turbulence, and pollutant dispersion in urban atmospheres. Phys Fluids 22(5):051301

Fingerson LM, Freymuth P (1983) Thermal anemometers. In: Goldstein RJ (ed) Fluid mechanics measurements, Washington DC, Hemisphere, pp 99–154

Fisher HB, List EJ, Koh RCY, Imberger J, Brooks NH (1979) Mixing in inland and coastal waters. Academic Press, San Diego

Fukui S, Kaneko R (1988) Analysis of ultra thin gas film lubrication based on linearized Boltzmann equation. First report: derivation of a generalized lubrication equation including thermal creep flow. J Tribol 110:253–262

Gad-El-Hak M (1999) The fluid mechanics of microdevices. J Fluids Eng 12(1):5–33

Garratt JR (1992) The atmospheric boundary layer. Cambridge University Press, Cambridge

Gingold RA, Monaghan JJ (1977) Smoothed particle hydrodynamics: theory and application to non-spherical stars. Mon Not R Astronom Soc 181:375–389

Graessley WW (2004) Polymer liquids and networks: structure and properties. Garland Science, New York

Graf WH, Mortimer CH (1979) Hydrodynamics of lakes. Elsevier Scientific Publishing Company, Amsterdam

Hanna SR, Tehranian S, Carissimo B, Macdonald RW, Lohner R (2002) Comparisons of model simulations with observations of mean flow and turbulence within simple obstacle arrays. Atmos Environ 36:5067–5579

Hanna SR et al (2006) Detailed simulation of atmospheric flow and dispersion in downtown Manhattan: an application of five computational fluid dynamics models. Bull Am Meteorol Soc 87:1713–1726

Hayes MA, Polson NA, García AA (2001) Active control of dynamic supraparticle structures in microchannels. Langmuir 17:2866–2871

Hedrick TL, Cheng B, Deng X (2009) Wingbeat time and the scaling of passive rotational damping in flapping flight. Science 324:252–255

von Helmholtz H (1868) Über discontinuierliche Flüssigkeits-Bewegungen. Monatsberichte der Königlichen Preussische Akademie der Wissenschaften zu Berlin 23:215–228

Hemond HF, Fechner EJ (1994) Chemical fate and transport in the environment. Academic Press, San Diego

Henriksen K, Kemp WM (1988) Nitrification in estuarine and coastal marine sediments. Chapter 10. In Blackburn TH, Sorensen J (eds) Nitrogen cycling in coastal marine environments. SCOPE. Wiley, New Jersey

Hille B (2001) Ion channels of excitable membranes. Sinauer Associates, Publisher Suderland, Massachusetts

Ho CM, Tai YC (1998) Micro-electro-mechanical systems (MEMS) and fluid flows. Annu Rev Fluid Mech 30:579–612

Hoskins M, Kunz R, Bistline J, Dong C (2009) Coupled flow-structure-biochemistry simulations of dynamic systems of blood cells using an adaptive surface tracking method. J Fluids Struct 25:936–953

Imberger J (1998) Physical processes in lakes and oceans. American Geophysical Union, Washington

Jain N, Ottino JM, Lueptow RM (2002) An experimental study of the flowing granular layer in a rotating tumbler. Phys Fluids 14(2):572–582

Jin X, Aluru NR (2011) Gated transport in nanofluidic devices. Microfluid Nanofluid 11:297–306

Karniadakis G, Beskok A, Aluru N (2005) Microflows and nanoflows. Fundamentals and simulations. Springer, New York

King J, Brown C, Sabet H (2003) A scenario-based holistic approach to environmental flow assessments for rivers. River Res Appl 19:619–639

Klessen RS, Peters T, Banerjee R, Mac Low M-M, Galván-Madrid R, Keto ER (2011) Modeling high-mass star formation and ultracompact H_{II} regions. In: Alves J, Elmegreen BG, Girart JM, Trimble V (eds) Computational star formation. Proceedings of the international astronomical union, IAU symposium vol 270, pp 107–114

Kline SJ, Reynolds WC, Schraub FA, Runstadler PW (1967) The structure of turbulent boundary layers. J Fluid Mech 30:741–773

Kolmogorov AN (1941) The local structure of turbulence in incompressible viscous fluid for very large Reynolds numbers. Proc USSR Acad Sci 30:299–303 (in Russian)

Korshunov VA, Berk BC (2004) Strain-dependent vascular remodeling: the "Glagov phenomenon" is genetically determined. Circulation 110:220–226

Kroger M (2004) Simple models for complex non-equilibrium fluids. Phys Rep 390:453–551

Kuo TC, Cannon DM, Shannon MA, Bohn PW, Sweedler JV (2003) Hybrid three-dimensional nanofluidic/microfluidic devices using molecular gates. Sens Actuators A: Phys 102:223–233

Lawrence GA, Browand FK, Redekopp LG (1991) The stability of a sheared density interface. Phys Fluids A 3:2360–2370

Lekakis I (1996) Calibration and signal interpretation for single and multiple hot-wire/hot-film probes. Measur Sci Technol 7:1313–1333

Leyton-Mange J, Sung Y, Henty M, Kunz RF, Zahn J, Dong C (2006) Design of a side-view particle imaging velocimetry flow system for cell-substrate adhesion studies. J Biomech Eng 128:271–278

Liang S, Slattery M, Wagner D, Simon S, Dong C (2008) Hydrodynamic shear rate regulates melanoma-leukocyte aggregation, melanoma adhesion to the endothelium, and subsequent extravasation. Ann Biomed Eng 36(4):661–671

List EJ (1982) Turbulent jets and plumes. Annu Rev Fluid Mech 14:189–212

Liu H (2005) Simulation-based biological fluid dynamics in animal locomotion. Appl Mech Rev 58(4):269–283

Löhner R, Cebral J, Soto O, Yim PJ, Burgess JE (2003) Applications of patient-specific CFD in medicine and life sciences. Int J Numer Methods Fluids 43:637–650

Lord Kelvin WT (1871) Hydrokinetic solutions and observations. Philos Mag 42:362–377

Lubin P, Glockner S, Chanson H (2010) Numerical simulation of a weak breaking tidal bore. Mech Res Comm 37(1):119–121

Lucy LB (1977) A numerical approach to the testing of the fission hypothesis. Astron J 82:1013–1024

Lueptow RM, Akonur A, Shinbrot T (2000) PIV for granular flows. Exp Fluids 28(2):183–186

Macosko CW (1994) Rheology: principles measurements and applications. Wiley, New York

Makowski MR et al (2011) Assessment of atherosclerosis plaque burden with an elastin-specific magnetic resonance contrast agent. Nat Med 17(3):383–388

Malek AM, Alper SL, Izumo S (1999) Hemodynamic shear stress and its role in atherosclerosis. J Am Med Assoc 282:2035–2042

Mehregany M, Nagarkar P, Senturia S, Lang JH (1990) Operation of microfabricated harmonic and ordinary side-drive motor. In: IEEE Micro electro mechanical system workshop, Napa Valley, CA (Feb, 1990)

Mellor GL, Herring HJ (1973) A survey of the mean turbulent field closure models. AIAA J 11:590–599

Mellor GL, Yamada T (1982) Development of a turbulence closure model for geophysical fluid problems. Rev Geophys Space Phys 20:851–875

Moeng CH, Sullivan PP (2002) Large eddy simulation. Encyclopedia of atmospheric sciences. Academic Press, San Diego, pp 1140–1150

Morrison FA (2001) Understanding rheology. Oxford University Press, Oxford

Morton BR, Taylor GI, Turner JS (1956) Turbulent gravitational convection from maintained and instantaneous sources. Proc R Soc A: Math Phys Eng Sci 234:1–23

Mulvany MJ, Baumbach GL, Aalkjaer C, Heagerty AM, Korsgaard N, Schiffrin EL, Heistad DD (1996) Vascular remodeling. Hypertension 28(505–506):1996

Nakane JJ, Akeson M, Marziali A (2003) Nanopores sensors for nucleic acid analysis. J Phys: Condens Matt 15:R1365–R1393

Orazzo A, Coppola G, de Luca L (2011) Numerical simulation of single-wave Kelvin-Helmholtz instability in two-phase channel flow. In: 24th European conference on liquid atomization and spray systems, Estoril, Portugal (in press)

Owens RG, Phillips TN (2002) Computational rheology. Imperial College Press, London

Pitot M (1732) Description d'une machine pour mesurer la vitesse des eaux courantes et le sillage des vaisseaux. Histoire de l'Académie Royale des Sciences avec les Mémoires de Mathématique et de Physique Tirés des Registres de cette Académie 363–376

Pitsch H (2006) Large-eddy simulation of turbulent combustion. Annu Rev Fluid Mech 38:453–482

Poff NL et al (2010) The ecological limits of hydrologic alteration (ELOHA): a new framework for developing regional environmental flow standards. Freshw Biol 55:147–170

Pope SB (2000) Turbulent flows. Cambridge University Press, Cambridge

Priestley CHB (1959) Turbulent transfer in the lower atmosphere. Chicago University Press, Chicago

Pudasaini SP, Hsiau S-S, Wang Y, Hutter K (2005) Velocity measurements in dry granular avalanches using particle image velocimetry-technique and comparison with theoretical predictions. Phys Fluids 17(9):093301

Pullin DI (1992) Contour dynamics methods. Annu Rev Fluid Mech 24:89–115

Raffel M, Willert C, Wereley S, Kompenhans J (2007) Particle image velocimetry: a practical guide. Springer, Berlin

Richter BD, Warner AT, Meyer JL, Lutz K (2006) A collaborative and adaptive process for developing environmental flow recommendations. River Res Appl 22:297–318

Rouse H, Yih C-S, Humphreys HW (1952) Gravitational convection from a boundary source. Tellus 4:201–210

Schnoor JL (1996) Environmental modeling: fate and transport of pollutants in air, water, and soil. Wiley, New Jersey

Scorer RS (1997) Dynamics of metereology and climate. Wiley, New York

Seitzman JM, Hanson RK (1993) Planar fluorescence imaging in gases. In: Taylor AMKP (ed) Instrumentation for flows with combustion. Academic Press, San Diego, pp 405–466

Sigalotti LDiG, Klapp J (2001) Protostellar collapse models of prolate molecular cloud cores. Astron Astrophys 378:165–179

Singh VP, Hager WH (1996) Environmental hydraulics. Kluwer Academic Publishers, Dordrecht

Smagorinsky J (1963) General circulation experiments with the primitive equations: I. The basic equations. Mon Weather Rev 91:99–164

Smith RB (1991) Kelvin-Helmholtz instability in severe downslope wind flow. J Atmos Sci 48:1319–1324

Springel V et al (2005) Simulations of the formation, evolution and clustering of galaxies and quasars. Nature 435:629–636

Springel V, Yoshida N, White SDM (2001) GADGET: a code for collisionless and gasdynamical cosmological simulations. New Astron 6:79–117

Steinman DA (2002) Image-based computational fluid dynamics modeling in realistic arterial geometries. Ann Biomed Eng 30:483–497

Steinmetz M (1996) Simulating galaxy formation. In: Bonometto S, Primack JR, Provenzale A (eds) Dark matter in the universe. Proceedings of the international school of physics Enrico Fermi, course CXXXII, Varenna, pp 479–503

Stoll R, Porté-Agel F (2008) Large-eddy simulation of the stable atmospheric boundary layer using dynamic models with different averaging schemes. Bound-Layer Metereol 126:1–28

Sturm TW (2001) Open channel hydraulics. MacGraw Hill Higher Education, New York

Suzuki YJ, Koyaguchi T (2007) Numerical simulations of turbulent mixing in eruption clouds. J Earth Simul 8:35–44

Tagawa N (1993) State of the art for flying head slider mechanisms in magnetic recording disk storage. Wear 168:43–47

Tanner RI (2000) Engineering rheology. Oxford University Press, Oxford

Tell JL, Maris HJ (1983) Specific heats of hydrogen, deuterium, and neon in porous Vycor glass. Phys Rev B 28:5122–5125

Telleman P, Larsen UD, Philip J, Blankenstein G, Wolf A (1998) Cell sorting in microfluidic systems. In: van den Berg H (ed) Micro total analysis systems '98. Kluwer Academic Publishers, Dordrecht, p 39–44

Terray A, Oakey J, Marr D (2002) Microfluidic control using colloidal devices. Science 296:1841–1843

Tharme RE (2003) A global perspective on environmental flow assessment: emerging trends in the development and application of environmental flow methodologies for rivers. River Res Appl 19:397–441

Thorpe SA (1971) Experiments on the instability of stratified shear flows: miscible fluids. J Fluid Mech 46:299–319

Thorsen T, Maerkl SJ, Quake SR (2002) Microfluidic large-scale integration. Science 298:580–584

Tobalske BW (2009) Symmetry in turns. Science 324:190–191

Trimmer W (1997) Micromechanics and MEMS, Classical and seminal papers to 1990 (IEEE Press)

Turner JS (1973) Buoyancy effects in fluids. Cambridge University Press, Cambridge

Udupa JK, Herman GT (2000) 3D imaging in medicine. CRC Press, Boca Ratón

Vargo SE, Muntz EP (1996) A simple micromechanical compressor and vacuum pump for flow control and other distributed applications. In: Thirty-fourth aerospace sciences meeting and exhibit, Jan 15–18, 1996, Reno, NV, AIAA 96–0310

Venkatakrishnan L, Bhat GS, Narasimha R (1999) Experiments on a plume with off-source heating: implications for cloud fluid dynamics. J Geophys Res 104(D12):14271–14281

Wagner C, Hüttl T, Sagaut P (2007) Large-Eddy simulation for acoustics. Cambridge University Press, Cambridge

Ward AD, Trimble SW (2004) Environmental hydrology. Lewis Publishers, CRC Press, Boca Ratón

Wei T, Willmarth WW (1991) Examination of v-velocity fluctuations in a turbulent channel flow in the context of sediment transport. J Fluid Mech 223:241–252

Westerweel J (1993) Digital particle velocimetry—theory and application. Delft University Press, Delft

Wittek A, Nielsen PMF, Miller K (eds) (2011) Computational biomechanics for medicine. Springer, Heidelberg

Yea Y, Cummings HZ (1964) Localized fluid flow measurements with an He-Ne laser spectrometer. Appl Phys Lett 4:176–178

Zabusky NJ, Hughes MH, Roberts KV (1979) Contour dynamics for the Euler equations in two dimensions. J Comput Phys 30:96–106

Waves and Instabilities in Rotating and Stratified Flows

Patrice Le Gal

Abstract This review intended primarily for Master degree students, presents the different types of classical waves that can occur in astro and geophysical flows. Inertial waves, caused by the rotation of the fluid, will first be introduced as well as their 2D version called Rossby waves. Then it will be shown how a density stratification of the fluid can make internal gravity waves appear. In each case and in the case where both rotation and stratification are present, the dispersion relations of the waves are derived. A differential rotation will then be added on the flow. The classical Rayleigh criterium for the centrifugal instability is recovered in the case of an homogeneous fluid but it will be shown that a new instability, called the strato-rotational instability (SRI), can occur when the fluid is stratified. Some experiments will be described. Finally, we will show how the application of a magnetic field can create Alfven waves in a rotating electrically conducting fluid and in which conditions the magneto-rotational instability (MRI) can grow.

1 Inertial Waves

First, let us consider an inviscid fluid of density ρ rotating around an axis $\overrightarrow{Oz}$ at a rate $\overrightarrow{\Omega}$. The linearized Euler equation for the perturbed velocity $\overrightarrow{u}$ and pressure field p describing the flow reads in the rotating frame of reference:

P. Le Gal (✉)
Institut de Recherche sur les Phénomènes Hors Equilibre, UMR 6594,
CNRS—Aix-Marseille Université, 49 rue F. Joliot Curie,
13384 Marseille, Cédex 13, France
e-mail: legal@irphe.univ-mrs.fr

J. Klapp et al. (eds.), *Fluid Dynamics in Physics, Engineering and Environmental Applications*, Environmental Science and Engineering,
DOI: 10.1007/978-3-642-27723-8_2, © Springer-Verlag Berlin Heidelberg 2013

$$\frac{\partial \vec{u}}{\partial t} + 2\vec{\Omega} \wedge \vec{u} = \frac{-\vec{\nabla}p}{\rho} \tag{1}$$

As can be seen, the Coriolis force appears here as a restoring force, forcing the displaced fluid particles to move on circles. The Coriolis force is the generating force of waves called "inertial waves" or "Kelvin waves" (Kelvin 1880). Searching for a local solution (in opposition to a global one where the geometry and boundary conditions are taken into account) of Eq. 1 under the form of a plane wave with a frequency ω and a wave vector $\vec{k} = (\alpha, \beta, \gamma)$: $(u, v, w, p/\rho) = (\hat{u}, \hat{v}, \hat{w}, \hat{p})e^{i(\alpha x+\beta y+\gamma z-\omega t)}$, we find easily the following algebraic system of equations:

$$\begin{cases} -i\omega\hat{u} \quad -2\Omega\hat{v} = -i\alpha\hat{p} \\ -i\omega\hat{v} + 2\Omega\hat{u} = -i\beta\hat{p} \\ -i\omega\hat{w} \quad = -i\gamma\hat{p} \end{cases} \tag{2}$$

This set of equations can be completed by the divergence free flow hypothesis that reads in the Fourier space as the following:

$$i\alpha\hat{u} + i\beta\hat{v} + i\gamma\hat{w} = 0 \tag{3}$$

Eliminating the pressure and velocity fields from these four equations leads to the well known dispersion relation of the Kelvin or inertial waves:

$$\omega^2 = \frac{4\gamma^2\Omega^2}{\alpha^2 + \beta^2 + \gamma^2} \tag{4}$$

This dispersion relation is special in the sense that the angle θ of propagation of a wave beam versus the rotation axis is simply given by the frequency of the waves:

$$\omega^2 = 4\Omega^2 cos^2(\theta) \tag{5}$$

Another characteristic of inertial waves is that their phase velocity is perpendicular to their group velocity. In order to illustrate the shape of the dispersion relation curve, we can fix the values of wavenumbers α and β and plot ω as functions of the axial wavenumber γ. Figure 1 shows that the frequencies are confined between $\pm 2\ \Omega$.

A classical way to produce these waves in the laboratory is to vibrate an object in a rotating tank at a given frequency. As explained before, the angle of propagation of the waves is determined by their frequency and thus the periodic fluid motions take place along cones emerging from the oscillating generator. Figure 2b is taken from an experiment by Messio et al. (2008) and Courtesy of University of Paris where the velocity field is measured by PIV in a plane perpendicular to the axis of rotation (which was here also the axis of vibration of a small cylinder).

Of course, as in a real fluid these waves are damped by viscosity, they need to be excited as we just saw in the previous experiment. However, the shape of the dispersion relations of inertial waves permits also parametric resonances to appear

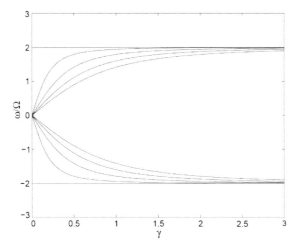

Fig. 1 Some dispersion relation curves of the Kelvin inertial waves as functions of axial wavenumber γ for various α and β

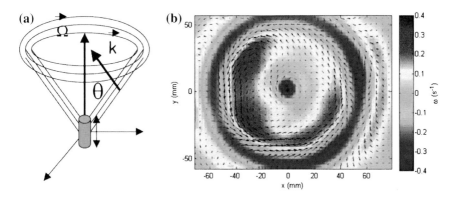

Fig. 2 **a** Schematic representation of the Kelvin wave cone generated by an oscillating cylinder in a rotating fluid. **b** Embarqued PIV measurements of the velocity field of a Kelvin wave conical beam generated by a vibrating cylinder, from Courtesy of University of Paris

under certain forcing conditions like those produced by precession, libration or tidal deformations of the rotating container. These resonances can indeed trigger the growth and interaction of inertial waves above a certain threshold. For instance, when a cylindrical rotating container is elliptically deformed in its cross section, two Kelvin waves can resonate with the elliptical deformation of the streamlines. An instability, called the elliptical instability can appear when these resonant conditions are met (Eloy et al. 2003). This instability is known to affect the vortices behind airplanes but can also play a role in astro and geophysics. Indeed, a planetary molten iron core flow could very well be destabilized by the

Fig. 3 Visualization of the elliptical instability of a tidally distorted deformable rotating cylindrical shell. In this illustration, the unstable mode is oscillating at a pulsation equal to Ω and corresponds to a resonance between two Kelvin waves with azimuthal wavenumbers 0 and 2

tidal distortions induced by a closed orbiting object (Lacaze et al. 2005; Cébron et al. 2010). Figure 3 presents a visualization of the elliptical instability in a rotating cylindrical shell where a Laser plane is sent in the meridional plane and illuminates Kalliroscope particles.

2 A Particular Case of Inertial Waves: The Rossby Waves

Laboratory observations of inertial waves are not easy and we saw in Sect. 1 that it was only relatively recently that these waves have been precisely measured when excited in a rotating tank. However, a particular case of these waves are known from a long time in meteorology. These waves called Rossby waves (Rossby 1939) can be described using first a bi-dimensional approximation of Eq. 1 ($\frac{\partial}{\partial z} = 0$ and $w = 0$), second when taking into account a linear variation of the Coriolis parameter 2Ω with the latitude (the β plane approximation). The flow is then described by the following equations:

$$\begin{cases} \frac{\partial u}{\partial t} \quad\;\; - 2\Omega(y)v = -\frac{1}{\rho}\frac{\partial p}{\partial x} \\ \frac{\partial v}{\partial t} + 2\Omega(y)u \quad\;\; = -\frac{1}{\rho}\frac{\partial p}{\partial y} \end{cases} \tag{6}$$

with $\frac{\partial \Omega(y)}{\partial y} = \beta_c$. The variation in time of the axial vorticity Ω_z of the two-dimensional motion can then easily be expressed as:

$$\frac{\partial \Omega_z}{\partial t} + 2\beta_c v = 0 \tag{7}$$

As in Sect. 1, this equation is completed by the incompressibility condition and taking their Fourier transforms leads to the following dispersion of the Rossby waves:

$$\omega = \frac{-2\beta_c \alpha}{\alpha^2 + \beta^2} \tag{8}$$

Fig. 4 Some dispersion relation curves of the Rossby waves as functions of the azimuthal wavenumber α for various β

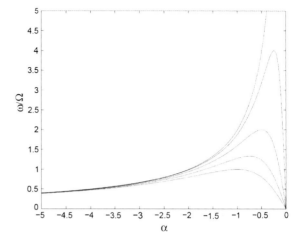

Fig. 5 Illustration of the propagation of a Rossby wave in the atmosphere, from Courtesy of University o Oregon (1999)

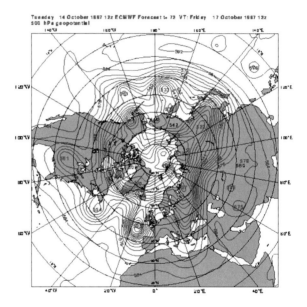

As can be seen from this dispersion relation, the phase velocity of the Rossby waves is $\frac{\omega}{\alpha} < 0$ and this is the reason why Rossby waves propagate westward in the Northern atmosphere. As in Sect. 1 we can fix the values of the wavenumbers β and plot in Fig. 4 the frequency ω as functions of the azimuthal wavenumber α. Classical meteorological images used for weather forecast give good illustrations of the propagation of the Rossby waves in the atmosphere. The succession of cyclonic and anticyclonic structures in the pressure field circumtravel around the North Pole as illustrated in Fig. 5.

Fig. 6 Illustration of the
propagation of a Rossby
wave in a rotating tank with a
tilted bottom, from Courtesy
of University of Washington

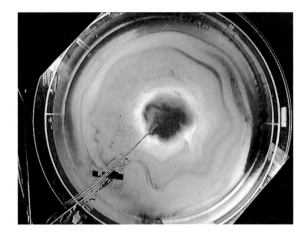

Fig. 7 Shadowgraph
visualizations of the
St Andrew cross formed by
the four internal wave beams
generated by a vertically
vibrated small cylinder in a
stratified layer of salt water,
from Courtesy of Sakai,
Iizawa, Aramaki (1997)

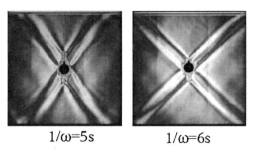

$1/\omega = 5s$ $1/\omega = 6s$

These waves can be reproduced in laboratory experiments using rotating tanks
at constant rate Ω that possess tilted bottoms at an angle $\hat{\theta}$. The three-dimensional
inviscid equation that describes the fluid motions in the rotating frame of reference
is Eq. 1. Taking the curl and then using the divergenceless property of the flow,
leads to the equation of the axial vorticity Ω_z:

$$\frac{\partial \Omega_z}{\partial t} = -2\Omega \frac{\partial w}{\partial z} \tag{9}$$

If we search solutions for u and v that are invariant along the z axis (the geo-
strophic hypothesis), the z derivative of the former equation gives: $\frac{\partial^2 w}{\partial z^2} = 0$. The
vertical component of the velocity is therefore a linear function of the axial
coordinate z: $w = az + b$. Applying the boundary conditions ($w = 0$) on the top
surface supposed at $z = 0$ and at the bottom at $z = -h$ ($v_\perp = w(t, x, -h)$
$cos\hat{\theta} + v(t, x, -h)sin\hat{\theta} = 0$) permits to link the axial velocity w and the velocity v
perpendicular to the axis of rotation in a neighborhood of x where h is supposed
constant:

$$\frac{\partial \Omega_z}{\partial t} = 2\frac{\Omega\, tg\hat{\theta}}{h} v(t, x, h) \tag{10}$$

Equation 10 is analogous to Eq. 7 where the β effect is given by $\beta_c = 2\frac{tg\hat{\theta}}{h}$. This analogy leads to the dispersion relation of Rossby waves in rotating tanks with tilted bottoms:

$$\omega = -2\frac{\Omega\, tg\hat{\theta}}{h\alpha} \tag{11}$$

Figure 6 illustrates this analogy where an oscillating cylinder generates wave trains visualized by dye in a rotating tank with a tilted bottom.

3 Gravity Waves

Often, planetary atmospheres or oceans are stably stratified in the vertical direction $\overrightarrow{Oz}$ leading to the existence of a new set of waves called gravity or internal waves. Starting with no rotation, the linearized equations of motion for a stratified flow are:

$$\begin{cases} \frac{\partial u}{\partial t} = & -\frac{1}{\rho}\frac{\partial p}{\partial x} \\ \frac{\partial v}{\partial t} = & -\frac{1}{\rho}\frac{\partial p}{\partial y} \\ \frac{\partial w}{\partial t} = -\frac{1}{\rho}\frac{\partial p}{\partial z} + b \\ \frac{\partial b}{\partial t} = & -N^2 w \end{cases} \tag{12}$$

where the buoyancy force is approximated by the first term of its Taylor expansion: $\overrightarrow{b} = \frac{1}{\rho}\frac{\partial \rho}{\partial z}\overrightarrow{g}z$. Traditionally, the vertical gradient of density is written in term of a Brunt-Väisälä frequency N which is defined by: $N^2 = -\frac{1}{\rho}\frac{\partial \rho}{\partial z}g$. As before, taking the Fourier transform of the linearized Euler equation leads to the algebraic set of equations:

$$\begin{cases} -i\omega\hat{u} = & -i\alpha\hat{p} \\ -i\omega\hat{v} = & -i\beta\hat{p} \\ -i\omega\hat{w} = -i\gamma\hat{p} + \hat{b} \\ -i\omega\hat{b} = & -N^2\hat{w} \end{cases} \tag{13}$$

which leads to the dispersion relation of internal gravity waves:

$$\omega^2 = N^2\frac{\alpha^2 + \beta^2}{\alpha^2 + \beta^2 + \gamma^2} \tag{14}$$

Or, if we note θ the angle of the wave vector versus the vertical axis of stratification:

$$\omega^2 = N^2 sin^2(\theta) \tag{15}$$

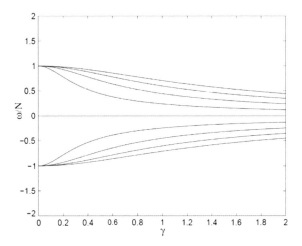

Fig. 8 Some dispersion relation curves of the internal waves as functions of axial wavenumber γ for various α and β values

Comparing both dispersion relations (5) and (15), the parallel between inertial and internal waves is striking. This confers to both systems of waves similar characteristics: for instance, their frequency determines the angle of propagation of the beams and their phase and group velocities are perpendicular one to the other respectively. Figure 7 shows two shadowgraph images of the St-Andrew cross generated by the oscillation of a small cylinder in a salt stratified layer of water (Courtesy of Sakai, Iizawa, Aramaki 1997). As explained before, it can be observed that changing the excitation frequency changes the angle of the wave beams.

The dispersion relation given by Eq. 14 can be illustrated by plotting ω as functions of γ for different values of α and β. Because of the form of this dispersion relation, Fig. 8 shows that the frequencies of internal waves are confined between $\pm N$.

4 Gravito-Inertial Waves

Rotation and stratification versus $\overrightarrow{Oz}$ can be taken into account together to study gravito-inertial waves. The linearization of the Euler equations written in the rotating frame of reference gives the following system of equations:

$$\begin{cases} \frac{\partial u}{\partial t} - 2\Omega v = -\frac{1}{\rho}\frac{\partial p}{\partial x} \\ \frac{\partial v}{\partial t} + 2\Omega u = -\frac{1}{\rho}\frac{\partial p}{\partial y} \\ \frac{\partial w}{\partial t} = -\frac{1}{\rho}\frac{\partial p}{\partial z} + b \\ \frac{\partial b}{\partial t} = -N^2 w \end{cases} \tag{16}$$

with the same notation as before. The Fourier transform of this system of differential equations leads naturally to the dispersion relation of the gravito-inertial waves:

Fig. 9 Conditions of existence of gravito-inertial waves in the (ω, N) plane. Two distinc domains of existence (*empty areas*) are defined by the *lines* $\omega = \pm N$ and $\omega = \pm 2\,\Omega$

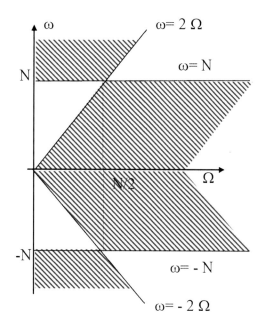

$$\omega^2 = \frac{4\gamma^2\Omega^2 + N^2(\alpha^2 + \beta^2)}{\alpha^2 + \beta^2 + \gamma^2} \tag{17}$$

which writes also as a function of the angle θ of the wave vector versus the vertical axis of rotation:

$$\omega^2 = 4\Omega^2\cos^2(\theta) + N^2\sin^2(\theta) \tag{18}$$

Note that this dispersion relation was only quite recently studied in Peacock and Weidman (2005). Because of the presence of the sine and cosine functions, these waves can only exist for certain frequencies ω. Figure 9 shows that two distinct regions of existence in the (ω, Ω) plane are defined by the lines $\omega = \pm N$ and $\omega = 2 \pm \Omega$. In each domain, a typical dispersion relation curve can be plotted. Figure 10 gives some example of these curves in both cases.

5 Waves in Differentially Rotating Flows

Evaluating the dispersion relation of waves when the rotation of the flow is not a solid-body rotation but depends upon the radius: $\Omega = f(r)$ is more tricky. The equation of motion must be written in the stationary frame of reference:

$$\frac{\partial \vec{u}}{\partial t} + (\vec{U}.\vec{\nabla})\vec{u} + (\vec{u}.\vec{\nabla})\vec{U} = -\frac{\vec{\nabla}p}{\rho} \tag{19}$$

Fig. 10 Some dispersion relation curves of the gravito-internal waves as functions of axial wavenumber γ for various α and β. In this case, N was chosen equal to $\pm 4\,\Omega$ for the curves outside the range $[-2\Omega, 2\Omega]$ (*solid curve*), and equal to $\pm 0.3\,\Omega$ for the curves inside the range $[-2\Omega, 2\Omega]$ (*dashed curves*)

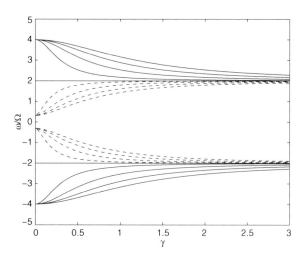

The difficulty in the calculation of the dispersion relations of these inertial waves comes from the nonlinear advective term that this time, possesses a new term coming from the variation of Ω with r. However, the problem can be solved if we search for solutions in the form of waves, but where the wave vector also rotates around the axis of rotation of the flow:

$$\begin{cases} \frac{\partial \alpha}{\partial t} + \Omega\beta = 0 \\ \frac{\partial \beta}{\partial t} + \Omega\alpha = 0 \\ \frac{\partial \gamma}{\partial t} = 0 \end{cases} \tag{20}$$

Using this property, it can be shown that $\frac{x}{r} = \frac{\alpha}{a}$ and $\frac{y}{r} = \frac{\beta}{a}$, where a is the norm of the projection of the wavevector in the plane perpendicular to the rotation axis: $a^2 = \alpha^2 + \beta^2$. The equations of motion can then be written under the form:

$$\begin{cases} -i\omega\hat{u} \qquad\quad - \Omega\hat{v} + \frac{y\gamma}{a^i}\frac{d\Omega}{dr}\hat{w} = -i\alpha\hat{p} \\ -i\omega\hat{v} + \Omega\hat{u} \qquad - \frac{x\gamma}{a^i}\frac{d\Omega}{dr}\hat{w} = -i\alpha\hat{p} \\ -i\omega\hat{w} \qquad\qquad\qquad\qquad = -i\gamma\hat{p} \end{cases} \tag{21}$$

Again, using that $\frac{x}{r} = \frac{\alpha}{a}$ and $\frac{y}{r} = \frac{\beta}{a}$, the y and x terms can be transformed. The dispersion relation of inertial waves (at radius r) in a differentially rotating flow can then be calculated after some algebra:

$$\omega^2 = \frac{2\Omega\left(2\Omega + r\frac{d\Omega}{dr}\right)\gamma^2}{\alpha^2 + \beta^2 + \gamma^2} \tag{22}$$

As can be seen from the above equation, we retrieve the original Kelvin wave dispersion relation when Ω is constant. Also, we can observe that ω is a complex number when $\Omega(2\Omega + r\frac{d\Omega}{dr}) < 0$ which is nothing else than the Rayleigh criterium

Fig. 11 The Strato-
Rotational Instability
between differentially
rotating cylinders (Le Bars
and Le Gal 2007)

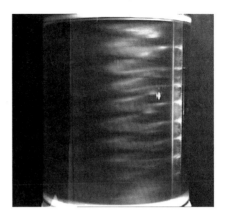

for the appearance of the centrifugal instability in a rotating flow. Any inviscid flow rotating with a differential rotation $\Omega = f(r)$, is unstable if its angular momentum is a decreasing function of r.

If the flow is furthermore axially stratified, the dispersion relation of these waves can also be calculated. It takes the following form:

$$\omega^2 = 2\Omega \left(2\Omega + r\frac{d\Omega}{dr} \right) cos^2(\theta) + N^2 sin^2(\theta) \tag{23}$$

where θ is as before the angle of the wavevector versus the rotation axis of the flow. As can be seen on this equation, the stratification term is positive and thus stabilizes the flow as soon as $\theta \neq 0$. However, if $\theta = 0$, as it is the case for Taylor-Couette vortices, stratification has no effect on the threshold of the centrifugal instability at least at this first order of calculation. A new instability has been however discovered analytically recently (Molemaker et al. 2001). This instability comes from a resonant interaction of the gravito-inertial waves and was named the SRI. It was then studied numerically by Shalybkov and Rüdiger (2005) and experimentally by Le Bars and Le Gal (2007). Figure 11 illustrates the SRI instability in a cylindrical Couette flow. As can be observed, two counter-rotating helices propagate in the gap between both cylinders and produce a braided pattern.

Note that this instability was further studied in the case where the flow is not confined between two walls (Riedinger et al. 2011). This case is particularly interesting for modelling the stability of accretion disks around stars as the laminar Kepler flow of the rotating gas cloud ($\Omega(r) \sim r^{-3/2}$) is stable versus the Rayleigh criterium. Another famous instability which is often invoked for the destabilization of accretion disks, is the so-called Magneto-Rotational Instability MRI (Balbus and Hawley 1998). As we will see in next paragraph, contrary to the SRI, the MRI directly affects the Rayleigh criterium of rotating flows.

6 Alfven and Magneto-Inertial Waves

Now the fluid is considered to be a perfect electric conductor. This framework is called the Ideal Magneto-Hydrodynamics limit. If the viscous effects are still neglected, the linearized Euler equation describes the fluid motions with an additional Lorentz force generated by an imposed axial homogeneous magnetic field $\vec{B}$. The second equation is the induction equation for the magnetic fluctuations $\vec{b}$ and is reduced in this case to the induction of the magnetic field by the flow perturbations:

$$\begin{cases} \frac{\partial \vec{u}}{\partial t} = -\frac{1}{\rho}\vec{\nabla}p + \frac{\vec{F}}{\rho} \\ \frac{\partial \vec{b}}{\partial t} = (\vec{B}.\vec{\nabla})\vec{u} \end{cases} \tag{24}$$

with $\vec{F} = \vec{J} \wedge \vec{b}$ where $\vec{J}$ is the electric current density given by $\vec{J} = \frac{\vec{\nabla} \wedge \vec{b}}{\mu}$, where μ is the magnetic diffusivity. Solving these equations with plane waves for $\vec{u}$ and $\vec{B}$ leads to:

$$\begin{cases} -\omega\hat{u} = \frac{B}{\rho\mu}(\gamma\hat{b}_y - \alpha\hat{b}_z) - \alpha\hat{p} \\ -\omega\hat{v} = \frac{B}{\rho\mu}(\gamma\hat{b}_x - \beta\hat{b}_z) - \beta\hat{p} \\ -\omega\hat{w} = \qquad\qquad -\gamma\hat{p} \end{cases} \tag{25}$$

and

$$\begin{cases} -\omega\hat{b}_x = B\gamma\hat{u} \\ -\omega\hat{b}_y = B\gamma\hat{v} \\ -\omega\hat{b}_z = B\gamma\hat{w} \end{cases} \tag{26}$$

The dispersion relation of Alfven waves with the associated Alfven phase speed V_A is then easily deduced (Alfven 1942):

$$\omega^2 = \frac{B^2}{\rho\mu}\gamma^2 \quad V_A = \frac{B}{\sqrt{\rho\mu}} \tag{27}$$

The linear form of Alfven waves dispersion relation shows that these waves are not dispersive and propagate at a constant Alfven speed V_A. Although these waves are known in astrophysics for a long time (Tsurutani et al. 2005), it is only recently that these waves have been observed in a laboratory experiment (Alboussière et al. 2011). Note that the Alfven speed is around 1 m/s for liquid Gallium under a magnetic field of 0.1 Tesla.

When rotation is added to the flow of conducting fluid, the Coriolis terms need to be incorporated into the equations of motion that become in the Fourier space:

Fig. 12 Some dispersion relation curves of the magneto-internal waves as functions of axial wavenumber γ for various α and β. In this case V_A was chosen equal to $\Omega\gamma$

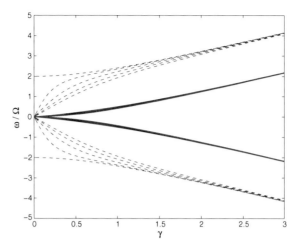

$$\begin{cases} \omega\hat{u} = 2i\Omega\hat{v} - \frac{B^2\gamma^2}{\rho\mu\omega}\hat{u} + \frac{B^2\gamma\alpha}{\rho\mu\omega}\hat{w} + \alpha\hat{p} \\ \omega\hat{v} = -2i\Omega\hat{u} + \frac{B^2\gamma^2}{\rho\mu\omega}\hat{v} - \frac{B^2\gamma\alpha}{\rho\mu\omega}\hat{w} + \beta\hat{p} \\ \omega\hat{w} = \gamma\hat{p} \end{cases} \tag{28}$$

The analytical formula for the dispersion relation can be calculated and takes the following form:

$$\omega^2 = \frac{\gamma^2}{k^2}\left(V_A^2 k^2 + 2\Omega^2 \pm 2\Omega\sqrt{k^2 V_A^2 + \Omega^2}\right) \tag{29}$$

where $k^2 = \alpha^2 + \beta^2 + \gamma^2$ is the square of the norm of the wavenumber and $V_A = \frac{B}{\sqrt{\rho\mu}}$ the Alfven speed as before. These waves are called Magneto-inertial waves or Magneto-Coriolis waves and their dispersion relation can be plotted the same way it was for pure inertial waves. As observed in Fig. 12, two kinds of branches (the magnetic (solid curves) or the hydrodynamic branches (dashed curves)) can be distinguished. As can be seen by the curvature of the curves, the magneto-inertial waves are dispersive except when the rotation of the flow is dominated by the magnetic field effect where the Alfven waves characteristics are recovered at large γ.

This dispersion relation formula can then be extended to non solid body rotation cases when introducing the axial vorticity $\Omega_z = \Omega + r\frac{d\Omega}{dr}$, already calculated in Sect. 5. The dispersion relation then becomes:

$$\omega^2 = \frac{\gamma^2}{k^2}\left(V_A^2 k^2 + \Omega\Omega_z \pm \Omega\sqrt{4k^2 V_A^2 + \Omega_z^2}\right) \tag{30}$$

This dispersion relation formula shows in particular that ω^2 can become negative and thus the flow be unstable if $V_A^2 k^2 < -2r\Omega\frac{d\Omega}{dr}$. This criteria replaces the Rayleigh criteria for the centrifugal instability and gives the threshold of the Magneto-Rotational Instability which has a tremendous importance in astrophysics in particular for

Fig. 13 Stability thresholds for the centrifugal (Rayleigh criterium) and the Magneto-Rotational Instability in the $(2\Omega, r\frac{d\Omega}{dr})$ plane. The *straight line* is the Rayleigh criterium and the hyperbolas correspond to the MRI threshold for given magnetic Alfven speed and non zero wave number. We observe that a stable centrifugal inviscid flow can become MRI unstable in the *colored* and *hatched* region in the limit of ideal MHD

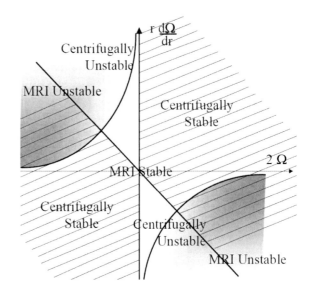

the destabilization of accretion disks (Balbus and Hawley 1998). We can plot for a given r both stability thresholds in the plane $(2\Omega, r\frac{d\Omega}{dr})$. Figure 13 shows these limits for given magnetic Alfven speed and non zero wave number. In particular, we can notice that a centrifugally stable flow can indeed be destabilized by the effect of an imposed magnetic field in the colored and hatched zones of the diagram. As can be seen on the figure the smallest Ω is given by the equality of both criteria. Supposing that the smallest wavelength in the device of typical length L is $k = \pi/L$, it is easy to see that the smallest velocity reached in the flow is $V = 2\pi\Omega R = \pi^2 V_A$. As seen before, the order of magnitude of the Alfven speed V_A is 1 m/s in Gallium liquid metal under a magnetic field of 0.1 Tesla. This gives a minimum experimental velocity of 10 m/s and shows the extreme difficulties of designing such a liquid metal experiment. Despite these technical difficulties, several experiments are attempting today to reproduce this instability in Taylor-Couette flow devices, using Sodium as working fluid (Sisan et al. 2004; Stefani et al. 2009).

7 Conclusion

Considering several generic cases of rotating flows, the dispersion relations of different types of waves have been derived from the linearized equations of motion. First inertial or Kelvin waves that propagate in solid body rotating flows were introduced. The 2D version of these waves, known as Rossby waves, that propagate in shallow layers when taking account the so called β effect, was then derived. Pure inertial waves were further enriched by the effect of first a density stratification, then by a non homogeneous rotation and finally by the application of a magnetic field. The Rayleigh criterium for the inviscid threshold of the centrifugal instability was recovered from these dispersion relations and compared to the Magneto-Rotational Instability threshold. Besides, it was moreover shown that the Strato-Rotational Instability is not caused by a modification of the Rayleigh discriminant inequality but by a resonant phenomenon of gravito-inertial waves that can destabilize a centrifugally stable Taylor-Couette flow.

Especial acknowledgements to Prof. Anne Cros for her kind help during my visit to Mexico. I also thank the Secretariá de Relaciones Exteriores, Dirección General de Cooperación Educativa y Cultural de México for their financial support. I am also grateful to S. Le Dizès for fruitful discussions in the course of preparing this lecture.

References

Alboussière T, Cardin P, Debray F, La Rizza P, Masson JP, Plunian F, Ribeiro A, Schmitt D (2011) Experimental evidence of Alfven wave propagation in a Gallium alloy. Phys Fluids 23:096601

Alfven H (1942) Existence of electromagnetic-hydrodynamic waves. Nature 150:405–406

Balbus SA, Hawley JF (1998) Instability, turbulence, and enhanced transport in accretion disks. Rev Mod Phys 70:1–53

Cébron D, Le Bars M, Leontini J, Maubert P, Le Gal P (2010) A systematic numerical study of the tidal instability in a rotating triaxial ellipsoid. Phys Earth Planet Inter 182:119–128

Courtesy of University of Paris-Sud. http://www.fast.u-psud.fr/ppcortet/inertialwave.php

Courtesy of University of Washington. http://www.ocean.washington.edu/people/faculty/rhines/rossbypbr.html

Courtesy of Sakai S, Iizawa I, Aramaki E (1997) Atmosphere and Ocean in a Laboratory: Hokusai. http://www.gfd-dennou.org/library/gfdexp/expe/exp/iw/1/res.htm

Courtesy of University of Oregon (1999). http://zebu.uoregon.edu/1999/ph161/images/rossbyfor.gif

Eloy C, Le Gal P, Le Dizès S (2003) Elliptic and triangular instabilities in rotating cylinders. J Fluid Mech 476:357–388

Kelvin L (1880) Vibrations of a columnar vortex. Phil Mag 10:155–68

Lacaze L, Le Gal P, Le Dizès S (2005) Elliptical instability in a rotating spheroid. J Fluid Mech 505:1–22

Le Bars M, Le Gal P (2007) Experimental analysis of the stratorotational instability in a cylindrical couette flow. Phys Rev Lett 99:064502

Messio L, Morize C, Rabaud M, Moisy F (2008) Experimental observation using particle image velocimetry of inertial waves in a rotating fluid. Exp Fluids 44(4):519–528

Molemaker MJ, McWilliams JC, Yavneh I (2001) Instability and equilibration of centrifugally stable stratified Taylor-Couette flow. Phys Rev Lett 86:5270–5273

Peacock T, Weidman P (2005) The effect of rotation on conical wave beams in a stratified fluid. Exp Fluids 39(1):32–37

Riedinger X, Le Dizès S, Meunier P (2011) Radiative instability of the flow around a rotating cylinder in a stratified fluid. J Fluid Mech 672:130–146

Rossby C-G (1939) Relation between variations in the intensity of the zonal circulation of the atmosphere and the displacements of the semi-permanent centers of action. J Mar Res 2(1):38–55

Shalybkov D, Rüdiger G (2005) Non-axisymmetric instability of density-stratified Taylor-Couette flow. J Phys 14:128–137

Sisan DR, Mujica N, Tillotson WA, Huang YM, Dorland W, Hassam AB, Antonsen TM, Lathrop DP
     (2004) Experimental observation and characterization of the magnetorotational instability. Phys
     Rev Lett 93:114502
Stefani F, Gerbeth G, Gundrum T, Hollerbach R, Priede J, Rüdiger G, Szklarski J (2009) Helical
     magnetorotational instability in a Taylor-Couette flow with strongly reduced Ekman pumping.
     Phys Rev E 80:066303
Tsurutani BT, Lakhina GS, Pickett JS, Guarnieri FL, Lin N, Goldstein BE (2005) Nonlinear
     Alfven waves, discontinuities, proton perpendicular acceleration, and magnetic holes/
     decreases in interplanetary space and the magnetosphere: intermediate shocks? Nonlinear
     Process Geophys 12:321–326

The Sloshing-Induced Motion of Free Containers

Patrick D. Weidman and Andrzej Herczynski

Abstract Experiments for the time-periodic liquid sloshing-induced sideways motion of free containers are compared with theory using standard normal mode representations for rectangular boxes, upright cylinders, wedges and cones of 90° apex angles, and cylindrical annuli. While the wedge and cone exhibit only one mode of oscillation, the boxes, cylinders and annuli have an infinite number of modes. In some cases we have been able to excite the second mode of oscillation. Frequencies ω were acquired as the average of three experimental determinations for every filling of mass m in the dry containers of mass m_0. Measurements of the dimensionless frequencies ω/ω_R over a range of dimensionless liquid masses $M = m/m_0$ are found to be in essential agreement with theoretical predictions. The frequencies ω_R used for normalization arise naturally in the mathematical analysis, different for each geometry considered.

This work, presented as a plenary talk for the international Enzo-Levi meeting held during May 5, 6 2011 in Guadalajara, Mexico, has now been published in the *Journal of Fluid Mechanics*. This article represents a condensed version of that *JFM* publication cited as Herczynski and Weidman (2012).

P. D. Weidman (✉)
Department of Mechanical Engineering, University of Colorado, Boulder, 80309-0427 CO, USA
e-mail: weidman@colorado.edu

A. Herczynski
Department of Physics, Boston College, Chestnut Hill, MA 02467-3811, USA

J. Klapp et al. (eds.), *Fluid Dynamics in Physics, Engineering and Environmental Applications*, Environmental Science and Engineering, DOI: 10.1007/978-3-642-27723-8_3, © Springer-Verlag Berlin Heidelberg 2013

1 Introduction

The hydrodynamic coupling of liquid sloshing in containers moving in some constrained manner has been the subject of scientific investigation for more than seven decades. Much of the original work was concerned with the effect of liquid propellant sloshing on the stability of ballistics and space vehicles in the 1960s; see Moiseev (1964) and Abramson (1966). Other studies have dealt with disturbances of trucks or ships transporting large partially-filled liquid containers to external forcings induced by a corrugated road on a moving truck or by periodic surface waves on a moving ship; see Dodge (2000), Ibrahim (2005) and Faltinsen and Timokha (2009) for an extensive summary of these parametrically-forced liquid transport problems.

Cooker (1994) studied the motion of partially filled containers suspended as a bifilar pendulum in the shallow-water limit. Weidman (1994) extended Cooker's theory to multi-compartment rectangular boxes and right circular cylinders suspended as bifilar pendula and later Weidman (2005) reported experimental data at different pendulum lengths and different liquid fillings for those geometries. Recently, Ardakani and Bridges (2010) provided an alternative derivation and a Lagrangian representation of Cooker's problem, still in the shallow-water limit. Yu (2010) reported finite-depth potential theory for the bifilar suspension of boxes and cylinders with two aims: (i) to eliminate the shallow-water restriction which assumes the pressure acting on the sidewalls is hydrostatic, and (ii) to explicitly display the effect of evanescent waves in the system.

The problem of interest here may be thought of as the infinite pendulum length limit of Cooker's bifilar suspended container. Yu (2010), having taken an interest in our experiments on freely moving containers driven by liquid sloshing (Herczynski and Weidman 2009), provided the finite-depth infinite pendulum length limit eigenvalue equations for boxes and cylinders.

The presentation begins with the problem formulation in Sect. 2. The experimental procedure is described in Sect. 3. The finite-depth eigenvalue equations for each geometry considered are given and results are compared with measurements in Sect. 4. A discussion and concluding remarks are presented in Sect. 5.

2 Problem Formulation

Consider a container of mass m_0 partially filled with liquid of mass m free to move in frictionless horizontal motion. If $X(t)$ denotes the horizontal position of the container in the stationary laboratory reference frame, then Newton's second law for the container takes the form

$$m_0 \ddot{X} = F_p \qquad (1)$$

where

$$F_p = \int_S p(\mathbf{n} \cdot \mathbf{i}) dS \qquad (2)$$

is the X-component of the pressure force acting on the container walls. Here p is the hydrodynamic pressure acting over the wetted surface S of the container, $\mathbf{n}$ is the unit normal to S pointing out of the fluid domain, and $\mathbf{i}$ is the unit vector directed along the X-axis. Surface tension is neglected, though capillary effects could easily be included in the calculation of the sloshing waveforms. It is convenient to calculate the pressure, velocity potential ϕ, and free surface displacement ζ in a coordinate system (x, y, z) attached to the container, with z pointing upwards and $z = 0$ the position of the quiescent free surface. Linearized potential motion is assumed so that the pressure in the frame of reference moving with the container is given by

$$p = -\rho(\phi_t + gz + x\ddot{X}). \qquad (3)$$

Here ρ is the liquid density, g is gravity, and $-\rho x\ddot{X}$ is the body force due to the acceleration of the container.

The velocity potential, pressure, and free surface displacement are determined from the solution of the linearized potential flow boundary-value problem

$$\nabla^2 \phi = 0 \quad (\text{in } \mathcal{D}) \qquad (4a)$$

$$\phi_t + g\zeta + x\ddot{X} = 0 \quad (z = 0) \qquad (4b)$$

$$\phi_z = \zeta_t \quad (z = 0) \qquad (4c)$$

$$\mathbf{n} \cdot \nabla \phi = 0 \quad (\text{on } S) \qquad (4d)$$

in which $\mathcal{D}$ is the domain of the quiescent liquid. For future reference we note that $\zeta(x, t)$ may be eliminated from (4b, 4c) yielding the combined kinematic and dynamic free surface condition

$$\phi_{tt} + g\phi_z + x\ddot{X} = 0 \quad (z = 0). \qquad (5)$$

We are interested in container shapes amenable to analytic solution which usually require some form of symmetry about the vertical axis or plane. Considered below are rectangular geometries, upright cylinders, cones and wedges with 90° apex angles, and cylindrical annuli. Experiments show that complicated rectilinear motions can arise depending on how the system is put into motion. With some practice we have learned how to manually excite the containers to yield damped periodic motion with little drift. With 'improper' excitation, the container was observed to oscillate while translating. Although these latter cases are certainly interesting from a dynamical systems point of view, we analyze here only periodic motions of the simplest form

$$X(t) = X_0 \cos \omega t. \tag{6}$$

The important dimensionless parameter for this study is the ratio of liquid mass m to dry container mass m_0, denoted by

$$M = \frac{m}{m_0}. \tag{7}$$

The goal is to calculate the dimensionless finite-depth frequency of periodic motion ω/ω_R as a function of M and compare with laboratory experiments. Natural choices for the reference frequency ω_R will avail themselves for each geometry considered.

The adopted classical modal method inherently includes the effect of evanescent waves but does not explicitly separate out their contribution to the composite travelling- and evanescent-wave solution as in Yu (2010). While our solutions can be cast in a more compact form compared to those presented in Yu (2010), this is at the expense of slower convergence of the series expressions obtained. Nevertheless, there is no hinderance for determining accurate solutions with the aid of Mathematica (Wolfram 1991).

3 Measurement Procedure

The experimental set-up consisted of a low-friction cart and a 1.2 m long aluminum track commercially available from PASCO specializing in physics apparatus for teaching laboratories. The cart has a mass of about 0.5 kg and is outfitted with four knife-edge wheels rotating freely on high quality ball bearings which are attached to the chassis via a suspension system with springs above each wheel. The cart's wheels fit snugly into two parallel groves along the track, which could be accurately leveled using four adjustment screws to assure one-dimensional, horizontal motion with minimal mechanical resistance.

A small plastic insert was fastened with two screws to the cart inside the hollow on its upper surface (designed to carry extra masses). The insert provided a flat surface on which each container could be attached using strong, double-sided adhesive tape. This mounting system proved reliably rigid and allowed us to attach and detach the containers with ease. However, we were limited in the maximum weight on the cart's wheels to less than 40 N since beyond this load the springs began to give-in and became extremely soft, making the cart wobbly and subject to transverse oscillations. Since our typical dry mass m_0 (cart + insert + dry container) was in the range 900–1,600 g, our containers could be filled with roughly 2–3 kg of water, depending on the container being tested. We were also limited by

Table 1 Container dimensions, dry masses m_0, and maximum percentage change in M due to evaporation

	L (cm)	W (cm)	H_b (cm)	R (cm)	R_1 (cm)	R_2 (cm)	m_0 (g)	$\%M_{loss}$
Large box	24.74	8.255	7.5	—	—	—	886.5	0.76
Tall box	14.67	9.59	14.7	—	—	—	922.4	0.33
Large cylinder	—	—	7.5	13.02	—	—	1129	0.76
Tall cylinder	—	—	17.6	7.335	—	—	1086	0.27
Wedge	—	24.75	12.3	—	—	—	1138	0.82
Cone	—	—	10.5	—	—	—	1228	1.48
Annulus ($\eta = 0.364$)	—	—	7.5	—	4.73	12.98	1207	0.76
Annulus ($\eta = 0.777$)	—	—	15.2	—	10.113	13.013	1644	0.32

the brimful heights H_b of the containers (see Table 1) and could fill them only up to about 2 cm below the rim in order to prevent spilling during the back-and-forth motion of the cart.

Each container was fabricated from transparent lucite to minimize the total weight m_0 of the tank plus the cart. The disadvantage is that the damping of standing waves in lucite containers is much greater compared to that in glass containers (Keulegan 1959). Boxes and wedges had sidewalls composed of 1/8 inch plate while the cylindrical and annular geometries were composed of 1/8 inch wall cylindrical stock. The bottom surfaces were fabricated from 3/16 to 1/4 inch plate. The wedge was cemented to a base platform 5.1×7.6 cm. The cone and its 7.6 cm diameter cylindrical base were machined as a single unit from solid cylindrical stock. Relevant details of the containers are given in Table 1 in which $L = 2D$ for the boxes, H_b is the brimful height of a container, and m_0 is the dry mass of the system, i.e. the mass of the cart and insert, double sticky adhesive and plexiglass container. To verify that the liquid volume remained reasonably constant, we have measured the evaporation rate of water from our containers. We found a very linear rate of evaporation per unit surface area with time, the value being $R_{evap} = 7.00 \times 10^{-3}$ g/hr-cm^2. For the maximum six hour period over which frequency measurements were made, we can estimate the maximum percentage change in M for each container, $\%M_{loss}$, and these data are also included in Table 1.

With the system set in motion, its position was recorded using a PASCO motion sensor aligned with, and located 30–40 cm from, the end of the cart. The motion sensor works by repeatedly sending bursts of 49 kHz ultrasonic pulses and measuring the time they take to reflect back from the moving cart. The sensor is connected to a computer via a PASCO universal interface (ScienceWorkshop 750) and the position versus time data can be saved in tabular form and/or displayed graphically on a computer screen. We used the sampling rate of 100 or 120 Hz and the position data was obtained with the nominal accuracy of ±0.001 m; in practice, the measurements are reliable at least to ±0.1 mm. To determine the frequency of cart's oscillations, we read the elapsed time over multiple cycles (peak-to-peak or trough-to-trough), and averaged the resultant periods over three separate runs for

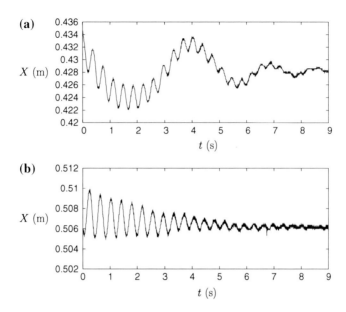

Fig. 1 Damping of *horizontal* motions initiated for **a** the *tall box* at $M = 1.66$ and **b** the *large box* at $M = 1.52$. Details for the *tall* and *large boxes* are given in Table 1

each liquid filling. We made sure that there was a few seconds delay between releasing the cart and the start of the recording so that most transients would attenuate. We also ignored the first few recorded cycles, and the ringing at the end of each run with very small amplitudes; see Fig. 1b.

Containers partially filled with water were set into motion by manually oscillating the system and setting it free. Two traces of recorded container motion are presented in Fig. 1. Figure 1a for the tall cylinder at $M = 1.66$ shows a complicated response of the system that can arise depending on the initial conditions, in this case a superposition of two distinct frequencies. More typical were traces resulting from the superposition of oscillations in the lowest mode with a nearly constant-speed translation of the center of mass (drift). None of these more complicated traces were deemed usable for our measurements. We relied on regular, single frequency oscillatory traces with minimal drift, such as that shown in Fig. 1b, taken using the tall box at $M = 1.52$. We recognize that a sinusoidally driven linear actuator could have been devised to place the cart supporting the container into motion at precisely the expected frequency for each liquid filling. Sophisticated control systems have been used, for example, to prevent sloshing of liquid moved in an open container as it is carried by a robotic arm; see Feddema et al. (1996). However, as Fig. 1b illustrates, the desired single-frequency oscillation mode can be obtained manually with some practice. The drawback of this approach was that many runs had to be discarded since they had unacceptable drift. For any particular configuration, our batting average for a good, usable run was about one out of three attempts when exciting the fundamental mode, and perhaps one out of ten when trying to excite the second mode. The cart's oscillation frequency was determined by measuring the elapsed time for three to ten cycles, always at low oscillation amplitude to stay within linear theory.

We used the local value of the gravitational constant $g = 980.366$ cm/s^2 provided to us by a colleague in the Department of Earth and Environmental Sciences at Boston College. The density was taken to be that for pure water at the average room temperature for the experiments, namely $\rho = 0.9977$ g/cm^3. Also, three digit accuracy in computing the oscillation frequencies was obtained by including twelve to fifteen terms in the series expressions in the eigenvalue equations.

4 Theory and Comparison with Experiment

The derivation of the eigenvalue equations for each geometry are given in Herczynski and Weidman (2012). Here we simply present the finite-depth eigenvalue equations and, when applicable, their shallow-water counterparts.

4.1 Rectangular Containers

Consider a rectangular box of length $L = 2D$, width W, and mass m_0 filled to depth H with liquid of density ρ. The finite-depth eigenvalue equation for this case is

$$1 + M\left(1 + 2\omega^2 \sum_{n=1}^{\infty} \frac{\tanh k_n H}{(k_n H)(k_n D)^2 (gk_n \tanh k_n H - \omega^2)}\right) = 0 \qquad (8)$$

where $k_n = (2n - 1)\pi/2D$. In solving the equation it must be remembered that $H = H(M)$.

The longwave limit $k_n H \to 0$ gives the shallow-water behavior for the box. In this case we make the approximation $\tanh k_n H \sim k_n H$ in Eq. 8 and insert the definition for $H = H(M)$ to obtain a series solution, the sum of which yields

$$Z + M \tan Z = 0, \qquad Z = \frac{1}{M^{1/2}}\left(\frac{\omega}{\omega_R}\right), \qquad \omega_R = \sqrt{\frac{m_0 g}{2\rho D^3 W}}. \qquad (9)$$

The measured frequencies for the first two modes of the large box are shown as solid circles in Fig. 2. The corresponding finite-depth results computed from Eq. 8 are shown by the solid lines and their shallow-water behaviors computed from Eq. 9 are shown by the dashed lines.

Visible in this figure is the Mode 2 maximum $\omega/\omega_R = 3.7269$ at $M = 1.900$ corresponding to a maximum frequency $\omega = 19.678$ rad/s. Mode 1 oscillations were recorded over a wide range of liquid fillings $0.5 \lesssim M \lesssim 1.5$; below $M \approx 0.5$ there was too little liquid mass to excite well-defined oscillations and above $M \approx 1.5$

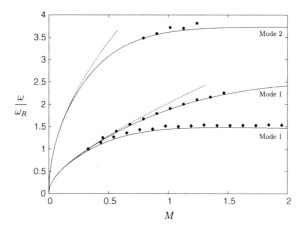

Fig. 2 Measured frequencies for the first two modes of the *large box* are shown by the *solid circles* and the reference frequency is $\omega_R = 5.280$ rad/s. Measured frequencies for the first mode of the *tall box* are shown by the *solid diamonds* and the reference frequency is $\omega_R = 10.943$ rad/s. The corresponding finite-depth theoretical results are show by *solid lines* and the shallow-water results are shown by the *dashed lines*

liquid sloshed out of the container. We were also able to initiate Mode 2 oscillations, but only over a narrow range centered about $M = 1$. As will be seen with the other geometries, the Mode 1 frequency measurements agree better with theory than the Mode 2 measurements. It was always the case that the amplitudes of the Mode 2 oscillations were considerably diminished compared with those of the fundamental mode and, as a result, a relatively small number of well defined damped oscillations were observed for this higher mode. Nevertheless, the agreement is considered very good for both modes.

In an effort to experimentally observe the frequency maximum we fabricated the tall box. The experimental data for Mode 1 oscillations of this container are displayed as solid diamonds in Fig. 2 and compared with the theoretical results calculated from Eq. 8 shown by the corresponding solid line. The maximum $\omega/\omega_R = 1.4796$ in the numerical calculation occurs at $M = 1.47$; this corresponds to a maximum frequency $\omega = 16.191$ rad/s. The relatively flat maximum is tracked by the data quite well. Owing to the relatively high frequencies associated with first mode in the tall box, the agreement with theory is not as good as with Mode 1 in the large box. Note that the shallow-water solution is the same for all Mode 1 box geometries as illustrated here for the large and tall boxes.

4.2 Cylindrical Containers

Now consider a circular cylinder of radius R and dry mass m_0 filled with liquid to depth H. The finite-depth eigenvalue equation in this case is given by

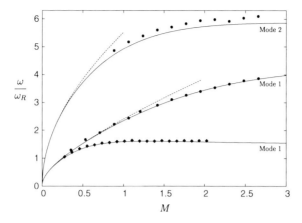

Fig. 3 Measured frequencies for the first two modes of the *large cylinder* are shown by the *solid circles* and the reference frequency is $\omega_R = 3.505$ rad/s. Measured frequencies for the first mode of the *tall cylinder* are shown by the *solid diamonds* and the reference frequency is $\omega_R = 10.833$ rad/s. The corresponding finite-depth theoretical results are show by *solid lines* and the shallow-water results are shown by the dashed lines

$$1 + M\left(1 + 2\omega^2 \frac{R}{H} \sum_{n=1}^{\infty} \frac{J_2(\alpha_n)}{(\alpha_n^2 - 1)J_1(\alpha_n)} \frac{\tanh k_n H}{(gk_n \tanh k_n H - \omega^2)}\right) = 0 \qquad (10)$$

where $J_1'(k_n R) = 0$ fixes the radial wave numbers k_n and J_1 is the Bessel function of the first kind. Again it must be kept in mind that $H = H(M)$.

The shallow-water limit is obtained from (10) by replacing $\tanh k_n H$ with $k_n H$ and incorporating the definition for $H = H(M)$. This yields a series solution which sums to the result

$$Z J_1'(Z) + MJ_1(Z) = 0, \qquad Z = \frac{1}{M^{1/2}}\left(\frac{\omega}{\omega_R}\right), \qquad \omega_R = \sqrt{\frac{m_0 g}{2\rho\pi R^4}}.$$
$$(11)$$

The measured frequencies for the first two modes of the large cylinder are shown as solid circles in Fig. 3. The corresponding finite-depth results computed from Eq. 3 of Chap. 4 are shown by the solid lines and their shallow-water behaviors computed from Eq. 4 of Chap. 4 are shown by the dashed lines.

The maxima for Modes 1 and 2 in the large cylinder occur beyond $M = 3$. In comparison to the large box we note that experimental data may be gathered over wider ranges of M for both the Mode 1 and Mode 2 oscillations. Again, the agreement with theory is better for the larger amplitude Mode 1 oscillations compared to Mode 2.

In order to try to capture a maximum in the frequency oscillation curve, we designed the tall cylinder. The experimental data for the Mode 1 response in this geometry shown as the solid diamonds in Fig. 3 are compared with the finite-depth

Fig. 4 Measured frequencies for the 90° wedge shown by the *solid circles* are compared with the theoretical results shown by the *solid line*. The small-M asymptote $M^{-1/4}$ is shown by the *dashed line* and the reference frequency is $\omega_R = 12.017$ rad/s

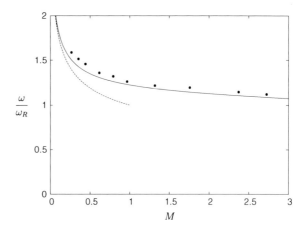

theoretical results shown as the corresponding solid line. The maximum in the frequency curve occurs at $M = 1.31$ with the value $\omega/\omega_R = 1.61244$, corresponding to $\omega = 17.46$ rad/s. It is clear that the measurements track the theoretical curve very closely. As with the box, owing to the relatively flat maximum, it is hard to distinguish the peak frequency in the experimental data. Again, note that the shallow-water Mode 1 solution is the same for all cylindrical containers.

4.3 A Wedge Geometry

The next geometry to be considered is that of wedge of apex angle $\pi/2$ radians having mass m_0 and width W. The apex angle is bisected by the vertical axis aligned with gravity. A simple analysis using the stationary eigenfunction solution of (Lamb (1932), Sect. 258) yields the exact solution

$$\frac{\omega}{\omega_R} = \frac{1}{M^{1/4}} \sqrt{\frac{1+M}{1+\frac{M}{3}}}, \qquad \omega_R = \left(\frac{\rho g^2 W}{m_0}\right)^{1/4}. \qquad (12)$$

Here it is pertinent to observe that this ω_R is not a shallow-water reference frequency, for there is no shallow-water limit for this wedge geometry; the solution for all M is a finite-depth solution.

Theoretical and experimental results for the wedge with 90° apex angle are shown in Fig. 4. The theory for the single mode possible in this geometry is that given in Eq. 12. For this container data could be obtained up to $M \approx 2.8$ above which fluid sloshed out of the wedge. The agreement between experiment and theory is considered good, but we now see a clear trend — theory and experiment are generally in better agreement for rotationally symmetric geometries compared to planar geometries (box and now the wedge). Note in this case that $\omega/\omega_R \sim M^{-1/4}$ as $M \to 0$ in distinct contrast to the box and cylinder geometries. But one must bear

Fig. 5 Measured frequencies for the 90° *cone* shown by the *solid circles* are compared with the theoretical result shown by the solid line. The small-M asymptote $M^{-1/6}$ is shown by the *dashed line* and reference frequency is $\omega_R = 9.638$ rad/s

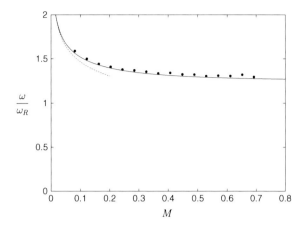

in mind that as $M \to 0$ the liquid mass m in the container tends to zero so there is no liquid to excite the sideways periodic container motion in this limit. For this experiment the reference frequency calculated from Eq. 12 is $\omega_R = 12.017$ rad/s.

4.4 A Cone Geometry

The rotationaly symmetric analogue of the wedge discussed in the previous section is a cone of apex angle $\pi/2$ radians and mass m_0. The analysis follows closely that for the wedge and in this case one finds

$$\frac{\omega}{\omega_R} = \frac{1}{M^{1/6}} \sqrt{\frac{1+M}{1+\frac{M}{4}}}, \qquad \omega_R = \left(\frac{\rho \pi g^3}{3 m_0}\right)^{1/6} \tag{13}$$

for the oscillation frequency of the freely moving cone.

Theoretical and experimental results for the cone are shown in Fig. 5. The theoretical frequencies for the only mode in this geometry is that given in Eq. 13. For this container data could be obtained only up to $M \approx 0.7$ above which fluid sloshed out of the cone. The agreement between experiment and theory is considered excellent and supports the trend observed previously that theory and experiment are generally in better agreement for axisymmetric geometries (cylinder and now the cone) compared to planar geometries (box and wedge). For the cone $\omega/\omega_R \sim M^{-1/6}$ as $M \to 0$, but again in this limit there is no liquid in the cone to excite the horizontal oscillations.

4.5 *Annular Containers*

Finally, consider sloshing in the region between two vertical concentric cylinders. The original work on this problem dates back to Sano (1913) who studied the seiching motion observed in a circular lake with central circular island; see also Campbell (1953) and Bauer (1960). The annulus of inner radius R_1, outer radius R_2, and dry mass m_0 is filled with liquid to depth H. We denote $\eta = R_1/R_2$ as the radius ratio. We use (r, θ, z) for the coordinate system attached to the sideways moving container and align the z-coordinate with the axis of the concentric cylinders. It is clear that any description of the free surface deflection for sloshing between the cylinders must include higher order azimuthal (e.g. $\cos n\theta$) terms, particularly in the narrow gap limit—the free surface cannot slosh back-and-forth in vertical planes as it does in the fundamental sloshing mode of a cylinder with a single nodal diameter. Nevertheless, one can compute the liquid-structure interaction for purely rectilinear motion of the annulus owing to the orthogonality of trigonometric functions. We thus proceed to determine the frequency of motion of the system by retaining only the lowest azimuthal dependence, $\cos \theta$, realizing that computation of the time-dependent free surface will not be available at this level of analysis. A rather long and tedious calculation gives the eigenvalue equation

$$1 + M\left(1 + \frac{\omega^2 \eta}{(1-\eta^2)} \sum_{n=1}^{\infty} \frac{B_n}{k_n R_1 H} [P(R_2) - \eta P(R_1)] \frac{\tanh k_n H}{(gk_n \tanh k_n H - \omega^2)}\right) = 0$$

(14a)

where the radial wavenumbers k_n are determined by solution of

$$J_1'(k_n R_1)Y_1'(k_n R_2) - J_1'(k_n R_2)Y_1'(k_n R_1) = 0,$$ (14b)

the coefficients B_n are given by

$$B_n = \frac{2k_n R_1^2 \{Y_1'(k_n R_1)[J_2(k_n R_2) - \eta^2 J_2(k_n R_1)] - J_1'(k_n R_1)[Y_2(k_n R_2) - \eta^2 Y_2(k_n R_1)]\}}{\eta^2 \{P^2(k_n R_2)[(k_n R_2)^2 - 1] - P^2(k_n R_1)[(k_n R_1)^2 - 1]\}}$$

(14c)

and

$$P(r) = Y_1'(k_n R_1)J_1(k_n r) - J_1'(k_n R_1)Y_1(k_n r).$$ (14d)

Another tedious calculation in the limit $R_1 \to 0$ shows that this result reduces to eigenvalue Eq. 10 for a cylinder, as expected.

There now arises the manner in which we normalize the frequencies computed. We have not determined the shallow-water behavior for the annulus, and since it likely depends on the radius ratio η, we choose to normalize all frequencies with the shallow-water value ω_R given in Eq. 11 for a cylinder of radius R_2 with dry mass m_0.

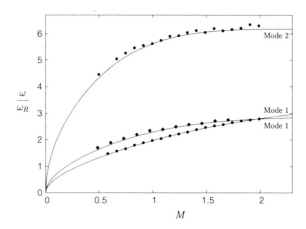

Fig. 6 Measured frequencies for the first two modes of the $\eta = 0.3644$ annulus are shown by the *solid circles* and the measured frequencies for the first mode of the $\eta = 0.777$ annulus are shown by the *solid diamonds*. The finite-depth theoretical results are shown by the *solid lines*. The reference frequency $\omega_R = 3.505$ rad/s computed from Eq. 11 is that for $\eta = 0.0$ for which $m_0 = 1129$ g

We now present results for the sloshing-induced motions of partially filled concentric cylindrical annuli. The experimental data plotted as solid circles and theoretical results calculated from Eq. 14 for the first two modes of sideways oscillation at $\eta = 0.3644$ are displayed in Fig. 6. While there is no maximum in the plotted range of M for Mode 1, Mode 2 displays a maximum at $M = 2.20$ with value $\omega/\omega_R = 6.1798$ corresponding to $\omega = 21.66$ rad/s. Agreement between theory and experiment for both modes is considered excellent. As mentioned above, we cannot determine the free surface deflection using only the first azimuthal mode in the analysis. However, we did observe damped oscillations in the anular region of wave sloshing and it was very interesting indeed. The free surface signature of the motion revealed waves propagating around opposite sides of the annulus that ultimately met in head-on collisions at $\theta = 0, \pi$. Collisions with splashing were observed only during the first couple of oscillations for which the wave amplitudes were relatively large. The splashing observed in our experiment is reminescent of that produced by the head-on collision of solitary waves reported by Maxworthy (1976).

Sample calculations revealed that successive frequency curves do not cross at sufficiently low values of η but that, above a critical value, the higher η curves cross some of the lower η curves. With this in mind, we fabricated a new annulus at $\eta = 0.777$ for which crossing is predicted. The experimental measurements for this case displayed as solid diamonds in Fig. 6 are compared with corresponding finite-depth theoretical results computed from Eq. 14 of Chap. 4. All experimental data agree very well with the theoretical predictions and indeed there is strong experimental evidence that the curves for $\eta = 0.364$ and $\eta = 0.777$ will indeed cross in the neighborhood of $M = 2$.

Fig. 7 A comparison of free
surface waveforms for the
large box, large cylinder,
wedge and *cone* computed at
$M = 1$. The normalized
horizontal coordinate is
$\xi = x/D$ for the box, $\xi = x/H$
for the *wedge*, $\xi = r/R$ for the
cylinder and $\xi = r/H$ for the
cone. Details for these
geometries are given in
Table 1

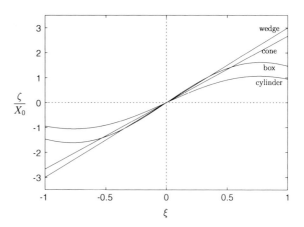

4.6 Waveforms and Amplification Ratios

The free surface wave profiles ζ/X_0 for the large box, large cylinder, wedge and cone are compared at the common value $M = 1$ in Fig. 7; see Herczynski and Weidman (2012) for detailed formulae for these waveform profiles.

The waveforms are plotted against the normalized coordinate $\xi = x/D$ for the box, $\xi = r/R$ for the cylinder, $\xi = x/H$ for the wedge and $\xi = r/H$ for the cone. Note that both the wedge and cone surfaces are flat, but that owing to the larger magnification ratio for the wedge, its rise height is larger than that for the cone. We have taken photographs, and in some instances videos, of the fundamental sloshing waveforms viewed from the side and find qualitative agreement between the surface profiles with those presented in Fig. 7. In particular, free surfaces appeared completely flat for sloshing waves in the cone and wedge, except for small capillary effects around the wetted perimeter of the containers.

We made one measurement of the amplification ratio for the large box. The amplification ratio is defined as ζ_0/X_0 where $\zeta_0 = \zeta(x = D)$. For a liquid filling $m = 815$ g corresponding to $M = 0.919$, we took a video of the oscillating wave near the end of the box on which was mounted a millimeter scale to estimate the vertical displacement of the liquid at the endwall $x = D$. The formula for the amplification ratio (Herczynski and Weidman 2012) for a box at fixed ω is given by

$$\left| \frac{\zeta_0}{X_0} \right| = \frac{D\omega^2}{g} \left[1 + 2\omega^2 \sum_{n=1}^{\infty} \frac{1}{(k_n D)^2 (gk_n \tanh k_n H - \omega^2)} \right]. \tag{15}$$

Evaluation of Eq. 8 at $H = m_0/\rho WL$ for the large box gives $\zeta_0/X_0 = 1.4192$. Our measured amplification ratio $\zeta_0/X_0 = 1.46 \pm 0.05$ is thus in excellent agreement with the theoretical prediction.

5 Discussion and Conclusions

Experiments on the horizontal, rectilinear, sloshing-induced motion of free containers oscillating over a nearly frictionless surface have been presented. The apparatus used, made by PASCO scientific, consisted of an aluminum, four-wheel cart with a fine suspension system which can move with very low friction on an aluminum track. Measured frequencies for the fundamental and second modes of transverse oscillation for box, cylinder, and annulus geometries were obtained over a range of dimensionless masses $M = m/m_0$, where m_0 is the dry mass of the system and m is the liquid mass inside the container. In addition, the frequency of the only sloshing mode available for a wedge and a cone with 90° apex angles were obtained over a range of M. Additional rectangular and cylindrical containers were designed in an attempt to capture the predicted maximum frequency that obtains for each geometry, with only partial success because of the relatively flat maxima in each case. More successful was an experiment devised to document the theoretical prediction that a large radius ratio annulus frequency curve will cross a lower radius ratio curve at some value of M. In all cases, measurements are considered to be in very good, if not excellent, agreement with the theoretical predictions.

We attempted to excite higher modes in all of our containers. In three of them (large box, large cylinder and the $\eta = 0.364$ annulus) we were able to observe the second mode, though sometimes over only a limited range of filling ratios M. In none of our containers could we observe the third (or any higher) harmonic, presumably because these oscillations would be at frequencies too high to excite manually. They would also have very small amplitudes, making them hard to discern.

Two trends in the data are apparent. First, measurements of the sloshing-induced frequency of the axisymmetric containers (cylinder, cone, annulus) were generally in better agreement with theory than those for the containers of planar symmetry (box, wedge). We tested to see if this might be some capillary effect by adding several drops of PhotoFlo to reduce the surface tension, but no discernible change in the frequency was noted for these long, damped standing waves. Second, while damping might be expected to reduce the oscillation frequency, our experimental results are sometimes slightly above, but almost never below, the theoretical predictions. Also, in all geometries except the wedge and the cone, evaporation would lower the observed frequencies, whereas our measurements are nearly always above the predicted values. We contend that this systematic trend is probably due to a slight restoring force provided by the cart's suspension system, especially at high fillings when the depressed springs are more susceptible to coupling with the oscillations of the liquid in the container.

Of particular interest is the fact that the transverse oscillation of sloshing-induced motion in an annulus can be determined using only the fundamental azimuthal mode, $\cos \theta$, with one nodal diameter. The shape of the oscillating free surface, however, cannot be determined unless higher modes are included. Observations of the free surface motion revealed waves propagating around opposite sides of the annulus that met in head-on collisions at $\theta = 0, \pi$. Collisions with splashing were observed during the first couple of oscillations during which the wave amplitudes were relatively large, reminiscent of those produced by the head-on collision of solitary waves.

Acknowledgments We have benefitted greatly from discussions with Dr. Mark Cooker and Professor Jie Yu during all phases of the work. We appreciate the precision work of John Butler at Colorado Plastic Products, Inc. in fabricating the boxes and the wedge, and of James Tucker of Tucker Precision Machining for turning the cone on a lathe from solid stock. Michael Sprague provided guidance in programming Mathematica. We thank Yun Peng who assisted in taking videos of our experiments and in some of the measurements.

References

Abramson HN (1966) The dynamical behavior of liquids in a moving container. Technical Report SP-106, NASA, Washington

Ardakani HA, Bridges TJ (2010) Dynamic coupling between shallow-water sloshing and horizontal vehicle motion. Eur J Appl Math 21:479–517

Bauer HF (1960) Theory of fluid oscillations in a circular ring tank partially filled with liquid. NASA TN-D-557

Campbell IJ (1953) Wave motion in an annular tank. Phil Mag Ser 7(44):845–853

Cooker MJ (1994) Waves in a suspended container. Wave Motion 20:385–395

Dodge FT (2000) The new dynamical behavior of liquids in moving containers. Southwest Research Institute, San Antonio

Faltinsen OM, Timokha AN (2009) Sloshing. Cambridge University Press, Cambridge

Feddema J, Dohrmann C, Parker G, Robinett R, Romero V, Schmitt D (1996) Robotically controlled slosh-free motion of an open container of liquid. In: IEEE Proceedings of the international conference on robotics and automation, pp 596–602

Herczynski A, Weidman PD (2009) Synchronous sloshing in a free container. APS division of fluid dynamics, 62nd annual meeting. Minneapolis, 22–24 Nov 2009

Herczynski A, Weidman PD (2012) Experiments on the periodic oscillation of free containers driven by liquid sloshing. J Fluid Mech 293:216–242

Ibrahim RA (2005) Liquid sloshing dynamics: theory and applications. Cambridge University Press, Cambridge

Keulegan GH (1959) Energy dissipation in standing waves in rectangular basins. J Fluid Mech 6:33–50

Lamb H (1932) Hydrodynamics, 5th edn. Cambridge University Press, Cambridge

Maxworthy T (1976) Experiments on collisions between solitary waves. J Fluid Mech 76:177–185

Moiseev NN (1964) Introduction to the theory of oscillations of liquid-containing bodies. Adv Appl Mech 8:233–289

Sano K (1913) On seiches of lake toya. Proc Tokyo Math Phys Soc 7(2):17–22

Weidman PD (1994) Synchronous sloshing in free and suspended containers. APS division of fluid dynamics, 47th annual meeting. Atlanta, 20–22 Nov 1994

Weidman PD (2005) Sloshing in suspended containers. APS division of fluid dynamics, 58th annual meeting. Chicago, 20–22 Nov 2005

Wolfram S (1991) Mathematica: a system for doing mathematics by computer. Addison-Wesley, Redwood City

Yu J (2010) Effects of finite depth on natural frequencies of suspended water tanks. Stud Appl Math 125:373–391

Experimental Investigation of the North Brazil Current Rings During Their Interaction with the Lesser Antilles

Raúl C. Cruz Gómez

Abstract Interaction of the North Brazil Current Rings (NBCR) with the Lesser Antilles Arc and their penetration into the Caribbean Sea was investigated by laboratory experiments. Self-propagating cyclonic vortices were made with barotropic fluid by one of two methods: suction and electromagnetic. The parameters causing the reflection, destruction and penetration of the rings into the Caribbean Sea were as follows: The Rossby number associated with the vortex, the width of the strait between islands, the incidence angle with the barrier, and the velocity of translation and intensity of interaction with the barrier. The principal observations were made by lagrangian measurements, using particle image velocimetry. It is concluded that the formation of bigs rings within the Caribbean Sea can be attributed to the interaction of vestiges of the NBCR with the chain of islands at the Antilles.

1 Introduction

Mesoscale oceanic vortices have the capacity to trap and transport mass, momentum and heat, as well as biochemical properties of the water, along hundreds of kilometers. During their propagation, these vortices can modify their characteristics, or can conserve them as coherent structures. In many cases, vortices can collide over the bottom topography or with islands during their trajectory. The dynamic, transport and auto-propagation of these structures have been the subject of observational and

R. C. Cruz Gómez (✉)
Departamento de Física, Instituto de Astronomía y Meteorología,
Universidad de Guadalajara, Guadalajara, Mexico
e-mail: rcruz@astro.iam.udg.mx
URL: rcruzx@gmail.com

J. Klapp et al. (eds.), *Fluid Dynamics in Physics, Engineering and Environmental Applications*, Environmental Science and Engineering,
DOI: 10.1007/978-3-642-27723-8_4, © Springer-Verlag Berlin Heidelberg 2013

theoretical and numerical studies (Adem 1956; Firing and Beardsley 1976; van Heijst 1994; Flór and Eames 2002; Fratantoni and Glikson 2002; Richardson 2005; Cruz Gómez and Bulgakov 2007; Cruz Gómez et al. 2008).

The motion of cyclonic vortices has been studied experimentally by Firing and Beardsley (1976), Carnevale et al. (1991), van Heijst (1994). Recently, Flór and Eames (2002) demonstrated by a qualitative experimental analysis that the initial direction of movement of isolated and non-isolated vortices depends on the decay of the vortex azimuthal velocity profile. In their experiment, vortices translate by the beta topographic effect.

To date, few studies have investigated the interaction between vortices and multiple straits. Studies made by analysis of satellite, theoretical and experimental data are controversial about the result of such interaction. In particular, there is no clear idea about the process that rings detaching from the North Brazil Current undergo when interacting with the chain of islands of the Lesser Antilles Arc. The present work, this question is investigated experimentally.

Cenedese et al. (2005), studying in the laboratory the evolution of a vortex that interacts with a pair of circular islands, found that only the smaller vortices completely cross the opening between the islands, a result expressed by the relationship $D/R > 3.6$ (where D is the opening width and R is the radius of maximum azimuthal velocity); vortices of intermediate size partially cross ($2.3 < D/R < 3.6$), and big vortices do not cross ($D/R < 2.3$).

Cenedese and Tanabe (2008) studied the evolution of a vortex interacting with a chain of circular islands in the laboratory where the vortices were impulsed by the beta topographic effect and generated by an ice cube. The results were reported in terms of the ratio G/d, where G is the diameter of the island and d the diameter of the vortex. For values $0.1 < G/d < 0.4$, a dipole formed on the other side of the barrier after the vortex interacted with the cylinders (islands), at the location of one of the straits, turning into the cyclonic-dominant part. The interaction was different for different configurations of the islands, with only a filament crossing at the southern island of the formation of the dipole.

Matias Duran and Velasco Fuentes (2008) studied the evolution of a vortex passing through a gap in laboratory and numerical experiments. Their results were based in: the intensity and initial position of vortex and the width of the gap. They found three different cases: the vortex cross totally through the gap; the vortex was split up into two and only one fraction passed through the gap; and a last case where the vortex was totally blocked.

In contrast with previous studies, the dynamic of barotropic vortices is analyzed in the present work by means of laboratory experiments using a chain of islands-straits. The first objective was to determine whether the chain of islands constitutes an impenetrable barrier for the vortices. The second objective was to determine the different scenarios of the rings interaction, before, during and after the collision. Three main factors were taken into account in the experiments: (a) The initial vortex profile of vorticity and velocity, (b) the width of the straits in relation to the vortex size, and (c) the dynamic by which the vortex approaches the barrier of islands.

2 The β-Effect

2.1 The Planetary β-Effect

In the shallow-water model for a fluid in a rotating and inviscid system, the potential vorticity is materially conserved:

$$\frac{D}{Dt}\left[\frac{f+\omega}{H}\right] = 0, \tag{1}$$

where ω is the relative vorticity, H is the local depth of the fluid, and $f = 2\Omega\sin\phi$ is the planetary vorticity or Coriolis parameter, with ϕ the geographic latitude and Ω is the angular velocity of the rotating system. Expanding f in Taylor's series: $f = 2\Omega\sin\phi_0 + 2\Omega\frac{y}{a}\cos\phi_0 + \cdots$ and keeping only the first two terms, we obtain $f = f_0 + \beta y$. where ϕ_0 the reference latitude, a the Earth's radius and y the northward pointing Cartesian coordinate (the Cartesian coordinates x and z are taken in local eastward and local vertical direction, respectively).

The parameter of the second term, $\beta = \frac{2\Omega\cos\phi_0}{a}$, corresponds to the latitudinal variation of f, or gradient of planetary vorticity. For a homogeneous fluid over a flat bottom (H = constant), the dynamic Eq. (1) is reduced to

$$\frac{D}{Dt}(\omega + \beta y) = 0. \tag{2}$$

This expression indicates that when a parcel of fluid is displaced to the north (y increases) its relative vorticity decreases, such that its potential vorticity is conserved. Conversely, displacements to the south imply an increase in the relative vorticity.

2.2 Laboratory Simulation of the β-Effect

If the horizontal scale of a vortex is much smaller than that of the topography over which it moves, then it is possible to model the β-effect in a laboratory (Carnevale et al. 1991). Considering this and under the assumption that the equation for conservation of potential vorticity is approximately valid, the spatial variations of planetary vorticity can be dynamically simulated with a false bottom in the tank (van Heijst 1994). In the laboratory case, the Eq. (1) is written as:

$$\frac{D}{Dt}\left[\frac{f_0+\omega}{H(x,y)}\right] = 0 \tag{3}$$

where $f_0 = 2\Omega$, and Ω is the angular velocity of the rotating table where the tank is placed. The depth $H(x,y)$ can be written according to Fig. 1 as:

Fig. 1 Diagram of the
laboratory setting for the
topographic β plane. The
false bottom simulates the β
effect

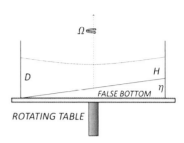

$H(x, y) = D - \eta(x, y)$, with D as the maximum depth, and η the local height of
the bottom. When the height of the topography is much smaller that the maximum
depth, we obtain: $\delta \equiv \eta/D \ll 1$.

For the particular case of a constant slope of the bottom in the y direction, $\eta = sy$,
with s as the slope, the following dynamic equation is then derived by neglecting the
terms of order $O(\delta^2)$ and $O(Ro\delta)$ where $Ro = \omega/f_0$ is the Rossby number.

$$\frac{D}{Dt}[\omega + \beta^* y] = 0 \tag{4}$$

where $\beta^* = sf_0/D$ is the topographic β parameter. In this way, the dynamics of a
geophysical fluid in the presence of a gradient of planetary vorticity (Eq. 2) is
identical to that of a fluid over an inclined plane with constant rotation (Eq. 4). By
rewriting the potential vorticity of a column of fluid for each case, we obtain:

$$q_{planet} = \omega + \beta y = constant \tag{5}$$

$$q_{lab} = \frac{f_0 + \omega}{H(x, y)} = constant \tag{6}$$

These relationships show that an increment of depth implies a decrease of y in the
ocean. Therefore, the shallow part of the tank is equivalent to the north and the
deep part is equivalent to the geographic south.

The non-dimensional number that measures the beta effect is: $\beta' = \frac{\beta L}{f_0}$, where L
is the scale of horizontal longitude of the flow under study; in our case, it
corresponds to that of the vortices generated in the laboratory. Given the characteristics of the oceanic rings in the region under study $(L \sim 150 \, km,$
$\beta \sim 2.2 \times 10^{-13} \, (cm \cdot s)^{-1}, f_0 \sim 0.3 \times 10^{-4} \, s^{-1})$ and of those generated in the
laboratory $(L \sim 5.5 \, cm, \beta \sim 0.0154 \, (cm \cdot s)^{-1}, f_0 = 1)$, the values of β' are 0.086
and 0.085, respectively. For the experimental case, $D = 13 \, cm$, and
$s = 12 \, cm/60 \, cm = 0.20$ is used.

3 Experimental Configuration

All experiments used a square tank, 60×60 cm, fixed on a rotating table with the axis of vertical rotation. The water depth in the middle of the tank was 13 cm. The rotation was anticlockwise. The experimental setting consisted of rotating the table at a constant angular velocity for 40 min. This ensured that the fluid reached the spin-up, as illustrated in Fig. 2.

In all experiments the Coriolis parameter was constant at $f_0 = 1$ s^{-1}, ($\Omega = 0.5$ s^{-1}) with a period of T $= 12.56$ s. Each experiment lasted 120 s. Considering the value of kinematic viscosity of the fluid $v \sim 0.0091$ cm^2/s (approximate value at 25 °C), the Ekman period $T_E = \frac{H}{(v\Omega)^{1/2}}$ was approximately 170 s, and hence the effects of bottom friction could be neglected for the time the experiments lasted (Zavala Sansón and van Heijst 2000).

To capture the information from the experiments, a digital video camera was installed above the rotation axis. The video was saved and the frames extracted for further analysis. In this method, the flow was visualized by injecting a colouring (potassium permanganate) to obtain qualitative information.

A second method measured the flow by using small particles of polyamide, with diameter of 50 μm. The motion of the particles was recorded with a video camera (CCD) that rotated along with the tank. The recorded area was 33×25 cm. The fluid was illuminated with a laser whose beam was horizontally projected in the x, y plane of the table. The displacement of the passive tracers was estimated from sequential images by cross correlation between pairs of them using Particle Image Velocimetry (PIV) system. With this technique it is possible to determine the position and velocity of a great number of tracers and make the interpolation to a rectangular grid with the objective of finding the vorticity fields.

3.1 Non-Isolated and Isolated Vortices

Flow was generated by one of two methods: suction and electromagnetic. Each produced monopolar circular vortices but with different profiles of tangential velocity and relative vorticity.

Vortices generated by the suction method are formed by means of the extraction of water with a siphon: the resulting low pressure zone produces an anticyclonic gyre for the rotation effects of the system (see Fig. 2). The intensity of these vortices depends on the amount of water extracted by the siphon (Hopfinger and van Heijst 1993). In the cases shown here, 2 l of water were extracted in each experiment, by using a 1 cm diameter siphon. The profiles of the resulting cyclonic vortex are shown in Fig. 3. The tangential velocity profile of these vortices reached a maximum and then slowly decayed, inversely proportional to the radial distance. On the other hand, the vorticity field had always the same sign and decayed exponentially from the center of the vortex. These vortices are called "non-isolated". The Rossby number associated with these vortices was $Ro = 1$.

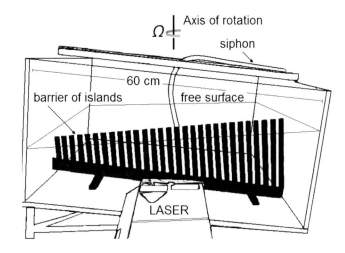

Fig. 2 Scheme of the laboratory setting for the experiments with the topographic β plane. The barrier represents the chain of islands and the siphon is at the center of the tank

The vortices generated by the electromagnetic method were forced by the interaction of electromagnetic fields. The intensity depends on the magnitude of both fields. In order to do this, salt (16 g/l) was added to increase the electrical conductivity. An electrode was placed on each corner of the tank and another one was put in the center, with different polarity. All the electrodes were set in a direction parallel to the axis of rotation of the table. When a potential difference is applied, an electric field is generated Fig. 4. A square magnet (5 cm^2) was put at the center, underneath the tank, with an intensity of 0.4 T. The magnet induces a continuous magnetic field in the vertical direction within the fluid, parallel to the rotation axis of the table. Owing to the interaction of the magnetic field with the electric field, a resulting force is produced (Lorentz's force) over the salt ions. This force acts on the fluid in a direction perpendicular to the electric current and magnetic field, producing cyclonic and anticyclonic vortices, depending on the polarity of the applied voltage.

The typical profiles of tangential velocity and vorticity are shown with the continuous lines in Fig. 3. The profile of tangential velocity reaches a maximum and then decays rapidly to zero. The vorticity field consists of a center with a defined sign, surrounded by a ring of vorticity with opposite sign. The integral area of vorticity is zero, and hence these are called "isolated" vortices. The Rossby number for these vortices was $Ro < 1$.

3.2 Islands and Straits

As an approximation to the Antilles Arc, two types of barriers between islands were implemented, representing wide and narrow straits with respect to the incident vortex. These barriers consisted of a series of obstacles with width I (islands)

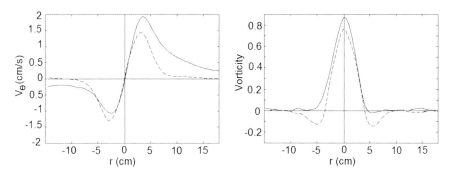

Fig. 3 Typical profile of tangential velocity (*left*) and vorticity (*right*) along a section, for a vortex generated by the suction method or *non-isolated* (*continuous line*). Vortex generated by the electromagnetic method or *isolated* (*dashed line*), both cyclonic, in experiments with flat bottom

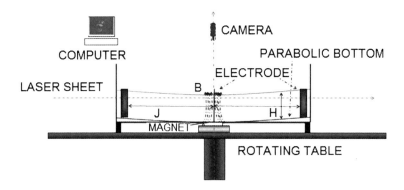

Fig. 4 Drawing of the experimental setting. J is the density of the electric current due to the electric field horizontally directed in both directions, depending on the polarity of the electrodes. B is the magnetic field generated by the magnet. H is the water layer depth

separated by a distance S (straits). In each case, $I = 1.2$ cm was used. The barrier differed in the separation between islands: one barrier (**A**) had narrow straits, $S = 0.8$ cm, and the other barrier (**B**) had wider straits, $S = 2.8$ cm. In each configuration, the islands covered vertically the whole water column (south to north of the tank). In the experimental model the barrier of islands was in a straight line, whereas the Antilles islands form an arc; however, the curvature ratio of the Antilles (~ 800 km) is much larger than that of the incident vortices (~ 150 km) and therefore the straight barrier is a reasonable approach.

Assuming that the vortices have an approximate radius of $R \sim 3$ cm, the important non-dimensional values are: $I' = \frac{I}{R} = 0.4$ for each barrier, $S' = \frac{S}{R} = 0.26$ for barrier **A**, and $S' = 0.93$ for **B**. These values are consistent with those corresponding to the NBCR in the area of study, characterized by $I \sim 60$ km, $S \sim 40$ km and $R \sim 150$ km, (Simmons and Nof 2002).

4 Dynamic in the Topographic β Plane

4.1 Interaction with the A Barrier

Figure 5 shows the sequence of vorticity fields of a non-isolated vortex colliding against barrier **A** and a qualitative sequence of another vortex generated under similar conditions. These vortices are initially very intense and move toward the barrier with an approximate translation velocity of 0.9 cm/s and direction to the northwest, the angle of incidence with the barrier is $\approx 20°$ at the moment of collision. The displacement and angle of incidence is shown with (+) symbols on center of vortex with a difference of time 1 s between pulses.

During the interaction, the vortex deforms slightly as it moves to the north with an approximate speed of 0.4 cm/s. A region with negative vorticity forms adjacent to the barrier that subsequently surrounds the vortex. After the vortex slightly separates from the barrier, it recovers its move to the west and collides again with the islands. Finally, the vortex stands and dissipates by viscous effects. The path followed by the vortex, whose increment between samples is 1 s, is shown in the left column. In many experiments of this type, when the vortex interacts with the barrier some weak filaments pass through some straits. This behavior is quite similar to that observed in the interaction of a vortex with a solid wall (Zavala Sansón et al. 1999), with the exception of the weak mass transfer to the other side of the barrier. Hence it is inferred that the separation of the islands is very small, so that the barrier acts as if it was a closed boundary.

The isolated vortices (Fig. 6) are less intense and their translation velocity toward barrier **A** is slower, approximately 0.3 cm/s. The angle of incidence with the barrier is $\approx 45°$ in a northwest direction. From its generation, the center of the vortex is surrounded by vorticity with opposite sign. When the vortex reaches the barrier, it stands moving slightly parallel to the barrier, towards the north, until it dissipates (Fig. 6). This pattern is also similar to the interaction with a solid wall, although with less intensity than the non-isolated vortex.

4.2 Interaction with the B Barrier

The sequence of a non-isolated vortex coming into contact with barrier **B** (Fig. 7) shows more interesting results from the dynamic and, probably, oceanographic point of view. Before the vortex comes into contact with the barrier, zones of vorticity with the opposite sign form between the straits, on the west side of the barrier. At the moment of incidence, the vortex displaces toward the north parallel to the barrier. However, in the quantitative experiment the cyclonic structures on the west side of the barrier fuse, creating a new vortex. This effect is also seen in the experiment with colouring, although more weakly.

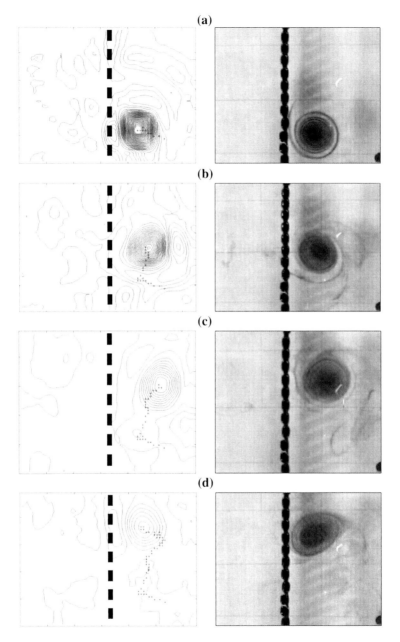

Fig. 5 *Left* vorticity contours for the non-isolated vortex, for times **a** 15 s, **b** 30 s, **c** 45 s, and **d** 60 s after the vortex has initiated. The *blue* and *red lines* represent cyclonic and anticyclonic vorticity. The contour increments are 0.02 of the maximum vorticity value. The *vertical line* represents barrier **A**. The displacement of the vortex center is shown with (+) symbols; the frequency between points is 1 s. *Right* sequence of images showing the evolution of a similar vortex, visualized with ink. The behavior is similar at the interaction with a solid wall

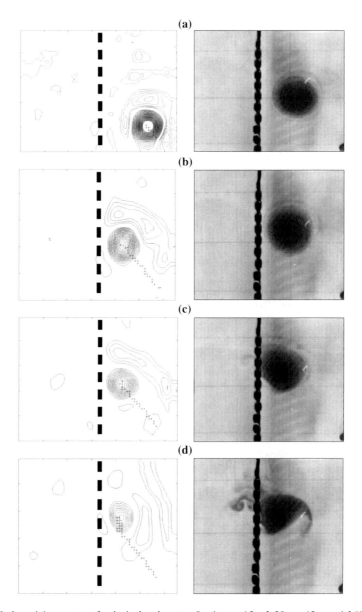

Fig. 6 *Left* vorticity contours for the isolated vortex, for times **a** 15 s, **b** 30 s, **c** 45 s, and **d** 60 s after the vortex has initiated. The *blue* and *red lines* represent cyclonic and anticyclonic vorticity. The contour increments are 0.02 of the maximum vorticity value. The *vertical line* represents barrier **A**. The displacement of the vortex center is shown with (+) symbols; the frequency between points is 1 s. *Right* sequence of images showing the evolution of a similar vortex, visualized with ink. The interaction with barrier was less intense and some weak filaments pass through some straits

(a)

(b)

(c)

(d)

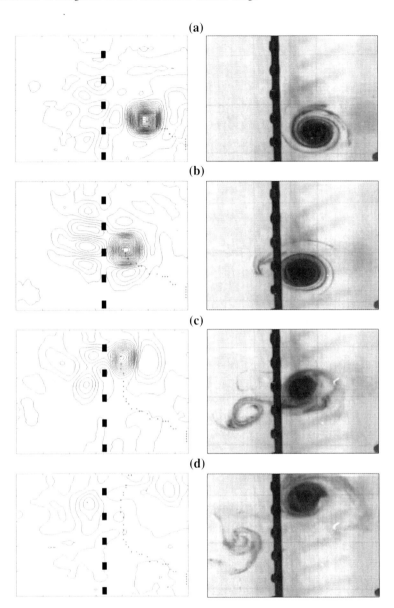

Fig. 7 *Left* vorticity contours for the non-isolated vortex, for the times **a** 15 s, **b** 30 s, **c** 45 s, and **d** 60 s after the vortex has initiated. The *blue* and *red lines* represent cyclonic and anticyclonic vorticity. The contour increments are 0.02 of the maximum vorticity value. The *vertical line* represents barrier **B**. The displacement of the vortex center is shown with (+) symbols; the frequency between points is 1 s. *Right* sequence of images showing the evolution of a similar vortex, visualized with ink. Zones of vorticity with the opposite sign form between the straits, on the west side of the barrier

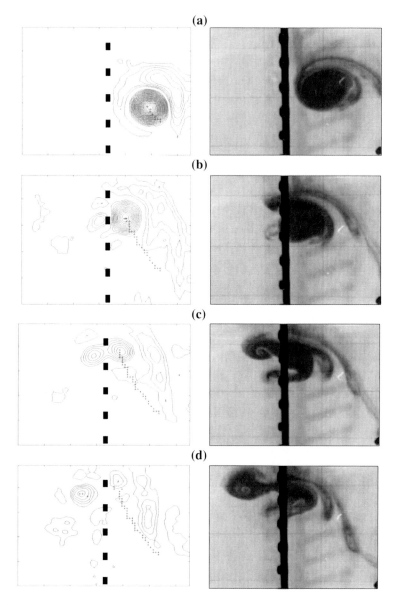

Fig. 8 *Left* vorticity contours for the isolated vortex, for times **a** 15 s, **b** 30 s, **c** 45 s, and **d** 60 s after the vortex has initiated. The *blue* and *red lines* represent cyclonic and anticyclonic vorticity. The contour increments are 0.02 of the maximum vorticity value. The *vertical line* represents barrier **B**. The displacement of the vortex center is shown with (+) symbols; the frequency between points is 1 s. *Right* sequence of images showing the evolution of a similar vortex, visualized with ink. This vortices do not generate opposite vorticity on the west side of the barrier, filament that crossed produces a new vortex on the west side

In figure (Fig. 8) The isolated vortex with barrier **B** is similar to the non-isolated ones. However, owing to their low intensity, isolated vortices do not generate opposite vorticity on the west side of the barrier. During its displacement parallel to the barrier, a great part of the colliding vortex begins to cross to the west side through the straits. The portion that has crossed produces a new vortex on the west side whose magnitude is similar to that of the original vortex. Not all the vortex crosses through the islands, and the part that stays on the east side moves parallel to the barrier, towards the north.

4.3 Formation of Vortices on the West Side of the Islands

The two types of vortices differ during their interaction with barrier **A** (Figs. 5 and 6). Mainly in terms of intensity; the non-isolated vortices are less intense than the isolated ones and for this reason their interaction with the barrier can be more violent. However, barrier **A** behaves as a solid wall for both types of vortex. Although some very weak filaments from the isolated vortices pass across the barrier, they are not energetic enough to form a sufficiently coherent structure.

On the other hand, the dynamic of the non-isolated vortices with barrier **B** is different and presents very particular characteristics: when the vortex is getting closer to the barrier, some vortices of opposite sign form on the west side of the barrier. A little later, vortices with the same sign of vorticity fuse and form a dipolar structure. Subsequently, when the original vortex comes into contact with the barrier, it displaces towards the north along the barrier and an intense filament begins to cross through a strait. This filament maintains the coherence of the incident vortex and fuses with the cyclonic part of the dipole, forming a coherent cyclone almost as intense as the original vortex.

The interaction of the two types of vortex with barrier **B** (Figs. 7 and 8) leads to the passage of either type through the barrier with a very similar dynamic. These experiments clearly show that new vortices can be formed on the west side of the barrier from the interaction of the vortices that reach the barrier on the east side.

5 Discussion and Conclusions

In this study we analyzed the dynamic of the interaction of cyclonic monopolar barotropic vortices with multiple straits, by experimental modeling. For the analysis of the experiments we used two techniques: (a) qualitative, that consisted of adding a colouring to the flow, and (b) quantitative, by making lagrangian measurements, using the PIV method.

The vortices were generated by two different methods: (1) suction, which has been extensively used in other studies, and (2) electromagnetic.

Analysis of the horizontal characteristics of the two types of vortex showed that those generated with suction were circular and their vorticity field always had the same sign (*non-isolated*) (Kloosterziel and van Heijst 1992). In contrast, the center of the electromagnetic vortices was surrounded by vorticity with opposite sign (*isolated*). In general, the vortices produced by suction were more intense ($Ro \sim 1$) than the electromagnetic vortices ($Ro \sim 0.5$). On the other hand, its *non-isolated* vortices are extremely stable to small perturbations (Kloosterziel and van Heijst 1992).

Two types of barrier were implemented to study the interaction of vortices with the chain of islands: one with small straits, and one with large straits, with respect to the size of the vortex. This interaction can be considered as a first approximation of the real case of the interaction vortex-strait at oceanic scales. However, the results obtained here can be compared with previous experimental studies (i.e. Cenedese and Tanabe 2008; Cenedese et al. 2005).

According to the results of the present work, the variation of the strait width, the intensity of the vortex and possibly the angle of collision with the barrier, are fundamental parameters in the dynamic of the interaction vortex-islands.

Although, in this study the real values of the NBCR and the chain of islands of Antilles Arc were adjusted with the values corresponding to the configuration of barrier **A**, the experiments showed that the barrier acts as a solid wall that obstructs the total or partial crossing of the vortices, independently of their intensity and angle of incidence when approaching the barrier. In this case, only some filaments are able to cross the strait, and these are so weak that they are not able to form coherent vortices on the west side of the islands. The generation of vorticity of opposite sign to that of the incident vortex due to the non-slid boundary condition, shows that barrier **A** is not an adequate model to represent the Antilles, since this behavior is not expected at oceanographic level, dominated by viscous effects. Even so, these cases show that if the straits are really narrow, the crossing of the vortices can be totally or partially blocked.

Another observed characteristic was the importance of the Rossby number in the interaction. For those vortices with *Ro* close to one (non-isolated vortices), their interaction with the barrier was violent, the incidence angle almost perpendicular and their translation velocity intense. On the other hand, the vortices with *Ro* less than one (isolated vortices) had a weak interaction with the barrier, an incidence angle $\sim 45°$ and a slow translation velocity.

In the configuration with barrier **B**, that the intrusion of part of the incident vortex was always by elongation of the original vortex into a filament. This filament crossed through one of the straits of the barrier, conserving its coherence and forming an other vortex on the west side. Mean while, the original vortex (on the east side of the barrier) continued to move parallel to the barrier towards the north.

The formation of alternating vorticity (cyclonic and anti-cyclonic) was observed with this same configuration. These vortices fused with other vortices having the same sign of vorticity. This dynamic can be a possible mechanism to form larger vortices on the other side of the Antilles in the Caribbean.

Although, the non-dimensional values were adjusted with those of the Antilles Arc the two configurations of barrier used here, can be considered as only a first approximation to the real situation, i.e. to the shape of the arc and the islands. A closer approximation of Antilles Arc and its islands, and of the bathymetry, are studies to be done in the future.

Nevertheless, it is possible to conclude that the important parameters causing the reflection, destruction and penetration of the rings toward the Caribean Sea are the Rossby number associated with the vortex and the width of the strait. Moreover, the formation of the large gyres in the Caribean maybe due to the intensity of the flow crossing through the straits between the islands, caused in part by the NBCR.

References

Adem J (1956) A series solution for the barotropic vorticity equation and its application in the study of atmospheric vortices. Tellus 8:364–372

Carnevale GF, Kloosterziel RC, van Heijst GJF (1991) Propagation of barotropic vortices over topography in a rotating tank. J Fluid Mech 233:119–139

Cenedese C, Tanabe A (2008) Laboratory experiments on mesoscale vortices colliding with an island chain. J Geophys Res 113:C04022

Cenedese C, Adduce C, Fratantoni DM (2005) Laboratory experiments on mesoscale vortices interacting with two islands. J Geophys Res 110:C09023

Cruz Gómez RC, Bulgakov SN (2007) Remote sensing observations of the coherent and non-coherent ring structures in the vicinity of Lesser Antilles. Ann Geophys 25:1–10

Cruz Gómez RC, Monreal-Gómez MA, Bulgakov SN (2008) Efectos de los vórtices en sistemas acuáticos y su relación con la Química, Biología y Geología. Interciencia 33:741–746

Firing E, Beardsley RC (1976) The behavior of a barotropic eddy on a Beta plane. J Phys Oceanogr 6:57–65

Flór JB, Eames I (2002) Dynamics of monopolar vortices on a topographic beta-plane. J Fluid Mech 456:353–376

Fratantoni DM, Glikson DA (2002) North Brazil current ring generation and evolution observed with SeaWiFS. J Phys Oceanogr 36(7):1241–1264

Hopfinger EJ, van Heijst GJF (1993) Vortices in rotating fluids. Annu Rev Fluid Mech 25: 241–289

Kloosterziel RC, van Heijst GJF (1992) The evolution of stable barotropic vortices in a rotating free-surface fluid. J Fluid Mech 239:607–629

Matías Durán M, Velasco Fuentes OU (2008) Passage of a barotropic vortex thorough a gap. J Phys Oceanogr 38:2817–2831

Richardson PL (2005) Caribbean current and eddies as observed by surface drifters. Deep-Sea Res II 52:429–463

Simmons HL, Nof D (2002) The squeezing of eddies through gaps. J Phys Oceanogr 32:314–335

van Heijst GJF (1994) Topographic effects on vortices in a rotating fluid. Meccanica 29:431–451

Zavala Sansón L, van Heijst GJF (2000) Interaction of barotropic vortices with coastal topography: laboratory experiments and numerical simulations. J Phys Oceanogr 30:2141–2162

Zavala Sansón L, van Heijst GJF, Janssen FJJ (1999) Experiments on barotropic vortex-wall interaction on a topographic beta-plane. J Geophys Res 104:10917–10932

Physical Processes of Interstellar Turbulence

Enrique Vázquez-Semadeni

Abstract This review discusses the role of radiative heating and cooling, as well as self-gravity, in shaping the nature of the turbulence in the interstellar medium (ISM) of our galaxy. The ability of the gas to radiatively cool, while simultaneously being immersed in a radiative heat bath, causes it to be much more compressible than if it were adiabatic, and, in some regimes of density and temperature, to become thermally unstable, and thus tend to spontaneously segregate into separate phases, one warm and diffuse, the other dense and cold. On the other hand, turbulence is an inherently mixing process, thus tending to replenish the density and temperature ranges that would be forbidden under thermal processes alone. The turbulence in the ionized ISM appears to be transonic (i.e, with Mach numbers $M_s \sim 1$), and thus to behave essentially incompressibly. However, in the neutral medium, thermal instability causes the sound speed of the gas to fluctuate by up to factors of ~ 30, and thus the flow can be highly supersonic with respect to the dense, cold gas. However, numerical simulations suggest that the supersonic velocity dispersion corresponds more to the ensemble of cold clumps than to the clumps' internal velocity dispersion. Finally, coherent large-scale compressions in the warm neutral medium (induced by, say, the passage of spiral arms or by supernova shock waves) can produce large, dense, and turbulent clouds that are affected by their own self-gravity, and begin to contract gravitationally. Because they are populated by the nonlinear turbulent density fluctuations, whose local free-fall times can be significantly smaller than that of the whole cloud, the fluctuations terminate their collapse earlier, giving rise to a regime of hierarchical gravitational fragmentation, with small-scale collapses occurring within larger-scale ones. Thus, the "turbulence" in the cold, dense clouds may actually consist primarily of gravitationally contracting motions at all scales within them.

E. Vázquez-Semadeni (✉)
Centro de Radioastronomía y Astrofísica, 58089 UNAM Campus Morelia, Mexico
e-mail: e.vazquez@crya.unam.mx

J. Klapp et al. (eds.), *Fluid Dynamics in Physics, Engineering and Environmental Applications*, Environmental Science and Engineering,
DOI: 10.1007/978-3-642-27723-8_5, © Springer-Verlag Berlin Heidelberg 2013

1 Introduction

Our galaxy, the Milky Way (or simply, the Galaxy) is a flattened conglomerate of stars, gas, dust, and other debris, such as planets, meteorites, etc., with a total mass $\sim 10^{12} M_\odot$, where $M_\odot = 2 \times 10^{33}$ g is the mass of the Sun. Most of this mass is believed to be in a roughly spherical dark matter halo, while $\sim 6.5 \times 10^{10} M_\odot$ are in stars (McMillan 2011), and $\sim 10^{10} M_\odot$ are contained in the gaseous component, mostly confined to the Galactic disk (Cox 2000).

The gaseous component may be in either ionized, neutral atomic or neutral molecular forms, spanning a huge range of densities and temperatures, from the so-called hot ionized medium (HIM), with densities $n \sim 10^{-2}$ cm^{-3} and temperatures $T \sim 10^6$ K, through the warm ionized and neutral (atomic) media (WIM and WNM, respectively), with $n \sim 0.3$ cm^{-3} and $T \sim 10^4$ K and the cold neutral (atomic) medium (CNM, $n \sim 30$ cm^{-3}, $T \sim 100$ K), to the *giant molecular clouds* (GMCs, $n \gtrsim 100$ cm^{-3} and $T \sim 10$–20 K). GMCs can span several tens of parsecs across, and, in turn, contain plenty of substructure, which is commonly classified into *clouds* ($n \sim 10^3$ cm^{-3}, size scales L of a few parsecs), *clumps* ($n \sim 10^4$ cm^{-3}, $L \sim 1$ pc), and *cores* ($n \gtrsim 10^5$ cm^{-3}, $L \sim 0.1$ pc). It is worth noting that the temperature of most molecular gas is remarkably uniform, ~ 10–30 K (e.g., Ferrière 2001).

The gaseous component, along with the dust, a cosmic-ray background, and a magnetic field of mean intensity of a few μG constitute what we know as the *interstellar medium* (ISM). This medium is in most cases well described by the fluid approximation (Shu 1992, Chap. 1). Moreover, the ISM is most certainly turbulent, as typical Reynolds numbers in it are very large. For example, in the cold ISM, $R_e \sim 10^5$–10^7 (Elmegreen and Scalo 2004, Sect. 4.1). This is mostly due to the very large spatial scales involved in interstellar flows. Because the ISM's temperature varies so much from one type of region to another, so does the sound speed, and the flow is often super- or trans-sonic (e.g., Heiles and Troland 2003; Elmegreen and Scalo 2004 and references therein). This implies that the flow is significantly compressible, inducing significant density fluctuations (Sect. 3.2).

In addition to being turbulent, the ISM is subject to a number of additional physical processes, such as gravitational forces exerted by the stellar and dark matter components as well as by the ISM itself (that is, the ISM's *self-gravity*), magnetic fields, cooling by radiative microscopic processes, and radiative heating due both to nearby stellar sources as well as to diffuse background radiative fields. All of this adds up to make the ISM an extremely complex and dynamical medium.

Finally, the self-gravity of the gas causes local (i.e., spatially intermittent) gravitational collapse events, in which a certain gas parcel within a molecular cloud goes out of equilibrium between its self-gravity and all other forces that oppose it, undergoing an implosion that leads the gas density to increase by tens of orders of magnitude, and whose end product is a star or a group ("cluster") of stars.

In this review, we focus on the interaction between turbulence, the effects of radiative heating and cooling, which effectively enhance the compressibility of the flow, and the gas' self-gravity. Their complex interactions have a direct effect on the star formation process. The plan of the paper is as follows: in Sect. 2 we first review the effects that the net heating and cooling have on the effective equation of state of the flow and, in the case of thermally unstable flows, on its tendency to spontaneously segregate in distinct phases. Next, in Sect. 3 we discuss a few basic notions about turbulence and the turbulent production of density fluctuations in the compressible case, to then discuss, in Sect. 4, the interplay between turbulence and the heating and cooling. In Sect. 5, we discuss the likely nature of turbulence in the diffuse (warm and hot) parts of the ISM, as well as in the dense, cold atomic and molecular clouds. We conclude in Sect. 6 with a summary and some final remarks. Due to space limitations, we do not discuss magnetic fields, although we refer the interested reader to the reviews by Vázquez-Semadeni et al. (2000b), Cho et al. (2003), Elmegreen and Scalo (2004), and McKee and Ostriker (2007).

2 ISM Thermodynamics: Thermal Instability

The ISM extends essentially over the entire disk of the Galaxy and, when considering a certain subregion of it, such as a cloud or cloud complex, it is necessary to realize that any such subregion constitutes an open system, whose interactions with its environment need to be taken into account. A fundamental form of interaction with the surroundings, besides dynamical interactions, is through the exchange of heat. Indeed, the ISM is permeated by a radiation field, due to the combined shine of the stars in the disk. Moreover, a bath of relativistic charged particles, mostly protons, known as *cosmic rays*, also exists in the ISM. These are believed to be accelerated in strong shocks produced by supernova explosions (Blandford and Eichler 1987). Both UV radiation and cosmic rays provide heating and ionization sources for the ISM at large (see, e.g., Dalgarno and McCray 1972; Wolfire et al. 1995). Finally, violent events, such as supernova explosions, can locally heat their surroundings to very high temperatures, causing local bubbles of hot, million-degree ionized gas.

On the other hand, the ISM can cool by emission from ions, atoms and molecules, and by thermal emission from dust grains (Dalgarno and McCray 1972; Sutherland and Dopita 1993; Wolfire et al. 1995). The balance between these radiative heating and cooling radiative processes, along with the heat due to mechanical work and thermal conductivity, determine the thermodynamic properties of the ISM. These properties will be the focus of the present section, rather than the detailed microphysical processes that mediate the radiative transfer, about which there exist plenty of excellent books and reviews (e.g., Dalgarno and McCray 1972; Sutherland and Dopita 1993; Osterbrock and Ferland 2006).

Globally, and as a first approximation, the ISM is roughly isobaric, as illustrated in the left panel of Fig. 1. As can be seen there, most types of regions, either dilute or dense, lie within an order of magnitude from a thermal pressure

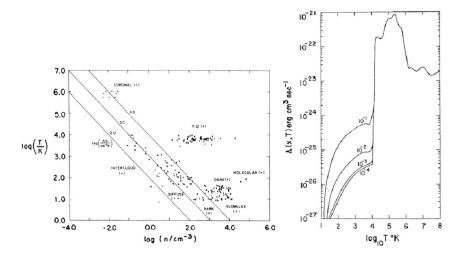

Fig. 1 *Left*: thermal pressure in various types of interstellar regions. The points labeled *coronal* correspond essentially to what we refer to as the HIM in the text; *intercloud* regions refer to the WIM and WNM; *diffuse*, to CNM clouds, and *dark*, *globule* and *molecular* to molecular gas. From Myers (1978). *Right*: temperature dependence of the cooling function. The labels indicate values of the ionization fraction (per number) of the gas. From Dalgarno and McCray (1972)

$P \sim 3000$ Kcm^{-3}.[1] The largest deviations from this pressure uniformity are found in HII regions, which are the ionized regions around massive stars due to the star's UV radiation, and molecular clouds, which, as we shall see in Sect. 5.4, are probably pressurized by gravitational compression.

The peculiar thermodynamic behavior of the ISM is due to the functional forms of the radiative heating and cooling functions acting on it, which depend on the density, temperature, and chemical composition of the gas. The right panel of Fig. 1 shows the temperature dependence of the cooling function Λ (Dalgarno and McCray 1972).

To understand the effect of the functional form of the radiative heating and cooling functions on the net behavior of the gas, let us write the conservation equation for the internal energy per unit mass, e. We have (e.g., Shu 1992)

$$\frac{\partial e}{\partial t} + \mathbf{u} \cdot \nabla e = -(\gamma - 1)e\nabla \cdot \mathbf{u} + \Gamma - n\Lambda, \tag{1}$$

where we have neglected thermal conduction and heating by magnetic reconnection. In Eq. (1), $\mathbf{u}$ is the velocity vector, γ is the ratio of specific heats, Γ is the heating rate per unit mass, and $n = \rho/\mu m_{\mathrm{H}}$ is the number density, with ρ being the mass density, μ the mean particle mass, and m_{H} the hydrogen atom's mass. Note

---

[1] It is customary in Astrophysics to express pressure in units of [Kcm^{-3}]. Strictly speaking, this corresponds to P/k, where k is the Boltzmann constant.

that $n\Lambda$ is the cooling rate per unit mass. The first term on the right hand side is the $P\,dV$ work per unit mass. We also assume an ideal-gas equation of state, $P = nkT$. This is a good approximation, given the very low densities of the ISM.

In the simplest possible case, that of a hydrostatic ($\mathbf{u} = 0$) and steady ($\partial/\partial t = 0$) state, Eq. (1) reduces to the condition of *thermal equilibrium*,

$$\Gamma(\rho, T) = n\Lambda(\rho, T), \tag{2}$$

where we have denoted explicitly the dependence of Γ and Λ on the density and temperature, and neglected the dependence on chemical composition.

Let us now assume we have a gas parcel in thermal and hydrostatic equilibrium, at a temperature somewhere in the range where the slope of Λ with respect to T is negative $\left(\text{i.e.,}(\partial\Lambda/\partial T)_\rho < 0\right)$, $10^{5.5} \lesssim T \lesssim 10^6$ K (see the right panel of Fig. 1). If we then consider a small isochoric (i.e., at constant density) increase in T of this fluid parcel, we see that Λ decreases. Because we had started from thermal equilibrium, a drop in Λ implies that now $\Gamma > n\Lambda$, a condition that increases the parcel's temperature even further, causing a runaway to higher temperatures, until the fluid parcel exits the temperature range where $(\partial\Lambda/\partial T)_\rho < 0$, at $T \gtrsim 10^6$K. This behavior is known as *thermal instability* (TI), and the condition $(\partial\Lambda/\partial T)_\rho < 0$ is known as the *isochoric criterion* for TI (Field 1965; see also the review by Vázquez-Semadeni et al. 2003).

Another, less stringent condition for the development of TI is the so-called *isobaric criterion*. This can be most easily understood as follows. Note that Eq. (2) provides a relation between the density and temperature of the medium. This can be inserted in the equation of state for the gas, allowing the elimination of the temperature from it and writing a *barotropic* equation of the form $P_{\text{eq}} = P_{\text{eq}}(\rho)$, where P_{eq} is the thermal pressure under conditions of thermal equilibrium, and is a function of the density only. Figure 2 shows the resulting density dependence of P_{eq} for the atomic medium under "standard" conditions (Wolfire et al. 1995). The region above the graph of $P_{\text{eq}}(n)$ corresponds to $n\Lambda > \Gamma$, and therefore to net cooling. Conversely, the area under the curve corresponds to net heating, $n\Lambda < \Gamma$. From this figure, we observe that the slope of the graph of P_{eq} versus ρ is *negative* for densities in the range $0.6 \lesssim n \lesssim 5$ cm^{-3} (i.e., $-0.2 \lesssim \log_{10}(n/1\,\text{cm}^{-3}) \lesssim 0.7$).

Let us now consider a fluid parcel in this density range, and apply to it a small, quasi-static compression (i.e., a volume reduction) to it, increasing its density. This displaces the parcel to the region above the thermal equilibrium curve, where net cooling occurs. Because the compression is quasi-static, the parcel has time to cool under the effect of the net cooling, which causes its pressure to decrease as it attempts to return to the thermal equilibrium curve. But thus it is now at a lower pressure than its surroundings, which then continue to further compress the parcel, and runaway compression sets in, until the parcel exits the regime where $dP_{\text{eq}}/dn < 0$. The density range where this condition is met is called the *unstable range*.

Thus, if the mean density of the medium is in the unstable range, this mode of TI tends to cause the medium to spontaneously segregate into a cold, dense phase and a

Fig. 2 Thermal-equilibrium pressure P_{eq} as a function of number density for "standard" conditions of metallicity and background UV radiation for the atomic medium. The *horizontal* axis gives $\log_{10}(n/cm^3)$. From Wolfire et al. (1995)

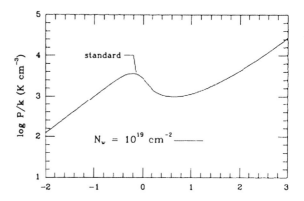

warm, diffuse one (Field et al. 1969). That is indeed the case of the ISM in the Galactic midplane in the Solar neighborhood, a fact which led Field et al. (1969) to propose the so-called *two-phase model* of the ISM: assuming dynamical and thermal equilibrium, the atomic ISM would consist of small dense, cold clumps (the CNM), immersed in a warm, diffuse background (the WNM). The clumps are expected to be small because the fastest growing mode of the instability occurs at vanishingly small scales in the absence of thermal conductivity, or at scales ~ 0.1 pc for the estimated thermal conductivity of the ISM (see, e.g., Audit and Hennebelle 2005). Note that the isochoric criterion for TI implies that the isobaric one is satisfied, but not the other way around. Technical and mathematical details, as well as other modes of TI, can be found in the original paper by Field (1965), and in the reviews by Meerson (1996) and by Vázquez-Semadeni et al. (2003).

It is important to note that, even if the medium is *not* thermally unstable, the balance between heating and cooling implies a certain functional dependence of $P_{eq}(\rho)$, which is often approximated by a *polytropic* law of the form $P_{eq} \propto \rho^{\gamma_e}$ (e.g., Elmegreen 1991; Vázquez-Semadeni et al. 1996), where it should be noted that γ_e, which we refer to as the *effective polytropic exponent*, is in general *not* the ratio of specific heats for the gas in this case, but rather a parameter that depends on the functional forms of Λ and Γ. The isobaric mode of TI corresponds to $\gamma_e < 0$.

3 Compressible Turbulence

3.1 Equations

In the previous section we have discussed thermal aspects of the ISM, whose main dynamical effect is the segregation of the medium into the cold and warm phases. Let us now discuss dynamics. As was mentioned in Sect. 1, the ISM is in general highly turbulent, and therefore it is necessary to understand the interplay between turbulence and the effects of the net cooling $(n\Lambda - \Gamma)$, which affects the compressibility of the gas (Vázquez-Semadeni et al. 1996).

The dynamics of the ISM are governed by the fluid equations, which, neglecting magnetic fields, comprise Eq. (1) and (e.g., Landau and Lifshitz 1959; Shu 1992)

$$\frac{\partial \rho}{\partial t} + \mathbf{u} \cdot \nabla \rho = -\rho \nabla \cdot \mathbf{u}, \tag{3}$$

$$\frac{\partial \mathbf{u}}{\partial t} + \mathbf{u} \cdot \nabla \mathbf{u} = -\frac{\nabla P}{\rho} - \nabla \varphi + v \left[\nabla^2 \mathbf{u} + \frac{\nabla (\nabla \cdot \mathbf{u})}{3} \right], \tag{4}$$

$$\nabla^2 \varphi = 4\pi G \rho, \tag{5}$$

where φ is the gravitational potential, and v is the kinematic viscosity. Equation (3) represents mass conservation, and is also known as the *continuity equation*. Equation (4) is the momentum conservation, or *Navier-Stokes* equation per unit mass, with an additional source term representing the gravitational force $\nabla \varphi / \rho$. In turn, the gravitational potential is given by *Poisson's equation*, Eq. (5). Equations (1), (3), (4), and (5) are to be solved simultaneously, given some initial and boundary conditions.

A brief discussion of the various terms in Eq. (4) is in order. The second term on the left is known as the *advective* term, and represents the transport of i-momentum by the j component of the velocity, where i and j represent any two components of the velocity. It is responsible for *mixing*. The pressure gradient term (first term on the right-hand side [RHS]) in general acts to counteract pressure, and therefore density, gradients across the flow. Finally, the term in the brackets on the RHS, the *viscous* term, being of a diffusive nature, tends to erase velocity gradients, thus tending to produce a uniform flow.

3.2 Reynolds and Mach Numbers Compressibility

Turbulence develops in a flow when the ratio of the advective term to the viscous term becomes very large. That is,

$$\frac{\mathcal{O}[\mathbf{u} \cdot \nabla \mathbf{u}]}{\mathcal{O}\left[v\left(\nabla^2 \mathbf{u} + \frac{\nabla(\nabla \cdot \mathbf{u})}{3}\right)\right]} \sim \frac{U^2}{L}\left[v\frac{U}{L^2}\right]^{-1} \sim \frac{UL}{v} \equiv R_e \gg 1, \tag{6}$$

where R_e is the *Reynolds number*, U and L are characteristic velocity and length scales for the flow, and $\mathcal{O}$ denotes "order of magnitude". This condition implies that the mixing action of the advective term overwhelms the velocity-smoothing action of the viscous term.

On the other hand, noting that the advective and pressure gradient terms contribute comparably to the production of density fluctuations, we can write

$$1 \sim \frac{\mathcal{O}(\mathbf{u} \cdot \nabla \mathbf{u})}{\mathcal{O}(\nabla P/\rho)} \sim \frac{U^2}{L} \left[\frac{\Delta P}{L\rho} \right]^{-1} \sim U^2 \left(\frac{c_s^2 \Delta \rho}{\rho} \right)^{-1} \equiv M_s^2 \left(\frac{\Delta \rho}{\rho} \right)^{-1}, \tag{7}$$

$$\Rightarrow \frac{\Delta \rho}{\rho} \sim M_s^2, \tag{8}$$

where $M_s \equiv U/c_s$ is the *sonic Mach number*, and we have made the approximation that $\Delta P/\Delta \rho \sim c_s^2$, where c_s is the sound speed. Equation (8) then implies that strong compressibility requires $M_s \gg 1$. Conversely, flows with $M_s \ll 1$ behave incompressibly, even if they are gaseous. Such is the case, for example, of the Earth's atmosphere. In the incompressible limit, $\rho = $ cst., and thus Eq. (3) reduces to $\nabla \cdot \mathbf{u} = 0$.

Finally, a trivial, but often overlooked, fact is that, in order to produce a density enhancement in a certain region of the flow, the velocity at that point must have a negative divergence (i.e., a *convergence*), as indicated by the continuity equation, Eq. (3). It is very frequent to encounter in the literature discussions of pre-existing density enhancements ("clumps") in hydrostatic equilibrium. But it should be kept in mind that these can only exist in multi-phase media, where a dilute, warm phase can have the same pressure as a denser, but colder, clump. But even in this case, the *formation* of that clump must have initially involved the convergence of the flow towards the cloud, and the hydrostatic situation is applicable in the limit of very long times after the formation of the clump, when the convergence of the flow has subsided.

3.3 Production of Density Fluctuations

According to the previous discussion, a turbulent flow in which the velocity fluctuations are supersonic will naturally develop strong density fluctuations. Note, however, that the nature of turbulent density fluctuations in a single-phase medium (such as, for example, a regular isothermal or adiabatic flow) is very different from that of the cloudlets formed by TI (cf. Sect. 2). In a single-phase turbulent medium, turbulent density fluctuations must be transient, as a higher density generally conveys a higher pressure,[2] and therefore the fluctuations must re-expand after the compression that produced them has subsided.

For astrophysical purposes it is important to determine the distribution of these fluctuations, as they may constitute, or at least provide the seeds for, what we

---

[2] An exception would be a so-called Burgers' flow, which is characterized by the absence of the pressure gradient term (Burgers 1974).

normally refer to as "clouds" in the ISM. Because of the transient nature of turbulent density fluctuations in single-phase media, however, this distribution refers to a time-stationary population of fluctuations, although the fluctuations themselves will appear and disappear on timescales that are short compared to the time over which the distribution is considered.

The probability density distribution (PDF) of the density field in turbulent iso-thermal flows was initially investigated through numerical simulations. Vázquez-Semadeni (1994) found that, in the isothermal case, the PDF possess a lognormal form. A theory for the emergence of this functional form was later proposed by Passot and Vázquez-Semadeni (1998), in which the production of density fluctuations was assumed to arise from a succession of compressive or expansive waves, each one acting on the value of the density left by the previous one. Because the medium contains a unique distribution of (compressible) velocity fluctuations, and because the density jumps in isothermal flow depend only on Mach number but not in the local density, the density fluctuations belong all to a unique distribution as well, yet each one can be considered independent of the others if the global time scales considered are much longer than the autocorrelation time of the velocity divergence (Blaisdell et al. 1993). Finally, because the density jumps are multiplicative in the density (cf. Eq. 8), then they are additive in $s \equiv \ln \rho$. Under these conditions, the Central Limit Theorem can be invoked for the increments in s, implying that s will be normally distributed. In consequence, ρ will have a lognormal PDF.

In addition, Passot and Vázquez-Semadeni (1998) also argued that the variance of the density fluctuations should scale linearly with M_s, a suggestion that has been investigated further by other groups (Padoan et al. 1997; Federrath et al. 2008). In particular, using numerical simulations of compressible turbulence driven by either solenoidal (or "vortical") or compressible (or "potential") forces, the latter authors proposed that the variance of s is given by

$$\sigma_s = \ln(1 + bM_s^2), \tag{9}$$

where b is a constant whose value depends on the nature of the forcing, taking the extreme values of $b = 1/3$ for purely solenoidal forcing, and $b = 1$ for purely compressible forcing. The lognormal density PDF for the one-dimensional, iso-thermal simulations of Passot and Vázquez-Semadeni (1998), with its dependence on M_s, is illustrated in the *left panel* of Fig. 3.

Finally, Passot and Vázquez-Semadeni (1998 see also Padoan and Nordlund 1999) also investigated the case where the flow behaves as a polytrope with arbitrary values of γ_e, by noting that in this case the sound speed is not constant, but rather depends on the density as $c_s \propto \rho^{(\gamma_e-1)/2}$, implying that the local Mach number of a fluid parcel now depends on the local density besides its dependence on the value of the flow velocity. Introducing this dependence of M_s on ρ in the expression for the lognormal PDF, Passot and Vázquez-Semadeni (1998) con-cluded that the density PDF should develop a power-law tail, at high densities when $\gamma_e < 1$, and at low densities when $\gamma_e > 1$. This result was then confirmed by numerical simulations of polytropic turbulent flows (Fig. 3, *right panel*).

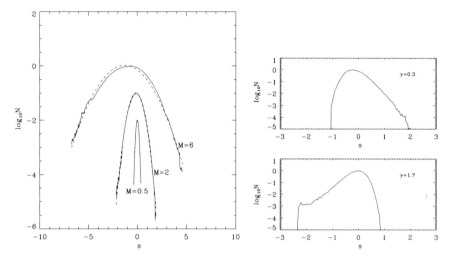

Fig. 3 *Left*: lognormal density PDFs for isothermal one-dimensional simulations at various Mach numbers, indicated by the labels. The independent variable is $s \equiv \ln \rho$. *Right*: density PDFs for polytropic cases (i.e., with $P \propto \rho^{\gamma_e}$), with effective polytropic exponent $\gamma_e = 0.3$ (*top*) and $\gamma_e = 1.7$ (*bottom*). From Passot and Vázquez-Semadeni (1998)

4 Turbulence and Thermodynamics

In the previous sections we have separately discussed two different kinds of physical processes operating in the ISM: radiative heating and cooling (to which we refer collectively as *net cooling*, $n\Lambda - \Gamma$), and compressible turbulence. However, since both operate simultaneously, it is important to understand how they interact with each other, especially because the mean density in the Solar neighborhood, $\langle n \rangle \sim 1 \text{ cm}^{-3}$, falls precisely in the thermally unstable range. This problem has been investigated numerically by various groups (e.g., Hennebelle and Perault 1999; Walder and Folini 2000; Koyama and Inutsuka 2000, 2002; Vázquez-Semadeni et al. 2000a; Vázquez-Semadeni et al. 2003; Vázquez-Semadeni et al. 2006; Gazol et al. 2001; Gazol et al. 2005; Kritsuk and Norman 2002; Sánchez-Salcedo et al. 2002; Piontek and Ostriker 2004, 2005; Audit and Hennebelle 2005; Audit and Hennebelle 2010; Heitsch et al. 2005; Hennebelle and Audit 2007).

4.1 Density and Pressure Distributions in the ISM

The main parameter controlling the interaction between turbulence and net cooling is the ratio $\eta \equiv \tau_c/\tau_t$, where $\tau_c \approx e/(\mu m_H n\Lambda)$ is the cooling time and $\tau_t \approx L/U$ is the turbulent crossing time. The remaining symbols have been defined above. In the limit $\eta \gg 1$, the turbulent compressions' dynamical evolution occurs much more rapidly than they can cool, and therefore the compressions behave nearly adiabatically. Conversely, in the limit $\eta \ll 1$, the fluctuations cool down essentially instantaneously while the turbulent compression is evolving, and thus they tend to reach the thermal equilibrium pressure P_{eq} as soon as they are produced (Elmegreen 1991; Passot et al. 1995; Sánchez-Salcedo et al. 2002; Vázquez-Semadeni et al. 2003; Gazol et al. 2005).[3] Because in a turbulent flow velocity fluctuations of a wide range of amplitudes and size scales are present, the resulting density fluctuations in general span the whole range between those limits, and the actual thermal pressure of a fluid parcel is not uniquely determined by its density, but rather depends on the details of the velocity fluctuation that produced it. This causes a scatter in the values of the pressure around the thermal-equilibrium value in the pressure-density diagram (Fig. 4, *left panel*), and also produces significant amounts of gas (up to nearly half of the total mass) with densities and temperatures in the classically forbidden thermally unstable range (Gazol et al. 2001; de Avillez and Breitschwerdt 2005; Audit and Hennebelle 2005; Mac Low et al. 2005), a result that has been encountered by various observational studies as well (e.g., Dickey et al. 1978; Heiles 2001). In any case, the tendency of the gas to settle in the stable phases still shows up as a multimodality of the density PDF, which becomes less pronounced as the *rms* turbulent velocity increases (Fig. 4, *right panel*).

4.2 The Formation of Dense, Cold Clouds

Another important consequence of the interaction of turbulence (or, more generally, large-scale coherent motions of any kind) and TI is that the former may *nonlinearly* induce the latter. Indeed, Hennebelle and Pérault (1999, see also Koyama and Inutsuka 2000) showed that transonic (i.e., with $M_s \sim 1$) compressions in the WNM can compress the medium and bring it sufficiently far from thermal equilibrium that it can then undergo a phase transition to the CNM (Fig. 5, *left panel*). This process amounts then to producing a cloud with a density up to $100\times$ larger than that of the WNM by means of only moderate compressions. This is in stark contrast with the process of producing density fluctuations by pure supersonic compressions in, say, an isothermal medium, in which such density contrasts would require Mach numbers $M_s \sim 10$. It is worth noting that the turbulent velocity dispersion of $\sim 8-11 \, \mathrm{km \, s^{-1}}$ in the warm Galactic ISM (Kulkarni and Heiles 1987; Heiles and Troland 2003) is, precisely, transonic.

Moreover, the cold clouds formed by this mechanism have typical sizes given by the size scale of the compressive wave in the transverse direction to the

---

[3] Note that it is often believed that fast cooling implies isothermality. However, this is an erroneous notion. While it is true that fast cooling is a necessary condition for isothermal behavior, the reverse implication does not hold. Fast cooling only implies an approach to the thermal equilibrium condition, but this need not be isothermal. The precise form of the effective equation of state depends on the details of the functional dependence of Λ and Γ on T and ρ.

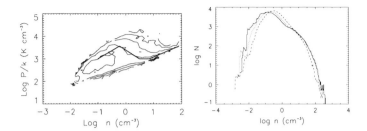

Fig. 4 *Left*: two-dimensional histogram of the grid cells in the pressure-density diagram for a two-dimensional simulation of turbulence in the thermally-bistable atomic medium, with *rms* velocity dispersion of $9 \, \mathrm{km \, s^{-1}}$, a numerical box size of 100 pc, and the turbulent driving applied at a scale of 50 pc. *Right*: density PDF in simulations like the one on the *left* panel, but with three different values of the *rms* velocity: $4.5 \, \mathrm{km \, s^{-1}}$ (*solid line*), $9 \, \mathrm{km \, s^{-1}}$ (*dotted line*), and $11.3 \, \mathrm{km \, s^{-1}}$ (*dashed line*). The peaks are seen to become less pronounced as the rms velocity increases. From Gazol et al. (2005)

compression, thus avoiding the restriction of having the size scale of the fastest growing mode of TI, which is very small (~ 0.1 pc; cf. Sect. 2). The initial stages of this process may produce thin CNM sheets (Vázquez-Semadeni et al. 2006), which are in fact observed (Heiles and Troland 2003). However, such sheets are quickly destabilized, apparently by a combination of nonlinear thin shell (NTSI; Vishniac 1994), Kelvin–Helmholtz and Rayleigh–Taylor instabilities (Heitsch et al. 2005), fragmenting and becoming turbulent. This causes the clouds to become a complex mixture of cold and warm gas, where the cold gas is distributed in an intricate network of sheets, filaments and clumps, possibly permeated by a dilute, warm background. An example of this kind of structure is shown in the *right panel* of Fig. 5.

5 Turbulence in the ISM

5.1 Generalities

As discussed in the previous sections, the ionized and atomic components of the ISM consist of gas in a wide range of temperatures, from $T \sim 10^6$ K for the HIM, to $T \sim 40$ K for the CNM. In particular, Heiles and Troland (2003) report temperatures in the range $500 < T < 10^4$ K for the WNM, and in the range $10 < T < 200$ K for the CNM. The WIM is expected to have $T \sim 10^4$ K. Additionally, those same authors report column density-weighted *rms* velocity dispersions $\sigma_v \sim 11 \, \mathrm{km \, s^{-1}}$ for the WNM, and $\sigma_v \sim 7 \, \mathrm{km \, s^{-1}}$ for the CNM. Since the adiabatic sound speed is given by (e.g., Landau and Lifshitz 1959)

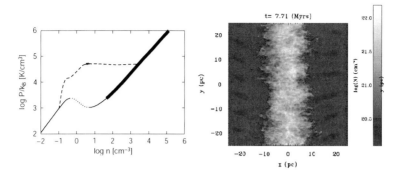

Fig. 5 *Left*: evolutionary path (*dashed line*) in the P versus ρ diagram of a fluid parcel initially in the WNM after suffering a transonic compression that nonlinearly triggers TI. The *solid* and *dotted lines* show the locus of $P_{eq}(\rho)$, the *solid* sections corresponding to linear stability and the *dotted* ones to linear instability. The *solid* section to the *left* of the *dotted line* corresponds to the WNM and the one at the *right*, to the CNM. The perturbed parcel evolves from *left* to *right* along the *dashed line*. From Koyama and Inutsuka (2000). *Right*: projected (or *column*) density of a numerical simulation of the formation of a dense cloud formed by the convergence of two large-scale streams of WNM. The projection is along lines of sight perpendicular to the direction of compression. The cloud is seen to have become turbulent and highly fragmented. From Hennebelle et al. (2008)

$$c_s = \sqrt{\frac{\gamma kT}{\mu m_H}} = 10.4 \, \text{km s}^{-1} \left(\frac{T}{10^4 \, \text{K}} \right)^{1/2}, \tag{10}$$

it is clear that the warm, or *diffuse*, gas is transonic ($M_s \sim 1$), while the cold, or *dense*, gas is strongly supersonic ($3 \lesssim M_s \lesssim 20$).

5.2 The Warm Ionized Medium

Collecting measurements of interstellar scintillation (fluctuations in amplitude and phase of radio waves caused by scattering in the ionized ISM) from a variety of observations, Armstrong et al. (1995) estimated the power spectrum of density fluctuations in the WIM, finding that it is consistent with a Kolmogorov spectrum, characteristic of incompressible turbulence (see, e.g., the reviews by Vázquez-Semadeni et al. 2000b; Mac Low and Klessen 2004; Elmegreen and Scalo 2004), on scales $10^8 \lesssim L \lesssim 10^{15}$ cm.

More recently, using data from the Wisconsin Hα Mapper Observatory, Chepurnov and Lazarian (2010) have been able to extend the spectrum to scales $\sim 10^{19}$ cm, suggesting that the WIM behaves as an incompressible turbulent flow over size scales spanning more than 10 orders of magnitude. This suggestion is supported also by the results of Hill et al. (2008) who, by measuring the distribution of Hα emission measures in the WIM, and comparing with numerical simulations of turbulence at various Mach numbers, concluded that the sonic Mach number of the WIM should be $\sim 1.4-2.4$. Although the WIM is ionized, and thus should be strongly coupled to the magnetic field, the turbulence then being magnetohydrodynamic (MHD), Kolmogorov scaling should still apply, according to the theory of incompressible MHD fluctuations (Goldreich and Sridhar 1995). The likely sources of kinetic energy for these turbulent motions are stellar energy sources such as supernova explosions (see, e.g., Mac Low and Klessen 2004).

5.3 The Atomic Medium

In contrast to the relatively clear-cut situation for the ionized ISM, the turbulence in the neutral (atomic and molecular) gas is more complicated, and is currently under strong debate. According to the discussion in Sect. 5.1, the temperatures in the atomic gas may span a continuous range from a few tens to several thousand degrees, and have velocity dispersions of $\sigma_v \sim 7-10\,\mathrm{km\,s^{-1}}$, suggesting that it should range from mildly to strongly supersonic. But because the atomic gas is *thermally bistable* (i.e., has two stable thermodynamic phases separated by an unstable one; Sect. 2), and because transonic compressions in the WNM can nonlinearly induce TI and thus a phase transition to the CNM (Sect. 4.2), the neutral atomic medium is expected to consist of a complex mixture of gas spanning over two orders of magnitude in density. Early models (e.g., Field et al. 1969; McKee and Ostriker 1977) proposed that the phases were completely separate, but the results reported in Sect. 4.1 suggest that significant amounts of gas exist as well in the unstable range, transiting between the stable phases. Numerical simulations of such systems suggest that the velocity disperion *within* the densest "clumps" is subsonic, but that the velocity dispersion of the clumps within the diffuse substrate is supersonic with respect to the clumps' sound speed (although subsonic with respect to the warmest gas; Koyama and Inutsuka 2002; Heitsch et al. 2005). Also, note that, contrary to earlier ideas (e.g. Kwan 1979; Blitz and Shu 1980), the clumps in the modern simulations actually form from *fragmentation* of large-scale clouds formed by large-scale compressive motions in the WNM, rather than the clouds forming from random collision and coagulation of the clumps. The complexity of this type of structure is illustrated in the *right panel* of Fig. 5.

5.4 The Molecular Gas

The discussion so far, involving mainly turbulence and thermodynamics, has referred to the ionized and atomic components of the ISM. However, molecular clouds (MCs) have long been known to be strongly self-gravitating (e.g.,

Goldreich and Kwan 1974; Larson 1981). In view of this, Goldreich and Kwan (1974) initially proposed that MCs should be in a state of gravitational collapse, and that the observed motions in MCs (as derived by the non-thermal linewidths of molecular lines) corresponded to this collapse. However, shortly thereafter, Zuckerman and Palmer (1974) argued against this possibility by noting that, if all the molecular gas in the Galaxy ($M_{mol} \sim 10^9 M_\odot$) were in free-fall, a simple estimate of the Galaxy's star formation rate (SFR), given by SFR $\sim M_{mol}/\tau_{ff} \sim 200\,M.\ \mathrm{yr}^{-1}$, where $\tau_{ff} = \sqrt{3\pi/32G\rho}$ is the free-fall time, would exceed the observed rate of $\sim 2\,M.\,\mathrm{yr}^{-1}$ (e.g., Chomiuk and Povich 2011) by roughly two orders of magnitude. This prompted the suggestion (Zuckerman and Evans 1974) that the non-thermal motions in MCs corresponded instead to small-scale (in comparison to the clouds' sizes) random turbulent motions. The need for these motions to be confined to small scales arose from the need of the turbulent (*ram*) pressure to provide an isotropic pressure that could counteract the clouds' self-gravity at large, maintaining them in near virial equilibrium (Larson 1981). Because turbulence is known to be a dissipative phenomenon (e.g., Landau and Lifshitz 1959), research then focused on finding suitable sources for driving the turbulence and avoiding rapid dissipation. The main driving source was considered to be energy injection from stars (e.g., Norman and Silk 1980; McKee 1989; Mac Low and Klessen 2004), and reduction of dissipation was proposed to be accomplished by having the turbulence being MHD, and consisting mostly of Alfvén waves, which do not dissipate as rapidly (e.g., Shu et al. 1987).

However, in the last decade several results have challenged the turbulent pressure-support scenario: (1) Turbulence is known to be characterized by having the largest-velocities occuring at the largest scales, and MCs are no exception, exhibiting scaling relations between velocity dispersion and size which suggest that the largest dispersions tend to occuring at the largest scales (Larson 1981; Heyer and Brunt 2004; Brunt et al. 2009, Fig. 6, *left and middle panels*). This is inconsistent with the small-scale requirement for turbulent support. (2) It was shown by several groups that MHD turbulence dissipates just as rapidly as hydrodynamic turbulence (Mac Low et al. 1998; Stone et al. 1998; Padoan and Nordlund 1999), dismissing the notion of reduced dissipation in "Alfvén-wave turbulence", and thus making the presence of strong driving sources for the turbulence an absolute necessity. (3) Clouds with very different contributions from various turbulence-driving mechanisms, including those with little or no star formation activity, such as the so-called *Maddalena's cloud*, show similar turbulence characteristics (Williams et al. 1994; Schneider et al. 2011), suggesting that stellar energy injection may not be the main source of turbulence in MCs.

Moreover, simulations of dense cloud formation have shown that, once a large cold CNM cloud forms out of a collision of WNM streams, it quickly acquires a large enough mass that it can begin to collapse gravitationally (Vázquez-Semadeni et al. 2007; Vázquez-Semadeni et al. 2010; Vázquez-Semadeni et al. 2011;Heitsch et al. 2008a; Heitsch et al. 2008b). The enhancement in its column density promotes the formation of molecular hydrogen (H_2) (Hartmann et al. 2001; Bergin

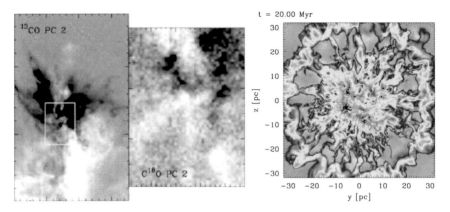

Fig. 6 *Left and middle panels*: second eigenimages obtained by principal component analysis of spectroscopic data of the star-forming region NGC 1333, showing the main contribution to the linewidth of molecular emission in this region (Brunt et al. 2009). The *middle* image shows the region enclosed in the *rectangle* in the *left* image. *Black and white colors* represent oppositely-signed components of the velocity. Brunt et al. (2009) describe the pattern as a "dipole", in which large-scale patches of alternating velocity direction are observed. This is seen in both the large-scale and the small-scale images. *Right panel:* image of the projected density field of a 3D numerical simulation with cooling, self-gravity, and magnetic fields, representing the formation of a dense atomic cloud by the collision of WNM streams in the direction perpendicular to the plane of the figure. The time shown is 20 Myr after the start of the simulation. *The black dots* denote "sink" particles, which replace local collapsing zones in the simulation. The whole cloud is also collapsing, although its collapse is not completed yet by the end of the simulation, at $t = 31$ Myr. From Vázquez-Semadeni et al. (2011)

et al. 2004; Heitsch and Hartmann 2008). Thus, it appears that the formation of a *molecular* cloud may require previous gravitational contraction (see also McKee 1989). In addition, according to the discussion in Sects. 4.2 and 5.3, the CNM clouds formed by converging WNM flows should be born turbulent and clumpy. The turbulent nature of the clouds further promotes the formation of molecular hydrogen (Glover and Mac Low 2007). The simulations by Vázquez-Semadeni et al. (2007); Vázquez-Semadeni et al. (2010); Vázquez-Semadeni et al. (2011) show that the nonlinear, turbulent density fluctuations can begin to collapse *before* the global collapse of the cloud is completed (Fig. 6, *right panel*), both because their densities are large enough that their free-fall time is significantly shorter than that of the whole cloud, and because the free-fall time of a flattened or elongated cloud may be much larger than that of an approximately isotropic clump within it of the same volume density (Toalá et al. 2012). In addition, the turbulent velocities initially induced in the clouds by the converging flows in the simulations of Vázquez-Semadeni et al. (2007) were observed to be small compared to the velocities that develop due to the subsequent gravitational contraction, while in the simulations of Banerjee R et al. (2009), the clumps with highest internal velocity dispersions were those that had already formed collapsed objects ("sink" parti-cles), although energy feedback from the sinks was not included.

All of the above evidence suggests that the observed supersonic motions in molecular clouds may have a significant, perhaps dominant, component of infalling motions, with a (possibly subdominant) superimposed random (turbulent) component remaining from the initial stages of the cloud (Bate et al. 2003; Ballesteros-Paredes et al. 2011a; Ballesteros-Paredes et al. 2011b), and perhaps somewhat amplified by the collapse (Vázquez-Semadeni et al. 1998). In this scenario of *hierarchical gravitational fragmentation*, the first structures that complete their collapse are small-scale, high-amplitude density fluctuations that are embedded within larger-scale, smaller amplitude ones, which complete their collapse later (Vázquez-Semadeni et al. 2009). The main role of the truly turbulent (i.e., fully random) motions is to provide the nonlinear density fluctuation seeds that will collapse locally once the global contraction has increased their density sufficiently for them to become locally gravitationally unstable (Clark and Bonnell 2005). Evidence for such multi-scale collapse has recently begun to be observationally detected (Galvan-Madrid et al. 2009; Schneider et al. 2010).

6 Summary and Conclusions

In this contribution, we have briefly reviewed the role of two fundamental physical processes (net radiative cooling and self-gravity) that shape the nature of interstellar turbulence. We first discussed the effects of the net thermal effects that arise from several microphysical processes in the ISM, such as the emission of radiation from various ions, atoms, and molecules, which carries away thermal energy previously stored in the particles by collisional excitation, thus cooling the gas, and photoelectric production of energetic electrons off dust grains by background stellar UV radiation, which heats the gas. The presence of radiative heating and cooling implies in general that the gas behaves in a non-isentropic (or non-adiabatic) way, and in particular it may become *thermally unstable* in certain regimes of density and temperature, where small (i.e., *linear*) perturbations can cause runaway heating or cooling of the gas that only stops when the gas exits that particular regime. This in turn causes the gas to avoid those unstable density and temperature ranges, and to settle in the stable ones, thus tending to segregate the gas into different phases of different densities and/or temperatures. In classical models of the ISM, only the stable phases were expected to exist in significant amounts.

We then discussed the interaction between trans- or supersonic turbulence, which produces large (i.e., *nonlinear*) density and velocity, and thermal instability (TI). We first discussed the probability density function (PDF) of the density fluctuations, which takes a lognormal form in isothermal regimes, and develops power-law tails in polytropic (i.e., of the form $P \propto \rho^{\gamma_e}$) ones, when $r_e \neq 1$. We then noted that, since turbulence is an inherently mixing phenomenon, it opposes the segregating effect of thermal instability, causing the production of gas parcels in the classically forbidden unstable regimes, which may add up to nearly half the mass of the ISM, although the density PDF in general still exhibits some multimodality due to the gas' preference to settle in the stable regimes. The existence of gas in the unstable ranges has been confirmed by various observational studies.

We next discussed the nature of the turbulence in the different ranges of density and temperature of the gas, noting that in the diffuse ionized regions, where the flow is transonic (i.e., with Mach numbers $M_s \sim 1$), the gas appears to behave in an essentially incompressible way, exhibiting Kolmogorov scalings over many orders of magnitude in length scale. However, in the neutral atomic regions, where the gas is expected to be thermally unstable under the so-called *isobaric criterion*, the flow is expected to exhibit large density and temperature fluctuations, by up to factors ~ 100, thus being highly fragmented. Numerical simulations of this process suggest that the gas is transonic with respect to the warm diffuse component, but supersonic with respect to the cold, dense one, although those supersonic motions seem to correspond more to the velocity dispersion of the dense clumps within the warm substrate, than to the internal velocity dispersion within the clumps themselves.

Finally, we pointed out that large-scale compressions in the warm neutral gas, which may be triggered by either random turbulent motions, or by yet larger-scale instabilities, may nonlinearly induce the formation of large regions of dense, cold gas; much larger, in particular, than the most unstable scales of TI, which have sizes ~ 0.1 pc. These clouds may easily be large enough to be *gravitationally* unstable, and numerical simulations of their evolution suggest that they rapidly engage in gravitational contraction. The latter may in fact promote the formation of molecules, so that the clouds are likely to become molecular only after they begin contracting. In addition, the clouds are born internally turbulent by the combined effect of TI and other dynamical instabilities, and the resulting nonlinear density fluctuations ("clumps") may themselves become locally gravitationally unstable during the contraction of the whole large-scale cloud. Because they are denser, they have shorter free-fall times, and can complete their local collapses before the global one does, thus producing a regime of *hierarchical gravitational fragmentation*, with small-scale, short-timescale collapses occurring within larger-scale, longer-timescale ones. It is thus quite likely that the flow regime in the dense molecular clouds corresponds to a dominant multi-scale gravitational contraction, with smaller-amplitude random (turbulent) motions superposed on it. Interstellar turbulence is seen to involve an extremely rich and complex phenomenology, even more so than the already-fascinating regimes of terrestrial turbulence.

References

Armstrong JW, Rickett BJ, Spangler SR (1995) Astrophys J 443:209

Audit E, Hennebelle P (2005) Astron Astrophys 433:1

Audit E, Hennebelle P (2010) Astron Astrophys 511:A76

Ballesteros-Paredes J, Hartmann LW, Vázquez-Semadeni E, Heitsch F, Zamora-Avilés MA (2011a) Mon Notices Royal Astro Soc 411:65

Ballesteros-Paredes J, Vázquez-Semadeni E, Gazol A et al (2011b) Mon Notices Royal Astro Soc 416:1436

Banerjee R, Vázquez-Semadeni E, Hennebelle P, Klessen RS (2009) Mon Notices Royal Astro Soc 398:1082

Bate MR, Bonnell IA, Bromm V (2003) Mon Notices Royal Astro Soc 339:577

Bergin EA, Hartmann LW, Raymond JC, Ballesteros-Paredes J (2004) Astron Astrophys 612:921

Blaisdell GA, Mansour NN, Reynolds WC (1993) J Fluid Mech 256:443

Blandford R, Eichler D (1987) Phys Rep 154:1

Blitz L, Shu FH (1980) Astron Astrophys 238:148

Brunt CM, Heyer MH, and Mac Low M (2009) Astron Astrophys 504:883

Burgers JM (1974) The nonlinear diffusion equation. Reidel, Dordrecht

Chepurnov A, Lazarian A (2010) Astron Astrophys 710:853

Cho J, Lazarian A, Vishniac ET (2003) In: Turbulence and magnetic fields in astrophysics. E. Falgarone, T. Passot eds. Lecture Notes in Physics, (Springer) 614:56

Chomiuk L, Povich MS (2011) Astron J 142:197

Clark PC, Bonnell IA (2005) Mon Notices Royal Astro Soc 361:2

Cox AN (2000) Allen's astrophysical quantities. Springer, New York. ISBN 0387987460

Dalgarno A, McCray RA (1972) Annu Rev Astron Astrophys 10:375

Dalgarno A, McCray RA (2005) Astron Astrophys 436:585

Dickey JM, Terzian Y, Salpeter EE (1978) Astrophys J 36:77

Elmegreen BG (1991) The physics of star formation and early stellar evolution. In: Proceedings of NATO ASIC 342, p 35

Elmegreen BG, Scalo J (2004) Annu Rev Astron Astrophys 42:211

Federrath C, Klessen RS, Schmidt W (2008) Astrophys J Lett 688:L79

Ferrière KM (2001) Rev Mod Phys 73:1031

Field GB (1965) Astrophys J 142:531

Field GB, Goldsmith DW, Habing HJ (1969) Astrophys J Lett 155:L149

Galván-Madrid R, Keto E, Zhang Q et al (2009) Astrophys J 706:1036

Gazol A, Vázquez-Semadeni E, Sánchez-Salcedo FJ, Scalo J (2001) Astrophys J Lett 557:L121

Gazol A, Vázquez-Semadeni E, Kim J (2005) Astrophys J 630:911

Glover SCO, and Mac Low MM (2007) Astrophys J 659:1317

Goldreich P, Kwan J (1974) Astrophys J 189:441

Goldreich P, Sridhar S (1995) Astrophys J 438:763

Hartmann L, Ballesteros-Paredes J, Bergin EA (2001) Astrophys J 562:852

Heiles C (2001) Astrophys J Lett 551:L105

Heiles C, Troland TH (2003) Astrophys J 586:1067

Heitsch F, Burkert A, Hartmann LW, Slyz AD, Devriendt JEG (2005) Astrophys J Lett 633:L113

Heitsch F, Hartmann L (2008) Astrophys J 689:290

Heitsch F, Hartmann LW, Burkert A (2008a) Astrophys J 683:786

Heitsch F, Hartmann LW, Slyz AD, Devriendt JEG, Burkert A (2008b) Astrophys J 674:316

Hennebelle P, Audit E (2007) Astron Astrophys 465:431

Hennebelle P, Banerjee R, Vázquez-Semadeni E, Klessen RS, Audit E (2008) Astron Astrophys 486:L43

Hennebelle P, Pérault M (1999) Astron Astrophys 351:309

Heyer MH, Brunt CM (2004) Astrophys J Lett 615:L45

Hill AS, Benjamin RA, Kowal G et al (2008) Astrophys J 686:363

Koyama H, Inutsuka SI (2000) Astrophys J 532:980

Koyama H, Inutsuka SI (2002) Astrophys J Lett 564:L97

Kritsuk AG, Norman ML (2002) Astrophys J Lett 569:L127

Kulkarni SR, Heiles C (1987) Interstellar Process 134:87

Kwan J (1979) Astrophys J 229:567

Landau LD, Lifshitz EM (1959) Course of theoretical physics. Fluid mechanics, Pergamon Press, Oxford

Larson RB (1981) Mon Notices Royal Astro Soc 194:809

Mac Low MM, Klessen RS, Burkert A, and Smith MD (1998) Physi Rev Lett 80:2754

Mac Low MM, Balsara DS, Kim J, and de Avillez MA (2005), Astrophys J 626:864

Mac Low MM, and Klessen RS (2004) Rev Mod Phys 76:125
McKee CF (1989) Astrophys J 345:782
McKee CF, Ostriker JP (1977) Astrophys J 218:148
McKee CF, Ostriker EC (2007) Annu Rev Astron Astrophys 45:565
McMillan PJ (2011) Mon Notices Royal Astro Soc 414:2446
Meerson B (1996) Rev Mod Phys 68:215
Myers PC (1978) Astrophys J 225:380
Norman C, Silk J (1980) Astrophys J 238:158
Osterbrock DE, Ferland GJ (2006) Astrophysics of gaseous nebulae and active galactic nuclei. In:
    Osterbrock DE, Ferland GJ (eds) 2nd edn. University Science Books, Sausalito
Padoan P, Nordlund Å (1999) Astrophys J 526:279
Padoan P, Nordlund A, Jones BJT (1997) Mon Notices Royal Astro Soc 288:145
Passot T, Vázquez-Semadeni E (1998) Phys Rev E 58:4501
Passot T, Vazquez-Semadeni E, Pouquet A (1995) Astrophys J 455:536
Piontek RA, Ostriker EC (2004) Astrophys J 601:905
Piontek RA, Ostriker EC (2005) Astrophys J 629:849
Sánchez-Salcedo FJ, Vázquez-Semadeni E, Gazol A (2002) Astrophys J 577:768
Schneider N, Bontemps S, Simon R et al (2011) Astron Astrophys 529:A1
Schneider N, Csengeri T, Bontemps S et al (2010) Astron Astrophys 520:A49
Shu FH (1992) Physics of astrophysics, vol II. University Science Books, Sausalito
Shu FH, Adams FC, Lizano S (1987) Annu Rev Astron Astrophys 25:23
Stone JM, Ostriker EC, Gammie CF (1998) Astrophys J Lett 508:L99
Sutherland RS, Dopita MA (1993) Astrophys J 88:253
Toalá JA, Vázquez-Semadeni E, Gómez GC (2012) Astrophys J 744:190
Vázquez-Semadeni E (1994) Astrophys J 423:681
Vázquez-Semadeni E (1999) Millimeter-wave astronomy. Mol Chem and Phys space 241:161
Vázquez-Semadeni E, Banerjee R, Gómez GC et al (2011) Mon Notices Royal Astro Soc
    414:2511
Vázquez-Semadeni E, Cantó J, Lizano S (1998) Astrophys J 492:596
Vázquez-Semadeni E, Colín P, Gómez GC, Ballesteros-Paredes J, Watson AW (2010) Astrophys
    J 715:1302
Vázquez-Semadeni E, Gazol A, Scalo J (2000a) Astrophys J 540:271
Vázquez-Semadeni E, Gazol A, Passot T et al (2003) In: Turbulence and magnetic fields in
    astrophysics. E. Falgarone, T. Passot eds. Lecture Notes in Physics, (Springer) 614:213
Vázquez-Semadeni E, Gómez GC, Jappsen AK et al (2007) Astrophys J 657:870
Vázquez-Semadeni E, Gómez GC, Jappsen A-K, Ballesteros-Paredes J, Klessen RS (2009)
    Astrophys J 707:1023
Vázquez-Semadeni E, González RF, Ballesteros-Paredes J, Gazol A, Kim J (2008) Mon Notices
    Royal Astro Soc 390:769
Vázquez-Semadeni E, Ostriker EC, Passot T, Gammie CF, and Stone JM (2000b) In: Protostars
    and planets IV. V. Mannings, AP Boss, SS Russell eds. (University of Arizona Press), 3
Vázquez-Semadeni E, Passot T, Pouquet A (1996) Astrophys J 473:881
Vázquez-Semadeni E, Ryu D, Passot T, González RF, Gazol A (2006) Astrophys J 643:245
Vishniac ET (1994) Astrophys J 428:186
Walder R, Folini D (2000) Astron Astrophys Space Sci 274:343
Williams JP, de Geus EJ, Blitz L (1994) Astrophys J 428:693
Wolfire MG, Hollenbach D, McKee CF, Tielens AGGM, Bakes ELO (1995) Astrophys J 443:152
Zuckerman B, Evans NJ (1974) Astrophys J 192:L149
Zuckerman B, Palmer P (1974) Annu Rev Astron Astrophys 12:279

Assessing Significant Phenomena in 1D Linear Perturbation Multiphase Flows

Alberto Soria and Elizabeth Salinas-Rodríguez

Abstract A procedure based on small perturbations linearization is developed for the assessment of relevant physical effects in fast fluidized beds. The fluid compressibility and wall interaction effects onto a main incompressible behavior are appreciated. A model by contributions is developed and the coefficients of all terms are evaluated in order to assess their significance. The process to get a lumped model is performed and the equivalence of lumped terms and variables under asymptotic conditions is developed. It is shown how wall effects are able to change a parabolic to a hyperbolic structure and how a third order waving structure collapses to a first order one onto a diffusive operator, under the limit of the incompressibility assumption.

1 Introduction

Fluidized beds are vessels or pipes with a mass of small solid particles which are lifted by the flow of a liquid or a gas. The fluid is injected at the vessel bottom with a flow rate high enough to lift the solid particles by viscous drag exerted on their surface against the gravity force. They are widely used in the food, chemical, pharmaceutical and metallurgical industries (Epstein 2003). In turn to control the involved processes, the comprehension of the different interaction mechanisms between particles, fluid and walls is required. Depending on their application they operate at several regimes: slug flow, fast fluidization regime, core annulus flow or

A. Soria (✉) · E. Salinas-Rodríguez
I.P.H. Department, Universidad Autónoma Metropolitana-Iztapalapa,
San Rafael Atlixco 186 Col. Vicentina, 09340 México, D.F., Mexico
e-mail: asor@xanum.uam.mx

J. Klapp et al. (eds.), *Fluid Dynamics in Physics, Engineering and Environmental Applications*, Environmental Science and Engineering,
DOI: 10.1007/978-3-642-27723-8_6, © Springer-Verlag Berlin Heidelberg 2013

pneumatic conveying regime (Grace et al. 1999). The fast fluidization regime, where solid particles are carried up by a fluid, is the regime under which the circulating fluidized beds (CFB) usually operate. This is a convenient way to maintain long time operation due to solid recirculation, moreover, it enhances both, the contact surface between phases and the interfacial heat and mass transfer, minimizing high changes in the state variables.

Van der Schaaf et al. (1998) measured locally the pressure propagation speed in CFB made of air and sand particles at several solid volume fractions, by placing a set of pressure transducers at several heights. Dense and dilute regimes may be present in fast fluidized beds. Ryzhkov and Tolmachev (1983) obtained the perturbation propagation speed and studied the resonance in vibrating fluidized beds.

Bi et al. (1995) studied the propagation speed of pressure waves by the pseudo-homogeneous compressible wave theory for A and B groups of particles, according to Geldart (1976) classification, finding insignificant differences in the results. They also studied separated flow compressible wave theory and reported a significant dependence on the wave frequency for the propagation wave speed in fluidized dense phase flows.

Musmarra et al. (1995) considered a pseudo-homogeneous simplified expression for pressure waves (Wallis 1969; Gregor and Rumpf 1976) and the propagation speed of pressure waves as a function of the frequency. For void fraction 0.35, they found that the dynamic waves observed in their experiments are more elastic than compression waves.

Sanchez et al. (2011), cited as SSS1 for short, studied the fluid compressibility effect on the waving structure in an unbounded fast fluidized bed, say infinite in extension (Joshi et al. 2001). They showed how the incompressible behavior is embedded into the compressible one extending Liu's criterion (1982). They found a decoupling structure for the model with explicit incompressible and compressible parts. They also obtained a lumped model where there is just one differential operator for each order of differentiation. In this operator all physical contributions of each order are embedded in just one term.

Soria et al. (2011), cited as SSS2 for short, extended SSS1 results by incorporating the presence of the wall in the model. This has an impact on the mathematical structure, changing a former parabolic structure to a hyperbolic one; this means that the wall presence causes an additional waving propagation in the bounded system. The sound propagation speed in the pure fluid is a parameter that multiplies the set of all terms arising from compressibility effects in the uncoupled model. On the other hand, in the lumped model the sound speed is incorporated, not only as a coefficient, but also affecting the expressions of the propagation speeds of these compressibility dependent terms.

In the present work, a methodology which allows to incorporate additional effects in a systematic way, beginning by a linear perturbation scheme as the basis for the effects separation and interactions is established. For this purpose, in Sect. 2, the theoretical development used to obtain linearized perturbation models is settled. Thus the compressibility versus incompressibility behavior, as well as the bounded versus unbounded systems can be assessed. In Sect. 3, the obtained models are used to make predictions on two physical systems, an FCC powder catalyst dragged by water vapor at isothermal conditions and a sand-air system (SAI) (van der Schaaf et al. 1998). Finally, the main conclusions of this work are summarized.

2 Two Fluids Model

Isothermal continuity and momentum equations for gas and solid powder are obtained in this section, regardless the particular method to get the working equations. The methods commonly used to get such equations are averaging procedures (Drew 1971; Soria and de Lasa 1991 and references therein) or a postulation approach (Gregor and Rumpf 1976; Ransom and Hicks 1984).

2.1 Model Equations

A dispersion of solid spherical small particles moving within a steam current through a vertical pipe, with an inner diameter D_t, at constant temperature is considered. The spheres mean diameter is d_p and its intrinsic density ρ_s; the fluid is considered a compressible Newtonian gas of intrinsic averaged density ρ_g and constant viscosity μ_g. The solid and gas volume fractions, ε_s and ε_g respectively satisfy $\varepsilon_g + \varepsilon_s = 1$, and ε_s is set between 0 and 0.5. The average mixture velocity is given by $U_m = \varepsilon_s V_s + \varepsilon_g V_g$, where V_g and V_s are the gas and the solid up-flow mass-weighted mean velocities, respectively. The mixture is considered to be highly turbulent, therefore, the viscous effects are significant close to the granule surfaces and to the walls. The particle Reynolds number is $\mathrm{Re}_p = d_p \rho_g U_t / \mu_g$, where $U_t = |V_s - U_m| = \varepsilon_g (V_g - V_s)$ is the terminal velocity. In this model, both viscosity effects in the bulk phases and turbulent stresses everywhere, are neglected.

A space–time averaging method is applied to the local-instantaneous continuity and momentum equations for the gas and solid phases (SSS2). A closed non-linear set of one-dimensional and transient averaged equations for the dependent variables $\{\varepsilon_g, p_g, V_g, V_s\}$ is obtained (Drew 1983; Soria and de Lasa 1991):

$$\frac{\partial}{\partial t}\left[\varepsilon_g \rho_g (p_g)\right] + \frac{\partial}{\partial z}\left[\varepsilon_g \rho_g (p_g) V_g\right] = 0 \tag{1}$$

$$-\rho_s \frac{\partial}{\partial t}\varepsilon_g + \rho_s \frac{\partial}{\partial z}\left[(1 - \varepsilon_g) V_s\right] = 0 \tag{2}$$

$$\frac{\partial}{\partial t}\left[\varepsilon_g \rho_g (p_g) V_g\right] + \frac{\partial}{\partial z}\left[\varepsilon_g \rho_g (p_g) V_g^2\right] + \varepsilon_g \frac{\partial}{\partial z} p_g + T_g + F_{gW} = 0 \tag{3}$$

$$\rho_s \frac{\partial}{\partial t}\left[\left(1 - \varepsilon_g\right)V_s\right]$$
$$+ \rho_s \frac{\partial}{\partial z}\left[\left(1 - \varepsilon_g\right)V_s^2\right] + \left(1 - \varepsilon_g\right)\frac{\partial}{\partial z}p_g - \Phi'\left(1 - \varepsilon_g\right)\frac{\partial}{\partial z}\varepsilon_g + T_s + F_{sW} = 0. \tag{4}$$

Here z and t are the vertical coordinate and the time, respectively; p_g is the averaged gas phase pressure.

2.2 Closures and Interaction Forces

The averaged solid phase pressure is $p_s = p_g + \Phi(\varepsilon_s)$, where the function $\Phi(\varepsilon_s)$ is the compressibility modulus and $\Phi' = d\Phi/d\varepsilon_s$ its derivative. The interaction forces are

$$T_g = F + \rho_g(p_g)g$$
$$T_s = -F + \left(1 - \varepsilon_g\right)\left[\rho_s - \rho_g(p_g)\right]g, \tag{5}$$

where F is the local average force exerted by the fluid on the particles and g is the gravitational force per unit mass.

The fluid and the solid particle drag forces, F_{gW} and F_{sW} respectively, as well F, are modeled as

$$F = \beta\varepsilon_g\varepsilon_s\left(V_g - V_s\right)$$
$$F_{gW} = \beta_{gW}\rho_g(p_g)V_g^2 \tag{6}$$
$$F_{sW} = \beta_{sW}\rho_s V_s^2$$

where $\beta = \left(3/4d_p\right)C_D\rho_g U_t$ is an interaction parameter. The virtual mass effect (Liu 1982) and the effective viscosity effect in the dense fluidization regime (Homsy et al. 1980; Liu 1982; Jackson 1985) were not taken into account. The coefficients, where $a = \{g, s\}$, are wall drag functions involving drag friction factors, f_{aW}.

2.3 Linearized Equations for Small Perturbations

Equations (1)–(4) are linearized around a uniform base state (zero sub index) plus a small perturbation such that $\varepsilon_g = \varepsilon_0 + \varepsilon$, $p_g = p_0 + p$, $V_g = v_{g0} + v_g$ and $V_s = v_{s0} + v_s$. Solid grains are incompressible, so $\rho_s = \rho_{s0}$ is a constant. Perturbation wave hierarchies for both the incompressible and the compressible models for bounded and unbounded systems are obtained by proposing waving solutions for linearized Eq. (1)–(4). These equations are represented in matrix form as

$$\mathbf{B}\frac{\partial \mathbf{u}}{\partial t} + \mathbf{C}\frac{\partial \mathbf{u}}{\partial z} + \mathbf{D}\mathbf{u} = \mathbf{0} \tag{7}$$

where $\mathbf{u}$ is the perturbation vector for the linearized PDE first order system,

$$\mathbf{u} = \begin{pmatrix} \varepsilon & p & v_g & v_s \end{pmatrix}^T \tag{8}$$

where the elements of matrices $\mathbf{B}$, $\mathbf{C}$ and $\mathbf{D}$ are defined elsewhere (Soria et al. 2008, SSS1 and SSS2).

From the continuity and momentum equations, a set of two equations for gas volume fraction (ε) and pressure (p) fluctuations is obtained:

$$
\begin{aligned}
(L_1 + \Lambda_1)\varepsilon + \left[L_3 + \left(\frac{d\rho_g}{dp_g}\right)_0 (K_1 + \Gamma)\right] p &= 0 \\
(L_2 + \Lambda_2)\varepsilon + \left[L_3 + \left(\frac{d\rho_g}{dp_g}\right)_0 K_2\right] p &= 0
\end{aligned}
\tag{9}
$$

where the incompressible behavior is represented by L waving operators defined by

$$
\begin{aligned}
L_1 &= -\frac{\rho_{g0}}{\varepsilon_0}\left(\frac{\partial}{\partial t} + v_{g0}\frac{\partial}{\partial z}\right)^2 - \beta\left(\frac{\partial}{\partial t} + v_{g0}\frac{\partial}{\partial z}\right) - \frac{1-\varepsilon_0}{\varepsilon_0}\beta\left(\frac{\partial}{\partial t} + v_{s0}\frac{\partial}{\partial z}\right), \\
L_2 &= \frac{\rho_{s0}}{1-\varepsilon_0}\left(\frac{\partial}{\partial t} + v_{s0}\frac{\partial}{\partial z}\right)^2 + \frac{\varepsilon_0}{1-\varepsilon_0}\beta\left(\frac{\partial}{\partial t} + v_{g0}\frac{\partial}{\partial z}\right) + \beta\left(\frac{\partial}{\partial t} + v_{s0}\frac{\partial}{\partial z}\right) \\
&\quad - \Phi_0'\frac{\partial^2}{\partial z^2} - \frac{(\rho_{s0} - \rho_{g0})}{1-\varepsilon_0}g\frac{\partial}{\partial z}, \\
L_3 &= \frac{\partial^2}{\partial z^2},
\end{aligned}
\tag{10}
$$

while the fluid compressibility effects are accounted for by K waving operators:

$$
\begin{aligned}
K_1 &= -\left(\frac{\partial}{\partial t} + v_{g0}\frac{\partial}{\partial z}\right)^2 - (1-\varepsilon_0)\frac{\beta}{\rho_{g0}}\left(\frac{\partial}{\partial t} + v_{g0}\frac{\partial}{\partial z}\right) + \frac{g}{\varepsilon_0}\frac{\partial}{\partial z}, \\
K_2 &= \varepsilon_0\frac{\beta}{\rho_{g0}}\left(\frac{\partial}{\partial t} + v_{g0}\frac{\partial}{\partial z}\right) - g\frac{\partial}{\partial z}
\end{aligned}
\tag{11}
$$

and finally, the wall effects are denoted by capital Greek symbols:

$$
\begin{aligned}
\Lambda_1 &= -\frac{2\beta_{gW}\rho_{g0}v_{g0}}{\varepsilon_0^2}\left(\frac{\partial}{\partial t} + v_{g0}\frac{\partial}{\partial z}\right) \\
\Lambda_2 &= \frac{2\beta_{sW}\rho_{s0}v_{s0}}{(1-\varepsilon_0)^2}\left(\frac{\partial}{\partial t} + v_{s0}\frac{\partial}{\partial z}\right) \\
\Gamma &= -2\frac{\beta_{gW}v_{g0}}{\varepsilon_0}\left(\frac{\partial}{\partial t} + \frac{v_{g0}}{2}\frac{\partial}{\partial z}\right)
\end{aligned}
\tag{12}
$$

It should be stressed that L waving operators are second order as well as K_1, while K_2 and capital Greek operators are first order. Thus, Eq. (9) is a set of two second order equations, where the gas compressibility, the wall drag interaction and the basic incompressible contribution can be identified as additive separated terms. This mathematical structure is a direct consequence of our linearization scheme. The incompressible fluid approximation can be retrieved from Eq. (9) in the limit $s \to \infty$; also, the unbounded behavior can be retrieved if β_{gW} and β_{sW} are set zero.Eq. (9) can be rewritten as a higher order equation as

$$
\left[(L_2 - L_1)L_3 - \frac{1}{s^2}(L_1 K_2 - L_2 K_1) \right] \varphi +
$$
$$
\left[(\Lambda_2 - \Lambda_1)L_3 - \frac{1}{s^2}(\Lambda_1 K_2 - \Lambda_2 K_1 - L_2 \Gamma - \Lambda_2 \Gamma) \right] \varphi = 0 \tag{13}
$$

where φ is either ε or p. It should be pointed out that Eq. (13) is obtained regardless the specific perturbation substituted. In this waving structure, the product of waving operators builds up higher order waving structures; being fourth order the highest and second order the lowest.

After algebraic manipulation, it is possible to rewrite the first bracket in Eq. (13) as

$$
(L_2 - L_1)L_3 - \frac{1}{s^2}(L_1 K_2 - L_2 K_1) =
$$
$$
\left[\tau \left(\frac{\partial}{\partial t} + c_1 \frac{\partial}{\partial z} \right) \left(\frac{\partial}{\partial t} + c_2 \frac{\partial}{\partial z} \right) + \left(\frac{\partial}{\partial t} + a \frac{\partial}{\partial z} \right) \right] \frac{\partial^2}{\partial z^2}
$$
$$
- \frac{1}{s^2}
\begin{bmatrix}
T \left(\frac{\partial}{\partial t} + \eta_{41} \frac{\partial}{\partial z} \right) \left(\frac{\partial}{\partial t} + \eta_{42} \frac{\partial}{\partial z} \right) \left(\frac{\partial}{\partial t} + \eta_{43} \frac{\partial}{\partial z} \right) \left(\frac{\partial}{\partial t} + \eta_{44} \frac{\partial}{\partial z} \right) \\
+ C \left(\frac{\partial}{\partial t} + \eta_{31} \frac{\partial}{\partial z} \right) \left(\frac{\partial}{\partial t} + \eta_{32} \frac{\partial}{\partial z} \right) \left(\frac{\partial}{\partial t} + \eta_{33} \frac{\partial}{\partial z} \right) \\
- gC \left(\frac{\partial}{\partial t} + \eta_{21} \frac{\partial}{\partial z} \right) \frac{\partial}{\partial z}
\end{bmatrix} . \tag{14}
$$

This equation was obtained by ordering all same order terms and expressing them as binomial products of waving operators. Thus a set of propagation speeds has arisen. In Eq. (14), coefficients $T = \frac{\varepsilon_0 \rho_{s0}}{\beta}$ and $\tau = \frac{(1-\varepsilon_0)\rho_{g0} + \varepsilon_0 \rho_{s0}}{\beta}$ are relaxation times, g is the gravitational constant, $C = \varepsilon_0 \frac{\varepsilon_0 \rho_{g0} + (1-\varepsilon_0)\rho_{s0}}{\rho_{g0}}$; a, c_i, η_{ij} are propagation speeds.

The second bracket contains all drag interaction effects due to the bounds and allows distinguishing how these effects affect the compressible and the incompressible behaviors. This bracket can be developed as

$$(\Lambda_2 - \Lambda_1)L_3 - \frac{1}{s^2}(\Lambda_1 K_2 - \Lambda_2 K_1 - L_2\Gamma - \Lambda_2\Gamma) = A\left(\frac{\partial}{\partial t} + \xi_f \frac{\partial}{\partial z}\right)\frac{\partial^2}{\partial z^2}$$

$$-\frac{1}{s^2}\left[C_w\left(\frac{\partial}{\partial t} + \xi_{31}\frac{\partial}{\partial z}\right)\left(\frac{\partial}{\partial t} + \xi_{32}\frac{\partial}{\partial z}\right)\left(\frac{\partial}{\partial t} + \xi_{33}\frac{\partial}{\partial z}\right) + B\left(\frac{\partial}{\partial t} + \xi_{21}\frac{\partial}{\partial z}\right)\left(\frac{\partial}{\partial t} + \xi_{22}\frac{\partial}{\partial z}\right)\right]$$

$$(15)$$

where $\xi_{3i}, i = 1, 2, 3$, $\xi_{2i}, i = 1, 2$ and ξ_f are third, second and first order propagation speed components accounting for interactions of the wall effects with the compressible and the incompressible basic structures. Here, A, C_w and B are given by

$$A = 2\frac{\beta_{gW}(1-\varepsilon_0)^2\rho_{g0}v_{g0} + \beta_{sW}\varepsilon_0^2\rho_{s0}v_{s0}}{\beta\varepsilon_0(1-\varepsilon_0)}, \quad C_w = \frac{2\rho_{s0}}{\beta(1-\varepsilon_0)}\left[\beta_{sW}\varepsilon_0 v_{s0} + \beta_{gW}(1-\varepsilon_0)v_{g0}\right]$$

and $B = 2\varepsilon_0\rho_{s0}\left[\dfrac{\beta_{gW}v_{g0}}{\rho_{s0}} + \dfrac{\beta_{sW}v_{s0}}{\rho_{g0}} + 2\dfrac{\beta_{gW}\beta_{sW}}{\beta}\dfrac{v_{g0}v_{s0}}{\varepsilon_0(1-\varepsilon_0)}\right].$

The first order propagation speed ξ_f due to the walls is defined as $\xi_f = \frac{\beta_{gW}(1-\varepsilon_0)^2\rho_{g0}v_{g0}^2 + \beta_{sW}\varepsilon_0^2\rho_{s0}v_{s0}^2}{\beta_{gW}(1-\varepsilon_0)^2\rho_{g0}v_{g0} + \beta_{sW}\varepsilon_0^2\rho_{s0}v_{s0}}$. When performing the summation of Eqs. (14) and (15), considering that coefficient T is much smaller than τ on the basis of a dimensionless Eq. (15), for most of the operating conditions (see SSS1), the whole wave equation with wall effects is found to be

$$\left[\tau\left(\frac{\partial}{\partial t} + c_1\frac{\partial}{\partial z}\right)\left(\frac{\partial}{\partial t} + c_2\frac{\partial}{\partial z}\right) + (1+A)\left(\frac{\partial}{\partial t} + a_w\frac{\partial}{\partial z}\right)\right]\frac{\partial^2}{\partial z^2}\varphi$$

$$-\frac{C}{s^2}\left[\begin{array}{c}\left(\frac{\partial}{\partial t} + \eta_{31}\frac{\partial}{\partial z}\right)\left(\frac{\partial}{\partial t} + \eta_{32}\frac{\partial}{\partial z}\right)\left(\frac{\partial}{\partial t} + \eta_{33}\frac{\partial}{\partial z}\right) + \frac{C_w}{C}\left(\frac{\partial}{\partial t} + \xi_{31}\frac{\partial}{\partial z}\right)\left(\frac{\partial}{\partial t} + \xi_{32}\frac{\partial}{\partial z}\right)\left(\frac{\partial}{\partial t} + \xi_{33}\frac{\partial}{\partial z}\right) \\ -g\left(\frac{\partial}{\partial t} + \eta_{21}\frac{\partial}{\partial z}\right)\frac{\partial}{\partial z} + \frac{B}{C}\left(\frac{\partial}{\partial t} + \xi_{21}\frac{\partial}{\partial z}\right)\left(\frac{\partial}{\partial t} + \xi_{22}\frac{\partial}{\partial z}\right)\end{array}\right]\varphi = 0$$

$$(16)$$

where the incompressible basic propagation speed modified by wall effects is

$$a_w = \frac{a + A\xi_f}{1 + A} \tag{17}$$

and ξ_{ij} are propagation speeds that merge from the search for first order waving operators, as discussed above.

Equation (16) can be called a *model by contributions*, since all effects can yet be identified, thus the first bracket involves all incompressible contributions. The coefficient A quantifies the wall importance summed up to the basic incompressible contribution of value one. The second bracket weighted by C, means the compressibility contribution. Inside this bracket the basic third order wave has a coefficient of one and the basic second order wave has a $-g$ coefficient. Here the wall effects are additional terms with coefficients C_w and B whose importance should be assessed with respect to C. Nevertheless, when a physical fluidized bed is tested, pressure or volume fraction signals should be taken by appropriate experimental devices and time series should be acquired. In these signals the information is lumped and contributions cannot be identified. Therefore a *lumped model* should be preferred for comparison with experimental results. This can be done by lumping Eq. (16) in a way that only one term for each order wave is obtained. When this is performed, it gives rise to

$$
\tau \left(\frac{\partial}{\partial t} + c_1 \frac{\partial}{\partial z} \right) \left(\frac{\partial}{\partial t} + c_2 \frac{\partial}{\partial z} \right) \frac{\partial^2}{\partial z^2} \varphi
$$

$$
- \frac{C}{s^2} \left[\begin{array}{c} \left(1 + \frac{C_w}{C} \right) \left(\frac{\partial}{\partial t} + \alpha_1^w \frac{\partial}{\partial z} \right) \left(\frac{\partial}{\partial t} + \alpha_2^w \frac{\partial}{\partial z} \right) \left(\frac{\partial}{\partial t} + \alpha_3^w \frac{\partial}{\partial z} \right) \\ + \frac{B}{C} \left(\frac{\partial}{\partial t} + \gamma_1 \frac{\partial}{\partial z} \right) \left(\frac{\partial}{\partial t} + \gamma_2 \frac{\partial}{\partial z} \right) \end{array} \right] \varphi = 0 \qquad (18)
$$

where, c_j, $j = 1, 2$, α_j^w, $j = 1, 2, 3$ and γ_j, $j = 1, 2$ are propagation speeds that should be coincident with experimental results.

3 Results and Discussion

Both the model by contributions and the lumped model, obtained in Sect. 2, were applied to an FCC catalyst carried up by turbulent water vapor and to a Sand-air CFB experimental system reported by van der Schaaf et al. (1998). In Table 1 the solid and fluid physical properties, the relevant geometrical parameters and the operating conditions in a homogeneous ground state for each system are given. In this Section, both systems are analyzed, emphasizing their similarities and differences. It is considered, as in SSS1, that the ground state solid volume fraction is given by $\varepsilon_{s0} = \frac{U_s}{U_m - U_t}$, where either, the superficial solid velocity U_s or the solid volume fraction ε_{s0} is assumed to be given. The Jiradilok's et al. (2006) solid compressibility modulus, $\varepsilon_{s0}\Phi_0' = 10^{6.837 - 2.475\varepsilon_0}$, was considered. The solid–fluid interaction parameter β should be selected for FCC grains according to the Stokes regime. For sand grains, the best drag coefficient is given by the Ergun's equation.

Table 1 Parameter values and operating conditions at ground states

Parameter	(FCC)	Sand-air (SAI)
Sound speed in pure fluid, s (m/s)	655	346
Solid density, ρ_{s0} (kg/m^3)	1,300	2650
Fluid density, ρ_{g0} (kg/m^3)	0.7065	0.9340
Solid grain diameter, d_p (μm)	60	310
Fluid friction factor[a], f_{gW}	0.0158	0.01670
Solid friction factor[b], f_{sW}	$0.0285\sqrt{gD_t}/v_{s0}$	$0.0285\sqrt{gD_t}/v_{s0}$
Terminal velocity, U_t (m/s)	0.085	2.231
Superficial fluid velocity, U_g (m/s)	7.000	3.200
Column diameter, D_t (m)	1.168	0.960
Column length, L (m)	22	9
Fluid Reynolds number, Re$_g$	1.93×10^5	1.43×10^5
Solid Reynolds number, Re$_s$	0.12	32.30
Stokes number, St	0.13	2.15

[a] according to friction factor formula, [b] according to Konno and Saito (1969) formula

3.1 Comparison of Wave Coefficients and Significance of Wave Hierarchies

Dimensionless equations equivalent to Eqs. (16) and (18) were derived to compare the significance of the coefficients with and without wall effects for both, the lumped and the contributions arrangements. For this purpose characteristic length L, velocity U_0 and time L/U_0 scales were defined and used to get Eqs. (19) and (20), the referred dimensionless expressions:

$$\left[\tau' \left(\frac{\partial}{\partial t'} + c_1' \frac{\partial}{\partial z'} \right) \left(\frac{\partial}{\partial t'} + c_2' \frac{\partial}{\partial z'} \right) + (1 + A) \left(\frac{\partial}{\partial t'} + a_w' \frac{\partial}{\partial z'} \right) \right] \frac{\partial^2}{\partial z'^2} \varphi$$

$$-C' \left[\begin{array}{c} \left(\frac{\partial}{\partial t'} + \eta_{31}' \frac{\partial}{\partial z'} \right) \left(\frac{\partial}{\partial t'} + \eta_{32}' \frac{\partial}{\partial z'} \right) \left(\frac{\partial}{\partial t'} + \eta_{33}' \frac{\partial}{\partial z'} \right) \\[2mm] + \frac{C_w}{C} \left(\frac{\partial}{\partial t'} + \xi_{31}' \frac{\partial}{\partial z'} \right) \left(\frac{\partial}{\partial t'} + \xi_{32}' \frac{\partial}{\partial z'} \right) \left(\frac{\partial}{\partial t'} + \xi_{33}' \frac{\partial}{\partial z'} \right) \\[2mm] + B' \left[\left(\frac{\partial}{\partial t'} + \xi_{21}' \frac{\partial}{\partial z'} \right) \left(\frac{\partial}{\partial t'} + \xi_{22}' \frac{\partial}{\partial z'} \right) - \frac{G'}{B'} \left(\frac{\partial}{\partial t'} + \eta_{21}' \frac{\partial}{\partial z'} \right) \frac{\partial}{\partial z'} \right] \end{array} \right] \varphi = 0$$

$$(19)$$

Fig. 1 Wall effects
compared to the basic
incompressible behavior

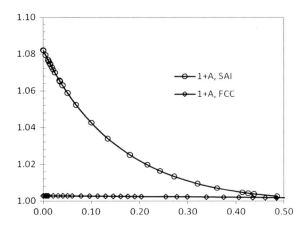

$$\tau'\left(\frac{\partial}{\partial t'}+c'_1\frac{\partial}{\partial z'}\right)\left(\frac{\partial}{\partial t'}+c'_2\frac{\partial}{\partial z'}\right)\frac{\partial^2}{\partial z'^2}\varphi$$

$$-C'\begin{bmatrix}\left(1+\dfrac{C_w}{C}\right)\left(\dfrac{\partial}{\partial t'}+\alpha_1^{w'}\dfrac{\partial}{\partial z'}\right)\left(\dfrac{\partial}{\partial t'}+\alpha_2^{w'}\dfrac{\partial}{\partial z'}\right)\left(\dfrac{\partial}{\partial t'}+\alpha_3^{w'}\dfrac{\partial}{\partial z'}\right)\\[2mm]+B'\left(\dfrac{\partial}{\partial t'}+\gamma'_1\dfrac{\partial}{\partial z'}\right)\left(\dfrac{\partial}{\partial t'}+\gamma'_2\dfrac{\partial}{\partial z'}\right)\end{bmatrix}\varphi=0 \qquad (20)$$

The dimensionless coefficients A, $\tau'\equiv\frac{U_0}{L}\tau$, $\frac{C_w}{C}$, $C'\equiv\frac{U_0^2}{s^2}C$, $B'\equiv\frac{BL}{CU_0}$ and $\frac{G'}{B'}\equiv\frac{gC}{U_0B}$ were estimated and plotted as functions of the solid volume fraction, ε_{s0}. All other primed variables are also dimensionless. In Fig. 1 the main incompressibility coefficient $1+A$ is shown to decrease from 1.082 to 1.002 as the solid volume fraction increases from 0 to 0.5. This coefficient is chosen as the basis for all further comparisons, since it depicts the basic incompressible contribution, as given by Eq. (19), enhanced by wall effects.

The fluid compressibility coefficient is depicted in Fig. 2 as the ratio C'/τ', in order to assess its relative importance with respect to the solid elasticity effect τ'. The ratio $C'/\tau' > 1$ for $\varepsilon_{s0} > 0.15$ (SAI system) and for $\varepsilon_{s0} > 0.015$ (FCC system). Then, the fluid compressibility effect is greater than the solid elasticity and should be considered as an important effect for modeling strategies. Nevertheless, the assumption of fluid incompressibility is frequently used without care. In Fig. 3 the dimensionless solid elasticity relaxation time, τ' is shown; its magnitude is below 5 % of the basic incompressible behavior.

The importance of B', the dimensionless coefficient of second order waving structures, is associated to interactions of the fluid compressibility with wall effects, as can be seen in Eq. (19). B' is also the coefficient of second order waving structures in the lumped model. Its magnitude should be comparable to the dominant wave components in the same bracket, say one, see Eqs. (19) and (20). In Fig. 4 it is seen that B' values go from 0.6 to 1.6 and therefore, second order

Fig. 2 Compressibility effects as a factor of the solid elasticity term

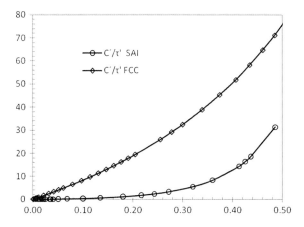

Fig. 3 Dimensionless solid elasticity relaxation time

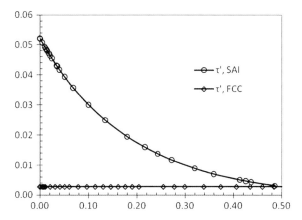

waves, associated to fluid compressibility, are as important as the third order ones (whose coefficient is around one). Moreover, in Eq. (19) B' is a factor of two terms enclosed by a bracket, being the second one due to gravity. This last term has a coefficient G'/B' and should be compared also to one, the coefficient of the first term inside the bracket. It is apparent that gravity is the dominant second order effect, since in Fig. 5 it is shown that $6.63 < G'/B' < 17.54$ for the SAI system and $2.10 < G'/B' < 4.20$ for the FCC system.

Coefficient C_w/C appears in Eq. (19) and in Eq. (20) as well. It means the relative importance of third order wall effects, with respect to the basic third order compressibility effect, whose coefficient inside the brackets is one. In Fig. 6 the fact that third order wall effects are always below 10 % of the main contribution is apparent. Also, for $\varepsilon_{s0} > 0.04$ wall effects decay below 0.10 % of main contribution.

Fig. 4 Second order
compressibility-wall
interaction effects

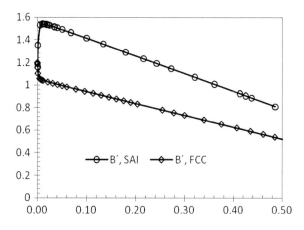

Fig. 5 Coefficient of gravity
effects

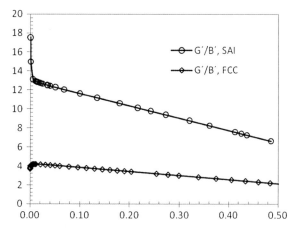

Fig. 6 Third order wall
effects coefficient

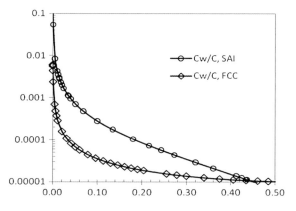

3.2 Analysis of Propagation Speeds

The propagation speed analysis is better suited from dimensional Eqs. (16) and (18) than from dimensionless ones. The main incompressible term, taken as a reference, is a first order wave structure acting onto a second order differential spatial operator, see Eq. (16). Its propagation speed, a_w results from modification of a, the unbounded propagation speed given in Eq. (14), once wall effects are accounted for, as can be observed in Eq. (17). In this expression the impact of wall effects on the propagation speed is two-folded, on one hand there is a propagation speed component ξ_f, merging from the wall dynamics by itself, Eq. (15). After ξ_f has been multiplied by coefficient A and summed up to the basic unbounded propagation speed a, coefficient A affects again the result, as an inverse factor added to one. This simple example clarifies how coefficients in the model by contributions, Eq. (16), become involved in lumped propagation speeds, as extensively done to get Eq. (18).

In Figs. 7 and 8 the second order propagation speeds resulting from the solid compressibility effect are shown. Their existence and importance for modeling strategies were established around 1970–80 and are plotted together with α_2^w for reference. The third order propagation speeds α_1^w, α_2^w and α_3^w, are plotted in Figs. 9 and 10. α_2^w is the middle third order propagation speed and can be taken as a further correction on the basic propagation speed a, once the fluid compressibility effect has been considered, additional to wall effects which formerly gave a_w in Eq. (17). Thus, variables a, a_w and α_2^w represent the same physical quantity, say the propagation speed of volume fraction waves, also known as the kinematic wave. Moreover, α_1^w and α_3^w can be understood as the adiabatic sound propagation speed in the dispersed medium, relative to the mass mean flow velocity, in upward and downward directions, respectively. Finally, Figs. 11 and 12 represent the interaction of gravity and wall effects with the fluid compressibility. Since there is not much literature concerning this aspect, let us address a discussion to this point and introduce an asymptotic analysis.

In fact, the lumping process performed to obtain Eq. (18) from Eq. (16) has to consider the equivalence on the second order terms as follows:

$$\frac{B}{C}\left(\frac{\partial}{\partial t} + \xi_{21}\frac{\partial}{\partial z}\right)\left(\frac{\partial}{\partial t} + \xi_{22}\frac{\partial}{\partial z}\right) - g\left(\frac{\partial}{\partial t} + \eta_{21}\frac{\partial}{\partial z}\right)\frac{\partial}{\partial z}$$
$$= \frac{B}{C}\left(\frac{\partial}{\partial t} + \gamma_1\frac{\partial}{\partial z}\right)\left(\frac{\partial}{\partial t} + \gamma_2\frac{\partial}{\partial z}\right) \tag{21}$$

which can be done by some algebra, in order to find speeds γ_1 and γ_2 as functions of the LHS variables and expressed in original variables.

Fig. 7 Solid compressibility propagation speeds c_1, c_2 and α_2^w for SAI system (m/s)

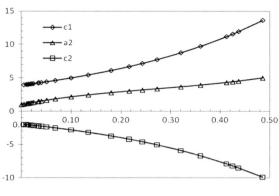

Fig. 8 Solid compressibility propagation speeds c_1, c_2 and α_2^w for FCC system (m/s)

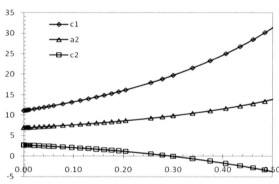

Fig. 9 Third order propagation speeds α_1^w, α_2^w and α_3^w, SAI system (m/s)

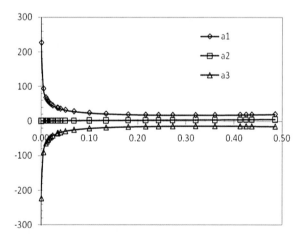

3.3 Asymptotic Analysis of Waving Structures

For an unbounded system $A = B = C_w = 0$ and the only existing term in Eq. (21) is the one of gravity effects. Then the following asymptotic limit occurs:

Fig. 10 Third order
propagation speeds
α_1^w, α_2^w and α_3^w, FCC system
(m/s)

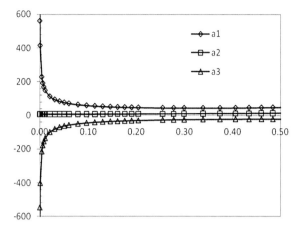

Fig. 11 Second order
propagation speeds
γ_1 and γ_2, SAI system (m/s)

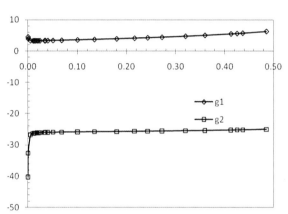

Fig. 12 Second order
propagation speeds,
γ_1 and γ_2, FCC system (m/s)

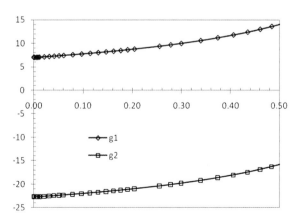

$$\lim_{B \to 0} \frac{B}{C} \left(\frac{\partial}{\partial t} + \gamma_1 \frac{\partial}{\partial z} \right) \left(\frac{\partial}{\partial t} + \gamma_2 \frac{\partial}{\partial z} \right) = -g \left(\frac{\partial}{\partial t} + \eta_{21} \frac{\partial}{\partial z} \right) \frac{\partial}{\partial z} \qquad (22)$$

which implies a change from hyperbolic to parabolic structures and also that $\gamma_2 \to -\infty$, $\gamma_1 \to \eta_{21}$, $\frac{B \gamma_2}{C} = -g$. Therefore γ_1 is a propagation speed mainly associated to gravity effects, while γ_2 merge from wall effects consideration and degenerates a hyperbolic structure to a parabolic one. Moreover, considering the unbounded system is equivalent to take the limit when $A \to 0$ in Eq. (17). It can be easily seen that $\lim_{A \to 0} a_w = a$, since ξ_f is bounded.

The lumped third order structure needs more elaboration since three terms contribute: (1) the main incompressible behavior modified by wall effects, (2) the main compressible third order structure and (3) the third order wall effects. These contributions are summed up and its rearrangement gives rise a lumped third order wave operator, see Eq. (18). For an unbounded system $A = C_w = 0$ and also $\alpha_j^w \to \alpha_j$, the third order propagation speeds of the unbounded system.

On the other hand, it is also interesting to analyze the incompressibility limit. This can be performed by equating both models, Eqs. (18) and (16), then taking the limit when the fluid propagation speed $s \to \infty$:

$$\lim_{s \to \infty} \frac{C}{s^2} \left[\begin{array}{c} \left(1 + \frac{C_w}{C} \right) \left(\frac{\partial}{\partial t} + \alpha_1^w \frac{\partial}{\partial z} \right) \left(\frac{\partial}{\partial t} + \alpha_2^w \frac{\partial}{\partial z} \right) \left(\frac{\partial}{\partial t} + \alpha_3^w \frac{\partial}{\partial z} \right) \\ + \frac{B}{C} \left(\frac{\partial}{\partial t} + \gamma_1 \frac{\partial}{\partial z} \right) \left(\frac{\partial}{\partial t} + \gamma_2 \frac{\partial}{\partial z} \right) \end{array} \right] \varphi$$
$$= -(1 + A) \left(\frac{\partial}{\partial t} + a_w \frac{\partial}{\partial z} \right) \frac{\partial^2}{\partial z^2} \varphi \qquad (23)$$

Since the second order wave remains bounded under the limit, the equality concerns just to the third order wave and is equivalent to the following set of limiting values: $\alpha_1^w \to \infty$, $\alpha_3^w \to -\infty$, $\alpha_2^w \to a_w$, $\lim_{s \to \infty} \frac{\alpha_1^w \alpha_3^w}{s^2} C = -1$, and it should also be needed that $C_w/C \to A$ upon the incompressibility limit in a bounded system.

4 Conclusions

On the foundation of the coefficient analysis and taking the unbounded incompressible behavior as the basis of comparisons, it should be concluded, for the physical systems chosen (see Table 1), that the (1) solid phase elastic effect is a small effect since it does not grows up a 5.2 % of main incompressible behavior, for SAI system. (2) The fluid compressibility effect is, for most operation conditions, a more important effect than the former one, growing up to a 10 or 20 % of main incompressible effect. (3) Wall effects enhance the main incompressible behavior up to a 8.2 % for small solid volume fraction. (4) Wall effects on the compressible part have two contributions, one on the third order wave and one on the second order one. (5) Second order impact of wall effects is same magnitude as the main compressibility term. (6) On the opposite, third order impact of wall effects is always below 10 % of the main compressibility contribution and for $\varepsilon_{s0} > 0.04$ decay below 0.10 %. (7) Despite the importance of wall effects on second order compressible part, the main second order contribution is due to gravity and is around three times greater than wall effects for FCC and around ten times for SAI system.

On the basis of the propagation speeds analysis it can be established that (8) the lumped model propagation speeds do hold physical significance and can be tested experimentally. (9) There is a set of seven propagation speeds for the lumped model, two of them are attached to the solid compressibility effect, two are the sound propagation speed relative to mass average velocity and one intermediate is the kinematic wave or volume fraction wave. A propagation speed is associated to gravity and the last one to wall drag interaction. (10) Each one of the wave hierarchies involves proper interactions between the considered effects, as can be followed up from operators in Eq. (13). (11) The asymptotic analysis performed allowed the association of lumped variables to their main physical meaning. (12) Within present approach, more meaning and structure is incorporated to the propagation speeds, as long as the lumping process is taking place. Thus the simplest first order incompressible unbounded propagation speed a, when combined with wall effects becomes a_w, then becomes α_2^w when fluid compressibility is allowed.

Acknowledgments The authors highly acknowledge the *Consejo Nacional de Ciencia y Tecnología (CONACyT)*, México, for financial support through Grant CB-2005-C01-50379-Y.

References

Bi HT, Grace JR, Zhu J (1995) Propagation of pressure waves and forced oscillations in gas-solid fluidized beds and their influence on diagnostics of local hydrodynamics. Powder Technol 82:239–253

Drew DA (1983) Mathematical modeling of two-phase flow. Ann Rev Fluid Mech 15:261–291

Drew DA (1971) Averaged field equations for two-phase media. Studies in Appl Math L, 133–166

Epstein N (2003) Applications of liquid-solid fluidization. Int J Chem React Eng 1:1–16

Grace JR, Issangya AS, Bai D, Bi H (1999) Situating the high-density circulating fluidized bed. AIChE J 45(10):2108–2116

Gregor W, Rumpf H (1976) Attenuation of sound in gas-solid suspensions. Powder Technol 15:43–51

Homsy GM, El-Kaissy MM, Didwinia A (1980) Instability waves and the origin of bubbles in fluidized beds-II comparison with theory. Int J Multiph Flow 6:305–318

Jackson R (1985) Hydrodynamic stability of fluid-particle systems. In: Davidson JF, Clift R, Harrison D (eds) Fluidization, 2nd edn, Academic Press, Inc, London, pp 47–72

Jiradilok V, Gidaspow D, Damronglerd S, Koves WJ, Mostofi R (2006) Kinetic theory based CFD simulation of turbulent fluidization of FCC particles in a riser. Chem Eng Sci 61:5544–5559

Joshi JB, Deshpande NS, Dinkar M, Phanikumar DV (2001) Hydrodynamic stability of multiphase reactors. Adv Chem Eng 26:1–130

Konno H, Saito S (1969) Pneumatic conveying of solids through straight pipes. J Chem Eng Jpn 2:211–217

Liu JTC (1982) Note on a wave-hierarchy interpretation of fluidized bed instabilities. Proc R Soc Lond A 380:229–239

Musmarra D, Poletto M, Vaccaro S, Clift R (1995) Dynamic waves in fluidized-beds. Powder Technol 82:255–268

Ransom VH, Hicks DL (1984) Hyperbolic two-pressure models for two-phase flow. J Comput Phys 53:124–151

Ryzhkov AF, Tolmachev EM (1983) Selection of optimal height for vibro-fluidized bed. Teor Osn Khim Tekhnol 17:140–147 [Theor Found Chem Eng (Engl. Transl.)]

Sánchez-López JRG, Soria A, Salinas-Rodríguez E (2011) Compressible and incompressible 1-D linear wave propagation assessment in fast fluidized beds. AIChE J 57:2965–2976

Soria A, de Lasa HI (1991) Averaged transport equations for multiphase systems with interfacial effects. Chem Eng Sci 46:2093–2111

Soria A, Salinas-Rodríguez E, Sánchez-López JRG (2012) submitted to AIChE Journal

Soria A, Sánchez-López JRG, Salinas-Rodríguez EM (2008) The incompressibility assumption assessment in fast fluidized bed wave propagation. In: Proceedings of the 14th WASCOM conference, Singapore, pp 530–535

van der Schaaf J, Schouten JC, van den Bleek CM (1998) Origin, propagation and attenuation of pressure waves in gas-solid fluidized beds. Powder Technol 95:220–233

Wallis GB (1969) One-dimensional two-phase flow. McGraw-Hill Inc, New York

Critical and Granular Casimir Forces: A Methodological Convergence from Nano to Macroscopic Scales

Y. Nahmad-Molinari, G. M. Rodríguez-Liñán, J. F. Reyes-Tendilla and G. Pérez-Ángel

Abstract Recent advances in colloidal science that have lead to directly measuring depletion and critical Casimir interaction potentials are reviewed. Methodological convergence of these studies is exploited and extended to granular systems in order to measure effective interactions among their constituent particles. Experimental evidence of depletion interactions as well as a novel "granular Casimir effect" is presented and subtle differences are discussed.

1 Introduction to Casimir Forces

Statistical mechanics concepts developed within the liquid state theory are nowadays powerful experimental tools for measuring interactions among particles in the sub-micron regime. At this scale, Casimir forces become important and they even may result in an irresistible attraction, for example in the vicinity of the consolute point, where a binary mixture separates into two components. Correlation functions measured by means of frustrated total internal reflection intensities made possible direct measurements of the so called Critical Casimir effect. Correlation functions of a pair of colloids confined to one dimensional optical trap allowed to determine the depletion interaction and in this work, for the first time we use similar methodologies to get measurements of these effects; being the mechanical analogue

Y. Nahmad-Molinari (✉) · G. M. Rodríguez-Liñán · J. F. Reyes-Tendilla
Instituto de Física, Universidad Autónoma San Luis Potosí, Av. Manuel Nava 6, Zona Universitaria, 78210 San Luis Potosí, Mexico
e-mail: yuri@ifisica.uaslp.mx

G. Pérez-Ángel
CINVESTAV-Mérida, Ap. Postal 73, Cordemex, Mérida, YUC, Mexico

J. Klapp et al. (eds.), *Fluid Dynamics in Physics, Engineering and Environmental Applications*, Environmental Science and Engineering,
DOI: 10.1007/978-3-642-27723-8_7, © Springer-Verlag Berlin Heidelberg 2013

of the quantum Casimir effect, the so called granular Casimir forces and the depletion forces.

Arising from electrodynamic quantum fluctuations of vacuum along with the boundary conditions imposed by two dielectric or conducting plates, there is an attraction between the considered surfaces first predicted by Casimir and Polder (1948). These forces can be measured in analogous systems in which electromagnetic fluctuations of the zero field in vacuum are replaced by fluctuations in the composition within a critical mixture close to its demixing transition (Beysens and Esteve 1985), the so called critical Casimir forces, firstly predicted by Fisher and de Gennes (1978) and then calculated in 2003 (Schlesener et al. 2003). The hunt for measuring and control this critical Casimir forces has guided the community to study colloidal particles suspended in critical mixtures (Tsori et al. 2004), and recently gave striking results when, for the first time, these forces were directly measured by Hertlein at Stuttgart University (Hertlein et al. 2008).

Since the strength of fluctuations in the critical mixtures diverges as temperature gets closer to the critical point, one can easily use temperature as a knob to tune the strength of critical Casimir effects among colloids or between a colloidal sphere and a flat boundary.

In general, when fluctuating fields are confined between two surfaces, long-range forces arise. Thus, one could search for acoustic analogues or even mechanical attraction when boundary conditions are imposed by these two surfaces in fluctuating acoustic or mechanical fields. The main aim of this paper is to describe the methodological convergence behind direct observations of two classical analogues of the quantum–mechanical Casimir, being the critical and the granular Casimir effects. This last analogue is been presented here for the first time.

2 Depletion Forces and Casimir Effect

2.1 Depletion Forces

Casimir forces resemble, and should be carefully discerned from depletion interactions. These depletion forces produce colloidal aggregation and colloidal phase separation and are induced by the osmotic pressure exerted by non adsorbing polymers on colloidal particles. Depletion interaction has an entropic origin, since large colloids effectively attract each other and they aggregate in order to make some room for the small tangles of polymer to mingle and roam with freedom, maximizing its own entropy and the system's entropy as a whole. However, depletion forces do not depend only on a difference among sizes, but mainly on how crowded the system is. Who has not experienced this tremendous attraction to his neighbors in the public transport at peak hours in a crowded city such as Tokyo or Mexico City?

Fig. 1 Picture showing the original Magdeburg hemispheres and the vacuum pump used in the demonstration (from Wikimedia commons)

Depletion interaction can induce nematic order in mixtures of tobacco viruses and spherical particles as Fraden has shown. It can even be used to separate the mixture into their constituents, as Cohen did in 1941 in order to separate Tobacco Mosaic from Tobacco Necrosis viruses by adding heparin as a depletant. Depletion interaction has been used as well as a fractionation method to separate emulsions and colloid by their size (Bibete) in order to produce monodisperse colloidal suspensions that are able to self-assemble into periodic structures that present the delicate phenomenon of iridescence (Fig. 1).

If a region in between two plates, or hemispheres, that are immersed in a sea of fluctuating particles or modes, is depleted from this particles or modes, a very strong attraction will be witnessed (Lekerkerker 2011). A famous example was the Magdeburg hemispheres pulled apart by two teams of horses in order to demonstrate the effects of atmospheric pressure. Air depleted from the interior region of the hemispheres is unable to compensate the momentum transfer of the exterior particles and both hemispheres suffer an almost irresistible attraction that the two teams of horses were unable to overcome. This attraction can induce order in thermally driven systems, which can be counterintuitive, since is commonly thought that thermal motion should drive systems towards disordered states. As an example, lets imagine a dinner in which guests are eagerly looking for dancing (this is usual in Latin-America). What is observed is that as soon as the dessert is

Fig. 2 Schematic
representation of **a** the
Casimir effect in which just
few modes are allowed to be
within the particles, and **b** the
depletion interaction in which
small Brownian particles are
excluded from the interior
region causing a unbalanced
osmotic pressure which
pushes the large particles
against each other

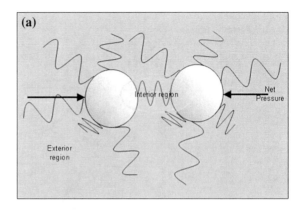

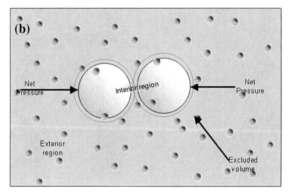

finished, tables will shortly end together pushed apart against the walls, freeing
space to dancers to perform their artistic evolutions. In this example, tables will
end in an ordered state in order to allow dancers to roaming around maximizing
their possible configurations. This is called entropic ordering, since closely
packing the tables in an ordered manner allows the dancers to maximize their
configurational entropy and thus, the total entropy of the system.

As we have stated above, depletion forces and Casimir forces should be carefully
distinguished from each other and in some cases, as we will see in our experiments,
they appear in conjunction. One may identify Casimir forces with objects immersed
in fluctuating oscillatory fields, while depletion forces with objects immersed within
a sea of thermally driven particles (Brownian motion) (Fig. 2).

In depletion forces, there is a region depleted from the constituent particles of
the Brownian Sea, and this density misbalance is the origin of the attraction. On
the other side, Casimir forces imply the inhibition of some oscillatory modes in the
region between objects, due to the fact that the objects themselves establish
boundary conditions that should be satisfied by the oscillating field. Analogously
to the case of depletion forces, in the case of Casimir attraction, the energy density
misbalance between the interior (in between the objects) and the exterior region of
the objects will produce an effective attraction. This Casimir effect is associated

with the radiation pressure exerted by the modes of the field in the exterior region, which will be higher since the boundary conditions for the interior region are more restrictive, and thus inhibit a larger number of modes.

3 Depletion and Critical-Casimir Forces Directly Observed

3.1 Methodological Tools

In 1999 Crocker and coworkers, at the University of Pennsylvania achieved direct measurements of entropic interaction between a single pair of colloidal PMMA (polymethilmethacrylate) 1.1 μm in diameter spheres, constrained to move in one dimension (Crocker et al. 1999). Depletion forces were achieved by means of a background of smaller (83 nm) polystyrene spheres at different concentrations. Attraction as well as repulsion was observed and even an oscillatory behavior of the pair potential was found. In order to explore the interaction mediated by the smaller spheres, the pair of larger spheres was trapped into a linear optical tweezers and the relative distance among the PMMA spheres was monitored by videomicroscopy. From the relative distance among particles the correlation function or the equilibrium probability $P(r)$ of finding the particles at a given distance r is calculated and thus inverted trough the Boltzmann relation, $P(r) = \exp[-F(r)/k_B T]$ to get the pair interaction energy $F(r)$. This is the kernel of the methodology shared by many clever experiments to unveil the pair potential among particles mainly at colloidal scales. Three features should be stressed, (i) the pair potential is inferred from statistically measuring distances, and (ii) an optical method is used to get the distance (videomicroscopy was used in depletion forces measurements and total internal reflection microscopy for determining the critical Casimir effect) and (iii) a soft and controllable constriction is imposed to the particles in order to limit their motion and at the same time be able to subtract the effects of this confinement.

In 2008, the Stuttgart team conducted experiments with polystyrene spheres and highly charged polystyrene colloidal particles which prefer lutidine or water respectively, being the solvent a critical mixture composed of 2,6-lutidine and water. By means of trapping with optical tweezers close to a plane surface, a colloidal particle, suspended on a binary mixture close to its critical point, is kept within the evanescent field of a total internally reflected EM Wave from below. The colloidal particle interacts with the evanescent field, partially frustrating the reflection and the transmitted intensity is collected by the same objective used to form the tweezers. This transmitted intensity is inversely proportional to the plate-sphere distance, and is recorded in order to probe the plate-sphere effective potential dependence on separation within the fluctuating field imposed by the critical mixture. This technique allows them to achieve femtonewton resolution.

Fig. 3 Aspect of our granular gasses as seen from the high speed camera

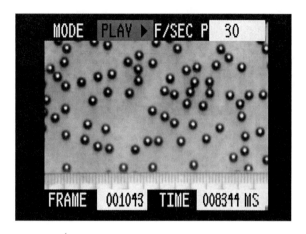

Fig. 4 Pair correlation function g (**r**) for five different restitution coefficients plotted as a function of normalized interparticle distance r with respect to particle diameter σ

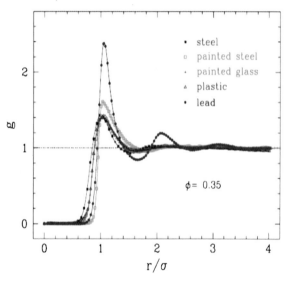

In San Luis Potosí, 2009, Bordallo-Favela et al., measured in quasi-two dimensional vertically vibrated granular gases, an attractive potential between dissipative hard spheres, about 3 mm in diameter (Bordallo-Favela et al. 2009). This attraction is produced by an increase in correlation due to inelastic collisions and to the depletion forces or osmotic pressure associated to area fraction. Again, the way of getting the interaction potential was by means of image analysis of pictures of the gas taken from above (Tata et al. 2000) as depicted in Fig. 3.

Radial distribution functions for several restitution coefficients (glass, steel, plastic, and lead) are shown in Fig. 4. A notorious increase of the correlation close to contact which will favor the formation of a deeper interaction potential (Fig. 5) is seen as the restitution coefficient of the spheres diminishes. These relations are plotted in Fig. 6.

Fig. 5 Interaction potentials U_{eff} obtained from pair correlation functions shown in the previous figure where β is a normalization constant

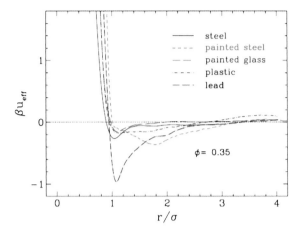

Fig. 6 Minimum of the interaction potentials U_{min} in *red squares* and maximum of the pair correlation functions in *blue triangles* as a function of the restitution coefficient

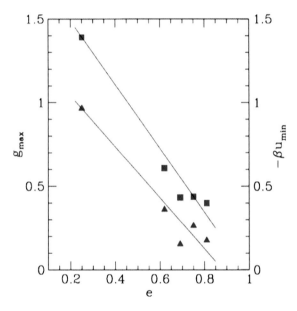

They have reported as well observations of the dependence of the attraction on the area fraction of their quasi-two dimensional granular gases. As we have seen, a methodological convergence on how to measure the interaction potential directly from particle positions is underneath these three experiments and the same kind of forces are measured.

The same team explored by means of turbidimetric measurements critical Casimir attractions among superparamagnetic nanoparticles suspended in a cyclohexane-methanol binary mixture close to phase separation. Remarkable magneto-optic and magneto-caloric effects were reported. Furthermore, irreversible aggregates that give account on double emulsions produced during the demixing transition (Hernández-Díaz 2010; Odenbach 2003).

Fig. 7 Segregated or final
state of a quasi-two-
dimensional granular gas
were the rods rich phase is
ordered into a nematic way

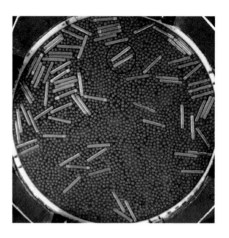

3.2 Granular Mixtures

In granular mixtures of spheres and rods conforming a quasi two-dimensional gas
a wealth of phenomena has been observed. Starting from an isotropic homogeneous mixture, the system evolves towards phase separation into a nematic phase
of rods coexisting with a homogeneous gas of spheres. First of all, a strong
depletion attraction between pairs of rods when they approach is witnessed and
they will form a stable couple that eventually separates and comes together again,
until they find a new member to do a trio or a new couple to form a quartet,
gradually segregating themselves and forming the nematically arranged new phase
of rods as can be seen in the Fig. 7. Once again, the example of dancers applies in
this case in which the tables could be thought as the rod like particles of our
experiments and the spheres as our guests.

A common configuration is seen in the same picture in which a chain of spheres
lays in between two rods making a configuration which is not as stable as a couple
of rods in contact, but nevertheless remains for large periods of time.

One may naively think that depletion forces are responsible as well for such
configurations. However, the region in between the two rods is not depleted from
small spheres, but instead this region is has a larger population density than the
exterior region and thus, the osmotic pressure misbalance should push apart both
roads (a repulsive depletion interaction is expected).

By zooming in and taking high speed video of such a configuration and
superimposing twenty frames, a pattern shows up, as can be seen in Fig. 8. Much
larger and intricate trajectories of the spheres in the exterior region and smaller
movements and displacements within the interior region can be clearly noted. This
is a signature of inhibition of mechanical modes transversal to rods and an
attractive force between rods at a distance larger than their thickness (equal to the
sphere diameter), at which cannot act.

Fig. 8 Twenty pictures superimposed of a rod-chain-rod configuration as it evolves. A clear suppression of motions transversal to road is seen for particles within both rods in comparison with particles outside this configuration

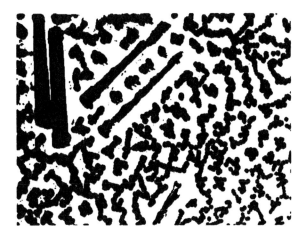

Radial or pair distribution functions among the rods centres of mass, angular distribution functions, and velocity distribution functions in and outside rod-chain-rod configurations as time goes on are in preparation to be published elsewhere.

4 Conclusions

We have reviewed a state of the art methodology to unveil subtle interactions among colloidal or granular particles. Pair or radial distribution function determination by optical means and inversion through the Boltzmann equation allows getting in a detailed way the interaction potential and helped us in determining the depletion attraction in granular materials. Furthermore, evidence of a novel "granular Casimir effect" is found in quasi 2D granular mixtures of rods and spheres. This evidence is related to the formation of rod-chain-rod stable structures that inhibit transversal momentum acquisition perpendicular to the rod axes.

References

Beysens D, Esteve D (1985) Adsorption phenomena at the surface of silica spheres in a binary liquid mixture. Phys Rev Lett 54:2123

Bordallo-Favela RA, Ramírez-Saíto A1 CA Pacheco-Molina1 JA, Perer Burgos1, 2 Y, Nahmad-Molinari1, Pérez G (2009) Effective potentials of dissipative hard spheres in granular matter. Eur Phys J E28:395–400

Casimir HBG, Polder D (1948) The Influence of Retardation on the London-van der Waals Forces. Phys Rev 73:360

Crocker JC, Matteo JA, Dinsmore AD, Yodh AG (1999) Entropic attraction and repulsion in binary colloids probed with a line optical tweezer. Phys Rev Lett 82:21

Fisher ME, De Gennes PG (1978) Phénomènes aux parois dans un mé-lange binaire critique. CR Acad Sci Paris B 287:207

Hertlein C, Helden L, Gambassi A, Dietrich S, Bechinger C (2008) Direct measurement of critical Casimir forces. Nature 451:172

Hernández-Díaz L, Hernández-Reta JC, Encinas-Oropesa A, Nahmad-Molinari Y (2010) Coupling of demixing and magnetic ordering phase transitions probed by turbidimetric measurements in a binary mixture doped with magnetic nanoparticles. J Phys Condens Matter 22:195101

Lekkerkerker HNW, Tuinier R (2011) Colloids and the depletion interaction. (Lecture Notes in Physics 833), Springer, New York, 233 pp

Odenbach S (ed) (2003) Ferrofluids: magnetically controllable fluids and their applications (Lecture notes in physics 594), Springer, New York, 253 pp

Schlesener F, Hanke A, Dietrich S (2003) Critical Casimir Forces in Colloidal Suspensions. J Stat Phys 110:981

Tata BVR, Rajamani PVJ, Chakrabarti AN, Wasan DT (2000) Gas-liquid transition in a two-dimensional system of millimeter-sized like-charged metal balls. Phys Rev Lett 84:3626–3629

Tsori Y, Tournilhac F, Leibler L (2004) Demixing in simple fluids induced by electric field gradients. Nature 430(544):02758

High Speed Shadowgraphy for the Study of Liquid Drops

José Rafael Castrejón-Pita, Rafael Castrejón-García
and Ian Michael Hutchings

Abstract The principles of shadowgraph photography are described in this work together with a few examples of its utilisation in the study of free liquid surfaces. Shadowgraph photography is utilized in combination with high speed imaging and image analysis to study the behaviour of sub-millimetre and millimetre-sized droplets and jets. The temporal and physical scales of these examples cover operational ranges of industrial, commercial, and academic interest. The aim of this work is to summarize the necessary optical and illumination properties to design an appropriate shadowgraph imaging system.

1 Introduction

Over the last two centuries, the shadowgraph technique has been extensively used in the study of fluid dynamics, the visualisation of objects in motion, in optical microscopy and in scientific photography. In fact, it can be said that several scientific breakthroughs could not have been achieved without this technique. Perfect examples of these are the early experiments in fluid mechanics carried out

J. R. Castrejón-Pita (✉) · I. M. Hutchings
Department of Engineering, University of Cambridge,
17 Charles Babbage Road, Cambridge CB3 0FS, UK
e-mail: jrc64@cam.ac.uk

I. M. Hutchings
e-mail: imh2@cam.ac.uk

R. Castrejón-García
Centro de Investigación de Energía, UNAM, Priv. Xochicalco s/n,
62580 Temixco, MOR, Mexico
e-mail: rcg@cie.unam.mx

J. Klapp et al. (eds.), *Fluid Dynamics in Physics, Engineering and Environmental Applications*, Environmental Science and Engineering,
DOI: 10.1007/978-3-642-27723-8_8, © Springer-Verlag Berlin Heidelberg 2013

by Savart and the study of the breakup of fluids by Raleigh, both in the nineteenth century (Savart 1833 and Strutt 1896). Savart developed a primitive but effective shadowgraph setup to visualize liquid jets emerging from a pressurized nozzle and his results were used some years later by Rayleigh and Plateau to lay the theoretical basis that describes the natural break up of jets (Eggers and Villermaux 2008). In simple terms, shadowgraphy is a visualisation technique in which the observed objects are illuminated from behind. Shadowgraph photography is often used to visualize liquids, but can also be used to observe solid objects and gases. Shadowgraphy is an advantageous visualisation technology due to its simplicity, low cost, quality and by the fact that a large range of optical components are readily commercially available. Furthermore, thanks to the advances in digital imaging, shadowgraphy is no longer a qualitative technique as the digitisation of images and their processing can provide information about the dynamics of the object studied (Castrejón-García 2011).

Shadowgraphy has a large range of potential applications: it is used in aerodynamics, in fluid dynamics and in ballistics. Although extensively used and mentioned in the scientific literature, the theoretical aspects of the technique are not generally discussed in detail or even generally available. As a result, the design and construction of shadowgraph systems is usually difficult and often carried out empirically. The aim of this work is to present and discuss the optical and technical concepts involved in shadowgraph photography, in particular the exposure time, field of view and depth of field. Several shadowgraph images of sprays and droplets are shown to exemplify the capabilities of the shadowgraph method for the study of fluids. These systems differ in their temporal and physical scales, components, and in the way they are designed and constructed; from custom-built electronics and optics to commercially available cameras, flash-lamps and lenses.

2 Shadowgraph Illumination

Shadowgraph photography or shadowgraphy is *de facto* a very simple technique that requires the optimisation of several parameters to obtain satisfactory results. Commonly, in conventional photography, the illumination of an object can be carried out in two ways: by front-illumination or by back-illumination. In the first case, illustrated in Fig. 1, the object is illuminated in such a way that the light dispersed or reflected by it forms the image on the camera sensor. This corresponds to conventional photography where a light source is always positioned in front of the object. This type of illumination is natural and necessary for the perception of colour in photography and in the image formation in the eye.

In the second case, illustrated in Fig. 2, the object is illuminated from behind by means of a focusing lens in such a way that the object's image formed on the sensor camera corresponds to its shadow or silhouette. On the recorded image, the object's shadow contrasts with a bright field formed by the light from the illumination system. This technique is known as *Shadowgraphy*. This type of

Fig. 1 Front illumination scheme, usually used in conventional photography

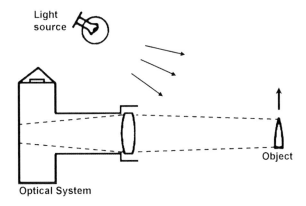

illumination does not usually record colour information as the light that interacts with the object is mostly refracted away from the sensor. However, this technique captures most of the light coming from the source and, as a result, is ideal in conditions of low light or with sensors with low sensitivity. This characteristic renders the shadowgraph technique the preferred type of visualisation in high speed imaging where the sensitivity is low at high frame rates. Due to this fact, shadowgraphy is once again becoming widely used. Examples of front and back (shadowgraph) illumination are presented in Figs. 3 and 4.

Figure 3 shows a sequence of images in which a sessile colourless droplet is being impacted by a coloured one. The impacting droplet is travelling at 1.1 m/s and has a radius of 1 mm. Both droplets consist of a 75 % glycerine and water mixture with a viscosity of 100 mPa s. The substrate is a flat transparent piece of acrylic sheet (polymethyl methacrylate). In this case, the illumination system is similar to that presented in Fig. 1, where the light source (a 50 W filament lamp) is oblique to the object and placed in front of it. For these experiments, a Phantom V640 high speed camera was used to capture the impact at a rate of 1,000 frames per second (fps) and at an exposure time of 0.3 ms.

Figure 4 shows the impact of a droplet into a sessile one (same conditions as in Fig. 3) but now with the illumination being produced from the back (i.e. shadowgraph). The uniform background is created by an optical diffuser placed in front of the 50 W lamp. For these experiments, a V310 Phantom camera was used at a frame rate of 7000 fps and with exposure times of 2 μs (only images at equivalent times to those in Fig. 3 are shown). The exposure time and the frame rate in these two experiments were chosen to acquire images with similar brightness and contrast. In high speed imaging, the brightness of the images depends mostly on the frame rate; the faster the frame rate is, the less the time there is to collect photons on the sensor. As a consequence, to create clear images, there is always a compromise between frame speed and brightness. In fact, in most high speed imaging experiments, the illumination is the principal factor limiting the speed of recording. Both these cameras (the V310 and the V640) share the same CCD (the Phantom 1,280 × 800 CMOS sensor) and sensitivity.

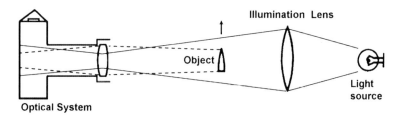

Fig. 2 Shadowgraph illumination

Fig. 3 Examples of high speed imaging using a front illumination scheme; the images correspond to a sequence of images with a 1 ms separation. In this case, only the light that interacts with the droplets is collected by the lens and the sensor of the camera allowing the production of colour. In this scheme, the reflected image of the light source often appears as a bright spot

The frame speed is different between the images presented in Figs. 3 and 4 (front and shadowgraph imaging) as it was adjusted to cope with the different type of illumination. The recording under shadowgraph conditions can be faster and with shorter exposure times due to a more efficient illumination. This is a useful characteristic of shadowgraphy as shorter exposure times are usually desired in the study of objects in motion in order to avoid the recording of blurred images. This experimental approach has been used in droplet deposition studies to validate theoretical models and to investigate the inner fluid dynamics during the deposition and coalescence of impacting droplets, (Castrejón-Pita et al. 2011).

Nowadays, the use of image analysis methods is becoming increasingly popular in studies of fluid dynamics. For the correct recognition of fluid or object boundaries, most image analysis algorithms require the objects to be sharply focused, frozen in time, and visualized on an even and homogeneous background.

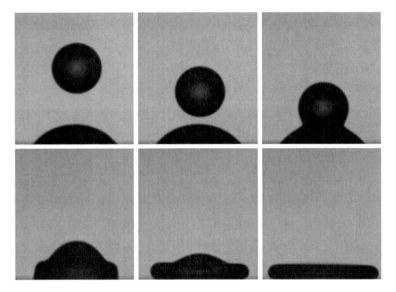

Fig. 4 The same experiments as in Fig. 3, but with shadowgraph illumination. In this case, only the light that does not interact with the droplets is captured by the sensor. All the other rays are mostly either reflected or refracted out of the camera lens. The brighter region at the center of the spherical drop is caused by light refracted through the drop that is not directed outside the CCD sensor

These characteristics are determined by the properties of the optical elements and the illumination arrangement. For objects in motion, the image sharpness is determined by the exposure time of the recording sensor and by the depth of field and focus of the optical system. In most cases, the blurring due to the motion is controlled by the use of short exposure and shutter times, or short light pulses. For example, let us assume that an object is moving at a speed of 50 m/s. If the object is observed through an optical system with a magnification factor of 3, the velocity of object's image on the camera sensor is actually 150 m/s. If the shutter time is set to 100 μs, the image would appear blurred on the sensor over a distance of 15 mm. On the other hand, if the object is moving at 1 m/s, the object would appear blurred only along 0.3 mm. In words used in conventional photography, the shutter time must be short enough to "freeze" the motion of the object. This effect is actually observed in Figs. 3 and 4. In Fig. 3, the impacting droplet appears blurred along the axial direction for about one-third of its radius. On the contrary, in Fig. 4, as the exposure time is shorter (150 times faster), the image is sharper, and its contour well-defined and perfectly circular whereas the image of the droplet in Fig. 3 has an elliptical shape due to the blurring.

The shutter times of most commercial single-lens reflex cameras (SLR cameras) are rarely below 500 μs, which is the key parameter to consider when considering photographing objects in motion with conventional CCD cameras. In contrast, in high speed imaging, exposure times are usually in the range of 1 ms–200 ns (subject to a change in resolution).

Fig. 5 Shadowgraph
snapshot of a 0.5 m/s droplet
of a 83 % glycerine and water
mixture. A Nikon SB-800
flash was used to produce the
24 μs pulse used to capture
this image

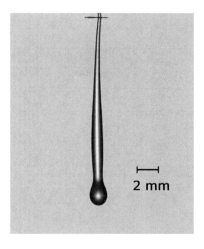

Alternatively, the use of a flash lamp or pulsed light source is recommended where the pulse duration is shorter than the shutter time. In the latter case, the effective exposure time is determined by the light duration and not restricted to the shutter time. Spark flashes or xenon flash lamps are the best options for shadowgraphy. Photographic flashes are readily commercially available, are not expensive, have a pulse duration in the range of 20–200 μs and are viable for low speed applications. These flashes are commonly used to study the jetting and behaviour of millimetre size droplets, (Castrejón-Pita et al. 2008). An example of this is shown in Fig. 5.

Laboratory xenon and spark lamps produce much faster light pulses but operate with high voltages which may restrict their application, based on safety issues, especially when used in studies involving volatile, conductive or explosive fluids. Specialist xenon systems produce light pulses with durations ranging from 120 ns to 1 μs with 100–200 mJ of pulse energy. In contrast, spark flashes have a pulse light duration of a few nanoseconds. These setups generally consist of a pair of electrodes connected to a capacitor which is charged until the stored energy is released as a luminous spark. The light intensity and the pulse duration can be adjusted by varying the separation and size of the electrodes and the surrounding medium. Many of the commercially available systems use argon gas, and produce pulse durations of the order of 5–20 ns, 9–25 mJ of electric flash energy, and utilize 3–5 kV power supplies. Although expensive, these systems are in many cases very convenient as they require little maintenance. Figure 6 shows an example of the use of a spark flash system to image a fast spray generator; this technology has been used in the past to study the fractal dimension of the aerosol formed and in the study of drop size distributions in sprays, (Castrejón-García et al. 2003) and (Le Moyne et al. 2008).

The exposure time is only one of the factors determining the sharpness of the objects in an image. Other important parameters are the depth of field and the working distance of the optical system used to visualize the objects. The correct

Fig. 6 A shadowgraph
snapshot of an oil spray. A
300 ns spark flash system was
used for this illumination.
The nozzle output has a
diameter of 2 mm and the
droplet terminal speed is
15 m/s

implementation of shadowgraph imaging requires the understanding not only of
the individual components of the imaging setup but also their overall effect on the
process of image formation. In most cases a few concepts of optics, ray tracing,
and photography are needed to set up a suitable working system. The following
section presents the theoretical basis required to design and implement a working
shadowgraph scheme.

3 Image Formation and Object Illumination

As mentioned above, the aim of shadowgraph photography is to obtain a sharp or
focused silhouette of an object on an even or homogeneous background. To fulfil
these conditions, any shadowgraph system should contain an appropriate optical
system capable of producing a focused image of the object on the plane of the optical
sensor (i.e. film or CCD) whilst producing simultaneously a uniformly illuminated
background. One of the best ways to understand and visualize the operation of many
optical systems is by drawing ray diagrams. In agreement with geometrical optics,
any shadowgraph setup contains two sets of optics, one for the illumination and
another one for the imaging of the object. Figure 7 represents a simplified setup
consisting of two lenses, the object, and a plane where a camera sensor is placed. The
illumination lens, labelled L_i in the figure, is placed between the light source and the
object. The imaging lens L_p is located between the object and the sensor plane, and
lens L_i focuses the image of the light source on the nodal plane of L_p. The converging
illumination rays from L_i intercept the plane of the object focused by L_p. This implies

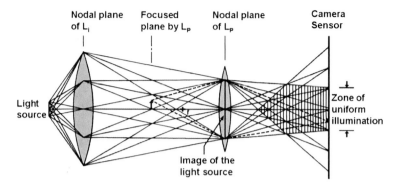

Fig. 7 Ray tracing diagram illustrating the formation of an image in the shadowgraph technique

that all the rays coming or passing through a particular point on the focused plane are collected by the lens L_p and form a corresponding point of the image on the camera sensor. In other words, all the rays emitted or reflected by an object placed on the focused plane form the object's image on the camera sensor plane. If the object is only illuminated from its back, as in shadowgraphy, the image formed on the camera plane corresponds to its silhouette or shadow. It must be borne in mind that the homogeneous illumination field is produced only by the rays converging to the camera sensor plane (shaded area in Fig. 7). Therefore, the camera sensor must be smaller than the area illuminated by these rays; otherwise, the illumination field would not be homogeneous or uniform.

A quick method to assess the size and position of the area of uniform illumination is to mount the camera lens without the camera body and form the image on a diffusive surface, e. g. a ground piece of glass. As optical components are usually more readily available than camera sensors this methodology is often useful to select a particular camera sensor.

4 Depth of Field of Shadowgraph Systems

The depth of field is an important parameter in the specification of a photographic system, as it quantitatively defines the observation volume or measuring volume of a given setup. The depth of field is defined as the region surrounding the focused plane (forward and backwards) where the loss of sharpness is negligible or imperceptible. This loss of sharpness is gradual and increases as the object moves away from the focused plane; therefore, is not possible to determine a given depth of field unless a fixed loss-of-sharpness is set. In conventional photography the depth of field is given by (Lambrecht and Woodhouse 2011):

$$d_F = 2c \frac{f}{D} \left(\frac{m+1}{m^2} \right), \tag{1}$$

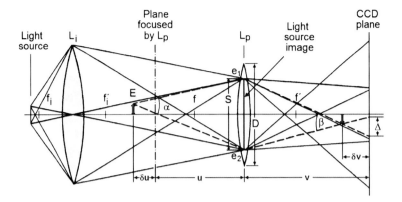

Fig. 8 Ray diagram for an object placed outside the focused plane in a shadowgraph system

where f is the focal length of the lens, m is the magnification of the optical system, D is the lens diameter and c is the so called *circle of confusion*, which is defined as the minimum loss-of-sharpness that is perceptible by the human eye. A typical value of the circle of confusion (c) for the average human eye is (Highton 2011):

$$c = \frac{\text{inch}}{1000} = 0.0254 \, \text{mm}.$$

A circle or dot with size equal or smaller than c will be seen as a point by the human eye at a comfortable reading distance (250 mm). That is to say, if the blurring of an image is less than or equal to 0.0254 mm, the image is seen by the human eye as in perfect sharpness.

Equation (1) is correct for conventional photography but not necessarily valid for a shadowgraph system. Consider Fig. 8, where an object is placed a small distance δu away from the plane focused by lens L_p. The lens diameter is D and the object's image is formed at a distance $-\delta v$ from the CCD or film plane. The illumination lens L_i is placed in such a way that the image S of a circular light-source is formed in the nodal plane of the imaging lens. The rays that pass grazing the edge E of the object can enter L_p only through the circle S formed by the image of the light source. The limit rays that can enter L_p are those coincident with points e_1 and e_2, and are separated by the angle β that is responsible for the image blurring Δ.

As seen in Fig. 8, the smaller the size of S, the smaller the blur Δ and larger the depth of field of the optical system. Consequently, the depth of field of a given shadowgraph system only depends on the size S of the image of the light source regardless of the diameter D of the lens L_p. The ray diagram in Fig 8 can be used to demonstrate that the depth of field in shadowgraph systems is (Castrejón-García and Milan 1982):

$$d_F = 2c\frac{f}{S}\left(\frac{m+1}{m^2}\right), \tag{2}$$

Fig. 9 Schematic view of an optical arrangement to control the depth of field in a shadowgraph system

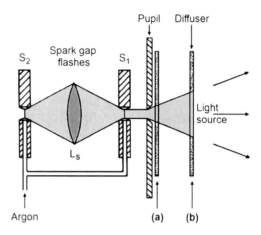

which is identical to Eq. (1), with the exception that S replaces D. Hence, the solely way to obtain a diaphragm-effect in a shadowgraph system to vary the depth of field is by changing the size of the image S of the light source. Figure 9 shows a variable size light source to control the depth of field of a shadowgraph optical system by changing the size of the light source. Light from the spark flash is stopped by a pupil where a diffuser plate (ground glass slide) is placed a given distance from the pupil. In position (a) near to the pupil, the diffuser works as a small light source, rendering a large depth of field; in position (b) the diffuser works as a larger light source, giving a small depth of field. Besides the device controlling the depth of field, the system also provides the same light energy, regardless of the position where the diffuser is placed. So there is no need to adjust other system parameters such as the sensitivity of the CCD sensor.

5 Applications of Shadowgraphy

Shadowgraph systems have been used in experiments in fluid mechanics for many years. Currently, the technique is used in industrial processes to assess the behaviour of liquid dispensers, droplet generators, fuel injectors, turbine burners, and aerosols. In many of these cases, this technique produces quantitative data that can be analysed and used to control the efficiency of certain processes. In this section some examples of the application of shadowgraphy in industry are explained.

5.1 Double-Exposure Shadowgraphy

One of the most utilized and often cited shadowgraph setups is the double-exposure system (Jones 1977), schematically shown in Fig. 10. As usual, the plane

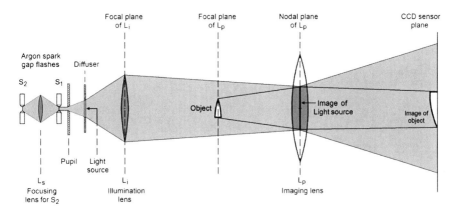

Fig. 10 A scheme of a double-exposure shadowgraph system

in which the object is moving is focused by the principal lens (L_p) forming the object image on the camera sensor. The illumination is produced by two spark light sources, namely; first spark flash (S_1) and second spark flash (S_2). The lens (L_S) forms a real image of the second spark flash (S_2) on the position where the first spark flash (S_1) is placed. This way, the light emitted by both spark flashes, comes optically from the same point. The illumination lens L_i focuses the light produced by S_1 and S_2, through the diffuser, on the nodal plane of the imaging lens L_p, and is sent towards the sensor (CCD or film) to produce the illumination background. The gray coloured area in Fig. 10shows the envelope of the light rays.

This system is known as a double exposure system because the light of each spark flash is recorded in the camera sensor on a single frame. The change of the position of the object's image (Δr) is determined by the time between flashes (Δt), the object speed (v), and the magnification of the optical system (m). Then, the actual velocity of the object is simply given by:

$$v = \frac{\Delta r}{m \Delta t}$$

Figure 11 shows a double-exposure shadowgraph image taken with an elapsed time (Δt) between flashes set to 100 μs. The second image produced by the second flash appears darker than the one produced by the first flash. The evolution of ligaments and droplets from sprayed oil moving at an average speed of 140 m/s can be observed.

5.2 Shadowgraphy in the Inkjet Industry

The shadowgraph technique is regularly used in the inkjet industry to monitor the delivery of fluid materials in a large variety of environments (Eggers and Villermaux

Fig. 11 A double-exposure
shadowgraph image of oil
droplets and ligaments. The
flash duration was 200 ns and
the images were formed in a
double spark flash rig

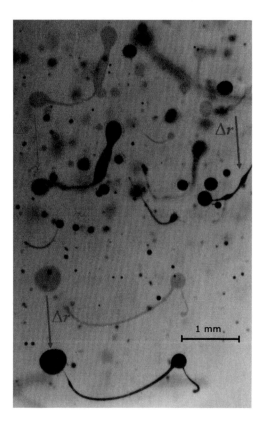

2008). In fact, some laboratory-based inkjet printers have built-in stroboscopic
visualisation setups to inspect the behaviour of the jetted droplets used to create
printed patterns, (i.e. Dimatix material printer DMP-2800). Although this capability
is not yet and may never be available in other printers, this approach is usually carried
out to test the performance of print-heads after manufacturing.

This section describes the use of a shadowgraph system to visualize a stream of
high speed droplets generated by a commercially available continuous ink jet (CIJ)
system by stroboscopic light. The setup utilized was a Domino Printing Sciences
Ltd A-series single nozzle CIJ printer with a reported nozzle diameter of 60 μm.
Essentially, in CIJ mode the liquid is pumped continuously into the head, generating an internal pressure which drives the liquid through the nozzle and creates a
jet of the desired speed. The CIJ technology relies on the creation of a stream of
droplets by the modulation (or periodic forcing) of a continuously running liquid
jet, (Kalaali et al. 2003 and Bruce 1976). During the process of jetting, the jet can
be charged by a nearby electrode set to a specific potential. After the jet break up
the droplet is no longer electrically connected to the rest of the fluid and so the
charge is retained by the droplet. The setting of the charging electrode determines
the amount of charge induced on the droplet. The direction of individual drops is

then controlled by a fixed electric field to form the printed pattern (Martin et al. 2007). The rate of droplet creation in these systems is equal to the frequency of the modulation that forces the breakup (Bruce 1976).

This technology is often used in the printing of caducity dates on plastic bottles and processed food containers. The printer was setup to jet a methyl ethyl ketone (MEK) based ink, at its standard modulation frequency of 64 kHz and a jet speed of 20 m/s. Stroboscopic methods are ideal and preferred in arrangements with a characteristic frequency or for a process that involves highly repeatable events. The basic principle is to illuminate the system in phase with the periodic phenomenon. This approach is applicable to a CIJ as the production of droplets is periodic and determined by the frequency of forcing. In this scheme, the exposure of the sensor can be arranged in such a way that single or multiple flashes can be recorded.

The shadowgraph stroboscopic system was built around the printhead to observe the jet behaviour. The imaging setup consisted of a spark flash system with a flash duration of 20 ns, a focusing system and an optical diffuser to produce a uniform short duration light background. The jet was visualized by a Navitar 12x ultra zoom microscope lens coupled to a CCD Prosilica (EC1380) camera with a resolution of 152.4 pixels/mm. The microscope lens was set to a depth of field of ≈ 200 μm which is large enough to contain the whole volume of the jet. The CCD camera and the spark flash were continuously triggered in phase with the printing frequency to obtain stroboscopic images; the recorded images are the result of the stroboscopic superposition of approximately 10 flashes.

Figure 12 shows some examples of shadowgraph images in terms of the modulation voltage used to drive the piezoelectric elements in the printhead. In a printhead, the drive modulation is delivered to the jet by a piezoelectric actuator located inside the printhead behind the nozzle. As the properties of both the piezoelectric element and the nozzle vary between suppliers, the optimisation of the jetting conditions is usually an empirical process that is carried out for each individual printhead. The jet break-up length, defined as the distance from the nozzle plane to the position of first break-up of the jet, is of commercial interest as this ultimately determines where the charging electrode is located (Curry and Portig 1976). In the CIJ industry, shadowgraph images like the ones presented in Fig. 12 are used to determine the breakup length of the jet in terms of the modulation drive voltage to produce a calibration curve. These calibrations are then utilized to find the optimum distance for the electrode. The breakup curve obtained by shadowgraphy using the MEK-based ink and the Domino printhead is shown in Fig. 13. The ideal point to set the printer is at the conditions producing the minimum breakup length because the jet is at it shortest length and so less susceptible to changes in environmental conditions. At this condition, the jet also exposes less surface, reduces evaporation of its components, has a minimum cross section and consequently has a minimum aerodynamic drag. The calibration curves can be derived automatically by the use of image analysis algorithms. The following section explains these techniques briefly.

Fig. 12 Shadowgraph imaging of a modulated jet of 50 μm diameter

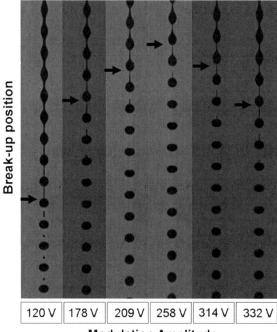

Fig. 13 Jet breakup length in terms of the driving amplitude

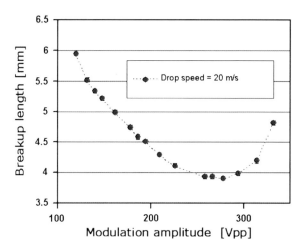

5.3 Shadowgraphy and Image Analysis Methods

The aim of this section is to briefly introduce some concepts of image analysis of shadowgraph digital pictures. Image analysis is a branch of computer science that deals with the detection of objects, still or in motion, from digital images and films (O'Gorman et al. 2008). This technology is currently used extensively in the study of

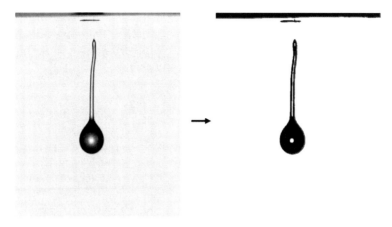

Fig. 14 Boundary detection (in *red*) by image analysis

```
Im = imread('image_name.jpg');
Im2 = rgb2gray(Im);
level = 0.8;
bww = im2bw(Im2,level);
bw=(~bww);
Im3 = imcomplement(bw);
imshow(Im3)
hold on
[B,L] = bwboundaries(bw,'noholes');
for k = 1:length(B)
    boundary = B{k};
    plot(boundary(:,2), boundary(:,1), 'r',
'LineWidth', 2)
end
```

Fig. 15 Basic algorithm (Matlab code) used to detect fluid boundaries. This code, without alterations, was used to produce Fig. 14

fluid dynamics; some common examples are velocimetry in flows by particle tracking, droplet directionality by drop detection and ligament size distribution studies in sprays, (Adrian1991; Hutchings et al. 2007 and Castrejón-García et al. 2003).

Most common algorithms for image analysis detect the boundary of objects by identifying local changes of contrast or colour levels in an image. In shadowgraphy, sudden changes of contrast only occur in the vicinity of abrupt changes of refractive index which are usually localized at the object's boundary (as seen in Fig. 14). As a consequence and as previously mentioned in Sect. 2, the shadowgraph technique is ideal for the use of image analysis.

Simple algorithms for boundary detection aim to convert colour images into black and white format where the object in study appears in black pixels and its background in white. In this type of algorithm, an image is first transformed into a grey-scale format so its pixel intensity levels can be discriminated by a threshold (RGB colour images are converted to grey scale by eliminating the hue and saturation information). In this way, the pixel intensity is individually compared with a pre-set intensity threshold. Any pixel of which the intensity level is above the threshold is converted to black while pixels in which the intensity is below the threshold are converted to white.

The object's boundary is then detected as the outer region of the black region. An example of this type of algorithm is shown in Fig. 15. In the code presented in Fig. 15, the image file is first loaded (line 1) and then converted into a grey-scale intensity image (line 2). The threshold level is set in line 3, line 4 compares it with the image pixels and line 5 records the resulting black and white (or binary) image (a value of 0 for black and 1 for white). The rest of the algorithm draws the image output as shown on the right in Fig. 14. Once the detection of boundaries and objects is performed many different studies can be carried out, e.g. the area, the center of mass, and volumes of objects can be measured (Hutchings et al. 2007). These algorithms can also be used to obtain fluid properties such as surface tension or viscosity by analyzing the oscillations of droplets or the profile of continuous jets, (Bellizia et al. 2003 and Castrejón-García et al. 2011).

6 Conclusions

A brief review of the shadowgraph technique has been presented in this paper. A summary of the basic concepts to design and construct useful visualisation systems and some examples has been presented. Although the shadowgraph is a rather old technique, it is still in use in laboratories around the world in conjunction with image analysis techniques. Shadowgraph systems are nowadays used as a quantitative and precise technique, no longer as a pure visualisation mean, and survive among modern competitors such as particle image velocimetry and laser Doppler anemometry. Further advances in electronics, optics, photography, and sensors development can only strengthen the place of shadowgraphy as a very useful technique in the study of fluid dynamics and objects in motion.

Acknowledgments This work was partially supported by the UK EPSRC and industrial partners in the Innovation in Industrial Inkjet Technology project. JRCP acknowledges support from the Grupo Santander Academic Travel Fund of the University of Cambridge. The authors are grateful for the assistance of J. Waldmeyer during the recording of the CIJ stroboscopic images.

References

Adrian RJ (1991) Particle-Imaging Techniques for Experimental Fluid Mechanics. Annu Rev Fluid Mech 23:261–304
Bellizia G, Megaridis GM, Mc Nallan M, Wallace DV (2003) A capillary-jet instability method for measuring dynamic surface tension of liquid metals. Proc R Soc Lond A 459:2195–2214

Bruce CA (1976) Dependence of ink jet dynamics on fluid characteristics. IBM J Res Dev 20:258–270

Castrejón-García R, Milan J (1982) Application of the shadowgraph technique to the analysis of mechanical sprayers. IX international meeting on boilers and pressure vessels (AMIME), Cuernavaca, Mexico

Castrejón-García R, Sarmiento-Galán A, Castrejón-Pita JR, Castrejón-Pita AA (2003) The fractal dimension of an oil spray. Fractals 11:155–161

Castrejón-García R, Castrejon-Pita JR, Martin G, Hutchings IM (2011) The shadowgraph imaging technique and its modern application to fluid jets and drops 57(3):266–275

Castrejón-Pita JR, Martin G, Hoath S, Hutchings IM (2008) A simple large-scale droplet generator for studies of inkjet printing. Rev Sci Instrum 79:075108

Castrejón-Pita JR et al (2011) The dynamics of the impact and coalescence of droplets on a solid surface. Biomicrofluidics 5:014112

Curry SA, Portig H (1976) Scale model of an ink jet. IBM J Res Dev 21:10–20

Eggers J, Villermaux E (2008) Physics of liquid jets. Rep Prog Phys 71:036601

Highton S (2011) Virtual reality photography: creating panoramic and object images. Library of congress. ISBN: 978-0-165-34223-8. p 38

Hutchings IM, Martin GD, Hoath SD (2007) High speed imaging and analysis of jet and drop formation. J Imaging Sci Technol 51(5):438–444

Jones AR (1977) A review of drop size measurement — the application of techniques to dense fuel sprays. Prog Energy Combust Sci 3:225–234

Kalaali A, Lopez B, Attane P, Soucemarianadin A (2003) Breakup length of forced liquid jets. Phys Fluid 15:2469–2479

Lambrecht RW, Woodhouse C (2011) Way beyond monochrome, 2nd edn. Focal Press, Elsevier, pp 134–136

Le Moyne L, Freire V, Conde DQ (2008) Fractal dimension and scale entropy applications in a spray. Chaos Solitons Fractals 38(3):696–704

Martin GD, Hoath SD, Hutchings IM (2007) Inkjet printing - the physics of manipulating liquid jets and drops. J Phys Conf Ser 105:01200

O'Gorman L, Sammon MJ, Seul M (2008) Practical algorithms for image analysis. Cambridge University Press, Cambridge, UK

Savart F (1833) Memoire sur la constitution des veines liquides lancees par des orifices circulaires en mince paroi. Ann Chim 53:337–386

Strutt (Lord Rayleigh) JW (1896) The theory of sound, vol II. Macmillan and Co. ltd, New York

Formation of Coherent Structures in a Class of Realistic 3D Unsteady Flows

Michel F. M. Speetjens and Herman J. H. Clercx

Abstract The formation of coherent structures in three-dimensional (3D) unsteady laminar flows in a cylindrical cavity is reviewed. The discussion concentrates on two main topics: the role of symmetries and fluid inertia in the formation of coherent structures and the ramifications for the Lagrangian transport properties of passive tracers. We consider a number of time-periodic flows that each capture a basic dynamic state of 3D flows: 1D motion on closed trajectories, (quasi-)2D motion within (approximately) 2D subregions of the flow domain and truly 3D chaotic advection. It is shown that these states and their corresponding coherent structures are inextricably linked to symmetries (or absence thereof) in the flow. Symmetry breaking by fluid inertia and the resulting formation of intricate coherent structures and (local) onset of 3D chaos is demonstrated. Finally, first experimental analyses on coherent structures and the underlying role of symmetries are discussed.

M. F. M. Speetjens (✉)
Department of Mechanical Engineering, Energy Technology
and J.M. Burgers Center for Fluid Dynamics,
P.O. Box 513 5600, Eindhoven, The Netherlands
e-mail: m.f.m.speetjens@tue.nl

H. J. H. Clercx
Department of Applied Physics, Fluid Dynamics Laboratory
and J.M. Burgers Center for Fluid Dynamics,
P.O. Box 513 5600, Eindhoven, The Netherlands
e-mail: h.j.h.clercx@tue.nl

J. Klapp et al. (eds.), *Fluid Dynamics in Physics, Engineering and Environmental Applications*, Environmental Science and Engineering,
DOI: 10.1007/978-3-642-27723-8_9, © Springer-Verlag Berlin Heidelberg 2013

1 Introduction

Lagrangian methods have proven their worth in transport studies on deterministic flows. Key to these methods is the notion that continuity organises fluid trajectories into coherent structures that geometrically determine the advective transport of material. Typical examples are the KAM islands and (un)stable manifolds of hyperbolic periodic points that constitute the flow topologies of 2D time-periodic flows (Ottino 1989). Most studies up to now focus on these 2D time-periodic (open and closed) flows and painted a fairly complete picture of its transport properties. Lagrangian transport in 3D flows, on the other hand, has received considerably less attention and remains a fairly unexplored field to date (Wiggins 2010). The present review demonstrates and discusses the rich dynamics of 3D Lagrangian transport by way of a representative and realistic class of 3D unsteady flows.

The following exposition concentrates on two key factors in the formation of coherent structures in 3D unsteady closed flows: symmetries and fluid inertia. We will address both analytical and numerical studies of coherent structures in 3D laminar flows and present first experimental evidence on their existence in real flows. To this end we consider a simple flow configuration: the time-periodic flow inside a cylinder $[r, \theta, z] = [0, 1] \times [0, 2\pi] \times [-1, 1]$ (Malyuga et al. 2002; Speetjens et al. 2004). The flow is driven by prescribed forcing protocols: a time-periodic repetition of a sequence of p piecewise steady translations of the bottom or top walls with unit velocity $U = 1$ and relative wall displacement $D = D_{wall}/R$ (with D_{wall} and R the in-plane wall displacement and cylinder radius, respectively). These forcing steps are of equal duration $T_{step} = T/p$, with T the period time of one cycle. A schematic of the configuration is shown in Fig. 1a and the specific forcing protocols are introduced below.

Highly-viscous flow conditions are assumed such that the viscous time scale $T_v = R^2/v$, with v the kinematic viscosity of the fluid, is much smaller than T_{step}, the duration of one forcing step (i.e. $T_v/T_{step} \ll 1$). As a consequence the transients during switching between forcing steps are negligible and the forcing steps then become reorientations of the base flow due to steady translation of the bottom wall in the x-direction. This base flow is governed by the non-dimensional steady Navier-Stokes and continuity equations,

$$Re\, \mathbf{u} \cdot \nabla \mathbf{u} = -\nabla p + \nabla^2 \mathbf{u}, \quad \nabla \cdot \mathbf{u} = 0, \tag{1}$$

with $Re = UR/v$ the well-known Reynolds number.

The motion of passive tracers is governed by the kinematic equation,

$$\frac{d\mathbf{x}(t)}{dt} = \mathbf{u}(t), \quad \mathbf{x}(0) = \mathbf{x}_0, \tag{2}$$

with initial and current passive tracer position $\mathbf{x}(0)$ and $\mathbf{x}(t)$, respectively. This equation has the formal solution $\mathbf{x}(t) = F_t(\mathbf{x}_0)$, which defines the continuous Lagrangian flow from the initial to the current tracer position along the trajectory

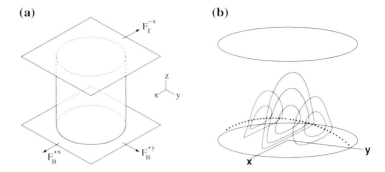

Fig. 1 Non-dimensional flow configuration. Reproduced from Speetjens et al. (2004). **a** Flow domain and forcing. **b** Streamline portrait base flow

$X(t; \mathbf{x}_0) = \{F_\xi(\mathbf{x}_0), 0 \leq \xi \leq t\}$. These trajectories for steady flows correspond with streamlines; Fig. 1b e.g. gives the streamline portrait for the base flow in the non-inertial limit $Re = 0$. Time-periodic flows admit representation by discrete maps, typically denoted "Poincaré maps" (Ottino 1989), defined as

$$\mathbf{x}_{k+1} = \Phi(\mathbf{x}_k), \quad \mathbf{x}_k = \mathbf{x}(kT), \tag{3}$$

where $\mathbf{x}_k$ is the position of the (passive) tracer after k periods of the time-periodic forcing protocol. The 3D dynamics will in the following be considered mainly in terms of such maps.

The first part of this review concerns three forcing protocols, denoted "protocols $\mathcal{A}$, $\mathcal{B}$, and $\mathcal{C}$" hereafter. They are respectively defined as

$$\Phi_\mathcal{A} = \mathbf{F}_B^{+y}\mathbf{F}_B^{+x}, \quad \Phi_\mathcal{B} = \mathbf{F}_T^{-x}\mathbf{F}_B^{+x}, \quad \Phi_\mathcal{C} = \mathbf{F}_B^{+y}\mathbf{F}_T^{-x}\mathbf{F}_B^{+x}, \tag{4}$$

where the subscripts in the forcing steps refer to top (T) and bottom (B) endwall and the superscripts indicate the translation direction (Fig. 1a). All forcing steps are transformations of the mapping $\mathbf{F}_B^{+x}$ corresponding with the base flow for interval $0 \leq t \leq T_{step}$ according to

$$\mathbf{F}_B^{+y} = \mathscr{F}_{\pi/2}(\mathbf{F}_B^{+x}), \quad \mathbf{F}_T^{-x} = S_z\mathscr{F}_\pi(\mathbf{F}_B^{+x}), \tag{5}$$

with $\mathscr{F}_\alpha : \theta \to \theta + \alpha$ and $S_z : (x, y, z) \to (x, y, -z)$. These forcing protocols each exhibit particular characteristics of 3D tracer dynamics, as will be demonstrated in Sects. 2 and 3.

Important to note is that protocols $\mathcal{A}-\mathcal{C}$ define so-called open forcing protocols, that is, have non-zero net wall displacement during each forcing cycle. This renders them, though constituting excellent case studies for demonstrating and investigation basic 3D dynamics, ill-suited for in-depth experimental studies, since finite size of endwalls in experimental set-ups limits the number of forcing cycles. Experimental studies therefore concern closed forcing protocols, i.e. with zero net wall displacement per forcing cycle, of the form

$$\Phi = \Phi_p \Phi_{p-1} \cdots \Phi_1, \quad \Phi_n = R^{n-1} \mathbf{F}_B^{+\times} R^{1-n}, \tag{6}$$

with $1 \leq n \leq p$ and $R : (r, \theta, z) \rightarrow (r, \theta + \theta_{step}, z)$ the reorientation operator. Step-wise reorientation angle $\theta_{step} = 2\pi/p$ and equal wall displacement D yields zero net wall motion. This ensures that these closed protocols can be repeated *ad infinitum* with a finite endwall, which facilitates long-term measurements. Section 4 discusses laboratory experiments for the closed three-step forcing protocol ("protocol $\mathcal{T}$") with the bottom wall translating along an equilateral triangle ($p = 3$ and $\theta_{step} = 2\pi/3$). It must be stressed that protocol $\mathcal{T}$—as well as any other closed protocol according to (6)—exhibits essentially the same dynamics as its open counterparts and thus is (qualitatively) representative for protocols $\mathcal{A}$–$\mathcal{C}$. This has been examined theoretically and numerically in Pouransari et al. (2010).

Coherent structures in the flow topology are spatial entities in the web of Lagrangian fluid trajectories that exhibit a certain invariance to the mapping Φ. Based on the classifications introduced by Guckenheimer and Holmes (1983) and Feingold et al. (1988) four kinds of such invariant structures can be distinguished in 3D time-periodic systems. They are defined by

$$\mathcal{P}^{(k)} = \Phi^k(\mathcal{P}^{(k)}), \quad \mathcal{L}^{(k)} = \Phi^k(\mathcal{L}^{(k)}), \quad \mathcal{C}^{(k)} = \Phi^k(\mathcal{C}^{(k)}), \quad \mathcal{S}^{(k)} = \Phi^k(\mathcal{S}^{(k)}), \tag{7}$$

constituting periodic points ($\mathcal{P}^{(k)}$), periodic lines ($\mathcal{L}^{(k)}$), invariant curves ($\mathcal{C}^{(k)}$) and invariant surfaces ($\mathcal{S}^{(k)}$) of order k (i.e. invariant with respect to k forcing cycles). Periodic lines consist of periodic points, and as a consequence each constituent point is invariant. Invariant curves and surfaces, however, are only invariant as entire entity. Any continuous mapping of a convex space[1] onto itself has, according to Brouwer's fixed-point theorem, at least one fixed (or periodic) point, see Speetjens et al. (2004). As a consequence, periodic points and associated coherent structures are the most fundamental building blocks of 3D flow topologies. Periodic points fall within one of the following categories: node-type and focus-type periodic points. Periodic lines are elliptic or hyperbolic or consist of elliptic and hyperbolic segments (Malyuga et al. 2002). Isolated periodic points and hyperbolic (segments of) periodic lines imply pairs of stable (W^s) and unstable (W^u) manifolds. They arise as surface-curve pairs ($W_{2D}^{s,u}, W_{1D}^{u,s}$) for periodic points and as surface-surface pairs ($W_{2D}^{s,u}, W_{2D}^{u,s}$) for periodic lines. Elliptic (segments of) periodic lines form the centre of concentric tubes. The 1D manifolds of isolated periodic points define invariant curves $\mathcal{C}^{(k)}$; 2D manifolds and elliptic tubes define invariant surfaces $\mathcal{S}^{(k)}$. Period-1 structures are the most important for the flow topology, as they determine the global organisation. We restrict the discussion below to period-1 structures.

---

[1] A space is termed convex if for any pair of points within the space, any point on the line joining them is also within the space.

2 Formation of Coherent Structures: The Role of Symmetries

Flows often accommodate symmetries due to the geometry of the flow domain and the mathematical structure of the governing conservation laws. Such symmetries, if present, play a central role in the formation of coherent structures. This is well-known for 2D time-periodic flows where this typically results in symmetry groups of coherent structures or physical separation of flow regions by symmetry axes (Feingold et al. 1989; Ottino et al. 1994; Meleshko and Peters 1996). The formation of coherent structures in 3D time-periodic flows may suppress truly 3D dynamics, see Feingold et al. (1988), Mezić and Wiggins (1994), Haller and Mezić (1998), Malyuga et al. (2002) and Speetjens et al. (2004). Such manifestations of symmetries are demonstrated below for the time-periodic cylinder flow subject to open forcing protocols $\mathscr{A}$, $\mathscr{B}$, and $\mathscr{C}$ in the non-inertial limit $Re = 0$.

Forcing by the rigid bottom wall in x-direction in the non-inertial limit $Re = 0$ yields a base flow $\mathbf{u}$ with symmetries

$$S_x u_x = u_x, \quad S_x u_{y,z} = -u_{y,z}, \quad S_y u_{x,z} = u_{x,z}, \quad S_y u_y = -u_y, \tag{8}$$

where $S_x : (x, y, z) \rightarrow (-x, y, z)$ and $S_y : (x, y, z) \rightarrow (x, -y, z)$, due to the linearity of the momentum Eq. (1) (Shankar 1997). This causes the base flow to adopt the simple form

$$u_r(\mathbf{x}) = \mathsf{u}_r(r, z) \cos \theta, \quad u_\theta(\mathbf{x}) = \mathsf{u}_\theta(r, z) \sin \theta, \quad u_z(\mathbf{x}) = \mathsf{u}_z(r, z) \cos \theta. \tag{9}$$

This implies closed streamlines in the base flow $\mathbf{F}_B^{+x}$ that are symmetric about the planes $x = 0$ and $y = 0$ (Fig. 1b) and two constants of motion (COM) of the generic form

$$F_1(\mathbf{x}) = f_1(r, z), \quad F_2(\mathbf{x}) = f_2(r, z) \sin \theta, \tag{10}$$

can be identified which both satisfy $dF_i/dt = \mathbf{u} \cdot \nabla F_i = 0$, with $i = \{1, 2\}$.

Essential is that the above properties are not specific to forcing by a rigid bottom wall; *any* bottom-wall boundary condition meeting (8) imparts those symmetries onto the base flow. This has the important implication that base flows with these conditions—and any time-periodic flow derived from them—constitute a family of flows with identical symmetries and equivalent flow topologies (Speetjens 2001; Znaien et al. 2012). Three kinds of bottom-wall conditions are relevant in the present context: (i) rigid bottom wall with uniform non-dimensional velocity $(u_x, u_y, u_z) = (1, 0, 0)$; (ii) smoothed non-dimensional bottom-wall velocity $(u_x, u_y, u_z) = ((r^2 - 1)^2, 0, 0)$; (iii) experimental forcing velocity with all components $u_{x,y,z}$ non-zero due to weak fluid exchange with an ambient fluid via a small gap between still cylinder and moving bottom wall. The smoothed conditions eliminate discontinuities that compromise the convergence properties of a spectral flow solver employed for simulations of inertial cases $Re > 0$ (Speetjens et al. 2006a, b; Pouransari et al. 2010); the gap prevents vibrations induced by

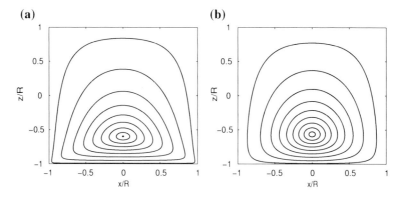

Fig. 2 Topological equivalence for flows subject to bottom-wall boundary conditions with symmetries (8) demonstrated for simulated streamline patterns in the plane $y = 0$. Reproduced from Znaien et al. (2012). **a** Rigid-wall conditions. **b** Smooth-wall conditions

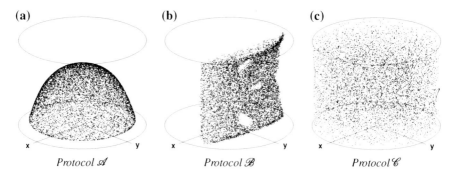

Fig. 3 Poincaré sections of single tracers for the time-periodic forcing protocols (rigid bottom-wall conditions). Reproduced from Speetjens et al. (2004). **a** Protocol $\mathscr{A}$. **b** Protocol $\mathscr{B}$. **c** Protocol $\mathscr{C}$

mechanical friction (Znaien et al. 2012). Figure 2 demonstrates the topological equivalence for the base flow by the streamline pattern in the plane $y = 0$ for rigid-wall and smooth bottom-wall conditions.

Properties (10) have essential consequences for the flow topologies of the forcing protocols (4). In Fig. 3 we have shown Poincaré sections of a single passive tracer for several time-periodic forcing protocols. Tracers released in protocol $\mathscr{A}$ are confined to invariant spheroidal surfaces (panel a) on which they perform effectively 2D (chaotic) dynamics. This occurrence of chaos on a sub-manifold of co-dimension one is an essentially 3D phenomenon. The reader may consult the studies by Gómez and Meiss (2002), Meier et al. (2007), Mullowney et al. (2008a, b), and Sturman et al. (2008) for dynamically similar systems. Protocol $\mathscr{B}$ restricts tracers to a quasi-2D (chaotic) motion within thin shells parallel to the yz-plane (panel b). Truly 3D (chaotic) dynamics covering the entire flow domain occurs only for protocol $\mathscr{C}$ (panel c). It clearly indicates dramatic differences in passive tracer dynamics due to the presence of geometric restrictions in protocols $\mathscr{A}$ and $\mathscr{B}$, similar to the KAM islands and cantori of 2D flows. This is a direct consequence of symmetries, as is demonstrated below.

2.1 Protocol $\mathscr{A}$

The restriction of tracers to invariant surfaces is an immediate result of the presence of "hidden" axisymmetry in the base flow. The COM F_1 according to (10) namely is invariant to the continuous transformation $\mathscr{F}_\alpha$, with $0 \leq \alpha \leq 2\pi$, i.e. $\mathscr{F}_\alpha(F_1) = F_1$. Hence, F_1 and the underlying continuous axisymmetry is retained by protocol $\mathscr{A}$. This means that passive tracers are entrapped on its level sets, which are defined by the surfaces of revolution of the trajectories $dr/dz = u_r(r,z)/u_z(r,z) = g(r,z)$ in the rz-plane (Speetjens et al. 2006a). Figure 4a shows a few members of the infinite family of these concentric spheroidal surfaces.

The symmetries of the base flow, induced by $\mathbf{F}_B^{+x}$, through $\Phi_\mathscr{A}$ following (4) and transformations (5) translate into the discrete symmetries

$$\Phi_\mathscr{A} = S_1 \Phi_\mathscr{A}^{-1} S_1, \quad \Phi_\mathscr{A} = \tilde{S} \Phi_\mathscr{A} \tilde{S}, \tag{11}$$

with $S_1 : (x,y,z) \rightarrow (-y,-x,z)$, $\tilde{S} = S_2 \mathbf{F}_B^{+x}$ and $S_2 : (x,y,z) \rightarrow (y,x,z)$. The time-reversal reflectional symmetry S_1 has the fundamental consequence that the flow must possess at least one period-1 line $\mathscr{L}^{(1)}$, which is located within the symmetry plane $I_1 = S_1(I_1)$ (plane $y = -x$), see Speetjens et al. (2004). Coexistence of S_1 with $\tilde{S}$ dictates that $\mathscr{L}^{(1)}$ be invariant to both discrete symmetries, i.e.

$$\mathscr{L}^{(1)} = S_1(\mathscr{L}^{(1)}) = \tilde{S}(\mathscr{L}^{(1)}) = S_1\tilde{S}(\mathscr{L}^{(1)}) = \tilde{S}S_1(\mathscr{L}^{(1)}). \tag{12}$$

This means that the symmetries essentially "shape" the period-1 line and its associated structures. The curve in Fig. 4a outlines the period-1 line for $D = 5$. The heavy and normal parts of the periodic line indicate elliptic and hyperbolic segments, respectively.

The intra-surface topologies within the invariant spheroids are organised by the periodic points defined by the intersection of the periodic lines with the spheroids. The segmentation of the periodic lines into elliptic and hyperbolic parts results in multiple kinds of intra-surface topologies. Figure 4b (left) shows an invariant spheroid that intersects with the hyperbolic segment of the above period-1 line, resulting in two hyperbolic period-1 points (indicated by the dots), and with elliptic segments of two period-2 lines. The latter results in two pairs of period-2 islands embedded in regions with chaotic intra-surface dynamics. Figure 4b (right) shows an invariant spheroid that intersects with hyperbolic segments of said periodic lines, leading to fully-chaotic intra-surface dynamics.

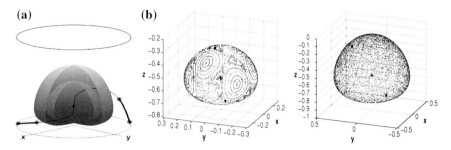

Fig. 4 Coherent structures and Hamiltonian intra-surface dynamics for protocol $\mathscr{A}$ (smooth bottom-wall conditions). Heavy/normal sections of the period-1 line indicate elliptic/hyperbolic segments. Intra-surface dynamics are visualised by Poincaré sections of a ring of tracers. Reproduced from Speetjens et al. (2006a). **a** Spheroids and period-1 line. **b** Intra-surface dynamics

2.2 Protocol $\mathscr{B}$

This forcing protocol consists of an alternating in-plane bottom and top plate translation. An immediate consequence of including both top and bottom walls in $\Phi_{\mathscr{B}}$ is the vanishing of COM F_1. Thus here passive tracers are, in contrast to protocol $\mathscr{A}$, no longer restricted to invariant surfaces. Protocol $\mathscr{B}$ nonetheless accommodates discrete symmetries, reading

$$\Phi_{\mathscr{B}} = \bar{S}\Phi_{\mathscr{B}}^{-1}\bar{S}, \quad \Phi_{\mathscr{B}} = \bar{S}'\Phi_{\mathscr{B}}^{-1}\bar{S}', \quad \Phi_{\mathscr{B}} = S_y\Phi_{\mathscr{B}}S_y, \tag{13}$$

with $\bar{S} = S_x\mathbf{F}_B^{+x}$, $\bar{S}' = S_z\bar{S}S_z$ and $S_{x,y,z}$ as before. Note that symmetries $\bar{S}$ and $\bar{S}$—and corresponding symmetry planes $\bar{I} = \bar{S}(\bar{I})$ and $\bar{I}' = S_z(\bar{I})$—are conjugate in that they relate via S_z; the latter in fact is a time-reversal symmetry "hidden" in $\bar{S}$ and $\bar{S}'$: $\Phi_{\mathscr{B}} = S_z\Phi_{\mathscr{B}}^{-1}S_z$. Time-reversal symmetry again implies at least one period-1 line $\mathscr{L}^{(1)}$ within the corresponding symmetry plane. However, coexistence of two such symmetries imposes an additional restriction compared to protocol $\mathscr{A}$ in that period-1 lines identify with the intersections $\mathscr{L}^{(1)} \in \bar{I} \cap \bar{I}'$ of the conjugate symmetry planes so as to simultaneously belong to both, see Speetjens et al. (2004). The symmetry of $\bar{I}$ and $\bar{I}'$ about $z = 0$ implies organisation of period-1 lines into the group

$$\mathscr{M}_B = \left[\{\mathscr{L}_{z,1}^{(1)}, \mathscr{L}_{z,2}^{(1)}, , \mathscr{L}_{z,n}^{(1)}\}, \{\mathscr{L}_1^{(1)}, S_z(\mathscr{L}_1^{(1)})\}, \{\mathscr{L}_2^{(1)}, S_z(\mathscr{L}_2^{(1)})\}, , \{\mathscr{L}_m^{(1)}, S_z(\mathscr{L}_m^{(1)})\} \right],$$

with $\mathscr{L}_{z,i}^{(1)} = S_z(\mathscr{L}_{z,i}^{(1)})$ $(i \in [1, n])$ period-1 lines within $z = 0$ and $\{\mathscr{L}_i^{(1)}, S_z(\mathscr{L}_i^{(1)})\}$ $(i \in [1, m])$ symmetry pairs of period-1 lines about $z = 0$. Note that all period-1 lines possess the self-symmetry $\mathscr{L}_{z,i}^{(1)} = S_y(\mathscr{L}_{z,i}^{(1)})$ and $\mathscr{L}_i^{(1)} = S_y(\mathscr{L}_i^{(1)})$ about the plane $y = 0$. Thus here a symmetry group of period-1 lines forms. Note that for protocol $\mathscr{A}$ a single period-1 line consisting of symmetric segments forms.

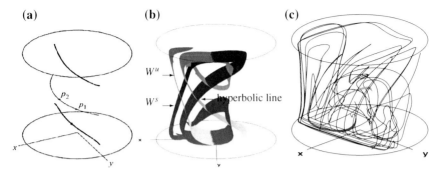

Fig. 5 Coherent structures for protocols $\mathcal{B}$ and $\mathcal{C}$ (rigid bottom-wall conditions). Reproduced from Speetjens (2001). **a** Period-1 lines protocol $\mathcal{B}$. **b** Manifolds segment $p_1 - p_2$. **c** Manifold pairs protocol $\mathcal{C}$

Mass conservation implies at least one period-1 line that, for given symmetries, must sit in the plane $z = 0$, which means that $\mathcal{L}_{z,1}^{(1)}$ always exists. Figure 5a shows a typical symmetry group $\mathcal{M}_B = \{\mathcal{L}_{z,1}^{(1)}, \mathcal{L}_1^{(1)}, S_z(\mathcal{L}_1^{(1)})\}$ of period-1 lines, with $\mathcal{L}_{z,1}^{(1)}$ fully hyperbolic and the conjugate pair $\{\mathcal{L}_1^{(1)}, S_z(\mathcal{L}_1^{(1)})\}$ fully elliptic. The stable/ unstable manifolds $W_{2D}^{s/u}$ of line segment $p_1 - p_2$ of $\mathcal{L}_{z,1}^{(1)}$, relating via $W_{2D}^u = S_z(W_{2D}^s)$, are shown in Fig. 5b and envelop the elliptic region comprising concentric tubes (the 3D counterpart to KAM islands) centred on the elliptic lines (not shown). The manifolds extend primarily in the direction normal to $\mathcal{L}_{z,1}^{(1)}$ (i.e. parallel to the yz-plane) and exhibit only marginal y-wise variation. Moreover, they exhibit transversal interaction, which is a "fingerprint" of chaotic dynamics in 2D systems (Ottino 1989). This manifold behaviour causes the quasi-2D chaotic tracer motion within a thin layer normal to the period-1 lines, as illustrated in Fig. 3b. Tracers released near the elliptic segments exhibit similar behaviour by remaining confined to thin "slices" of elliptic tubes (not shown). Hence, passive tracer dynamics within each layer basically is of a Hamiltonian nature and thus intimately relates to the intra-surface dynamics of protocol $\mathcal{A}$. Primary difference with the latter is that tracers are not strictly confined to an invariant surface. Whether these less restrictive conditions may be of any consequence is an open question. Protocols $\mathcal{A}$ and $\mathcal{B}$ thus reveal that time-reversal symmetries, through their link with periodic lines, imply effectively (quasi-)2D dynamics. This suppression of truly 3D dynamics is an essentially 3D manifestation of this kind of symmetries.

2.3 Protocol $\mathcal{B}$

Inclusion of a third forcing step results in a flow that is devoid of global symmetries. In particular the absence of time-reversal symmetries is of fundamental consequence in that periodic lines must thus no longer be present. However, the current flow must, according to Brouwer's fixed-point theorem, accommodate at least one isolated period-1 point. Two node-type period-1 points indeed exist, and have associated manifold pairs (W_{2D}^u, W_{1D}^s) with essentially 3D foliations that densely fill the entire flow domain (Fig. 5c). Here the stable and unstable manifolds, in contrast with those associated with periodic lines, are not related via a time-reversal symmetry. This asymmetry in time results in essentially 3D transport and is a key element in the truly 3D dynamics demonstrated in Fig. 3c. The intrinsic hyperbolicity of the isolated period-1 points strongly suggests that this tracer motion, irrespective of manifold interactions, must always be chaotic (Speetjens et al. 2004). This underscores the fundamental difference between periodic lines and isolated periodic points. The former are the 3D counterpart to periodic points in 2D systems and the associated symmetry in time gives rise to (quasi-)2D dynamics. The latter, on the other hand, are truly 3D entities and prime indicators of an efficient 3D mixing flow, see Feingold et al. (1988) and Malyuga et al. (2002).

3 Formation of Coherent Structures: The Role of Fluid Inertia

Fluid inertia ($Re > 0$) is always present in realistic flows and may have a strong impact upon the transport properties. Inertia introduces a secondary circulation to the base flow $\mathbf{F}_B^{+x}$ transverse to its primary circulation parallel to the wall motion (Fig. 6a). This breaks the symmetry about $x, y = 0$, causing the closed streamlines of $Re = 0$ to become non-closed and wrapped around concentric invariant tori. These tori, in turn, undergo progressive disintegration into tori with winding numbers[2] larger than one with increasing Re. This manifests itself in the characteristic island chains and chaotic seas shown in Fig. 6b, c. This cross-sectional behaviour of perturbed 3D steady flows is similar to the Hamiltonian response scenario for 2D time-periodic systems subject to perturbations (Ottino 1989).

Studies on response scenarios of invariant surfaces to perturbations are almost exclusively restricted to tori. This may be largely attributed to the prominent role of Hamiltonian mechanics in (2D) mixing studies. However, the classification theorem for closed surfaces states that any orientable closed surface (the category including level sets of a COM in bounded flows) is topologically equivalent to a sphere or a connected sum of tori (Alexandroff 1961). This puts forth invariant spheroids, besides invariant tori, as second fundamental form of invariant surfaces relevant in the present context. Note that tori and spheres are in fact the only two kinds of invariant surfaces that may occur in Euler flows (Mezić and Wiggins 1994). The cylinder flow offers a way to investigate the response of invariant

---

[2] The winding number W represents the number of revolutions around the axis of rotation required for completing a full loop on closed trajectories. The closed streamlines in Fig. 1b have unit winding number.

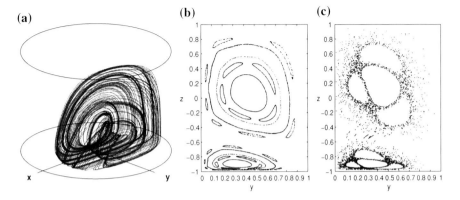

Fig. 6 Inertial effects ($Re > 0$) in the base flow $\mathbf{F}_B^{+x}$ (smooth bottom-wall conditions). The cross-sections correspond with plane $x = 0.12$. Reproduced from Speetjens (2001). **a** $Re = 100$. **b** $Re = 50$. **c** $Re = 100$

spheroids to (inertial) perturbations under realistic conditions. This may contribute to a more complete picture of the fate of invariant surfaces subject to (inertial) perturbations.

The effect of fluid inertia upon invariant spheroids is demonstrated for protocol $\mathscr{A}$. This case is representative for generic forcing protocols with such an unperturbed topology (Pouransari et al. 2010). Inertia breaks both the time-reversal reflectional symmetry S_1 in (11) and the continuous axisymmetry due to COM F_1; only symmetry $\tilde{S}$ is preserved for $Re > 0$ (Speetjens et al. 2006a). This causes the period-1 line $\mathscr{L}^{(1)}$, shown in Fig. 4a, to give way to a focus-type isolated period-1 point with a (W_{2D}^s, W_{1D}^u) manifold pair that for sufficiently high Re completely destroys the invariant spheroids and, in consequence, yields 3D chaotic motion of passive tracers. Figure 7 demonstrates this for $Re = 100$. However, even minute departures from the non-inertial limit (with $Re \ll 1$) may change the flow topology drastically, which is investigated below.

The secondary circulation causes progressive drifting of tracers transverse to the invariant spheroids that grows stronger with increasing Re. This is demonstrated in Fig. 8 by means of the rz-projection of a Poincaré section of a single tracer tracked for approximately 10,000 forcing periods. The drifting tracers remain confined within thin shells centred upon the invariant spheroids (provided $Re \lesssim 0.1$) for time spans of $\mathcal{O}(20,000)$ periods. Thus invariant spheroids survive in an approximate way as so-called "adiabatic shells" (Speetjens et al. 2006a). However, this survival occurs only for regions with chaotic intra-surface dynamics. Quite remarkably, here chaos in fact promotes the persistence of (partial) transport barriers or, in perhaps more intriguing words, 2D chaos suppresses the onset of 3D chaos. The formation of a complete adiabatic shell, as the one shown in Fig. 8b, thus signifies an underlying invariant spheroid with fully-chaotic tracer motion. Chaotic and non-chaotic regions of invariant spheroids (e.g. shown in Fig. 4b) transform into incomplete adiabatic shells and elliptic tubes,

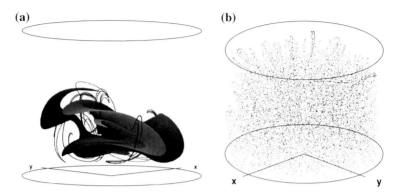

Fig. 7 Coherent structures for protocol $\mathscr{A}$ at $Re = 100$ (smooth bottom-wall conditions). Reproduced from Speetjens et al. (2006a). **a** Manifold pair (W_{2D}^s, W_{1D}^u). **b** Poincaré section of a single tracer

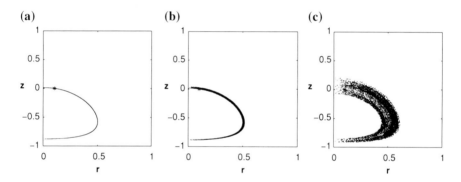

Fig. 8 Inertia-induced drifting of tracers transverse to invariant spheroids visualised by the Poincaré section of a single tracer (10,000 periods; smooth bottom-wall conditions). The *star* represents the initial tracer position. Reproduced from Speetjens et al. (2006a). **a** $Re = 0$. **b** $Re = 0.1$. **c** $Re = 1$

respectively, that merge into intricate adiabatic structures by a mechanism termed "resonance-induced merger" (RIM). This is illustrated in Fig. 9 for $Re = 0.1$. Shown is an adiabatic structure formed by RIM, comprising an inner and outer adiabatic shell, connected via an elliptic tube on each elliptic segment of the period-1 line. Both tubes, similar as the underlying elliptic segments of the period-1 line of the non-inertial limit, form a symmetry pair. RIM thus results in a family of nested closed adiabatic structures that are topologically equivalent to tori. Refer to Speetjens et al. (2006a, b) for a more detailed treatment of RIM.

The dynamics of the (perturbed) invariant spheroids remains largely an open problem. Fully-chaotic spheroids survive weak inertia as complete adiabatic shells and constitute transport barriers akin to those of the non-inertial limit. Non-chaotic regions on (sub-)families of invariant spheroids have fundamental consequences by causing the formation of intricate adiabatic structures through RIM. Its occurrence

(a)   **(b)**

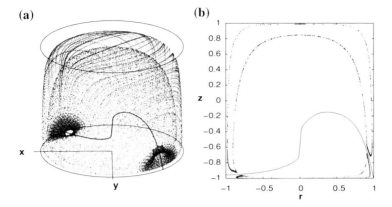

Fig. 9 Formation of adiabatic structures by resonance-induced merger (RIM) of adiabatic shells and elliptic tubes emanating from the elliptic segments of the period-1 line (*curve*) for small departures from the non-inertial limit ($Re = 0.1$; smooth bottom-wall conditions). Reproduced from Speetjens et al. (2006a). **a** Perspective view. **b** Slice centred upon symmetry plane $y = -x$

for a wide range of forcing protocols strongly suggests that RIM is a universal phenomenon and part of an essentially 3D route to chaos. Refer to Speetjens et al. (2006a, b) and Pouransari et al. (2010) for a more extensive discussion.

4 Experimental Analysis of Periodic Lines

The above theoretical and numerical analyses exposed symmetries and periodic lines as two key determinants for the dynamics of (perturbed) 3D unsteady flows. Recent studies by Znaien et al. (2012) have shifted the focus to laboratory experiments so as to investigate their existence and properties in real flows. This experimental analysis has been carried out by 3D particle tracking-velocimetry (3D-PTV) using the 3D-PTV algorithm developed at ETH, Zürich, Switzerland (Luethi et al. 2005). Exploratory measurements in the cylinder flow demonstrated the great potential of this technique for 3D quantitative experimental studies on Lagrangian fluid trajectories and corresponding coherent structures (Speetjens et al. 2004).

This constitutes a first important step towards experimental validation of the 3D transport phenomena, such as RIM, observed in the cylinder flow.

The experimental study is performed for protocol $\mathscr{T}$ introduced in Sect. 1 on grounds of their suitability for long-term measurements enabled by the zero net wall motion of the driving wall. These alternative flows are nonetheless representative for protocols $\mathscr{A}$ and $\mathscr{B}$ investigated in Sect. 2 and 3 in that they exhibit essentially the same dynamics. The non-inertial limit $Re = 0$ accommodates a time-reversal reflectional symmetry

$$\Phi = S_\beta \Phi^{-1} S_\beta, \quad S_\beta : (r, \theta, z) \rightarrow (r, 2\beta - \theta, z), \tag{14}$$

with $\beta = (\pi - \theta_{step})/2$. This implies, similar to protocols $\mathscr{A}$ and $\mathscr{B}$, a period-1 line $\mathscr{L}^{(1)}$ in the symmetry plane $I_\beta = S_\beta I_\beta = \{\mathbf{x} \in \mathscr{D}|_{\theta=\beta}\}$. The latter coincides with $\beta = \pi/6$ for protocol $\mathscr{T}$ (Pouransari et al. 2010). Moreover, both closed protocols involve only forcing by the bottom wall and thus also posses the invariant spheroids of protocol $\mathscr{A}$ shown in Fig. 4a.

The experimental set-up imposes different boundary conditions on the bottom wall due to a small gap between still cylinder and moving wall. However, this is inconsequential for the present discussion in that symmetries (8) are retained. Hence, the experimental flow thus belongs to the same family as those deriving from base flows subject to rigid and smooth bottom-wall conditions (Sect. 2). Symmetry (14) of the closed protocols thus holds for *any* of these bottom-wall conditions; differences are entirely quantitative in that the shape of the period-1 line $\mathscr{L}^{(1)}$ may vary slightly between these conditions.

Figure 10 demonstrates the topological equivalence of the period-1 lines for numerical simulations of protocol $\mathscr{T}$ using the rigid-wall and smooth boundary conditions on the bottom wall for increasing wall displacement D. This directly reveals a close correlation between both sets of period-1 lines. Moreover, this exposes three generic topological properties that are essentially independent on the particular boundary conditions: (i) progressive convolution of the period-1 line with increasing D; (ii) a common attachment point $(r/R, z/R) = (0, 1)$ at the top wall; (iii) a common interior stagnation point $(r/R, z/R) = (0, z_0)$ on the cylinder axis, with $z_0 \approx -0.56$ (Znaien et al. 2012). The interior stagnation point acts as a "pivot" around which period-1 lines curl up with increasing D_{wall}. The period-1 lines for rigid-wall and smooth conditions approximately coincide for displacements $D_{smooth}/D_{rigid} \approx \bar{U}_{smooth}/\bar{U}_{rigid} = 8/15$ the ratio of the mean wall velocities of the base flow ($\bar{U} = \int \int u_x dx dy / \pi$) on the bottom wall (Speetjens 2001). Hence, period-1 lines of both cases, besides the qualitative equivalence, undergo quantitatively comparable progressions yet at different wall displacements D.

The experimental flow due to forcing protocol $\mathscr{T}$ belongs to the same family as the flows simulated above and thus must exhibit topologically equivalent behaviour. This implies formation of a period-1 line in the symmetry plane I_β according to the above scenario. Figure 11 shows the period-1 lines $\mathscr{L}^{(1)}$ (symbols) within the symmetry plane I_β for displacements $D_{wall} = [1.4, 2.9, 4.3]$ obtained from direct 3D-PTV experiments via the procedure outlined above. Shape and dependence upon displacement D are entirely consistent with the generic progression shown in Fig. 10. Note in particular the intersection of the period-1 lines at the common interior stagnation point on the cylinder axis, which again sits at $z_0 \approx -0.56$. Comparison with the simulations for the rigid-wall conditions reveals that the employed displacement D_{sim} roughly correlates with D as $D_{sim}/D \approx 2/3$. The curves in Fig. 11 indicate the simulated period-1 lines using this rule of thumb. This exposes a close quantitative agreement, signifying a formation process that, save the particular displacement D, is relatively insensitive to the

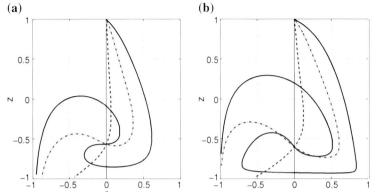

Fig. 10 Generic progression of period-1 lines within the symmetry plane $I_\beta = \pi/6$ versus displacement D for protocol $\mathcal{T}$. *Blue/red/black* corresponds with $D = 1, 6, 14$. Reproduced from Znaien et al. (2012). **a** Smooth bottom-wall conditions. **b** Rigid bottom-wall conditions

Fig. 11 Experimental period-1 lines for protocol $\mathcal{T}$ within the symmetry plane $I_\beta = \pi/6$ versus wall displacement: $D = 1.4$ (*blue* +), $D = 2.9$ (*red* ∘) and $D = 4.3$ (*black* ∗). *Continuous curves* are period-1 lines obtained by numerical simulations for the rigid-wall boundary conditions at $D = 1.0$, $D = 1.9$ and $D = 2.9$. Reproduced from Znaien et al. (2012)

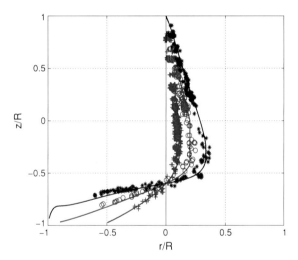

particular boundary conditions. Moreover, this further demonstrates the essential topological equivalence of period-1 lines in the family of flows under investigation here. Recall in this respect that a similar scaling rule connects the period-1 lines of the rigid-wall and smooth boundary conditions. Hence, comparison of experimental period-1 lines with those simulated for the smooth conditions yields identical qualitative agreement and equal quantitative resemblance.

The range of the experimental study has been extended by a hybrid particle-tracking method for numerical simulation of Lagrangian tracer dynamics using an Eulerian flow field constructed from 3D-PTV data. This method is based on the approach proposed by Voth et al. (2002) and is explained in more detail in Znaien et al. (2012). The hybrid method admits investigation of wall displacements beyond the operating limit $D \leq 4.3$ of the laboratory set-up. Figure 12 gives the

Fig. 12 Experimental
period-1 lines for protocol $\mathscr{T}$
within the symmetry plane
$I_\beta = \pi/6$ versus wall
displacement obtained with
the hybrid particle-tracking
method: $D = 4.3$ (*blue* +),
$D = 8.6$ (*red* ∘) and $D = 17.1$
(*black* ∗). *Continuous curves*
are period-1 lines obtained by
numerical simulations for the
rigid-wall boundary
conditions at $D = 2.9$,
$D = 5.7$ and $D = 11.4$.
Reproduced from Znaien
et al. (2012)

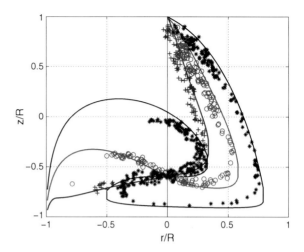

period-1 lines (symbols) for displacements $D = [4.3, 8.6, 17.1]$ computed with this method. Curves correspond with simulated period-1 lines, again using $D_{sim}/D \approx 2/3$. The semi-experimental period-1 lines, similar to their simulated counterparts in Fig. 10, undergo progressive convolution upon augmenting D while remaining fixated at the two designated positions on the cylinder axis. Moreover, the close agreement with the simulated period-1 lines is retained, notwithstanding the considerably greater geometric complexity. Shown results thus substantiate the previous findings in that the experimental period-1 lines behave entirely in accordance with the generic scenario following Fig. 10.

5 Outlook

Great progress has been made on Lagrangian transport phenomena in 3D unsteady flows since the first pioneering studies from the mid-1980s, for example, Dombre et al. (1986) and Feingold et al. (1987, 1988). However, many challenges remain, mainly on grounds of two complicating factors. First, 3D spaces admit far greater topological complexity of coherent structures compared to their counterparts in 2D spaces (Alexandroff 1961). The diversity in flow topologies of Protocols $\mathscr{A}$, $\mathscr{B}$, and $\mathscr{C}$ clearly illustrates this. Second, the 3D equations of motion (2) lack the well-defined Hamiltonian structure of 2D configurations. In particular the latter continues to be a major obstacle to fundamental advances in insights on 3D Lagrangian dynamics.

Development of a comprehensive mathematical framework thus is imperative for a systematic description and analysis of 3D Lagrangian transport. Theoretical developments primarily expand on the classical Hamiltonian concept of action-angle variables by (local) representation of coherent structures as invariant surfaces and curves defined by constants of motion (see e.g., Feingold et al. (1987,

1988), MacKay (1994), Mezić and Wiggins (1994), Cartwright et al. (1996), and Mezić (2001)). A promising recent concept with Hamiltonian foundation and devised specifically for mixing applications is found in the linked twist map (Sturman et al. 2006; Meier et al. 2007; Sturman et al. 2008).

The development of a comprehensive Hamiltonian-like theoretical framework for 3D Lagrangian transport—in particular response scenarios to perturbations and routes to chaos—is nonetheless in its infancy. The most important generalisation of classical Hamiltonian mechanics to 3D systems is perhaps the 3D counterpart to the well-known KAM theorem, describing the fate of invariant tori under weak perturbations. However, similar universal response scenarios for coherent structures of different topology, most notably the important case of invariant spheroids, remain outstanding. Moreover, the scope must be widened to include the effect of strong perturbations on the flow topology. These topics are largely unexplored, though bifurcations similar to those studied in Mullowney et al. (2008a, b) are likely to play a pivotal role. Further reconciliation with concepts from mathematical physics, for example Arnol'd (1978), Arnol'd and Khesin (1991), Mezić and Wiggins (1994), Haller and Mezić (1998), and Bennet (2006), and magnetohydrodynamics (Biskamp 1993; Moffatt et al. 1992) is essential in strengthening the current framework.

Our studies on the 3D cylinder flow, a cross-section of which has been given in this review, seek to contribute to deepening knowledge of 3D Lagrangian transport phenomena. We believe this system constitutes an excellent testbed for 3D transport studies. The various forcing protocols encompass basic dynamic states of 3D flows: 1D motion on closed streamlines, (quasi-)2D motion due to presence of invariant manifolds and periodic lines and truly 3D chaotic advection caused by isolated periodic points. The present cylinder flow thus captures the essence of a wide range of 3D unsteady flows including generic 3D volume-preserving maps with non-toroidal invariant surfaces, as e.g. investigated in Gómez and Meiss (2002) and Mullowney et al. (2008a, b), the 3D lid-driven cube considered in Anderson et al. (1999, 2006), and 3D granular flows inside spherical tumblers, as studied by Meier et al. (2007) and Sturman et al. (2008). Furthermore, the simple yet realistic configuration is amenable to laboratory experiments and in fact has already been the subject of investigation in first experimental studies on key aspects of 3D transport (Speetjens et al. 2004; Znaien et al. 2012). The 3D cylinder flow may thus serve as bridge between theoretical and numerical studies and real flows.

References

Alexandroff P (1961) Elementary concepts of topology. Dover, New York

Anderson PA, Galaktionov OS, Peters GWM, van de Vosse FN, Meijer HEH (1999) Analysis of mixing in three-dimensional time-periodic cavity flows. J Fluid Mech 386:149

Anderson PA, Ternet TJ, Peters GWM, Meijer HEH (2006) Experimental/numerical analysis of chaotic advection in a three-dimensional cavity flow. Int Polym Process 4:412

Arnol'd VI (1978) Mathematical methods of classical mechanics. Springer, New York

Arnol'd VI, Khesin BA (1991) Topological methods in hydrodynamics. Springer, New York

Bennet A (2006) Lagrangian fluid dynamics. Cambridge University Press, Cambridge

Biskamp D (1993) Nonlinear magnetohydrodynamics. Cambridge University Press, Cambridge

Cartwright JHE, Feingold M, Piro O (1996) Chaotic advection in three-dimensional unsteady incompressible laminar flow. J Fluid Mech 316:259

Dombre T, Frisch U, Greene JM, Hénon M, Mehr A, Soward AM (1986) Chaotic streamlines in the ABC flows. J Fluid Mech 167:353

Feingold M, Kadanoff LP, Piro O (1987) A way to connect fluid dynamics to dynamical systems: passive scalars. In: Hurd AJ, Weitz DA, Mandelbrot BB (eds) Fractal aspects of materials: disordered systems. Materials Research Society, Pittsburgh, pp 203–205

Feingold M, Kadanoff LP, Piro O (1988) Passive scalars, three-dimensional volume-preserving maps and chaos. J Stat Phys 50:529

Franjione JG, Leong C-W, Ottino JM (1989) Symmetries within chaos: a route to effective mixing. Phys Fluids A 11:1772

Gómez A, Meiss JD (2002) Volume-preserving maps with an invariant. Chaos 12:289

Guckenheimer J, Holmes P (1983) Nonlinear oscillations, dynamical systems and bifurcations of vector fields. Springer, New York

Haller G, Mezić I (1998) Reduction of three-dimensional, volume-preserving flows by symmetry. Nonlinearity 11:319

Luethi B, Tsinober A, Kinzelbach W (2005) Lagrangian measurement of vorticity dynamics in turbulent flow. J Fluid Mech 528:87

Malyuga VS, Meleshko VV, Speetjens M (2002) Mixing in the Stokes flow in a cylindrical container. Proc R Soc Lond A 458:1867

MacKay RS (1994) Transport in 3D volume-preserving flows. J Nonlinear Sci 4:329

Meier SW, Lueptow RM, Ottino JM (2007) A dynamical systems approach to mixing and segregation of granular materials in tumblers. Adv Phys 56:757

Meleshko VV, Peters GWM (1996) Periodic points for two-dimensional Stokes flow in a rectangular cavity. Phys Lett A 216:87

Mezić I, Wiggins S (1994) On the integrability and perturbation of three-dimensional fluid flows with symmetry. J Nonlinear Sci 4:157

Mezić I (2001) Break-up of invariant surfaces in action-angle-angle maps and flows. Physica D 154:51

Moffatt HK, Zaslavsky GM, Comte P, Tabor M (1992) Topological aspects of the dynamics of fluids and plasmas. Kluwer Academic Publishers, Dordrecht

Mullowney P, Julien K, Meiss JD (2008) Blinking rolls: chaotic advection in a three-dimensional flow with an invariant. SIAM J Appl Dyn Sys 4:159186

Mullowney P, Julien K, Meiss JD (2008) Chaotic advection and the emergence of tori in the Küppers–Lortz state. Chaos 18:033104

Ottino JM (1989) The kinematics of mixing: stretching, chaos and transport. Cambridge University Press, Cambridge

Ottino JM, Jana SC, Chakravarthy VS (1994) From Reynolds stretching and folding to mixing studies using horseshoe maps. Phys Fluids 6:685

Pouransari Z, Speetjens MFM, Clercx HJH (2010) Formation of coherent structures by fluid inertia in three-dimensional laminar flows. J Fluid Mech 654:5

Shankar PN (1997) Three-dimensional eddy structure in a cylindrical container. J Fluid Mech 342:97

Speetjens MFM (2001) Three-Dimensional chaotic advection in a cylindrical domain. PhD thesis, Eindhoven University of Technology, The Netherlands

Speetjens MFM, Clercx HJH, van Heijst GJF (2004) A numerical and experimental study on advection in three-dimensional Stokes flows. J Fluid Mech 514:77

Speetjens MFM, Clercx HJH, van Heijst GJF (2006) Inertia-induced coherent structures in a time-periodic viscous mixing flow. Phys Fluids 18:083603

Speetjens MFM, Clercx HJH, van Heijst GJF (2006) Merger of coherent structures in time-periodic viscous flows. Chaos 16:043104

Sturman R, Ottino JM, Wiggins S (2006) The mathematical foundation of mixing. Cambridge University Press, Cambridge

Sturman R, Meier SW, Ottino JM, Wiggins S (2008) Linked twist map formalism in two and three dimensions applied to mixing in tumbled granular flows. J Fluid Mech 602:129

Voth GA, Haller G, Gollub JP (2002) Experimental measurements of stretching fields in fluid mixing. Phys Rev Lett 88:254501

Wiggins S (2010) Coherent structures and chaotic advection in three dimensions. J Fluid Mech 654:1

Znaien JG, Speetjens MFM, Trieling RR, Clercx HJH (2012) On the observability of periodic lines in 3D lid-driven cylindrical cavity flows. Phys Rev E 85(6):066320–1/14

Jets in Symbiotic Stars: The R Aqr Case

Silvana G. Navarro and Luis J. Corral

Abstract In this paper we analyse the jet like features observed in some symbiotic systems, we present the objects in which such features were detected. One of this objects is R Aqr. In this object high collimated outflows emerging from the central region has been observed since the 1970s. We present the possible explanation for this outflows and analyse some new kinematic data of this object comparing them with previous observations.

1 Introduction

The term "symbiotic star" were developed from the peculiar spectra of this objects which shows spectral lines characteristic of low temperature stars plus emission nebular lines produced in regions with temperatures in the range 10^4–10^5 K and densities near 10^4 cm^{-3}. This type of spectra and the observed brightness and spectral variations, suggested that this objects consist in evolved binaries with a giant member and a compact companion. Nowadays, the binary nature of these interacting systems is well established, they are long period binaries some of them with very elongated orbits. The emission lines observed in the spectra of these objects suggest the existence of an accretion disk which is formed with the accreted material from the companion.

S. G. Navarro (✉) · L. J. Corral
Instituto de Astronomía y Meteorología, Universidad de Guadalajara,
Vallarta 2602, Col, Arcos, Guadalajara, Jal, Mexico
e-mail: silvana@astro.iam.udg.mx

L. J. Corral
e-mail: lcorral@astro.iam.udg.mx

J. Klapp et al. (eds.), *Fluid Dynamics in Physics, Engineering and Environmental Applications*, Environmental Science and Engineering,
DOI: 10.1007/978-3-642-27723-8_10, © Springer-Verlag Berlin Heidelberg 2013

Following the evolution of a binary system where both components are low mass stars, there are two scenarios in which such type of spectra could be observed: (1) when the more massive member evolves to the Asymptotic Giant Branch (AGB) and transfer mass to the main sequence companion, and (2) when the primary component turn into a white dwarf (WD) and the second component evolve and transfer mass to it.

In both cases an accreting disc could be formed, either by the accretion of the AGB wind material onto the compact companion or by overflow of its Roche lobe, but is in the second scenario where the accreting object has the necessary physical conditions to produce the ejection events, like jets, that we are observing in some symbiotic systems (Frankowski and Jorissen 2007). Until now highly collimated outflows (jets) were observed in few symbiotic systems, 10 of the almost 200 known (Munari et al. 2005). R Aqr, CH Cyg and MWC 560 are between the best studied.

R Aqr was the first symbiotic in which jet-like emission was detected, it appeared initially (latest 1970s) like a small spike at the NE (Wallerstein and Greenstein 1980), followed by a dramatic brightening near 1980 (Sopka et al. 1982).

CH Cyg, together with R Aqr, are the nearest symbiotic systems: 200 and 250 pc respectively (van Leeuwen et al. 1997; Perryman et al. 1997) they have been extensively studied in all frequencies. CH Cyg system consists of a M6 giant star transferring mass to a white dwarf (Hinkle et al. 2009). A powerful radio jet was first detected in 1984 (Taylor et al. 1986), this jet followed a sudden dimming in V magnitude; a similar behaviour were observed during the 1998 radio jet detection (Karovska et al. 1998).

R Aqr and CH Cyg are between the few symbiotic systems presenting also highly collimated X-ray emission, it was first detected in CH Cyg with Chandra in 2001 (Galloway and Sokoloski 2004). Karovska et al. (2010), observed this object with Chandra, HST and VLA in 2008, they detected the jet with Chandra and HST and discovered a new NE–SW jet with clumps along the jet, very similar to the observed in R Aqr (Kellogg et al. 2001; Navarro et al. 2003).

MWC 560 was classified as a peculiar M4ep star by Sanduleak and Stephenson (1973), they detected the TiO bands and Balmer emission lines in its spectra. Due to the drastic spectral variations in the blue shifted absorption lines, showing radial velocities from 500 km/s to near 4,000 km/s, Tomov (1990) proposed a jet outflow seen along the line of sight. Schmid et al. (2001) presented the results of an intensive monitoring program of this object from November 1998 to January 1999. They present a high resolution optical spectra with high S:N ratio and proposed a geometric model for this object.

Hen 2-104 (Corradi et al. 2001), RS Oph (Taylor et al. 1986), Hen 3-1341 (Tomov et al. 2000; Munari et al. 2005), StHa 190 (Munari et al. 2001), AG Dra (Mikolajewska 2002), V1319 Cyg, HD149427 (Brocksopp et al. 2003), and Z And (Brocksopp et al. 2004; Skopal et al. 2011) are other examples of jets in symbiotic systems. In R Aqr, the jet like outflows are detected in the optical, UV, X-ray and radio spectral regions (Sopka et al. 1982; Kafatos et al. 1986); although the respective emitting areas are not always coincident.

The MHD models for the production of jets on accreting systems have been developed for objects with higher mass, like black holes or neutron stars, although the acceleration and collimation mechanisms are probably the same for all the astrophysical objects exhibiting jets (Livio 2004). In their analytical work, Blandford and Payne (1982) explained the jet formation by magneto-centrifugal forces acting on the inner disc plasma. Soker and Regev (2003) proposed another form to accelerate the plasma close to the central object by "spatio-temporal localized accretion shocks" in the boundary layer of the WD, in which bubbles on the surface of the WD are heated, they expand and accelerate the plasma to the local escape velocity. In these models the accreting mass rate is a crucial factor for the jet ejection, Soker and Lasota (2004) determined a minimum accretion rate to produce jets of 10^{-6} M $\odot$ yr^{-1}.

In the case of symbiotic systems the determination of this parameter is difficult due to the companion and, in many cases due to the view angle. R Aqr and CH Cyg are symbiotics in which the jet is near perpendicular to the line of sight, while in MWC 560 is nearly parallel to it. This fact allow Stute et al. (2005) to obtained accurate determinations of the velocity and column density of the out-flowing gas in the jet and construct an hydrodynamical simulation of it [see also Stute (2006), and Stute and Sahai (2007)].

2 The Jet in R Aqr

R Aqr is the nearest symbiotic system with a distance near 200 pc (van Leeuwen et al. 1997). It consist of a Mira variable (spectral type M7 III) with a pulsation period of 287 days (Mattei and Allen 1979) and a white dwarf. Determinations of the orbital period estimations varies from 17 years (Nichols et al. 2007) to the more recently determination of Gromadzki and Mikolajewska (2009) of 43.6 years. These authors determined spectroscopically the orbital period, the eccentricity of the orbit: $e = 0.25$, and the probable mass of their components: $1-1.5$ M$\odot$ for the Mira and $0.6-1.0$ M $\odot$ for the WD.

Around R Aqr have been observed two coaxial bipolar shells (hourglass like), with the binary located in the centre of their waist. Sopka et al. (1982) detected variations in the brightness of the nebular features, and described the observation of a bright jet-like feature near to the central star. The kinematic study of the expanding outer and inner shells indicate an age of 650 and 185 years respectively (Solf and Ulrich 1985). In their kinematic model Solf and Ulrich proposed that these hour-glass shells were ejected in nova-like outbursts from the symbiotic. They noted that the spatio-kinematic structure of the inner nebulae resembles that of the outer nebulae but at smaller scale.

R Aqr was the first symbiotic in which jet-like emission was detected, it turns observable in the optical in the latest 1970s like a small spike at the NE direction (Wallerstein and Greenstein 1980), followed by a dramatic brightening near 1980 (Sopka et al. 1982). Subsequent VLA observations resolved this spike in two components (Hollis et al. 1985) and detected a SW knot (Kafatos et al. 1989). Paresce et al. (1988) observed a new knot farther to the NE. Subsequent observations detected more "knots" located along the same direction (NE and SW).

Variations in the position and brightness of the knots are registered since earlier studies. Solf (1992) detected remarkable changes in the kinematical properties of the jet-like features when they observed this object in 1987 and compared with previous observations (Solf and Ulrich 1985). The velocity profile of the knots are broadened and appears a notorious asymmetry, showing a blue wing that extend to more than 200 km/s. Variability have been detected also in other spectral regions: Kellogg et al. (2001), detected X-ray emission from the NE and SW jet. In subsequent 2004 observations the X-ray emission at the SW was not detected (Kellogg et al. 2007). Nichols et al. (2007) detected on their X-ray observations a 1,734 s periodic oscillation, which suggest that the compact component is a magnetic white dwarf.

R Aqr is an outstanding symbiotic system due the wide type of physical processes occurring on it and the many observable tracers of them. The jet in R Aqr is also remarkable due to its physical extension: more than 1400 AU for the NE jet, and near 2800 AU for the farthest SW knot.

Gonçalves et al. (2003) observed this object between 1991 and 1999, and study the evolution of the jet (Navarro et al. 2003). The location of the knots in diagnostic diagrams (Cantó 1981) were obtained with data of the 1999 epoch and showed that the regions on the outer shell are mainly excited by shocks while in the regions closer to the symbiotic the main excitation process are photoionization by the central object. Nichols and Slavin (2009) analyse the structure of OVI $\lambda\lambda$ 1,038 line and adjust bow shock models to the jet. They calculated full radiative, full non-radiative and hybrid models for the shock. They found that the model that best fits the data is an hybrid bow shock with a shock speed between 235 and 285 km/s and a turbulent velocity near 60 km/s. They found similar parameters for both sides of the jet (NE and SW), suggesting a common origin. They found preshock densities in the range $10^3 - 10^4$ cm^{-3} at the position of X-ray maximum emission.

3 Observations

We observed R Aqr with the MES spectrograph at San Pedro Mártir Observatory (Meaburn et al. 2003) obtaining high resolution spectra in the Hα - [NII] and in [OIII] λ 5007 Å spectral regions. The observations were obtained on September 24 and 25, 2010. The spectral resolution at Hα is near 10 km/s with the 150 μm slit; while the plate scale correspond to 0.35 arcsec/pix, both for a binning of 2×2. In this paper we discuss the first results obtained from our observations.

In Fig. 1 we draw over an Hα image of R Aqr, some of the slit positions that were observed. Nearly all of them were positioned to pass trough two or three

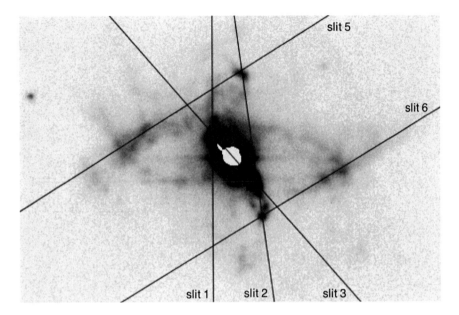

Fig. 1 Position of the slits observed on R Aqr. Due to its high brightness, only slit 3 pass trough the central object. The slits passing through the fainter knots avoid the centre

features avoiding the central source due to its high brightness; only one of the slits (number 3) pass through it. Slits number 1, 2, and 6 go through the jet and number 3 is nearly aligned to it.

In Fig. 2 we present the appearance of the spectra obtained with slit number one (we present only the H alpha and [NII] $\lambda\lambda6583$ due to scale convenience). In this figure the principal central emission correspond to the cut of the slit trough the NE jet. The upper and lower emissions comes from the expanding lobes mentioned in previous section, we are observing the emission from the expanding material, with projected velocities on the line of sight, toward and away us, this is the reason of the "X" like shape of the emission observed above and below the principal jet emission.

In Fig. 2 we add four lines at the position where we obtained the profile of [NII] line shown in Fig. 3a, b. On these figures is notorious the higher value of the velocity component along the line of sight when we move away from the centre (specially in Fig. 3a).

4 Results and Conclusions

In Fig. 4 we present the X-Ray emission observed by Kellogg et al. in 2000 (Kellogg et al. 2001) as contours over our Hα image. We observed a nearly coincidence of the knots in the NE and SW with the X-ray emission. The position

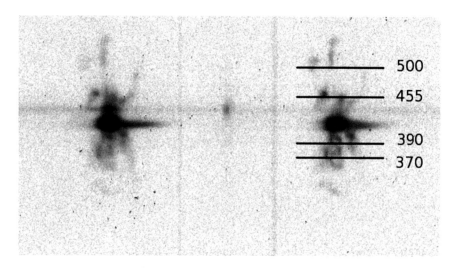

Fig. 2 Hα and [NII] λλ6583 emission obtained with the slit number 1 positioned trough R Aqr

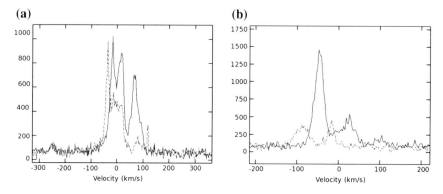

Fig. 3 Profile of the [NII] λλ6583 emission. In Fig. 3a we compare the spectra extracted adding 7 rows around 390 (*continuous line*) with the profile of the same line extracted from 7 rows around 370 (*dashed line*). In Fig. 3b we compare the profile centred at row 455 (*continuous line*) with the profile obtained from 367 to 373 rows (*dashed line*). The position of the row numbers mentioned here are indicated in Fig. 2

of the central source coincide with the space between the two principal maxima in the X-ray emission. The SW X-ray emitting region nearer the central source could coincide with the "A" SW knot observed by Navarro et al. (2003). Subsequent observations by Kellogg et al. 2001. in 2001 shows no emission in the far SW region (labeled S in Fig. 4). According with these authors, such behaviour could be explained by the cooling due to adiabatic expansion of the emitting volume.

According to spectroscopic observations in the optical region, it is possible to estimate electronic densities around $10^3 - 10^4$ cm^{-3}, with this value, the mean free path, adopting a Maxwell velocity distribution, is $3 \times 10^{10} - 3 \times 10^{11}$ cm.

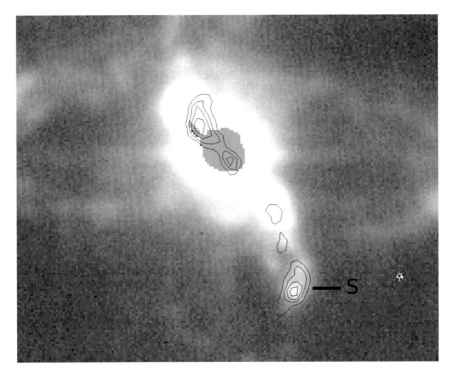

Fig. 4 Hα+ [NII] image the contours correspond to the X-ray emission detected by Kellogg et al. (2001). We note coincidence of the X-ray maxima with the jets observed at the NE and SW. In subsequent X-ray observations the emission on the S knot decreased until almost disappear (Kellogg et al. 2007)

Fig. 5 Velocity profile of the λ 6583 [NII] line for the slit number one. The *red wing* is clearly observed, the FWHM of the emission is 35 km/s wile the FWZI = 225 km/s

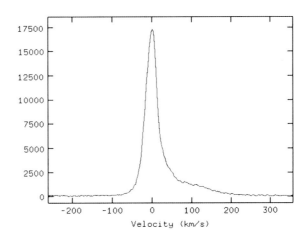

Comparing this value with the size of the emitting region (near 3 arcsec) Kellogg et al. (2007) concluded that is justified to consider the jet as a fluid. These authors take γ = 5/3 for a monatomic gas, they calculated the cooling time and found that the

inclusion of an adiabatic expansion explains a temperature decrease near 40 % for the 3.3 years between their observations (from 1.4×10^6 to 8.4×10^5 K). At september 2010, we observed emission in Hα and [HII] in the same regions, implying a temperature around few tens of thousands Kelvin. As it is expected that the adiabatic expansion continue, this fact contribute to the cooling of the plasma. The X-ray emission can be interpreted as a bow-shock produced when the collimated outflow encounters the relic material from the previous outbursts that produced the shells or with the wind material from the Mira. This scenario remembers the conditions present in HH objects. The high similarity in the geometry of the observed shocks and in the nebular line profiles observed in both type of objects support this coincidence.

In Fig. 5 we observed the velocity profile of the [NII] line λ 6583 for the slit position number one. In this figure we observed a high asymmetry in the line, presenting a red wing of more than 200 km/s (FWHM = 35 km/s, and HWZI $\geq$ 200 km/s). Solf (1992) observed a similar asymmetry of this line for knots B and D, but presenting a blue wing in his 1987 observations, in contrast with the 1982/83 observations (Solf and Ulrich 1985), when no asymmetry was detected.

5 Concluding Remarks

Due to the high similarity of the velocity profiles of the [NII] lines in R Aqr with the observed profiles in HH objects Solf (1992) proposed to compare with the Raga and Bohm (1986) and Hartigan et al. (1987) models for a bow-shock in HH objects. Solf obtained the post-shock velocity and the ϕ angle of the shock axis with the line of sight. Comparing our velocity profiles with Hartigan et al. models (their figures h and s), we conclude that our observations correspond to ϕ between 150 and 180 degrees, i.e. we are observing the shock from back. This is the first observation of R Aqr jet that are obtained at such ϕ angles and proof the model of a precessing jet that has been proposed previously by other authors (Hollis et al. 1999; Gonçalves et al. 2003). Actually we are working in the search of periodic profile variations, collecting previous observations [Solf and Ulrich (1985), Solf (1992), Navarro et al. (2003), etc.], and our new 2010 and 2011 observations.

Acknowledgments We acknowledge the help of SPM staff during the observation run and specially to Roberto Vazquez for the introduction to MEZCAL and its observation techniques.

References

Blandford RD, Payne DG (1982) Mon Notices Royal Astron Soc 199:883
Brocksopp C, Bode MF, Eyres S (2003) Mon Notices Royal Astron Soc 344:1264
Brocksopp C, Sokoloski JL, Kaiser C et al (2004) Mon Notices Royal Astron Soc 347:430
Cantó J (1981) In investigating the universe. Dordrecht, Reidel

Corradi RLM, Munari U, Livio M, Mampaso A, Gonçalves D, Schwarz H (2001) Astrophys J 560:912

Frankowski A, Jorissen A (2007) BaltA 16:104

Galloway DK, Sokoloski JK (2004) Astrophys J 613L:61

Gonçalves DR, Mampaso A, Navarro SG, Corradi RLM (2003) Am Soc Prev Cardiol 303:423

Gromadzki M, Mikolajewska J (2009) Astron Astrophys 495:931

Hartigan P, Raymond J, Hartmann L (1987) Astrophys J 316:323

Hinkle KH, Fekel FC, Joyce RR (2009) Astrophys J 692:1360

Hollis JM, Kafatos M, Michalitsianos AG, Mc Allister HA (1985) Astrophys J 289:765

Hollis JM, Vogel SN, van Buren D, Strong JP, Lyon RG, Dorband JE (1999) Astrophys J 522:297

Kafatos M, Michalitsianos AG, Hollis JM (1986) ApJS 62:853

Kafatos M, Hollis JM, Yusef-Zadeh F, Michalitsianos AG, Elitzur M (1989) Astrophys J 346:991

Karovska M, Carilli CL, Mattei JA (1998) J Am Assoc Var Star Obs 26:97

Karovska M, Gaetz TJ, Carilli ChL, Hack W, Raymond JC, Lee NP (2010) Astrophys J 710:L132

Kellogg E, Anderson C, Korreck K, DePasquale J, Nichols J, Sokoloski JL (2007) Astrophys J 664:1079

Kellogg E, Pedelty J, Lyon R (2001) Astrophys J 563:L151

Livio M (2004) BaltA 13:273

Mattei JA, Allen J (1979) J Royal Astron Soc Can 73:173

Meaburn J, López JA, Gutiérrez L, Quiróz F, Murillo JM, Valdéz J, Pedrayes M (2003) RMAA 39:185

Mikolajewska J (2002) Mon Notices Royal Astron Soc 335:L33

Munari U, Tomov T, Yudin B et al (2001) Astron Astrophys 369:L1

Munari U, Siviero A, Henden A (2005) Mon Notices Royal Astron Soc 360:1257

Navarro SG, Gonçalves DR, Mampaso A, Corradi RLM (2003) Am Soc Prev Cardiol 303:486

Nichols JS, DePasquale J, Kellogg E, Anderson CS, Sokoloski J, Pedelty J (2007) Astrophys J 660:651

Nichols J, Slavin JD (2009) Astrophys J 699:902

Paresce F, Burrows C, Horne K (1988) Astrophys J 329:318

Perryman MAC et al (1997) Astron Astrophys 323:L49

Raga AC, Bohm KH (1986) Astrophys J 308:829

Sanduleak N, Stephenson CB (1973) Astrophys J 185:899

Schmid HM, Kaufer A, Camenzind M, Rivinius T, Stahl O, Szeifert T, Tubbesing S, Wolf B (2001) Astron Astrophys 377:206

Skopal A, Tarasova TN, Pribulla T, Vanko M, Dubovsky PA, Kudzej I (2011) APN5procE 71

Soker N, Lasota JP (2004) Astron Astrophys 422:1039

Soker N, Regev O (2003) Astron Astrophys 406:603

Solf J (1992) Astron Astrophys 257:228

Solf J, Ulrich H (1985) Astron Astrophys 148:274

Sopka RJ, Herbig G, Kafatos M, Michalitsianos AG (1982) Astrophys J 258:L35

Stute M, Camenzind M, Schmid HM (2005) Astron Astrophys 429:209

Stute M (2006) Astron Astrophys 450:645

Stute M, Sahai R (2007) Astrophys J 665:698

Taylor AR, Seaquist ER, Mattei JA (1986) Nature 319:38

Tomov (1990) IAU Circ 4955

Tomov T, Munari U, Marrese P (2000) Astron Astrophys 354:L25

van Leeuwen F, Feast MW, Whitelock PA, Yudin B (1997) Mon Notices Royal Astron Soc 287:955

Wallerstein G, Greenstein JL (1980) Publ Astron Soc Pac 92:275

Granular Hydrodynamics

L. Trujillo, L. Di G. Sigalotti and J. Klapp

Abstract Sand flowing through the constriction of an hourglass or jumping on a vibrating plate is fluidized in the sense that it moves analogously to a fluid. Dense flows of grains driven by gravity down inclines occur in nature and in industrial processes. Natural examples include rock avalanches and landslides. Applications are found in the chemical, pharmaceutical and petroleum industry. Grain flow can be modeled as a fluid-mechanical phenomenon. However, granular fluids teach us about an astounding complexity that emerges from simple, macroscopic particles. For example, starting from an homogenous fluidized system, structures evolve and a dilute granular fluid co-exists with much denser solid-like clusters. Another example is the so-called Brazil nut effect, whereby larger and heavier particles placed into an agitated granular bed rise to the top. We present an outlook of the hydrodynamic description of granular materials. Our purpose is to outline a theory of grain flow which is based upon the description of continuous matter fields derived from the kinetic theory for dense gases, as is usually encountered in fluid dynamics.

L. Trujillo · L. D. G. Sigalotti
Instituto Venezolano de Investigaciones Científicas, IVIC, Centro de Física, Apartado Postal 20632 Caracas, 1020-A, Venezuela
e-mail: leonardo.sigalotti@gmail.com

L. Trujillo (✉)
The Abdus Salam, International Centre for Theoretical Physics, ICTP, Trieste, Italy
e-mail: leonardo.trujillo@gmail.com

J. Klapp
Instituto Nacional de Investigaciones Nucleares ININ, Carretera Mexico- Toluca Km 36.5, La Marquesa, 52750 Estado de Mexico, Mexico
e-mail: jaime.klapp@hotmail.com

J. Klapp
Departamento de Matemáticas, Cinvestav del I.P.N., San Pedro Zacatenco, 07360 Mexico D.F., Mexico

J. Klapp et al. (eds.), *Fluid Dynamics in Physics, Engineering and Environmental Applications*, Environmental Science and Engineering, DOI: 10.1007/978-3-642-27723-8_11, © Springer-Verlag Berlin Heidelberg 2013

1 Introduction

Granular materials consist of a collection of discrete macroscopic solid particles interacting via short-range repulsive contact forces. Classical examples are sand, powders, sugar, salt, and gravel. No limitation is imposed on the size of the particles, which may range from nanometer scale, as in pigments or aerosols, to the scale of mined or quarried materials and rocks. Their physical behavior involves complex nonlinear phenomena, including non-equilibrium static configurations, energy dissipation, nonlinear elastic response, and peculiar flow dynamics (Jaeger et al. 1996).

Why are granular materials interesting? *A new state of matter.*—Beyond their practical importance, granular materials interest physicists because they are an unusual form of matter with interesting properties that are still not yet fully understood. *Interdisciplinary effort.*—Investigations in granular materials gather teams from civil, chemical and mechanical engineers, food technologists, powder metallurgists, materials scientists, pharmaceutical scientists, applied mathematicians, and physicists. *Fascinating experiments.*—They exhibit a wealth of interesting phenomena like heaping under vibration, segregation, convection, fluidization, pattern formation, anomalous sound propagation, compaction, and density waves. *Computer applications.*—Physics and engineering on high performance computers modeling granular flows gives a great insight towards the understanding of the dynamics of particulate systems. *A theoretical challenge.*—There is no accepted set of universal governing equations describing the physics of granular materials. Important contributions have been made to improve our understanding of many new aspects using modern tools from statistical to fluid mechanics. *Industrial productivity.*—They are important since a large fraction of the materials handled and processed in the chemical, pharmaceutical, construction, metallurgical, and food-processing industries are granular in nature.

Granular materials can mimic several features of condensed matter: gas, liquid or solid states, plastic flow, and glassy behavior. But in many ways, a granular material is like an ordinary fluid. Insofar as possible, individual grains are treated as the molecules of a *granular fluid*. One example is the motion of sand on a vibrating plate. At sufficiently high vibrations the individual grains randomly jump up and down colliding between them. Other examples are the displacements inside a shear-cell or the flows of grains driven by gravity down inclines. These observations have inspired several authors to use continuum balance equations for mass, momentum, and energy analogous to fluid dynamics. Whereas classical fluids are well described by the Navier-Stokes equations, no constitutive law can reproduce the diversity of behavior observed with a granular material. This difficulty originates from fundamental characteristics of granular matter such as negligible thermal fluctuations, highly dissipative interactions, and a lack of separation between the microscopic grain scale and the macroscopic scale of the flow.

Here, we focus our presentation on the liquid regime and introduce a hydrodynamic description of dense granular flows. We restrict the discussion to rigid dry grains and do not consider soft particles, cohesive, and friction effects as well as the interaction with a surrounding fluid.

2 Some Analogies Between Fluidized Granular Materials and Fluid-Mechanical Phenomena

Now, we enumerate five representative examples of the analogies between fluidized granular materials and liquids.

- *Convection.*—An experimental setup particularly suited to study granular fluidization is putting sand on a loudspeaker or on a vibrating plate. Under gravity the sand jumps up and down and although kinetic energy is strongly dissipated, collisions between its grains reduce its density, thereby allowing it to flow. Under certain circumstances flow between top and bottom can occur in the form of convection cells reminiscent of those seen in fluids (Hayakawa et al. 1995). Such cell flows have been observed experimentally (Knight et al. 1996) and modelled via Molecular Dynamics simulations (Gallas et al. 1992).
- *Hydrostatic and Archimedes' buoyancy law.*—Buoyancy belongs to those physical phenomena that have fascinated people since the time of Archimedes of Syracuse.[1] In 2005 the Mexican group headed by Prof. Jesús Carlos Ruiz, designed a novel experimental setup where the granular system is fluidized by injecting energy through the lateral walls, perpendicular to the gravity field (Huerta et al. 2005). They measured buoyancy forces of a larger intruder particle immersed in the bed and found that they obey an Archimedes-like principle (i.e., these forces are proportional to the displaced volume). A similar mechanism was proposed previously by Trujillo and Herrmann (2003) based on a hydrodynamic theory for segregation. Very recently, Maes and Thomas (2011) studied the origin of buoyancy forces via an asymmetric exclusion process verifying the Archimedes' principle in the fluid limit of a dissipative lattice gas dynamics.
- *Leidenfrost effect.*—The Leidenfrost effect (Leidenfrost 1966) is a phenomenon in which a liquid, in near contact with a plate significantly hotter than the liquid's boiling point, produces an insulating vapor, which prevents direct heat transfer from the plate to the liquid, causing the fluid to hover and survive for a long time. When granular materials confined in a Hele-Shaw cell (quasi 2D container) are vertically vibrated a crystalline cluster is elevated and supported by a diluted "gaseous" layer of fast grains at the bottom. In 2005 the group headed by Prof. Detlef Lohse at the University of Twente, The Netherlands, demonstrated experimentally for the first time the *granular Leidenfrost effect*

---

[1] The "sand laptop": He has even said to have carried a small wooden tray filled with sand, which he used to draw his figures and work on his mathematical problems. This tray would have been Archimedes' version of the modern lap top computer.

and explained by a (granular) hydrodynamic model the parameters which control the effect (Eshuis et al. 2005). The agreement between experiment and theory was quite remarkable.

- *Hydraulic jump.*—The circular hydraulic jump following the impact of a liquid jet on a plate is a very common example of a free boundary problem in fluid mechanics: It is observable in everyday life as one opens a water tap into a sink. When the vertical jet hits the horizontal plate, it first spreads out radially into a thin layer. At some distance from the jet, however, the height of liquid suddenly jumps to a higher value. A granular jet impinging on a solid surface also gives rise to several features reminiscent of the hydraulic jump. This analogy between the *granular jump* and the water hydraulic jump was studied in detail by Boudet et al. (2007). This experiment brings forth the interplay between continuum hydrodynamics properties and the granular and discrete nature of the medium.
- *Granular jets.*—The impact of a solid object on the surface of a liquid creates a splash forming a surface cavity which immediately starts to collapse due to the hydrostatic pressure of the surrounding liquid. Then, a jet shoots up from the liquid surface. A spectacular analogy with the jet formation in liquids is observed when a heavy sphere impacts onto a bed of very loose fine sand (Lohse et al. 2004a, b). Experiments reveal that when the heavy solid plunges into the sand a splash is created. Then, a jet of grain shoots upwards at the position of impact. The free surface of the sand bed develops patterns analogous to the surface tension driven Rayleigh instability of a water jet, even though there is no surface tension in dry granular materials (Lohse et al. 2004a). The influence of air pressure on the jet and the rate of penetration of the heavy solid was studied by Caballero et al. (2007), showing a dynamical coupling between gas and the granular medium.

3 Hydrodynamic Theory for Granular Materials

Under suitable conditions, a noncohesive granular system can be maintained in a fluid-like state by continuous vibration of the container. Previous studies have established that this fluidized state is well described by hydrodynamic theories for inelastic hard particles (Jenkins and Savage 1983; Brey et al. 1998, 2001; Garzó and Dufty 1999).

In contrast to a molecular gas in equilibrium state, where the mean kinetic energy (mean–square velocity) of a molecule is proportional to the gas tempera-ture (thermodynamic temperature), the natural equilibrium state of a granular material is a static configuration due to the inelastic nature of particle collisions. Therefore, a stationary fluidized state needs a constant energy flow of energy into the system. This steady state driven by the energy flux can be assumed as a condition of "thermal equilibrium" (Herrmann 1993). On the other hand, in analogy to a molecular gas, we can define a "granular temperature" T_g, which is proportional to the mean kinetic energy E of the particles' velocity. Let us remark that this generalized notion of temperature is introduced for theoretical convenience to take advantage of a thermodynamical analogy for granular materials.

In fact, the definition of thermodynamic variables for nonequilibrium states is straightforward theoretically. The thermodynamic of nonequilibrium states has always been a matter of debate. Alternatives for a thermodynamic formulation for granular materials have been proposed by Edwards et al. (e.g., Edwards and Oakeshott (1989)), Herrmann (1993), and Hong and Hayakawa (1997). Certainly, the problem of a thermodynamic formulation for granular materials deserves a deeper analysis (cf. Kadanott (1999)), but to keep the scope of this work manageable we will limit the theoretical analysis into the framework of the kinetic theory of granular gases (Brilliantov and Pöschel 2004) (i.e., a Maxwell–Boltzmann statistics).

The development of statistical descriptions for granular systems involves averaging over the 'microscopic' laws for the particles motion to obtain 'macroscopic' balance equations for the hydrodynamic fields. A basis for the derivation of granular hydrodynamic equations (analogous to the Navier–Stokes equations) and detailed expressions for the constitutive relations (transport coefficients) is provided by the kinetic theory of gases conveniently modified to account for inelastic binary collisions. See the book (Brilliantov and Pöschel 2004) which provides an excellent introduction of the current knowledge about granular gases.

The balance laws for a granular fluid can be obtained on the basis of a mean field kinetic equation, like the Boltzmann or Enskog–Boltzmann equation. The only "nonclassical" term is the collision rate of dissipation per unit volume per unit time due to inelastic collisions. Many calculations are based on the kinetic theory of nonuniform gases (Chapman and Cowling 1970) using the Chapman–Enskog procedure.

In 1987–1989 Jenkins and Mancini (1987, 1989) introduced a remarkable extension of the kinetic theory of nonuniform gases to bidisperse granular mixtures. In the context of the so–called Revised Enskog Theory (RET) for multicomponent mixtures (López de Haro et al. 1983), they derived balance laws and constitutive relations for a dense binary mixture of smooth, nearly elastic particles by assuming a Maxwellian velocity distribution. They investigated in detail the use of the theory to a steady rectilinear shearing flow induced by the relative motion of parallel boundaries in the absence of external forces. Important modifications in these models were introduced by Arnarson et al. (Arnarson and Willits 1998; Willits and Arnarson 1999; Alam et al. 2002; Arnarson and Jenkins 2004). It is important to recall that there has been also an attempt to extend the kinetic theory of multicomponent mixtures to systems consisting of an arbitrary number of inelastic particles (Garzó et al. 2007a; b).

The point of departure is the conservation equations for the mixture density, momentum, and energy:

$$\frac{D\rho}{Dt} = -\rho \nabla \cdot \mathbf{u}, \tag{1}$$

$$\rho \frac{D\mathbf{u}}{Dt} = -\nabla \cdot \hat{\mathbf{P}} + \sum_i n_i \mathbf{F}_i, \tag{2}$$

$$\rho \frac{d}{2} \frac{DT}{Dt} = \frac{d}{2} T \nabla \cdot \mathbf{J} - \nabla \cdot \mathbf{Q} - \hat{\mathbf{P}} : \nabla \mathbf{u} + \sum_i \mathbf{J}_i \cdot \mathbf{F}_i - \mathcal{D}, \tag{3}$$

where ρ is the total mixture mass density, $\mathbf{u}$ is the mass average velocity of the mixture, $\hat{\mathbf{P}}$ is the pressure tensor, n_i is the number density of species i, $\mathbf{F}_i$ is the external force acting on the particle, T is the mixture *granular temperature*, $\mathbf{J}$ is the diffusive mass-flux, $\mathbf{Q}$ is the mixture energy flux, and $\mathcal{D}$ is the total inelastic dissipation rate. Also, $\frac{D(\cdot)}{Dt} = \frac{\partial(\cdot)}{\partial t} + \mathbf{u} \cdot \nabla(\cdot)$, is the substantial derivative (or material derivative), the symbol : denotes the full tensor contraction, and d is the dimension ($d = 2, 3$). Equations (1–3) are rigorous consequences of the Boltzmann–Enskog kinetic equation and must be supplemented with the respective constitutive relations for $\hat{\mathbf{P}}$, $\mathbf{J}$, $\mathbf{Q}$, and $\mathcal{D}$.

3.1 A Brief Introduction to the Kinetic Theory of Inelastically Colliding Particles

Here we sketch the structure of the kinetic theory for a binary mixture and focus on the species balance equations and fluxes. In the following we shall specialize to the case of a granular fluid with a single species component. The generalization to a multicomponent mixture is straightforward.

3.1.1 Hydrodynamic Fields

As a mechanical model for granular fluids we consider a binary mixture of slightly inelastic, smooth particles (disks/spheres) with radii r_i ($i = A, B$) and masses m_i in two and three dimensions ($d = 2, 3$). The system contains a number N_i of particles of species i in a volume V. The volume has a constant regular shape, and no particles are allowed to flow across the surface, so that the total number of particles is constant.

Mean values are calculated in terms of the single particle velocity distribution function $f_i(\mathbf{c}, \mathbf{r}, t)$ for each species. By definition $f_i(\mathbf{c}, \mathbf{r}, t) d\mathbf{c} d\mathbf{r}$ is the number of particles which, at time t, have velocities in the interval $d\mathbf{c}$ centered at $\mathbf{c}$ and positions lying within a volume element $d\mathbf{r}$ centered at $\mathbf{r}$. The information that there are a number N_i of particles of species i in the volume V of the system is expressed by means of the normalization condition

$$\int d\mathbf{c} d\mathbf{r} f_i(\mathbf{c}, \mathbf{r}, t) = N_i(\mathbf{r}, t). \tag{4}$$

If the particles are uniformly distributed in space, so that f_i is independent of $\mathbf{r}$, then the number density $n_i(\mathbf{r}, t)$ of species i is

$$n_i(\mathbf{r}, t) = \int d\mathbf{c} f_i(\mathbf{c}, \mathbf{r}, t). \tag{5}$$

The total number density n is the sum over both species, $n = n_A + n_B$. The species mass density ρ_i is defined by the product of n_i and m_i, and the total mixture density is $\rho = \rho_A + \rho_B = \rho_A \phi_A + \rho_B \phi_B$, where ρ_i is the material density of species i and ϕ_i is the d—dimensional volume fraction for species i: $\phi_i = \frac{\Omega_d}{d} n_i r_i^d$, where Ω_d is the surface area of a d—dimensional unit sphere. The mean value of any quantity $\psi_i = \psi_i(\mathbf{c})$ of a particle species i is

$$\langle \psi_i(\mathbf{c}) \rangle \equiv \frac{1}{n_i} \int d\mathbf{c} \psi_i(\mathbf{c}) f_i(\mathbf{c}). \tag{6}$$

The mean velocity of species i is $\mathbf{u}_i = \langle \mathbf{c}_i \rangle$. The mass average velocity (barycentric velocity) $\mathbf{u}$ of the mixture is defined by

$$\mathbf{u} \equiv \frac{1}{\rho} (\rho_A \mathbf{u}_A + \rho_B \mathbf{u}_B). \tag{7}$$

The peculiar velocity $\mathbf{C}_i$ of particle i is its velocity relative to the barycentric velocity, so that

$$\mathbf{C}_i \equiv \mathbf{c}_i - \mathbf{u}, \tag{8}$$

and the diffusion velocity $\mathbf{v}_i$ of species i is its relative mean motion, thus

$$\mathbf{v}_i \equiv \langle \mathbf{C}_i \rangle = \mathbf{u}_i - \mathbf{u}. \tag{9}$$

The species granular temperature is defined proportional to the mean kinetic energy of species i

$$T_i \equiv \frac{1}{d} m_i \langle \mathbf{C}_i \cdot \mathbf{C}_i \rangle = \frac{1}{d} m_i \langle C_i^2 \rangle, \tag{10}$$

and the mixture temperature is

$$T \equiv \frac{1}{n} (n_A T_A + n_B T_B). \tag{11}$$

The coefficient of restitution for collisions between particles i and j is denoted by e_{ij}, with $e_{ij} \leq 1$ and $e_{ij} = e_{ji}$.

3.1.2 The Boltzmann–Enskog Equation for a Mixture

The distribution functions $f_i(\mathbf{c}_i, \mathbf{r}, t)$ for the two species are determined from the set of nonlinear Boltzmann–Enskog equations

$$\left(\frac{\partial}{\partial t} + \mathbf{c}_i \cdot \nabla + \frac{\mathbf{F}_i}{m_i} \cdot \frac{\partial}{\partial \mathbf{c}_i} \right) f_i(\mathbf{c}_i, \mathbf{r}, t) = \sum_j J_{ij} \left[\mathbf{c}_i | f_i(\mathbf{c}_i), f_j(\mathbf{c}_j) \right]. \tag{12}$$

The Boltzmann–Enskog collision operator $J_{ij} \left[\mathbf{c}_i | f_i, f_j \right]$ describing the scattering of pairs of particles is

$$J_{ij} \left[\mathbf{c}_1 | f_i(\mathbf{c}_1), f_j(\mathbf{c}_2) \right] \equiv g_{ij} r_{ij}^{d-1} \int \int d\mathbf{c}_2 d\hat{\sigma} \Theta(\hat{\sigma} \cdot \mathbf{c}_{21})(\hat{\sigma} \cdot \mathbf{c}_{21})$$
$$\times \left[\frac{1}{e_{ij}^2} f_i(\mathbf{c}_1') f_j(\mathbf{c}_2') - f_i(\mathbf{c}_1) f_j(\mathbf{c}_2) \right], \tag{13}$$

where g_{ij} is the radial distribution function, $r_{ij} = r_i + r_j$, $\hat{\sigma}$ is the unit vector directed from the center of the particle of type i to the center of particle j separated at contact by r_{ij}, and $\Theta(x)$ is the Heaviside step function ($\Theta(x) = 0$, for $x < 0$ and $\Theta(x) = 1$, for $x > 0$). The post–collisional velocities $\mathbf{c}_i'$ are given in terms of the pre–collisional velocities $\mathbf{c}_i$ by

$$\mathbf{c}_i' = \mathbf{c}_i + M_{ji}(1 + e_{ij})(\hat{\sigma} \cdot \mathbf{c}_{ji})\hat{\sigma}, \tag{14}$$

where $M_{ji} = m_j/m_{ij}$, $m_{ij} = m_i + m_j$, and $\mathbf{c}_{ji} \equiv \mathbf{c}_j - \mathbf{c}_i$ is the relative velocity between particles.

Also, in the Boltzmann–Enskog collision operator we have used the Enskog assumption for dense gases, i.e.,

$$f_{i,j}^{(2)}(\mathbf{c}_1, \mathbf{c}_2) \rightarrow g_{ij} f_i^{(1)}(\mathbf{c}_1) f_j^{(1)}(\mathbf{c}_2), \tag{15}$$

for the complete pair velocity distribution function $f_{i,j}^{(2)}(\mathbf{c}_1, \mathbf{c}_2)$.

3.1.3 Analysis Based on the Boltzmann–Enskog Equation

To find the equation satisfied by $\langle \psi_i \rangle$, as defined in (6), we multiply both sides of Eq. (12) by ψ_i and then integrate over all velocities $\mathbf{c}$. Thus we get

$$\frac{\partial}{\partial t} \langle n_i \psi_i \rangle = -\nabla \cdot \langle n_i \mathbf{c}_i \psi_i \rangle + n_i \frac{\mathbf{F}_i}{m_i} \cdot \left\langle \frac{\partial \psi_i}{\partial \mathbf{c}_i} \right\rangle + \sum_j \left[\chi_{ij}(\psi_i) - \nabla \cdot \theta_{ij}(\psi_i) \right], \tag{16}$$

where $\chi_{ij}(\psi_i)$ is the collisional source integral

$$\chi_{ij}(\psi_i) \equiv g_{ij} r_{ij}^{d-1} \int \int \int d\mathbf{c}_1 d\mathbf{c}_2 d\hat{\sigma} \Theta(\hat{\sigma} \cdot \mathbf{c}_{21})(\hat{\sigma} \cdot \mathbf{c}_{21}) [\psi_i' - \psi_i]$$
$$\times \left[1 + (1/8)r_{ij}^2(\hat{\sigma} \cdot \nabla)^2 + \cdots\right] f_i(\mathbf{c}_1) f_j(\mathbf{c}_2), \tag{17}$$

and $\theta_{ij}(\psi_i)$ is the collisional flux integral

$$\theta_{ij}(\psi_i) \equiv \frac{1}{2} g_{ij} r_{ij}^d \int \int \int d\mathbf{c}_1 d\mathbf{c}_2 d\hat{\sigma} \hat{\sigma} \Theta(\hat{\sigma} \cdot \mathbf{c}_{21})(\hat{\sigma} \cdot \mathbf{c}_{21}) [\psi_i' - \psi_i]$$
$$\times \left[1 + (1/8)r_{ij}^2(\hat{\sigma} \cdot \nabla)^2 + \cdots\right] f_i(\mathbf{c}_1) f_j(\mathbf{c}_2). \tag{18}$$

When $\psi_i(\mathbf{c}_i) \to \psi_i(\mathbf{C}_i)$, the balance equation may be written in the form

$$\frac{\partial}{\partial t} \langle n_i \psi_i \rangle = -\nabla \cdot \langle n_i \mathbf{c}_i \psi_i \rangle + \left\langle n_i \frac{\partial \mathbf{C}_i}{\partial t} \cdot \frac{\partial \psi_i}{\partial \mathbf{C}_i} \right\rangle$$
$$+ \sum_j \left[\chi_{ij}(\psi_i) - \nabla \cdot \theta_{ij}(\psi_i) - \theta_{ij} \left(\frac{\partial \psi_i}{\partial \mathbf{C}_i} \right) : \nabla \mathbf{u} \right]. \tag{19}$$

To investigate transport processes, we must solve the Boltzmann–Enskog equation (12), with given initial conditions, to obtain the velocity distribution function. Some rigorous properties of any solution of (12) may be obtained from the fact that in any particle collision there are dynamical quantities that are rigorously conserved. For a granular fluid the independent conserved properties are mass and momentum. The transformation (14) conserves momentum but, when $e < 1$, does not conserve energy. Therefore, in a binary collision the change of kinetic energy $\Delta E \neq 0$. In the lowest–order approximation we assume that the fluid has a local Maxwell–Boltzmann (Maxwellian) distribution

$$f_i(\mathbf{c}_i) \approx f_i^{(0)}(\mathbf{c}_i) = \pi^{d/2} n_i \left(\frac{m_i}{2T_i}\right)^{d/2} \exp\left[-\frac{m_i}{2T_i}(\mathbf{c}_i - \mathbf{u})^2\right]. \tag{20}$$

3.1.4 Species Balance Equations

The equation of continuity for species i is obtained by taking $\psi_i = m_i$ in Eq. (16):

$$\frac{\partial}{\partial t}(n_i m_i) + \nabla \cdot (n_i m_i \mathbf{u}_i) = 0. \tag{21}$$

If $\psi_i = m_i \mathbf{c}_i$ in Eq. (16), the balance of linear momentum for species i is obtained as

$$\frac{\partial}{\partial t}(n_i \mathbf{u}_i) + \nabla \cdot (n_i \mathbf{u}_i \mathbf{u}_i) = -\frac{1}{m_i} \nabla \cdot \hat{\mathbf{P}}_i + \frac{n_i}{m_i} \mathbf{F}_i + \Gamma_i, \tag{22}$$

where Γ_i is the momentum source vector which arises due to the interaction between unlike particles and $\hat{\mathbf{P}}_i$ is the species stress tensor.

Taking $\psi_i = \frac{1}{2}m_i\mathbf{C}_i^2$ and using the balance law (19), we obtain the balance of fluctuation energy (granular temperature) for each species:

$$\frac{\partial}{\partial t}(n_iT_i) + \nabla \cdot (n_i\mathbf{u}_iT_i) = -\nabla \cdot \mathbf{Q}_i - \hat{\mathbf{P}}_i : \nabla\mathbf{u} + \mathbf{J}_i \cdot \mathbf{F}_i + \frac{\rho_i}{\rho}\mathbf{v}_i \cdot (\nabla \cdot \hat{\mathbf{P}} - n\mathbf{F}) + \mathscr{D}_i.$$

$$(23)$$

3.1.5 Constitutive Relations

The Chapman–Enskog method is a perturbation method based on a power series expansion of $f_i(\mathbf{c}_i)$ in terms of the Knudsen number $K_n = l/L$, where l is the mean free path and L a characteristic length. The method provides different expressions for the pressure tensor $\hat{\mathbf{P}}_i$, the momentum source term Γ_i, the energy (heat) flux $\mathbf{Q}_i$, and the kinetic energy dissipation rate $\mathscr{D}_i$.

The species stress tensor has at the Navier–Stokes level, the standard Newtonian form

$$\hat{\mathbf{P}}_i = p_i\hat{\mathbf{I}} + \mu_i(\nabla\mathbf{u} + \nabla\mathbf{u}^T),$$

$$(24)$$

where p_i is the partial pressure of species i, μ_i is the viscosity of species i, and $\hat{\mathbf{I}}$ denotes the unit tensor. The equation of state for the partial pressure of species i can then be written as: $p_i = n_iZ_iT_i$, where $Z_i := (1 + \sum_{j=l,s} K_{ij})$ is the *compressibility* factor of species i and $K_{ij} := \phi_jg_{ij}(1 + R_{ij})^d/2$, with g_{ij} being the radial distribution function and $R_{ij} = r_i/r_j$ the size-ratio.

At the zeroth-order approximation for the distribution function, the momentum source term can be written as:

$$\Gamma_i = n_iK_{ij}T\left[\left(\frac{m_j - m_i}{m_{ij}}\right)\nabla(\ln T) + \nabla\left[\ln\left(\frac{n_i}{n_j}\right)\right] + \frac{4}{r_{ij}}\left(\frac{2m_im_j}{\pi m_{ij}T}\right)^{1/2}(\mathbf{u}_j - \mathbf{u}_i)\right],$$

$$(25)$$

which arises due to the interaction of unlike particles and satisfies the relation $\sum_i \Gamma_i = \mathbf{0}$.

The energy flux is given by

$$\mathbf{Q}_i = -\kappa_i\sqrt{T_i}\nabla T_i,$$

$$(26)$$

where κ_i is the analog of the thermal conductivity of species i. Finally, the rate of kinetic energy dissipation of species i is given by

$$\mathscr{D}_i = \sum_j \frac{\pi^{\frac{d-1}{2}}}{d\Gamma(d/2)} g_{ij} r_{ij}^{d-1} m_i n_i n_j M_{ji} \left[M_{ji}(1 - e_{ij}^2) \left(\frac{2T_i}{m_i} + \frac{2T_j}{m_j} \right) \right.$$

$$\left. +4(1 + e_{ij}) \frac{T_i - T_j}{m_i + m_j} \right] \left(\frac{2T_i}{m_i} + \frac{2T_j}{m_j} \right)^{1/2} . \tag{27}$$

In the elastic limit $e_{ij} = 1$ the above constitutive relations reduce to the "classical" theory of non-uniform dense gases (Chapman and Cowling 1970). On the other hand, keeping the next order gradient terms corresponds to the *Burnett* or *super-Burnett* hydrodynamics (García-Colín et al. 2008; Serero et al. 2006). Let us mention that a complete treatment of the energy flux $\mathbf{Q}_i$ should include a "non-Fourier" term which relates the energy flux with the density gradient ∇n. This term does not have an analogue in the hydrodynamics of molecular fluids and should be included in the calculation of temperature and density profiles of a vertically vibrated granular system. The early studies of granular mixtures, cited above, used a nearly elastic formalism. Recently, an important step was taken by Serero et al. (2007), Garzó et al. (2009), who proposed a new Sonine approach for the calculation of higher order expansions to treat strong inelasticity. These contributions open a new perspective to increase the accuracy of hydrodynamics calculations for granular binary mixtures.

4 Illustrative Example: The Brazil Nut Segregation Phenomenon

When a mixture of nuts is shaken they are segregated according to their size and/or mass; the dynamics can results in an upward or downward movement of the larger particles (the Brazil nuts). The segregation on top is known as the Brazil Nut Problem (BNP), while the segregation towards the bottom is termed the Reverse Brazil Nut Problem (RBNP). This is a clear example where theoretical granular hydrodynamics have been applied to study the mechanisms driving the segregation phenomenon in vibro-fluidized granular mixtures.

Recently, scientists of the Department of Theoretical and Applied Mechanics, at the Cornell University, Ithaca, New York, USA, have developed kinetic theories for particle segregation in collisional flows of dry granular materials (Arnarson and Willits 1998; Willits and Arnarson 1999; Jenkins and Yoon 2002; Yoon and Jenkins 2006). In the works carried out by Jenkins (1998) and Arnarson and Jenkins (2000) the species diffusion velocity has been calculated in the context of the Chapman–Enskog procedure, and following the procedure introduced by Jenkins and Mancini (1987), they obtained the weighted difference of the species momentum balances and the gradients of partial pressure to outline expressions for the interparticle force. However, they did not integrate the equations to determine the variations of the fields and the segregation dynamics. On the other hand, for binary mixtures of inelastic spheres in the presence of a temperature gradient, Arnarson and Willits (1998) found that larger, denser particles tend to be more concentrated in cooler regions. This result was confirmed by numerical simulations (Henrique et al. 2000). However, this mechanism of segregation is a natural consequence of the imposed gradient of temperature and it is not related to the nature of the grains (Henrique et al. 2000). In 2002 Jenkins and Yoon investigated the upward ⇔ downward transition employing the hydrodynamic equations for a binary mixture of elastically colliding spheres. Their theory agrees with the phase diagram for the BNP/RBNP transition introduced by Hong et al. (2001). However, the driving mechanism for segregation in the hydrodynamic framework is presumably different from that of the percolation–condensation transition (Hong et al. 2001). On the other hand, most of the above mentioned kinetic models apply only for nearly elastic particles so that non-equipartition effects on segregation are not accounted for.

Since granular systems are dissipative, energy equipartition does not generally hold (Ippolito et al. 1995; McNamara and Luding 1998). Such lack of energy equipartition has been recently investigated in several theoretical and numerical works (Alam and Luding 2002, 2003), and also confirmed in air-table experiments (Ippolito et al. 1995) and vibrofluidized beds (Feitosa and Menon 2002; Wildman and Parker 2002). The influence of the non-equipartition of kinetic energy on segregation in binary granular mixtures was considered by Trujillo et al. (2003). They used constitutive relations for partial pressures, energy flux, and inelastic dissipation rate that take into account the breakdown of kinetic energy equipartition between the two species. However, up to now the non-equipartition seems to have no discernible influence on experimental observations (Schöter et al. 2006). On the other hand, the evolution equation for the relative velocity of the intruders was shown to be coupled to the inertia of the smaller particles. Let us remark that an application of the theoretical framework by Trujillo et al. (2003) illustrates the onset of the reverse buoyancy effect and the non-monotonic ascension dynamics reported by the experimentalists (Shinbrot and Muzzio 1998; Möbius et al. 2001) (See Alam et al. (2006) for a detailed analysis of the unsteady forces and the added mass effect). An extension of the works reported by Trujillo et al. (2003) and Alam et al. (2006), which incorporates the effects due to thermal diffusion was carried out recently by Garzó (2008). This work completes the theoretical landscape towards a unified mathematical description of the segregation dynamics based on granular hydrodynamics.

5 Conclusions

We have presented a brief introduction of the physics of granular materials in the fluidized state and described some remarkable analogies with different fluid-mechanical phenomena, such as convection, buoyancy, the Leidenfrost effect, the hydraulic jump and liquid jets.

We derived a phenomenological continuum description for granular fluids based on hydrodynamic equations conveniently modified to take into account important features such as the inelastic nature of the collisions between the grains. The granular hydrodynamics equations can be formally derived from the Boltzmann–Enskog kinetic equation for a dense gas of hard spheres. This kinetic theory description incorporates a sink term in the balance equation for the kinetic energy as a signature for granular fluids. These equations can be worked out via the Chapman–Enskog procedure. We calculated the constitutive relations up to the lowest-order approximation for a Maxwellian distribution. Our main purpose was to present a brief pedagogical introduction to the kinetic theory of inelastically colliding particles.

The collection of theoretical ideas presented above and their congruence with experimental outcomes opens the possibility to construct a global phenomenological understanding of the segregation dynamics in terms of granular hydrodynamics. Another promising feature is that the onset of granular convection can be calculated from a linear stability analysis of a hydrodynamic-like model of the granular flow (Eshuis et al. 2010). However, up to now the coexistence between solid and fluid phases in granular systems is not fully described by the hydrodynamic theory. On the other hand, the glassy behavior of granular materials can not be described within the theoretical approach presented in this work. Nevertheless, the theoretical granular hydrodynamics represent a good starting point towards the construction of a general theory for granular materials. In passing, we recall the new recent ideas introduced by Yimin Jiang and Mario Liu towards the construction of a granular solid hydrodynamics (Jiang and Liu 2009).

Acknowledgments L.T. acknowledges the organizer of the XVII Annual Meeting of the Fluid Dynamics Division (XVII-DDF) of the Mexican Physical Society, with special mention to Anne Cros. J. Klapp thank ABACUS, CONACyT grant EDOMEX-2011-C01-165873.

References

Alam M, Luding S (2002) How good is the equipartition assumption for the transport properties of a granular mixture? Granul Matter 4:139–142

Alam M, Luding S (2003) Rheology of bidisperse granular mixtures via event-drivent simulations. J Fluid Mech 476:69–103

Arnarson BÖ, Willits JT (1998) Thermal diffusion in binary mixtures of smooth, nearly elastic spheres with and without gravity. Phys Fluids 10:1324–1328

Alam M, Trujillo L, Herrmann HJ (2006) Hydrodynamic theory for reverse Brazil nut segregation and the non-monotonic ascension dynamics. J Stat Phys 124:587623

Alam M, Willits JT, Arnarson BÖ, Luding S (2002) Kinetic theory of a binary mixture of neraly elastic disks with size and mass disparity. Phys Fluids 14:4085–4087

Arnarson BÖ, Jenkins JT (2000) Particle segregation in the context of the species momentum balances. In: Helbing D, Herrmann HJ, Schreckenberg M, Wolf DE (eds) Traffic and granular flow'99: social, traffic and granular dynamics. Springer, Berlin, pp 481–487

Arnarson BÖ, Jenkins JT (2004) Binary mixtures of inelastic spheres: simplified constitutive theory. Phys Fluids 16:4543–4550

Boudet JF, Amarouchene Y, Bonnier B, Kellay H (2007) The granular jump. J Fluid Mech 572:413–431

Brey JJ, Dufty JW, Kim CS, Santos A (1998) Hydrodynamics for granular flow at low density. Phys Rev E 58:4638–4653

Brey JJ, Ruiz-Montero MJ, Moreno F (2001) Hydrodynamics of an open vibrated granular system. Phys Rev E 63:061306

Brilliantov NV, Pöschel T (2004) Kinetic theory of granular gases. Oxford University Press, Oxford

Caballero G, Bergmann R, van der Meer D, Prosperetti A, Lohse D (2007) Role of air in granular jet formation. Phys Rev Lett 99:018001

Chapman S, Cowling TG (1970) The mathematical theory of nonuniform gases. Cambridge University Press, Cambridge

Edwards SF, Oakeshott RBS (1989) Theory of powders. Physica A 157:1080–1090

Eshuis P, van der Weele K, van der Meer D, Lohse D (2005) Granular Leidenfrost effect: experiment and theory of floating particle clusters. Phys Rev Lett 95:258001

Eshuis P, van der Meer D, Alam M, van Gerner HJ, van der Weeke K, Lohse D (2010) Onset of convection in strongly shaken granular matter. Phys Rev Lett 104:038001

Feitosa K, Menon N (2002) Breakdown of energy equipartition in a 2D binary vibrated granular gas. Physical Review Letters 88:198301

Gallas JAC, Herrmann HJ, Sokołowski S (1992) Convection cells in vibrating granular media. Phys Rev Lett 69:1371–1373

García-Colín LS, Velasco RM, Uribe FJ (2008) Beyond the Navier-Stokes equations: Burnett hydrodynamics. Phys Rep 465:149–189

Garzó V (2008) Brazil-nut effect versus reverse Brazil-nut effect in a moderately dense granular fluid. Phys Rev E 78:020301

Garzó V, Dufty JW (1999) Dense fluid transport for inelastic hard spheres. Phys Rev E 59:5895–5911

Garzó V, Dufty JW, Hrenya CM (2007) Enskog theory for polydisperse granular mixtures. I. Navier-Stokes order transport. Phys Rev E 76:031303

Garzó V, Dufty JW, Hrenya CM (2007) Enskog theory for polydisperse granular mixtures. II. Sonine polynomial approximation. Phys Rev E 76:031304

Garzó V, Vega-Reyes F, Montanero JM (2009) Modified Sonine approximation for granular binary mixtures. J Fluid Mech 623:387–411

Hong DC, Hayakawa H (1997) Thermodynamic theory of weakly excited granular systems. Phys Rev Lett 78:2764–2767

Hayakawa H, Yue S, Hong DC (1995) Hydrodynamic description of granular convection. Phys Rev Lett 75:2328–2331

Henrique C, Batrouni G, Bideau D (2000) Diffusion as a mixing mechanism in granular materials. Phys Rev E 63:011304

Herrmann HJ (1993) On the thermodynamics of granular media. J de Physique II (France) 3:427–433

Hong DC, Quinn PV, Luding S (2001) Reverse Brazil nut problem: competition between percolation and condensation. Phys Rev Lett 86:3423–3426

Huerta DA, Sosa V, Vargas MC, Ruiz-Suárez JC (2005) Archimedes' principle in fluidized granular systems. Phys Rev E 72:031307

Ippolito I, Annic A, Lemaître J, Oger L, Bideau D (1995) Granular temperature: experimental analysis. Phys Rev E 52:2072–2075

Jaeger HM, Nagel SR, Behringer RP (1996) Granular solids, liquids, and gases. Rev Mod Phys 68:1259–1273

Jenkins JT (1998) Particle segregation in collisional flows of inelastic spheres. In: Herrmann HJ, Holvi J-P, Luding S (eds) Physics of dry granular media. Kluwer, Dordrecht, p 658

Jenkins JT, Mancini F (1987) Balance laws and constitutive relations for plane flows of a dense, binary mixture of smooth, nearly elastic, circular disks. J Appl Mech 54:27–34

Jenkins JT, Mancini F (1989) Kinetic theory for binary mixtures of smooth, nearly elastic spheres. Phys Fluids A 1:2050–2057

Jenkins JT, Savage SB (1983) A theory for the rapid flow of identical, smooth, nearly elastic, spherical particles. J Fluid Mech 130:187–202

Jenkins JT, Yoon DK (2002) Segregation in binary mixtures under gravity. Phys Rev Lett 88:194301

Jiang L, Liu M (2009) Granular solid hydrodynamics. Granul Matter 11:139–156

Kadanoff LP (1999) Built upon sand: theoretical ideas inspired by granular flows. Rev Mod Phys 71:435–444

Knight JB, Ehrichs EE, Kuperman V, Flint JK, Jaeger HM, Nagel SR (1996) Experimental study of granular convection. Phys Rev E 54:5726–5738

Leidenfrost JG (1966) On the fixation of water in diverse fire. Int J Heat Mass Transf 9:1153–1166

Lohse D, Bergmann R, Mikkelsen R, Zeilstra C, van der Meer D, Versluis M, van der Weele K, van der Hoef M, Kuipers H (2004) Impact on soft sand: void collapse and jet formation. Phys Rev Lett 93:198003

Lohse D, Rauhé R, Bergmann R, van der Meer D (2004) Creating a dry variety of quicksand. Nature 432:689–690

López de Haro M, Cohen EGD, Kincaid JM (1983) The Enskog theory for multicomponent mixtures. I Linear transport theory. J Chem Phys 78:2746–2759

Maes C, Thomas SR (2011) Archimedes' law and its corrections for an active particle in a granular sea. J Phys A Math Theor 44:285001

McNamara S, Luding S (1998) Energy non-equipartition in systems of inelastic, rough spheres. Phys Rev E 58:2247

Möbius ME, Lauderdale BE, Nagel SR, Jaeger HM (2001) Size separation of granular particles. Nature 414:270

Schöter M, Ulrich S, Kreft J, Swift JB, Swinney HL (2006) Mechanisms in the size segregation of a binary granular mixture. Phys Rev E 74:011307

Serero D, Goldhirsch I, Noskowicz SH, Tan M-L (2008) Hydrodynamics of granular gases and granular mixtures. J Fluid Mech 554:237–258

Serero D, Noskowicz SH, Goldhirsch I (2007) Exact results versus mean field solutions for binary granular gas mixtures. Granul Matter 10:37–46

Shinbrot T, Muzzio FJ (1998) Reverse buoyancy in shaken granular beds. Phys Rev Lett 81:4365–4368

Trujillo L, Alam M, Herrmann HJ (2003) Segregation in a fluidized binary granular mixture: competition between buoyancy and geometric forces. Europhys Lett 64:190–196

Trujillo L, Herrmann HJ (2003) Hydrodynamic model for particle size segregation in granular media. Physica A 330:519–542

Wildman RD, Parker DJ (2002) Coexistence of two granular temperatures in binary vibrofluidized beds. Phys Rev Lett 88:064301

Willits JT, Arnarson BÖ (1999) Kinetic theory of a binary mixture of nearly elastic disks. Phys Fluids 11:3116–3122

Yoon DK, Jenkins JT (2006) The influence of different species' granular temperature on segregation in a binary mixture of dissipative grains. Phys Fluids 18:073303

Efficient Neighborhood Search in SPH

Juan Pablo Cruz Pérez and José Antonio González Cervera

Abstract One of the main problems found during the implementation of an N-body algorithm, is its inefficiency when the number of points to evaluate is increased. This is a consequence of the order $O(N^2)$ of these methods. With this in mind, when we use the method of Smoothed Particle Hydrodynamics (SPH), it is necessary to find an algorithm that allows us to make the computation in an efficient way. The method presented in this article is of order $O(N)$, being more efficient as well as easy to implement, reducing the computing time.

1 Introduction

In order to study a physical system, we require to solve partial differential equations to extract and analyze the physical quantities required to understand the problem. Depending on the complexity of the system, the equations could be extremely complicated and may not be solved analytically. It is a common practice to use numerical tools to solve them.

One of the most used and well known methods used to analyze scenarios containing fluids is the Smoothed Particle Hydrodynamics method. Even though it is a powerful method to solve the equations describing the evolution of the fluids, in its simplest form it can be very inefficient. The reason behind this is that if we

J. P. Cruz Pérez (✉) · J. A. G. Cervera
Instituto de Física y Matemáticas, Universidad Michoacana de San Nicolás de Hidalgo.
Edificio C-3, Cd. Universitaria, 58040 Morelia, MICH, Mexico
e-mail: dirak3d@ifm.umich.mx

J. A. G. Cervera
e-mail: gonzalez@ifm.umich.mx

J. Klapp et al. (eds.), *Fluid Dynamics in Physics, Engineering and Environmental Applications*, Environmental Science and Engineering,
DOI: 10.1007/978-3-642-27723-8_12, © Springer-Verlag Berlin Heidelberg 2013

divide the physical system in N small parts, we require $O(N^2)$ operations to obtain the new values of the physical quantities describing the fluid. There are methods reducing the number of operations required for the evolution. For example, tree methods require a number of $O(NlogN)$ operations (Gafton and Rosswog 2011; Press et al. 2007; Hernquist and Kats 1988). In this work, we analyze another algorithm that requires $O(N)$ operations (Onderik et al. 2007). The paper is organized as follows: in Sect. 2 we introduce the SPH method, in Sect. 3 we introduce an efficient search of neighbors method, in Sect. 4 we describe in detail the Cell Index method, in Sect. 5 we describe the implementation of the algorithm an finally in Sect. 6 we conclude.

2 Smoothed Particle Hydrodynamics

SPH is a Lagrangian method and it can be used to solve differential equations describing the behavior of fluids in a physical system. The ideas behind the SPH method are the following:

- Approximation of the functions describing the fluid using the identity

$$f(x_0) = \int f(x)\delta(x - x_0)dx, \tag{1}$$

  then we substitute the Dirac's delta with a function W called *Kernel* such that

$$f(x_0) \approx \int f(x)W(x - x_0, h)dx. \tag{2}$$

  W is a function of the distance between x and x_0. The parameter h is known as the smoothing length, it satisfies $lim_{h \to 0} W(x - x_0, h) = \delta(x - x_0)$, $\int W(x - x_0, h)dx = 1$ and it is a compact support function.
- Discretize the fluid en N elements (called particles). With this approach we can substitute the integrals over all the space with sums over all the particles of the fluid. Then, substitute the volume element of the particle with the coefficient between its mass and its density:

$$f(x_b) \approx \sum_{a=1}^{N} f(x_a)W(x_a - x_b, h_b)\frac{m_a}{\rho_a}. \tag{3}$$

  The subscripts, name each one of the particles of the volume. For instance, m_a is the mass of the particle located at x_a. In this way, it is possible to compute any function related with the system. For example, for the density:

$$\rho(x_b) \approx \sum_{a=1}^{N} m_a W(x_a - x_b, h_b). \tag{4}$$

- Finally, we discretize the derivatives of the functions:

$$\partial_i f(\boldsymbol{x}_b) \approx \sum_{a=1}^{N} f(\boldsymbol{x}_a) \partial_i W(\boldsymbol{x}_a - \boldsymbol{x}_b, h_b) \frac{m_a}{\rho_a}. \tag{5}$$

With these recipes, it is possible to discretize the evolution equations of the fluids to solve them numerically. It is easy to notice from these equations that in order to evolve each one of the particles of the fluid, it is required to sum over all the other particles of the system. This implies that the simplest method to implement it is also the most inefficient one because it requires $O(N^2)$ operations. In the next section we introduce a method that increases the efficiency using only $O(N)$ operations.

3 Efficient Search of Neighbors

In order to increase the efficiency of the computations we use a method called *Index of Cells*. The general idea behind this method is the following: a numerical grid is created and each of the particles of the fluid is located inside of one of the *cells* of the grid. We associate each particle with an index representing this cell. Each of these numbers is encrypted in one number using a *key* function which returns an integer. Finally, the list of the keys is ordered providing an idea of the real spatial order of the particles in an efficient way. Lets explain in detail the procedure.

4 Cell Indexing

The Cell indexing method requires four steps: *classification*, *ordering*, *census* and *search* (Onderik et al. 2007).

4.1 Classification

A key is assigned to each one of the N particles with coordinates $\boldsymbol{r}_a = (x_a, y_a, z_a)$ where $a = 1, \ldots, N$. A virtual box containing all the particles is constructed such that its size is the minimum required and the sizes of the box are parallel to the axes of the coordinates:

$$x_{\max} = \max_{a=1,\ldots,N} \{x_a\},$$

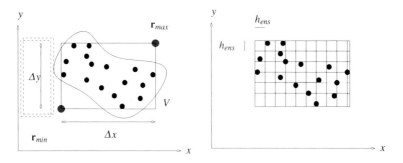

Fig. 1 *Left* in this figure we show in (2D) how to compute $\Delta x = x_{max} - x_{min}$ and $\Delta x = y_{max} - y_{min}$, comparing both the smallest one is elected and it is divided in N_p parts. *Right* in this example, $h_{ens} = \frac{\Delta y}{N_p}$ and we obtain squares of dimension h_{ens}

$$x_{min} = \min_{a=1,...,N} \{x_a\}$$

and similar for y_{max}, z_{max}, y_{min} and z_{min}. With $\boldsymbol{r}_{max} = (x_{max}, y_{max}, z_{max})$ and $\boldsymbol{r}_{min} = (x_{min}, y_{min}, z_{min})$ it is possible to visualize the box containing all the particles, Fig. 1 (in order to simplify the explanations all the plots in the article are in two spatial dimensions).

Once the box is constructed, we choose the smallest side and divide it in N_p parts. We choose the dimension of the *fundamental cell* as the size of the cube (a square in two spatial dimensions) with length h_{ens}

$$h_{ens} = \frac{1}{N_p} \min\{x_{max} - x_{min}, y_{max} - y_{min}, z_{max} - z_{min}\}. \tag{6}$$

Once the box has been divided in cubes of size h_{ens} called *cells*, each of the particles is inside of one of these cells. And we can see that N_p is a parameter that adjust the number of cells in which the main box will be divided. Lets define a *class* as the set of all the particles inside of one cell. We assign to each class a triad of integers (i_a, j_a, k_a) indicating the position of the cell with the following formula

$$i_a = 1 + \left[\frac{x_a - x_{min}}{h_{ens}}\right], \; j_a = 1 + \left[\frac{y_a - y_{min}}{h_{ens}}\right], \; k_a = 1 + \left[\frac{z_a - z_{min}}{h_{ens}}\right]. \tag{7}$$

The index i_a runs from 1 to n_x with $n_x = 1 + \left[\frac{x_{max} - x_{min}}{h_{ens}}\right]$.[1] The indices j_a and k_a have similar expressions.

---

[1] $\left[\frac{a}{b}\right] = c$, where a and b are real numbers and c is the integer such that $a = b \cdot c + r$, here r is a real too.

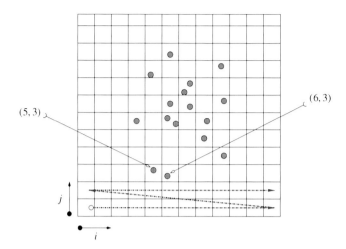

Fig. 2 The keys generated for the particles are: for the particle 1 with coordinates $(5,3)$, $key(1) = 5 + n_x \cdot 3$ and for the particle 2 with coordinates $(6,3)$, $key(2) = 6 + n_x \cdot 3$, if we know n_x and n_y the key gives an integer. The *dashed line* describes graphically describe (in $(2D)$) the way the indices increase such that the key function increases monotonically

Now, it is necessary to build a function such that it increases uniformly as the indices i_a, j_a, k_a increase. Lets define the key function depending on i_a, j_a and k_a in the following way:

$$key(a) = i_a + n_x \cdot j_a + n_x \cdot n_y \cdot k_a. \tag{8}$$

Every cell has a unique key fixing the class of the particles that can be found there. For instance, the cell with coordinates $(i, j, k) = (1, 1, 1)$ has the unique key $key_{(1,1,1)} = 1 + n_x \cdot 1 + n_x \cdot n_y \cdot 1$.

The key is a monotonically increasing function as long as the cells are listed in the following way: starting with $i_a = 1$, $j_a = 1$ and $k_a = 1$

1. The index i_a is increased by 1 until it reaches n_x.
2. The index j_a is increased by one unit.
3. Repeat (1) and (2), until j_a is equal to n_y.
4. Once $j_a = n_y$ increase k_a in one unit.
5. Repeat from (1) to (4) until $k_a = n_z$. See Fig. 2.

If the position of the particles r_a is generated using the *acceptance and rejection method* (Press et al. 2002) it is not possible to assume that the particles have an order in its spatial position. For example, in Fig. 2 we identify two particles in $(2D)$ the particle with index $a = 1$ and coordinates $(i, j) = (5, 3)$ and the particle with index $a = 2$ and coordinates $(i, j) = (6, 3)$. The key of the first one is $key(1) = 5 + b_x \cdot 3 = 5 + 12 \cdot 3 = 41$ and the key of the second particle is $key(2) = 6 + b_x \cdot 3 = 6 + 12 \cdot 3 = 42$.

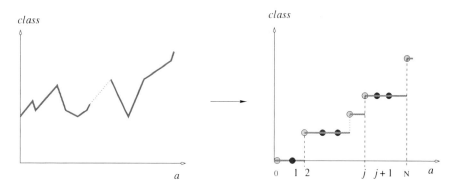

Fig. 3 *Left* if the particles of the configuration of interest have positions elected with a random process, something similar to this plot can be obtained. *Right* after ordering, it is important to store the information of the first particle of each class. These are the reference particle is the first on each stair

In general, the index of the particles could be disordered such that the key function it is not increasing monotonically. This is the main reason behind the ordering of the N particles using the keys of their classes. We use a method known as *radix sort* of order $O(N)$ and the efficiency of all the algorithm is inherited from it (Terdiman 2000).

4.2 Radix Sort

Assume we have N integer numbers. Now, we will consider the binary expression of these numbers. For example, consider the binary representation of the following four numbers: 23,48,25 y 6:

$$
\begin{array}{ll}
010111 & 23 \\
110000 & 48 \\
011001 & 25 \\
000110 & 6
\end{array}
$$

Checking the first digit in the binary expression from right to left: the number 23 has the number 1, the 48 has a 0, the 25 a 1 and the 6 a 0.

Take all the elements of the list with a 0 and put them in the top of the list and the numbers having a 1 on the bottom of the list. The result is the following

$$
\begin{array}{ll}
110000 & 48 \\
000110 & 6 \\
010111 & 23 \\
011001 & 25
\end{array}
$$

The next step is to repeat the procedure but now with the second digit from right to left. The number 48 has a 0, 6 a 1, 23 a 1 and 25 a 0. This means that now we have to put the 48 and the 25 on the top of the list and 6 and 23 on the bottom (in this order):

$$
\begin{array}{ll}
110000 & 48 \\
011001 & 25 \\
000110 & 6 \\
010111 & 23
\end{array}
$$

It is worth noticing that the order of the previous figure is altered and it does not matter. Continuing with the procedure, we obtain:

step 3		step 4		step 5		step 6	
110000	48	110000	48	000110	6	000110	6
011001	25	000110	6	110000	48	010111	23
000110	6	010111	23	010111	23	011001	25
010111	23	011001	25	011001	25	110000	48

The final result (as expected) is 6, 23, 25 y 48. The procedure required 6 steps related with the number of digits used in the binary representation, this means that the number of steps is independent of the size of the list; the number of internal operations, when the *zeros* and *ones* are stored in the lists, are the one changing depending on the size of the elements to order. In this example we used 6-bits integers, if 32-bits integers are required, then the number of operations increases to 32.

4.3 Census

Once we have the list of the particles and the class $(a, key(a))$, the next task is to classify and identify the class of each particle: starting with $a = 2$ we check the condition $key(a) = key(a - 1)$. If it is *true* the particle a is in the same class (i.e. the same cell) as the particle $a - 1$, if it is *false* then we have found that the particle is part of another class. We repeat this procedure with all the particles. During the process, every time the condition $key(a) = key(a - 1)$ is not satisfied, the following data are stored:

- $pr(i) = a$, the index of the ith class found in the ath particle (*reference particle*).
- $class(i) = key(a)$ is the ith class with value $key(a)$.

Once the N particles are checked, the number of times that the condition was not satisfied is the number of different classes $N_{classes}$ found in the configuration. This means $i = 1, \ldots, N_{classes}$ (Fig. 3).

4.4 Search

In order to obtain the averages, it is necessary to know how many neighbors each particle has. All the previous steps (*classification, ordering and census*) were done in order to improve this part of the process. Given the ath particle with key

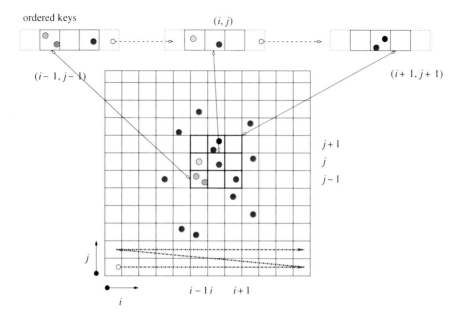

Fig. 4 In this plot we show the search processes for a particle with coordinates (i,j). It is easy to notice that we only search in the cells next to the cell containing the particle $(i+s, j+t)$ with $s, t = -1, 0, 1$ (9 cells in total). On one hand, for all the cells (with exception of $(i-1, j-1)$, $(i-1, j+1)$ and $(i+1, j+1)$) we will obtain from the *ordered keys* a reference particle and the number of particles in that class. On the other, we will no obtain any information from the cells $(i-1, j-1)$, $(i-1, j+1)$ and $(i+1, j+1)$. The one-dimensional array of cells, represents the function given in the Eq. (8): change the problem of assigning an order to the particles in $(2D)$ or $(3D)$ for the ordering in $(1D)$

$key(a) = q$, it is possible to find the cell occupied by this particle using the equation:

$$i = mod(q, n_x), \quad i = 0, \ldots, n_x - 1 \tag{9}$$

$$j = mod\left(\frac{q-i}{n_x}, n_y\right), \quad j = 0, \ldots, n_y - 1 \tag{10}$$

$$k = \frac{q - i - n_x \cdot j}{n_x \cdot n_y}, \quad k = 0, \ldots, n_z - 1. \tag{11}$$

If the indices of the cells (i, j, k) start with 1 instead of 0, the a readjust is necessary in the previous equations with $i = 0 \rightarrow n_x$, $j = 0 \rightarrow n_y$.

Once the coordinates of the ath particle are obtained (i_a, j_a, k_a), the neighbor cells are obtained:

$$\mathbf{N}_{cells}(a) = \{(i, j, k) | i = i_a + s, j = j_a + t, k = k_a + w, \ i, j, k = -1, 0, 1\} \tag{12}$$

each one of these cell has a key and using the ordered list built during the census, it is possible to check if the cells contain particles. If the ath particle is in the boundary, then it is not possible to explore the neighbor cubes. That is the reason why we have to confirm for each (i, j, k) of the $\mathbf{N}_{cells}$ the following condition

$$i \leq n_x, \, j \leq n_y, \, k \leq n_z, \tag{13}$$

$$i \geq 1, \, j \geq 1, \, k \geq 1. \tag{14}$$

4.5 Periodic Boundary Conditions

For cosmological or some astrophysical processes could be interesting to assign periodic boundary conditions. With this in mind, let us explain how to implement the algorithm for these boundaries. If the inequalities (13) and (14) are satisfied this means that the cell is part of the boundary of our domain. In two dimensions this means that the cell is on the side or the corner of the square of the domain. Consider a box containing all the particles in two dimensions:

- If $i = 1$, the cell is part of the side parallel to the x-axis closest to the origin. we use the box in $i = n_x$ because we do not have a box in $i = 0$,
- If $j = 1$, the cell is part of the side parallel to the y-axis. We identify $j = 0$ with $j = n_y$.
- If $i = j = 1$ the cell is in the lower left corner of the box. We identify the cell $i = j = 0$ with the cell $i = n_x$ and $j = n_y$. See Fig. 5.
- If $i = n_x$, we identify the cell $i = n_x + 1$ with the cell in $i = 1$.
- If $j = n_y$, we identify the cell $j = n_y + 1$ with the cell in $j = 1$.
- If $i = n_x$ and $j = n_y$ identify $i = n_x + 1, j = ny + 1$ with $i = j = 1$.
- For the rest of the corners, we have: $i = 1, j = n_y$ (search in $i = n_x$ and $j = 1$) and $i = n_x, j = 1$(search in $i = 1$ and $j = n_y$).

4.6 Big O-Notation

To estimate the time the algorithm consumes during the calculations lets use the big O-notation (Rodríguez-Meza et al. 2011).

1. **Classification:** In the first step of the implementation, it is required to compute x_{min}, y_{min} and z_{min}. The required time to do this is $T_{min} = \alpha_{min}N$, similarly for x_{max}, y_{max} and z_{max} we obtain $T_{max} = \alpha_{max}N$. To find (i, j, k) we have $T_{index} = \alpha_{index}N$ and finally $T_{keys} = \alpha_{keys}N$. From now on all the α's will represent an adhoc scalar to describe a time process proportional to the number of particles

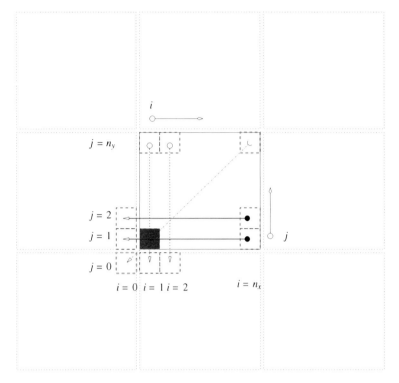

Fig. 5 As an example for the periodic boundary conditions we show the cell $(i = 1, j = 1)$ with its neighbors: $(i = 0, j = 0) \leftarrow (i = n_x, j = n_y)$, $(i = 0, j = 1) \leftarrow (i = n_x, j = 1)$, $(i = 0, j = 2) \leftarrow (i = n_x, j = 2)$, $(i = 1, j = 0) \leftarrow (i = 1, j = n_y)$, $(i = 2, j = 0) \leftarrow (i = 2, j = n_y)$

N. This process is done only once, this means that the total time required to build the virtual box is

$$T_{classification} = (\alpha_{min} + \alpha_{max} + \alpha_{index} + \alpha_{keys})N = \alpha_{classification}N \qquad (15)$$

this means

$$O(T_{classification}) = O(\alpha_{classification}N) = O(N). \qquad (16)$$

2. **Radix Sort:** We have a list of N integers obtained using n-bits. Since $n \ll N$ the total time spent ordering all the numbers is $T = n\alpha_{bit}N$, where $\alpha_{bit}N$ is the time consumed ordering the N numbers for each one of the n digits in the binary respresentation, obtaining

$$T_{radix} = \alpha_{radix}N \rightarrow O(T_{radix}) = O(N) \qquad (17)$$

3. **Census:** For each one of the N numbers two tasks are required: decide if the class is the same or different to the next one. Depending on the answer, it stores

a number or pass to the next analysis. The time consumed in the process could be expressed as

$$T_{census} = \alpha_{census} N \rightarrow O(T_{census}) = O(N) \tag{18}$$

4. **Search:** For each one of the N particles. The indices (i_a, j_a, k_a) are calculate using $key(a)$. The time for this step is T_{key}. Next, we compute N_{cells}, in T_{cells} units of time. Finally we search between the $N_{classes}$ with an estimated time of $T_{clases} = \alpha_{clases} N_{clases}$. For this step we have:

$$T_{search} = T_{key} + T_{cells} + \alpha_{class} N_{class}, \tag{19}$$

obtaining

$$O(T_{search}) = O(T_{key}) + O(T_{cells}) + O(N_{class}), \tag{20}$$

Adding up everything

$$O(T_{total}) = O(T_{classification}) + O(T_{radix}) + O(T_{census}) + N \cdot O(T_{search}), \tag{21}$$

where $N > N_{class}$ (we have multiplied by N because we search neighbors for each one of the particles). It is possible to reduce this estimation considering $O(T_{key}) \ll O(N)$ and $O(T_{census}) \ll O(N)$ obtaining

$$O(T_{total}) = O(N) + O(N) + O(N) + N \cdot O(N_{class}) = O(N). \tag{22}$$

5 Implementation

As an example of the efficiency of the implementation, we will compute the density of a particle using the SPH Eq. (3) (Liu and Liu 2009)

$$\rho_a = \sum_{b=1}^{N} m_b W_{ab}, \tag{23}$$

where $\rho_a = \rho(x_a)$ and $W_{ab} = W(x_a - x_b, h_b)$ then, for each a it is necessary to perform N summations. If we perform this operation over the N particles the final method would be of order $O(N^2)$. Lets keep in mind that the Kernel is different from zero only for some of the particles around the calculation point, implying that we are performing a huge number of useless operations. To avoid this problems, we use the cell indexing method. In this way, it is easy to know the cell where the point of interest is in and also all the neighbor cells. The algorithm is the following one:

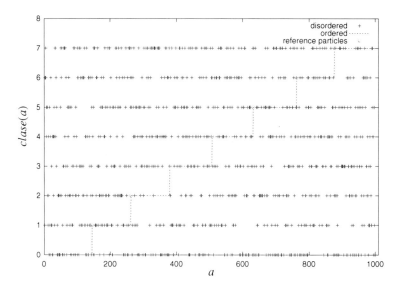

Fig. 6 This graphic shows the set of (N=1000) particles corresponding to a constant density distribution. The *dashed line* represents the ordered classes withe the radix sort

1. Construct the virtual box→ h_{ens},
2. Construct the grid and index the particles→ (i_a, j_a, k_a), $a = 1, \ldots, N$.
3. Classification→ $key(a) = i_a + n_x j_a + n_x n_y k_a$ and ordering (Fig. 6),
4. Census→ $pr(i), key(i)$, $i = 1, \ldots, n_{class}$.
5. For each of the N particles

    a. With the key compute the coordinates (i_a, j_a, k_a)

    b. Compute the coordinates of the neighbor cells $\mathbf{N}_{cells}(a)$ and the corresponding keys

    c. Search for the particles in the lists generated during the *Census*

    d. This subset of particles is used to compute the averages around the ath particle.

With this algorithm, we construct the initial data of a distribution of particles obeying a radial profile (Press et al. 2007; Hernquist and Kats 1988)

$$\rho(r) = \frac{M}{2\pi R^2 r}, \tag{24}$$

using the acceptance-rejection method, where M is the total mass of the configuration and R is the radius. The three probability distributions required to be satisfied to allocate the particles are:

$$\int f_r(r)dr = 1, \quad \int f_\theta(\theta)d\theta = 1, \quad \int f_\phi(\phi)d\phi = 1, \tag{25}$$

Fig. 7 The line $N1$ is the one approximating the analytic profile with 10^4 particles, $N2$ uses 10^5 and $N3$, 10^6. Observe that the analytic profile (*pink line*) is better approximated as the number of particles is increased

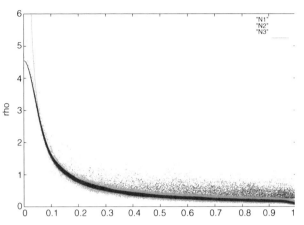

$$f_r(r) = \frac{4\pi\rho(r)}{M}r^2, \quad f_\theta(\theta) = \frac{\sin(\theta)}{2}, \quad f_\phi(\phi) = \frac{1}{2\pi}. \tag{26}$$

Finding three numbers (r, θ, ϕ) satisfying f_r, f_θ and f_ϕ, and the repeating N times, we find the positions to perform the averages using *Cell Indexing*.

The first test presented here is obtained increasing the number of particles N, see Fig. 7 for three different resolutions $N = 10^4$, 10^5, 10^6, the dispersion of the averaged values is smaller as we increase the resolution, with respect to the analytic value $\rho(r)$. This means that the acceptance-rejection method improves as we increase the number of particles describing the configuration.

The second test measures the time that the code requires to compute the averages using two methods: the particle-particle and the cell indexing methods.

In Fig. 8 number of particles N in a log-log scale, $N = 10^2, 10^3, 5 \times 10^3, 8 \times 10^3, 10^4, 1.5 \times 10^4, 1.7 \times 10^4, 10^5, 10^6$. It is worth to noticing that for $N = 100$ the first method is faster that the second one. But as we increase the number of particles we notice the advantage of the cell indexing method. For example, for $N = 10^3$, the time required by the first method is $T = 0.16$ s, and for the second one $t = 0.13$ s. If we keep increasing the number of particles, the difference becomes bigger. For $N = 17 \times 10^3$ particles (where the first method starts failing due to memory issues) we obtain $T = 133$ s and $t = 8.15$ s. The second one been 16 times faster than the first one.

In a log-log plot, the slope of the adjusted line for both methods represents m in $O(N^m)$. We found that the slope for $f1(x)$ is $m \approx 2$ while for $f2(x)$ is $m \approx 1$.

It is possible to improve different things in this algorithm, for example instead of only reporting the elements of the neighbor cell it is possible also to report all the neighbors inside a given radius h_{ens}.

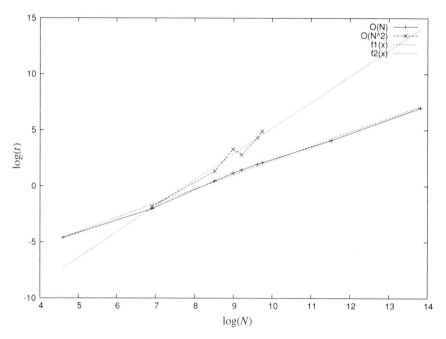

Fig. 8 We show the times required to compute $\rho_a = \sum_b m_b W_{ab}$ with $a = 1, \ldots, N$. The slope m of the *straight line* corresponds to the exponent in $O(N^m)$. We used $N = 10^2, 10^3, 5 \times 10^3$, $8 \times 10^3, 10^4, 1.5 \times 10^4, 1.7 \times 10^4, 10^5, 10^6$. For $N = 17 \times 10^3$, $t_{ens} = 8.15$s y $T = 133.06$s

6 Conclusions

In this manuscript we showed that the *Cell Index* method is lineal in time. The method shows convergence as we increase the number of particles. This allowed us to show that the SPH calculation of the density converges to the analytic value.

The standard particle to particle method limits the number of particles that can be used to the memory capacity of the computer. This is not a problem at all using the cell indexing method.

We tested the method, showing that it can be 16 times faster than the other one. This method can be a tool to substitute other methods with difficult implementations (TreeCode, K-trees, etc).

The parallel implementation of this algorithm and the fusion of this method with an SPH code is work in progress, in order to be used to solve physical problems.

References

Gafton E, Rosswog S (2011) A fast recursive coordinate bisection tree for neighbor search and gravity. Mon Not R Astron Soc 418(2):770–781

Hernquist L, Kats N (1989) TREESPH: a unification of SPH with the hierarchical tree method. Astrophys J Suppl Ser 70:419–446

Liu GR , Liu MB (2009) Smoothed Particle Hydrodynamics: a mesh free particle method. World Scientific Publishing Co Pte Ltd, New Jersey

Onderik J, Durikovic R (2007) Efficient neighbour search for particle-based fluids. J Appl Math Stat Inform 4(1):29–43

Paper online. URL: http://www.codercorner.com/RadixSortRevisited.htm

Press WH, Teukolsky SA, Vettering WT, Flannery BP (1997) Numerical recipes in fotran. Cambridge University Press, New york

Press WH, Teukolsky SA, Vetterling WT, Flannery BP (2007) Numerical recipes: the art of scientific computing, 3rd edn. Cambridge University Press, New York, ISBN 978-0-521-88068-8

Rodríguez-Meza MA, Suárez-Casino J, Matos T (2011) Métodos Numéricos en Astrofísica, Estado de México: Inovación Editorial Lagares de México, S.A. de C.V.

Part II
Multiphase and Granular Flow

On the Film Thickness Between a Bubble and the Wall in Liquids in Vertical Tubes

Abel López-Villa and Abraham Medina Ovando

Abstract We study numerically the film thickness that is formed between the free surface of a bubble and the inner wall in a vertical tube. The bubble is formed by gas injection in a tube filled with a viscous fluid. The computations were performed through the use of the Boundary element method (BEM) to solve the Stokes equations and a fourth order Runge–Kutta scheme to build the bubble shape. After the computation of the bubble shape, the thickness of the annular film was calculated for low Bond numbers, Bo, and a wide range of Capillary numbers, Ca. For the case $Ca \ll 1$ (inviscid approximation) it is found that the film actually touches the wall, meanwhile for the viscous case we found that the film thickness, scaled by radius of the tube, is a function of Ca and Bo. We also discuss experiments that validate the numerical results.

1 Introduction

The growth and detachment of bubbles generated by the continuous injection of gas into a quiescent liquid has been very much studied in conditions where the viscosity of the liquid plays no important role. Different models were given by Kumar and Kuloor (1970), Longuet-Higgins et al. (1991), Oguz and Prosperetti (1993), Corchero et al. (2006), Higuera and Medina (2006). The results of these studies are of interest in metallurgical and chemical industries, for example,

A. López-Villa (✉) · A. M. Ovando
SEPI ESIME Azcapotzalco, Instituto Politécnico Nacional, Av. de las Granjas 682, Col. Santa Catarina, 02250 Azcapotzalco D.F., Mexico
e-mail: abelvilla77@hotmail.com

J. Klapp et al. (eds.), *Fluid Dynamics in Physics, Engineering and Environmental Applications*, Environmental Science and Engineering,
DOI: 10.1007/978-3-642-27723-8_13, © Springer-Verlag Berlin Heidelberg 2013

where liquids of low viscosity, such as liquid metals and aqueous solutions, need to be handled.

The generation and dynamics of bubbles in very viscous liquids is also of interest but has not been so much studied. Thus, while many aspects of the dynamics of bubbles in unbounded viscous liquids are well understood (Manga and Stone 1994; Wong et al. 1998; Higuera 2005) the formation and detachment of bubbles in confined systems has received much less attention (Bretherton 1961; Ajaev and Homsy 2006; Coutanceau and Thizon 1981; López-Villa et al. 2011). Bubbles in very viscous liquids are commonly found when dealing with polymers in their liquid phases, in the flows of lava, and in processes of oil extraction from production pipelines, among others. In the present work we consider the problem of the film thickness that is formed between the free surface of a single bubble and the wall when the bubble reaches their critical size in the vertical circular tube filled with a quiescent liquid of high viscosity.

2 Theoretical Study of Bubbles Growing in Circular Cylinders

The model used here is valid to understand the bubble formation in a very viscous liquid in confined axi-symmetric geometries (López-Villa et al. 2011). We assume a fluid confined in a cylinder of inner radius R^*. The equations of continuity and Stokes have the following dimensionless forms, respectively

$$\nabla \cdot \mathbf{v} = 0, \tag{1}$$

$$0 = -\nabla p + \nabla^2 \mathbf{v} - Bo\,\mathbf{i}, \tag{2}$$

where p is the pressure, $\mathbf{v}$ is the velocity field, $\mathbf{i}$ is the normal vector pointing in the upward vertical direction (x is the vertical coordinate) and Bo is the Bond number given by

$$Bo = \frac{\rho g a^2}{\sigma}, \tag{3}$$

where ρ is the liquid density, g is the acceleration due to gravity, a is the radius of the orifice of gas injection and σ is the surface tension. When the bubble grows we assume that the dimensionless gas flow rate is $dV_0/dt = Ca = constant.$; i.e., the dimensionless flow of gas that forms the bubble is exactly the capillary number, given by

$$Ca = \frac{\mu Q}{\sigma a^2}. \tag{4}$$

where μ is the viscosity of the liquid and Q is a constant flow rate of gas.

When a bubble is formed in the liquid there exist a free surface of the form,
$f(\mathbf{x},t) > 0$

Equations (1) and (2) must be solved with the boundary conditions

$$\frac{Df_i}{Dt} = 0, \tag{5}$$

$$-p\mathbf{n}_i + \tau' \cdot \mathbf{n}_i = (\nabla \cdot \mathbf{n}_i - p_{g_i})\mathbf{n}_i. \tag{6}$$

in the surfaces of the ith bubbles and

$$\mathbf{v} = \mathbf{0} \tag{7}$$

on the inner cylinder's surface, located at $r = R^*$, and at infinity, because the fluid does not move there. Moreover, the pressure far from the bubble must satisfy

$$p + Bo\, x = constant \tag{8}$$

The problem contains the five dimensionless parameters which are a Bond number, a capillary number, the dimensionless radius of the base and height of the tube, and the contact angle of the liquid with the base. The set of equations given above will satisfy the boundary conditions of quit flow at infinity, the non slip conditions on walls and the quasi static pressure balance. The evolution of the free surface (bubble shapes) is given by the solution of Eq. (5), under a fourth order Runge–Kutta scheme, which is attained after the hydrodynamic problem, imposed by Eqs. (1) and (2), has been solved by using the BEM method (López-Villa et al. 2011).

Good predictions were found for the description of the bubble formation in cylinders for small Bond numbers and values above $Ca/(R^2) > 0.03$ where $R = R^*/a$ is the inner tube radius normalized by the radius of the injection orifice because the liquid film seems to develop a corrugation, due to lack of numerical resolution when the film becomes very thin (López-Villa et al. 2011). Thus, results in this study will be limited by the aforementioned capillary number.

3 The Film Thickness

In Fig. 1a it is shown that during the bubble growth in the pipe there exist an annular film of thickness b^*. In a classical study, Bretherton (1961) showed, by using the lubrication theory, that the dimensionless thickness of the film $b = b^*/a$ scales as

$$\frac{b}{R} \sim \left(\frac{Ca}{R^2}\right)^{\frac{2}{3}}, \tag{9}$$

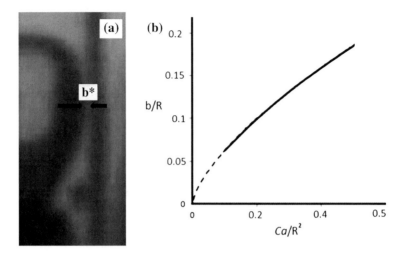

Fig. 1 **a** Bubble in a vertical tube, notice the existence of a film of thickness b^*. **b** Plot of the scaled dimensionless film thickness b/R as a function of the scaled capillary number Ca/R^2

which is valid whenever $Ca \to 0$ and R and he Bond number are also small. In Fig. 1b a plot obtained from the numerical computations that obeys the relation (9) is given. In such a plot the continuous curve was obtained through our numerical solution meanwhile the dashed part of the curve only shows the trend given by Eq. (9) but was numerically inaccessible. Despite it, in this case, clearly $b \to 0$ if $Ca \to 0$. Physically the condition $Ca \to 0$ implies that the bubble in an inviscid liquid touches the inner solid wall.

Figure 2a shows some bubble profiles: in this case they were obtained for low capillary numbers and is evident that the film thickness tends towards zero for small values of Ca and R. Also it can see that the profiles show some "corrugations", this is because they become unstable when the height of the tubes is very large compared with the radius tube, i.e. in this case height tube is 30 times the radius tube. In experiments it was observed that when $Ca \ll 1$ the small bubble profiles will be unstable.

Moreover, very different results were obtained when the film thickness was computed for very viscous liquids, i.e., $Ca \gg 1$ in the limit of low Bond number. In Fig. 2b it is possible to notice that the film thickness tend towards a constant value when the capillary number increases. In the Fig. 3 is sketched how $b \to$ constant for $Ca \gg 1$. In dimensional terms the actual thickness of the annular film, $b^* \to 1.5a^*$, i.e., the lower value of b^* is 1.5 times the radius of the gas injection orifice. Physically this condition is attained in very viscous liquids or at very large gas flow rates.

Fig. 2 Bubble shapes in cylinders filled with (a) liquids of low viscosity and (b) with very viscous liquids. In (a) the height and the film thickness between the bubble and the wall diminishes when $Ca \rightarrow 0$ ($Ca = 0.4$, 0.3, 0.2 and 0.1). In (b) the film thickness $b \rightarrow$ constant, for $Ca \gg 1$. The larger bubble corresponds to $Ca = 35$ other cases are $Ca = 20$, 10 and 5, the dimensionless pipe radius was $R = 5$

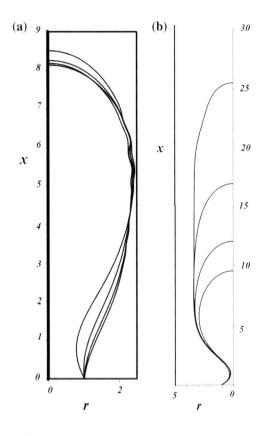

Fig. 3 Plot of the thickness of the annular film, b, as a function of the capillary number, Ca. Notice that $b \rightarrow constant$ for $Ca \gg 1$

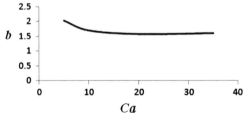

4 Conclusions

In this work we have analyzed numerically the behavior of the thin fluid film around a bubble growing in a tube when the Bond number is small (dominant surface tension effects on the gravity ones) and in the extreme cases: $Ca \ll 1$ and $Ca \gg 1$. In the first case the film thickness, b, obeys a power law relation in terms of the capillary number Eq. (9) which implies that b can be vanished if $Ca \rightarrow 0$. Conversely, in the limit case $Ca \gg 1$ film thickness tends to a constant value that depends on the radius of the tube.

Finally, the previous results can be of essential importance, for instance, to understand the bubble train motion and the foam flow in tubes (Kornev et al. 1999), among others. Work along these lines is in progress.

References

Ajaev VS, Homsy GM (2006) Modeling shapes and dynamics of confined bubbles. Annu Rev Fluid Mech 38:277–307

Bretherton FP (1961) The motion of long bubbles in tubes. J Fluid Mech 10:166–188

Corchero G, Medina A, Higuera FJ (2006) Effect of wetting conditions and flow rate on bubble formation at orifices submerged in water. Colloids and surfaces A. Physicochem Eng Aspects 290:41–49

Coutanceau M, Thizon P (1981) Wall effect on the bubble behaviour in highly viscous liquids. J Fluid Mech 107:339–373

Higuera FJ (2005) Injection and coalescence of bubbles in a very viscous liquid. J Fluid Mech 530:369–378

Higuera FJ, Medina A (2006) Injection and coalescence of bubbles in a quiescent invicid liquid. Eur J Mech B/Fluids 25:164–171

Kornev KG, Neimark AV, Rozhkov AN (1999) Foam in porous media: thermodynamic and hydrodynamic peculiarities. Adv Colloid Interface Sci 82:127–187

Kumar R, Kuloor NR (1970) The formation of bubbles and drops. Adv Chem Eng 8:255–368

Longuet-Higgins MS, Kerman BR, Lunde K (1991) The release of air bubbles form and underwater nozzle. J Fluid Mech 230:365–390

López-Villa A, Medina A, Higuera FJ (2011) Bubble growth by injection of gas into viscous liquids in cylindrical and conical tubes. Phys Fluids 23:102102

Manga M, Stone HA (1994) Interactions between bubbles in magmas and lavas: Effects of the deformation. J Vulcanol Res 63:269–281

Oguz HN, Prosperetti A (1993) Dynamics of bubble growth and detachment from a needle. J Fluid Mech 257:111–145

Wong H, Rumschitzki D, Maldarelli C (1998) Theory and experiment on the low-Reynolds number expansion and contraction of a bubble pinned at a submerged tube tip. J Fluid Mech 356:93–124

Mathematical Model for "Bubble Gas-Stratified Oil" Flow in Horizontal Pipes

C. Centeno-Reyes and O. Cazarez-Candia

Abstract A one-dimensional, isothermal, transient model for bubble gas-stratified oil flow is presented. Bubble gas- stratified oil flow pattern of heavy oil, water and gas, in horizontal pipelines, consists of two regions: (1) a stratified region with a gross water layer in the bottom, an oil layer in the middle and a thin water layer in the top, (2) a region with a water layer in the bottom, an oil layer in the middle and a gas bubble in the top. The two-fluid mathematical model consists of mass, momentum and energy conservation equations for every phase, considering the hydrostatic gradient. The model takes into account: (1) wall shear stress, (2) interfacial shear stress, and (3) the non-Newtonian oil behavior. The model is able to predict pressure, volumetric fraction, temperature, and velocity profiles for every phase. The numerical solution is based on the finite difference technique in an implicit scheme. The model was validated using experimental data reported in literature, for a heavy crude oil (14 °API), and it was observed that the pressure drop calculated by the model was reasonably close to the experimental data.

1 Introduction

Heavy oil–water–gas flow is commonly found in the oil and gas industry, mainly in the transport of produced fluids from wells. Trevisan (2003) observed nine flow patterns when heavy oil, gas and water flow simultaneously through a horizontal

C. Centeno-Reyes (✉) · O. Cazarez-Candia
Instituto Mexicano del Petróleo, Eje Central Lázaro Cárdenas, No. 152, Col. San Bartolo Atepehuacan, C.P 07730, México, D.F., Mexico
e-mail: ccenteno@imp.mx

O. Cazarez-Candia
e-mail: ocazarez@imp.mx

J. Klapp et al. (eds.), *Fluid Dynamics in Physics, Engineering and Environmental Applications*, Environmental Science and Engineering,
DOI: 10.1007/978-3-642-27723-8_14, © Springer-Verlag Berlin Heidelberg 2013

pipe, the presence of two immiscible liquids give rise to a wider variety of flow patterns, and this imply more discontinuities and greater complexity in the hydrodynamic models. Particularly the bubble gas- stratified oil flow pattern has not been modeled until this work.

2 Mathematical Model

The bubble gas- stratified oil is illustrated in Fig. 1, where l_1 is the length of the bubble region, l_2 is the length of the liquid region, A_k is the cross sectional area of the k phase (k = oil, water, gas), τ_{kwall} are the fluid-wall shear stress and τ_{kk} are the interfacial shear stress.

It was assumed: (1) a full-developed, one-dimensional, adiabatic flow with flat interfaces, (2) ideal gas, and (3) incompressible flow for liquid phases. The hydrostatic pressure was used to take into account the unequal phase pressure.

Taking as an starting point the work of Lahey and Drew (1989) the governing averaged (space-time) equations for a one-dimensional adiabatic heavy oil-water-gas flow in a constant area duct without interfacial mass, including the averaged hydrostatic pressure, were developed and can be written as:

Gas, oil and water mass conservation equations

$$\frac{\varepsilon_g}{C_g^2 + g(D - h_L)}\left(\frac{\partial P_i}{\partial t} + U_g \frac{\partial P_i}{\partial z}\right) + \frac{\varepsilon_g \rho_g}{T_g}\left(\frac{g(D - h_L)}{C_g^2 + g(D - h_L)} - 1\right)\left(\frac{\partial T_g}{\partial t} + U_g \frac{\partial T_g}{\partial z}\right)$$

$$+ \varepsilon_g \rho_g \frac{\partial U_g}{\partial z} + \left(\frac{\varepsilon_g g \rho_g}{\left(C_g^2 + g(D - h_L)\right)}\frac{A}{2\sqrt{h_L(D - h_L)}} + \rho_g\right)\left(\frac{\partial \varepsilon_g}{\partial t} + U_g \frac{\partial \varepsilon_g}{\partial z}\right) = 0 \tag{1}$$

$$\frac{\partial \varepsilon_o}{\partial t} + U_o \frac{\partial \varepsilon_o}{\partial z} + \varepsilon_o \frac{\partial U_o}{\partial z} = 0 \tag{2}$$

$$\frac{\partial \varepsilon_w}{\partial t} + U_w \frac{\partial \varepsilon_w}{\partial z} + \varepsilon_w \frac{\partial U_w}{\partial z} = 0 \tag{3}$$

Gas, oil and water energy conservation equations

$$\left(Cp_g - \frac{C_g^2 g(D - h_L)}{\left(C_g^2 + g(D - h_L)\right)T_g}\right)\frac{\partial T_g}{\partial t} + U_g\left(\frac{\partial U_g}{\partial t} + U_g \frac{\partial U_g}{\partial z}\right) + Cp_g U_g \frac{\partial T_g}{\partial z}$$

$$- \frac{C_g^2}{\rho_g\left(C_g^2 + g(D - h_L)\right)}\frac{\partial P_i}{\partial t} - \frac{C_g^2 g}{\left(C_g^2 + g(D - h_L)\right)}\frac{A}{2\sqrt{h_L(D - h_L)}}\frac{\partial \varepsilon_g}{\partial t} = 0 \tag{4}$$

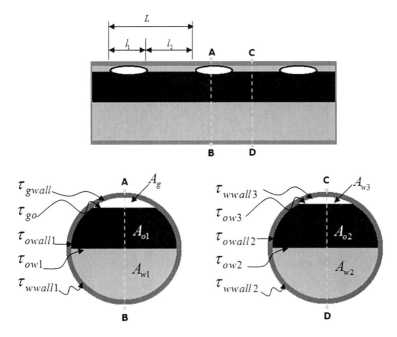

Fig. 1 Schematic description of bubbly gas-stratified oil flow

$$\frac{\partial T_o}{\partial t} + U_o \frac{\partial T_o}{\partial z} + \frac{U_o}{Cp_o}\left(\frac{\partial U_o}{\partial t} + U_o \frac{\partial U_o}{\partial z}\right) + \frac{U_o}{Cp_o\rho_o}\frac{\partial P_i}{\partial z} + \frac{U_o g}{Cp_o}\left(\frac{A}{4\sqrt{h_{L1}(D-h_{L1})}} + \frac{A\phi L}{4\sqrt{h_{L2}(D-h_{L2})l_1}}\right)$$

$$-\frac{R^3(1-\phi)l_2}{Al_1}\frac{\Lambda_1}{\left(\varepsilon_o - (1-\phi)\varepsilon_g(l_2/l_1)\right)^2} + \frac{R^3(1-\phi)}{A}\frac{\Lambda_2}{\left(\varepsilon_o + (1-\phi)\varepsilon_g\right)^2}\Bigg)\frac{\partial \varepsilon_g}{\partial z}$$

$$+\frac{U_o g R^3}{Cp_o A}\left(\frac{\Lambda_1}{\left(\varepsilon_o - (1-\phi)\varepsilon_g(l_2/l_1)\right)^2} + \frac{\Lambda_2}{\left(\varepsilon_o + (1-\phi)\varepsilon_g\right)^2}\right)\frac{\partial \varepsilon_o}{\partial z} = 0 \tag{5}$$

$$\frac{\partial T_w}{\partial t} + U_w \frac{\partial T_w}{\partial z} + \frac{U_w}{Cp_w}\left(\frac{\partial U_w}{\partial t} + U_w \frac{\partial U_w}{\partial z}\right) + \frac{U_w}{Cp_w\rho_w}\frac{\partial P_i}{\partial z} + \frac{U_w}{3Cp_w\rho_w}\left[\frac{\rho_o g A}{2\sqrt{h_{L1}(D-h_{L1})}} + \frac{\rho_o g A}{2\sqrt{h_{L2}(D-h_{L2})}}\frac{\phi L}{l_1}\right.$$

$$-\frac{g(\rho_w - \rho_o)A}{\sqrt{h_w(D-h_w)}}((l_1 + \phi l_2)/l_1) - \frac{4\rho_w g R^3\Theta}{A\left(1 - \varepsilon_g((l_1 + \phi l_2)/l_1) - \varepsilon_o\right)^2}((l_1 + \phi l_2)/l_1) - \frac{2\rho_w g R^3\Phi}{A\varepsilon_g^2}\frac{l_1}{\phi L}\Bigg]\frac{\partial \varepsilon_g}{\partial z}$$

$$-\frac{U_w}{3Cp_w\rho_w}\left(\frac{g(\rho_w - \rho_o)A}{\sqrt{h_w(D-h_w)}} + \frac{4\rho_w g R^3\Theta}{A\left(1 - \varepsilon_g((l_1 + \phi l_2)/l_1) - \varepsilon_o\right)^2}\right)\frac{\partial \varepsilon_o}{\partial z} = 0 \tag{6}$$

Gas, oil and water momentum conservation equations

$$\varepsilon_g \rho_g \left(\frac{\partial U_g}{\partial t} + U_g \frac{\partial U_g}{\partial z}\right) + \frac{\varepsilon_g C_g^2}{\left(C_g^2 + g(D - h_{L1})\right)}\frac{\partial P_i}{\partial z} + \frac{\varepsilon_g C_g^2 g \rho_g}{\left(C_g^2 + g(D - h_{L1})\right)}\frac{A}{2\sqrt{h_{L1}(D-h_{L1})}}\frac{\partial \varepsilon_g}{\partial z} +$$

$$\frac{\varepsilon_g C_g^2 g(D - h_{L1})\rho_g}{\left(C_g^2 + g(D - h_{L1})\right)T_g}\frac{\partial T_g}{\partial z} = -\frac{\tau_{gwall}S_g}{A}\frac{l_1}{L} - \frac{\tau_{go}S_{go}}{A}\frac{l_1}{L} \tag{7}$$

$$\varepsilon_o \rho_o \left(\frac{\partial U_o}{\partial t} + U_o \frac{\partial U_o}{\partial z} \right) + \varepsilon_o \frac{\partial P_i}{\partial z} + \varepsilon_o \rho_o g \left(\frac{A}{4\sqrt{h_{L1}(D - h_{L1})}} + \frac{A\phi L}{4\sqrt{h_{L2}(D - h_{L2})}l_1} \right)$$

$$- \frac{R^3(1 - \phi)l_2}{Al_1} \frac{\Lambda_1}{\left(\varepsilon_o - (1 - \phi)\varepsilon_g(l_2/l_1) \right)^2} + \frac{R^3(1 - \phi)}{A} \frac{\Lambda_2}{\left(\varepsilon_o + (1 - \phi)\varepsilon_g \right)^2} \right) \frac{\partial \varepsilon_g}{\partial z} \qquad (8)$$

$$+ \frac{\varepsilon_o \rho_o g R^3}{A} \left(\frac{\Lambda_1}{\left(\varepsilon_o - (1 - \phi)\varepsilon_g(l_2/l_1) \right)^2} + \frac{\Lambda_2}{\left(\varepsilon_o + (1 - \phi)\varepsilon_g \right)^2} \right) \frac{\partial \varepsilon_o}{\partial z}$$

$$= -\frac{\tau_{owall1}S_{o1}}{A}\frac{l_1}{L} + \frac{\tau_{go}S_{go}}{A}\frac{l_1}{L} - \frac{\tau_{owall2}S_{o2}}{A}\frac{l_2}{L} + \frac{\tau_{ow3}S_{ow3}}{A}\frac{l_2}{L} - \frac{\tau_{ow1}S_{ow1}}{A}\frac{l_1}{L} - \frac{\tau_{ow2}S_{ow2}}{A}\frac{l_2}{L}$$

$$(1 - \varepsilon_o - \varepsilon_g)\rho_w \left(\frac{\partial U_w}{\partial t} + U_w \frac{\partial U_w}{\partial z} \right) + (1 - \varepsilon_o - \varepsilon_g)\frac{\partial P_i}{\partial z}$$

$$+ \frac{(1 - \varepsilon_o - \varepsilon_g)}{3} \left[\frac{\rho_o g A}{2\sqrt{h_{L1}(D - h_{L1})}} + \frac{\rho_o g A}{2\sqrt{h_{L2}(D - h_{L2})}} \frac{\phi L}{l_1} - \frac{g(\rho_w - \rho_o)A}{\sqrt{h_w(D - h_w)}}((l_1 + \phi l_2)/l_1) \right.$$

$$\left. - \frac{4\rho_w g R^3 \Theta}{A\left(1 - \varepsilon_g((l_1 + \phi l_2)/l_1) - \varepsilon_o \right)^2}((l_1 + \phi l_2)/l_1) - \frac{2\rho_w g R^3 \Phi}{A\varepsilon_g^2} \frac{l_1}{\phi L} \right] \frac{\partial \varepsilon_g}{\partial z} \qquad (9)$$

$$- \frac{(1 - \varepsilon_o - \varepsilon_g)}{3} \left(\frac{g(\rho_w - \rho_o)A}{\sqrt{h_w(D - h_w)}} + \frac{4\rho_w g R^3 \Theta}{A\left(1 - \varepsilon_g((l_1 + \phi l_2)/l_1) - \varepsilon_o \right)^2} \right) \frac{\partial \varepsilon_o}{\partial z}$$

$$= -\frac{\tau_{wwall1}S_{w1}}{A}\frac{l_1}{L} - \frac{\tau_{wwall2}S_{w2}}{A}\frac{l_2}{L} + \frac{\tau_{ow3}S_{ow3}}{A}\frac{l_2}{L} + \frac{\tau_{ow1}S_{ow1}}{A}\frac{l_1}{L} + \frac{\tau_{ow2}S_{ow2}}{A}\frac{l_2}{L}$$

where T is temperature, U is velocity, S_k is the wetted perimeter, C_g is gas sound velocity, Cp is heat capacity, P_i is gas-oil interfacial pressure, ρ is density, g is acceleration due to gravity, D is pipe diameter, ε is volumetric fraction, A is the cross sectional area, h_L and h_w are liquid and water height respectively, t and z are temporal and spatial coordinates, respectively. The subscripts g, o and w represent gas, oil and water, respectively.

ϕ is the ratio between the water film area (A_3) in the liquid region and the gas cross sectional area (A_g) in the bubble region (Fig. 1):

$$\phi = A_{w3}/A_g \qquad (10)$$

Θ, Φ, Λ_1, and, Λ_2 are all geometrical parameters defined by

$$\left[\vartheta/4 - \text{sen}\,\vartheta/4 - 1/3\text{sen}^3(\vartheta/2) \right]_{\lambda 2}^{\lambda 1} \qquad (11)$$

where $\lambda 1$ is $\pi - 2\beta$, $\pi + 2\alpha_1$, $\pi + 2\alpha_2$, 2π, $\lambda 2$, is 0, $\pi - 2\beta$, $\pi - 2\beta$, $\pi + 2\alpha_2$, for Θ, Λ_1, Λ_2, and, Φ respectively. Where α and β are the angles between the interfaces and the centerline.

2.1 Shear-Stress Relationships

As was stated at the beginning of Sect. 2, different shear stresses are needed to solve the equation system. The shear stresses were correlated as follows.

Gas-wall and water-wall shear stresses.

For the shear stresses between the water or gas and the pipe surface, the gas- and water- wall friction factors were calculated using the correlations proposed by Taitel et al. (1995).

$$f_k = C\mathrm{Re}_k^{-n}, \quad k = w, g \tag{12}$$

where $C = 0.046$, $n = 0.2$ for turbulent flow, and $C = 16$, $n = 1$ for laminar flow.

Oil-wall shear stress.

The crude oil was considered to exhibit a non-Newtonian behavior, using the next power law model (Hasan 2007).

$$\tau = K\gamma^n, \quad n \leq 1 \tag{13}$$

where K is the power law consistency constant and n is a measure of the non-newtonian behavior. The friction factor can be calculated analytically for laminar flow (Govier and Aziz 1972) and is given by:

$$f_o = 16\mathrm{Re}_{oMR}^{-1} \tag{14}$$

For turbulent flow, the explicit Kawase correlation was used (Kawase et al. 1994):

$$f_o = \left\{ 3.57 \log_{10} \left[\frac{\mathrm{Re}_{oMR}^{1/n^{0.615}}}{\mathrm{Re}_{oMR}^{1/n^{0.615}} \left[10^{(3.75-8.5n)/5.756} \left(\frac{2\varepsilon}{D}\right) \right]^{1.14/n} + 6.5^{1/n^{(1+0.75n)}}} \right] \right\}^{-2} \tag{15}$$

where Re_{oMR} is Metzner-Reed Reynolds number (Metzner-Reed 1955)

$$\mathrm{Re}_{oMR} = \rho_o D h_o^n U_o^{2-n} / K((3n+1)/4n)^n 8^{n-1} \tag{16}$$

Water–oil and gas-oil interfacial shear stress.

The empirical values of interfacial friction factor were taken to be same as proposed by Taitel et al. (1995).

$$f_{ow} = 0.014 \tag{17}$$

$$f_{go} = \begin{cases} f_g & \text{for } f_g > 0.014 \\ 0.014 & \text{for } f_g \leq 0.014 \end{cases} \tag{18}$$

3 Method of Solution

The numerical method implemented was finite differences with a first order downstream implicit scheme for spatial derivatives and a first order upstream implicit scheme for time derivatives. Equations (1)–(9) are discretized and then can be written in a matrix form as:

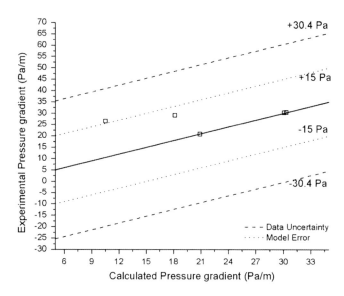

Fig. 2 Predicted pressure gradient compared with Trevisan data (Trevisan 2003)

$$\mathbf{D}_i\left(\mathbf{x}_i^0\right)\mathbf{x}_i^{t+\Delta t} = \mathbf{E}_i\left(\mathbf{x}_i^0, \mathbf{x}_i^t, \mathbf{x}_{i-1}^{t+\Delta t}\right) \tag{19}$$

$\mathbf{x}$ is a column vector of dependent variables given by:

$$\mathbf{x} = \left[\varepsilon_o, \varepsilon_g, P_i, U_g, T_g, U_o, U_w, T_o, T_w\right]^T \tag{20}$$

where the superscript T indicates the transpose.

The methodology followed is similar to that of Cazarez and Vazquez (2005). The numerical solution was obtained using the LINPACK (Dongarra et al. 1990) package of numerical routines for solving simultaneous algebraic linear equations. The algorithm used in this study is based on the factorization of a matrix using Gaussian elimination with partial pivoting.

4 Results and Discussion

The model was validated using the experimental data given by Trevisan (2003). The experiments took place in an acrylic tube, at ambient temperature and pressure. Heavy oil (14°API), with a density of 971Kgm^{-3}, and a viscosity of 5,040 mPas was used.

Figure 2 shows the comparison between the experimental pressure gradient and the pressure gradient calculated with the model, as can be seen the pressure drop predicted has a good agreement with the experimental data. The average error was 5 Pa/m. The huge uncertainty in the experimental data is because for horizontal

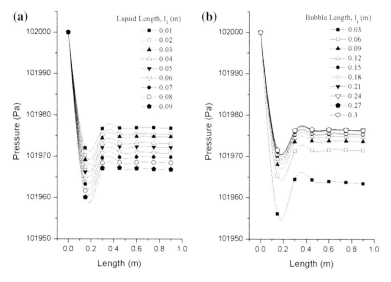

Fig. 3 Effect of: **a** Liquid length and **b** Bubble length on pressure profiles

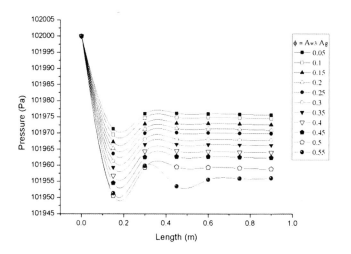

Fig. 4 Effect of gas/water cross sectional areas ratio on pressure profiles

pipelines, the pressure drop is related only to the friction, for this reason the pressure gradients are small, and they can be within the experimental data uncertainty.

A parametric study was performed using the model, Fig. 3a shows the effect of liquid region length on pressure profile predictions and it can be observed that as the liquid region length increases, pressure drop increases too. On the other hand, Fig. 3b shows the effect of bubble region length on pressure profile predictions, for bubble lengths greater than 0.15 m pressure profiles get close to the same value, which means that the lower pressure drop is achieved.

In Fig. 4 is shown the pressure profiles for different gas/water cross sectional areas, and as this ratio is bigger, the pressure drop is bigger too.

5 Conclusions

A mathematical model for the hydrodynamic of the bubble gas- stratified flow of heavy oil, water and gas in horizontal pipelines was proposed and validated. The pressure gradients predicted by the three-phase model were within the range of experimental data uncertainty.

Increases in the liquid region length increase the pressure drop in the pipe, as well as an increase of the bubble lengths diminish the pressure drop, but for bubble lengths greater than 0.15 m bubbles length does not impact any more the pressure profile prediction.

References

Cazarez O, Vazquez M (2005) Prediction of pressure, temperature and velocity distribution of two-phase flow in oil wells. J Petrol Sci Eng 46:195–208

Dongarra JJ, Bunch JR, Moler CB, Stewart GW (1990) LINPACK user guide. Soc Ind Appl Math

Govier GW, Aziz K (1972) The flow of complex mixtures in pipes. Van Norstrand -Reinhold, New York

Hasan S (2007) Rheology of heavy crude oil and viscosity reduction for pipeline transportation. M.Sc. Thesis. Concordia University Quebec, Canada

Kawase Y, Shenoy AV, Wakabayashi K (1994) Friction and heat and mass transfer for turbulent pseudoplastic non-newtonian fluid flows in rough pipes. Can J Chem Eng 72:798–804

Lahey RT Jr, Drew DA (1989) The three-dimensional time and volume averaged conservation equations of two-phase flow. Adv Nucl Sci Tech 20(1):1–69

Metzner AB, Reed JC (1955) Flow of non-newtonian fluids correlation of the laminar, transition, and turbulent-flow regions. AIChE J 1:434–440

Taitel Y, Barnea D, Brill JP (1995) Stratified 3-phase flow in pipes. Int J Multiphase Flow 21(1):53–60

Trevisan FE (2003) Padrões de Fluxo e Perda de Carga em Escoamento Trifásico Horizontal de Óleo Pesado, Água e Ar. Thesis. Universidade Estadual de Campinas, Brasil

Pseudoturbulence in Bubbly and Transition Flow Regimes

Santos Mendez-Diaz, Roberto Zenit, Sergio Chiva Vicent,
José Luis Muñoz-Cobo and Arturo Morales-Fuentes

Abstract Computational fluid-dynamics codes use the two-fluid model to simulate transport phenomena in liquid–gas flows. Numerical instabilities arise when flow regime transition occurs, i.e. bubbly flow to slug flow. One mechanism that produces the transition between regimes is the fluctuation induced over the liquid phase velocity due the relative movement of ascending bubbles. Using multi-tip impedance probe and Laser Doppler Anemometry (LDA), the main local flow parameters of the two-phase flow were obtained. The flow conditions selected were single phase flow up to liquid–gas flow at 30 % gas concentration. The power spectral energy distributions obtained show that turbulence intensity and the energy exponent decay have a nearly constant value when wall peak is observed (wall peak distribution occurs when the radial void fraction distribution shows a

S. Mendez-Diaz (✉) · A. Morales-Fuentes
Universidad Autónoma de Nuevo León, Av. Pedro de Alba s/n,
66451 San Nicolás de los Garza, Nuevo Leon, Mexico
e-mail: santos.mendezdz@uanl.edu.mx

A. Morales-Fuentes
e-mail: arturo.moralesfn@uanl.edu.mx

R. Zenit
Instituto de Investigación en Materiales, UNAM, Circuito Exterior s/n.
Apartado Postal 70-360, 04510 México, Distrito Federal, Mexico
e-mail: zenit@unam.mx

S. C. Vicent
Departamento de Ingeniería Mecánica y Construcción, Universidad Jaime I,
Castellón de la Plana, Spain
e-mail: schiva@emc.uji.es

J. L. Muñoz-Cobo
Departamento de Ingeniería Química y Nuclear, Universidad Politécnica de Valencia,
Camino de Vera s/n, Valencia, Spain
e-mail: jlcobos@iqn.upv.es

J. Klapp et al. (eds.), *Fluid Dynamics in Physics, Engineering and Environmental Applications*, Environmental Science and Engineering,
DOI: 10.1007/978-3-642-27723-8_15, © Springer-Verlag Berlin Heidelberg 2013

peak concentration near to the channel wall). Under the flow conditions tested, pseudo-turbulence effects can be observed at low gas concentrations in bubbly flow regime and the decay energy has a nearly constant slope; whereas at transition and slug regimes the exponent decay in PSD's shows a non-constant value, and the lift force would be less important.

1 Introduction

Liquid–gas flow can be found in many industrial processes as oil recovery, pharmaceutical applications and nuclear energy. At experimental laboratory facilities, bubble columns can be classified as laminar and turbulent according to the bubblance parameter (Rensen et al. 2005) defined as:

$$b = \frac{\frac{1}{2}\alpha U_R^2}{u_0'^2} \tag{1}$$

where α, U_R and $u_0'^2$ are respectively the void fraction, the relative velocity between phases, and the variance of velocity at single phase conditions. This dimensionless number relates the bubble energy due its relative movement respect to the liquid phase, and the turbulent energy due to the effects of liquid viscosity.

For a turbulent regime, when $b > 1$, the liquid agitation is produced by the induced velocity fluctuations produced by the ascending movement of the bubbles, and the turbulence due to liquid viscous effects:

$$\overline{u_E'^2} = \overline{u'^2} - \overline{u_0'^2} \tag{2}$$

The total fluctuations on the liquid velocity field (u'^2) are exclusively due to the bubble relative motion under laminar flow condition when the bubblance parameter is equal to ø and liquid velocity in stagnant conditions:

$$\overline{u_E'^2} = \overline{u'^2} \tag{3}$$

where $u_E'^2$ is the total velocity fluctuation. Many experimental and numerical works have been reported on bubbly columns under laminar (b = ø) and moderated turbulent conditions (b < 1). Some authors observe under pseudoturbulence flow conditions an energy decay slope of -3 in the power spectral distribution (Martinez-Mercado et al. 2007; Rensen et al. 2005; Lance and Bataille 1991; Risso 2011). Panidis (2011) reported a power spectrum distribution slope near to -2 for fully turbulent flows.

In co-current flow, liquid phase has non-zero mean velocity and the relative movement of the bubble produces an important influence on liquid velocity field, however, the pseudo-turbulence flow conditions are easily identified when the liquid phase has zero mean velocity and the only source of the liquid agitation is

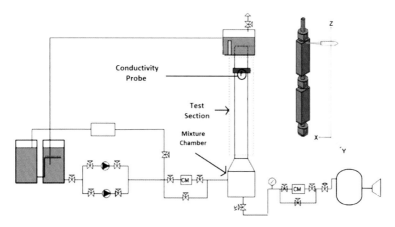

Fig.1 Experimental arrangement

the relative movement between phases. This work studies the pseudo-turbulence effects under turbulent conditions.

2 Experimental Setup

The experimental setup is shown in Fig. 1. It consists of a cylindrical test section with an internal diameter of 52 and 3,340 mm of length; it is made of Plexiglas. The liquid phase is tap water circulated by two centrifugal pumps. The gaseous phase is compressed air. The air and water temperature were nearly constant during the test. The air mass flow rate was measured with a thermal mass flow meter, and the liquid flow rate was measured with an electromagnetic flow meter. The liquid–gas mixture was formed employing a porous sintered element with an average pore size of 10 μm, located in the lower plenum.

The local flow parameters as bubble velocity and Sauter mean diameter were measured employing a four sensor impedance probe. It consists of four sensors made of stainless steel, with a diameter of 0.22 mm. The vertical distance between both tips was about 1.5 mm. Due to the large difference in electrical conductivity of the liquid phase and the gas phase, the impedance signal rises sharply when a bubble passes through one of the sensor tips, obtaining a stepping signal. From the time lag, between the impedance signals of the front and back tips, a measurable value of the bubble velocity is obtained. This sensor was located at $z/D = 52$.

The liquid phase velocity was measured using Laser Doppler Anemometry. The axial location of the LDA device is $z/D = 50$. From the experimental data obtained the autocorrelation function and the power spectra were calculated using the Welch's method.

The liquid superficial velocities tested were 0.5 and 1.0 m/s, and the gas concentration was changed from 0 to 30 %, approximately. The Reynolds number

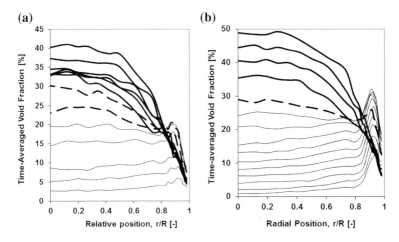

Fig. 2 Time-averaged void fraction at two liquid superficial velocities. Thin, dashed and tick lines represents wall-peak, transition and core-peak distributions. The gas concentration was changed from 4 to 30 %. Data taken from Mendez-Diaz (2008). **a** $j_l = 0.5$ m/s, **b** $j_l = 1$ m/s

for the single phase case was greater than 15,000, clearly a turbulent flow regime. The Reynolds bubble number was adjusted to values above 1,500. In Fig. 2 the radial void fraction is showed at 1 m/s of liquid velocity, thin lines represent the bubbly regime. When the void fraction is increased, the flow regime changes from bubbly to slug, the last are showed with thicker lines, whereas transition regime is represented by broken ones.

The initial diameter was not controlled in this experiment; it was measured by the multi-tip impedance probe. At the liquid velocities tested, it was observed that bubble disintegration is not produced by turbulence effects, and the initial bubble size determines the developing pattern in the radial void fraction distribution. The two-phase flow obtained using the sintered porous plate is poly-dispersed in size

3 Experimental Results

Figure 3 shows the radial distribution of the time-averaged velocity for both liquid and gas phases. The superficial liquid velocity was 0.5 m/s and the continuous thick line represents the single phase case. When the void fraction is increased from 3.7 to 9.6 %, a change on the local liquid velocity is observed. For this flow conditions, the bubble presence locally overrides the liquid velocity and a reduction occurs, in agreement with the reports of Hibiki et al. (2001) and Serizawa and Kataoka (1988). At void fractions greater than 10 %, the induced agitation increases until the change of flow regime.

Figure 4 shows the radial distribution of the time-averaged velocity of both liquid and gas phase. The superficial liquid velocity at the entrance was 1 m/s;

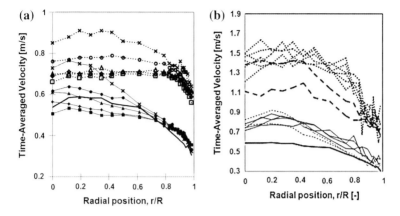

Fig. 3 Radial distribution of liquid (*thin continuous* line) and superficial gas velocity (*dotted* line) by $j_l = 0.5$ m/s. In figure a, the gas concentrations are identified as: 3.7 % (■), 6.7 % (♦), 9.6 % (▲), 13.3 % (●) and 18.1 % (x), all of them by bubbly flow regime. Data taken from Mendez-Diaz (2008). **a** Bubbly regime. **b** Transition and slug regime

Fig. 4 Radial distribution of liquid (*thin continuous* line) and superficial gas velocity (*dotted* line) by $j_l = 1$ m/s. The gas concentrations are identified as: 4 % (■), 7.8 % (♦), 8.6 % (▲), 12.6 % (●), 17.5 % (x) and 24 %(+). Data taken from Mendez-Diaz, 2008

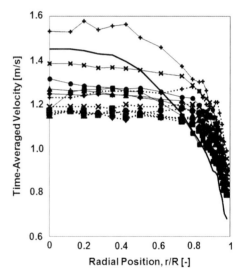

the continuous thick line in the figure represents the single phase case. When bubbles were introduced into the flow, a reduction in the local liquid velocity was observed. Under these flow conditions the effect of the bubble on the liquid velocity is not significant. However, when the gas concentration is greater than 10 %, the radial profile changed.

The change in the radial liquid velocity profile is produced by a redistribution of the shear rate originated by the presence of bubbles. For low gas concentrations and when the bubble size is within the range between 2 and 4 mm, the bubbles concentrate near the wall. Due to wake interaction, the shear rate from the wall is

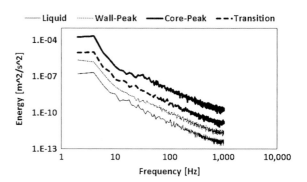

Fig. 5 Power spectral distribution at distinct flow regimes. Continuous, thin, dashed and tick lines represent single phase flow conditions, bubbly (8.6 %), transition (24.5 %), and slug (25.6 %) flow conditions

smooth, and the flow seems highly turbulent. At 0.5 m/s liquid velocity, the slip ratio is relatively high, whereas at 1 m/s the slip velocity is close to zero or negative

4 Discussion

From the radial void fraction distribution, there are flow conditions at low void fraction by which pseudoturbulence effects are present and dominate the phase distribution. These flow conditions are bubbly flow regime, at which the lift force dominates the bubble distribution. Figure 4 shows that the radial distribution on liquid phase is quite different from the single phase case due to the bubble induced agitation. Figure 5 shows the power spectral distribution for different gas concentrations and flow regimes. This result agree well with those reported by other authors: an increment in the liquid agitation is produced when void fraction is increased too.

The energy decay exponent is nearly constant when a bubbly flow regime is present; however, when the transition regime occurs, the value of the exponent decay is not constant, the same is observed at slug flow regime. The change in the exponent decay of energy reveals that the energy dissipation occurs faster at non-bubbly flow regimes due mainly to big bubbles present in the flow. This suggests that the shear rate induced by small bubbles is nearly homogeneous, and that the dissipation mechanisms are more efficient. On the other hand, the presence of big bubbles in slug regime results in a preferential concentration near the center of the channel. Also, the power spectral distribution can be observed to have less pronounced energy decay, as is showed in Fig. 6.

Fig. 6 Energy decay
exponent on different flow
conditions. Solid and empty
markers show bubbly and
slug flow regimes,
respectively. (■), $j_l = 0.5$ m/
s; (▲), $j_l = 1$ m/s

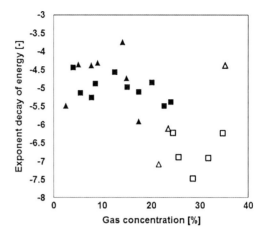

5 Conclusions

One mechanism that produces the transition between two phase flow regimes is the induced liquid velocity fluctuation level. Employing multi-tip impedance probe and laser Doppler anemometry, the main local flow parameters of both liquid and gas phases can be measured. The gas concentration was changed from 0 to 30 %.

From the data obtained, for bubbly flow regime, the turbulence intensity of the liquid phase is nearly constant, as well as the energy decay exponent. For this flow condition a wall peak pattern is present in void fraction distribution. At low gas concentration (less than 10 %) pseudo-turbulence effects are important and dominate the phase distribution. For the transition and slug regimes, the energy decay exponent shows a non-constant value. In the transition regime, a special redistribution of the shear rate occurs. A critical Reynolds number can be suggested to distinguish the bubbly regime; above this critical value, it can be argued that the lift force dominates the phase distribution and wall peak is observed in the radial void fraction distribution. This critical value is near 2,000.

References

Hibiki T, Ishii M, Xiao Z (2001) Axial interfacial area transport of vertical bubbly flows. Int J Heat Mass Transfer 44:1869–1888

Lance M, Bataille J (1991) Turbulence in the liquid phase of a uniform bubbly air-water flow. J Fluid Mech 222:95–118

Martinez-Mercado J, Palacios-Morales CA, Zenit R (2007) Measurements of pseudoturbulence intensity in monodispersed bubbly liquids for $10 < Re < 500$. Phys fluid 19

Mendez-Diaz S (2008) Medida experimental de la concentracion de área interfacial en flujos bifasicos finamente dispersos y en transicion. Ph.D. Thesis, Universidad Politecnica de Valencia

Panidis Th (2011) The development of the structure of water-air bubble grid turbulence. Int J Multiph Flow 37:565–575

Risso F (2011) Theorical model for k^{-3} spectra in dispersed multiphase flows. Phys Fluid 23:011701

Serizawa A, Kataoka I (1988) Phase distribution in two-phase flow. Transient phenomena in multiphase flow, hemisphere, Washington DC 1, pp. 179–224

Rensen T, Luther S, Lohse D (2005) The effect of bubbles on developed turbulence. J fluid Mec 538:153–187

Single- and Two-Phase Flow Models for Concentric Casing Underbalanced Drilling

J. Omar Flores-León, Octavio Cazarez-Candia
and Rubén Nicolás-López

Abstract Underbalanced Drilling process (UBD) is applied to oil fields where the original pressure has been depleted as a consequence of an intensive exploitation. In such technique, due to the incorporation of nitrogen into the Drilling Fluid (DF), the downhole pressure is close to the reservoir pressure. This work presents a single-phase, one-dimensional and transient mathematical model for the flow of: (1) nitrogen in an annulus, and (2) DF in a drill pipe. It is also presented a one-dimensional, transient, mathematical drift-flux model for the DF-gas two-phase flow in an annulus. The mathematical models are based on the mass, momentum and energy conservation equations, which are solved numerically using the finite difference technique in an implicit scheme. The models allow predicting pressure, velocity and temperature profiles. The predictions are in agreement with experimental and field data reported in the literature.

J. O. Flores-León (✉)
Facultad de Ingeniería, Universidad Nacional Autónoma de México,
C.P. 04510, Mexico, D.F., Mexico
e-mail: joomar35@hotmail.com

O. Cazarez-Candia · R. Nicolás-López
Instituto Mexicano del Petróleo, Eje Central Lázaro Cárdenas No. 152,
Col. San Bartolo Atepehuacan, C.P. 07730, Mexico, D.F., Mexico
e-mail: ocazarez@imp.mx

R. Nicolás-López
e-mail: rnlopez@imp.mx

J. Klapp et al. (eds.), *Fluid Dynamics in Physics, Engineering and Environmental Applications*, Environmental Science and Engineering,
DOI: 10.1007/978-3-642-27723-8_16, © Springer-Verlag Berlin Heidelberg 2013

1 Introduction

The Underbalanced Drilling (UBD) technique allows drilling fields where the original pressure has been depleted as a consequence of an intensive exploitation through the years. However, in literature there are few works on UBD, for instance, Urbieta et al. (2009) reported the first application of concentric nitrogen injection technique in Mexico. The reservoir pressure was originally 7500 psi, however due to the production rates and exploitation, nowadays the formation pressure is around 2200 psi. Then, concentric casing UBD is one option for drilling depleted reservoirs. This technique has all the UBD advantages and the good performance of the directional tools for the trajectory control (Urbieta et al. 2009).

The concentric casing UBD consists in the injection of an oil-based DF, at the standpipe through the drill pipe, which returns to surface in the principal annulus. Nitrogen is injected at surface through the concentric secondary annulus, and it meets the DF in the principal annulus at a certain depth. The incorporation of a gas phase in the principal annulus allows reducing the bottomhole pressure, reaching the underbalanced condition in the well. Figure 1 shows a pressure profiles scheme for a typical concentric casing gas injection UBD.

For the design and suitable operation of a concentric casing UBD, it is necessary to have theoretical tools, then this work presents a single-phase, one-dimensional and transient mathematical model for the flows of: (1) nitrogen in an annulus, and (2) drilling-fluid in a drill pipe. It is presented also a one-dimensional, transient, mathematical drift-flux model for the DF-gas two-phase flow in an annulus.

2 Single-Phase Flow Model

In the single-phase model, gas phase is real and compressible, however liquid phase is incompressible. Both fluids have a Newtonian behavior and they flow under adiabatic conditions. The forces presented are due to gravity and friction. The mathematical model is based on mass, momentum and energy conservation equations described respectively as follows (Bird et al. 2002):

$$\frac{\partial}{\partial t}\rho_k + \frac{\partial}{\partial z}\rho_k v_k = 0 \tag{1}$$

$$\frac{\partial}{\partial t}\rho_k v_k + \frac{\partial}{\partial z}\rho_k v_k v_k = -\frac{\partial}{\partial z}P_k \pm \rho_k g_z \cos\theta \pm \frac{\tau_{kw}}{D_h} \tag{2}$$

$$\frac{\partial}{\partial t}\rho_k e_k + \frac{\partial}{\partial z}\rho_k e_k v_k = -\frac{\partial}{\partial t}P_k \pm \rho_k v_k g \cos\theta \tag{3}$$

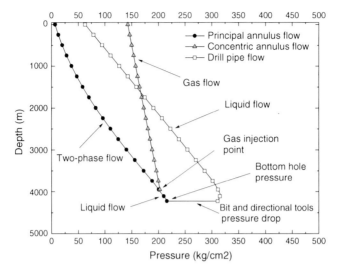

Fig. 1 Typical pressure profiles in a concentric casing UBD operation

where ρ is density, v is velocity, P is pressure, τ_w is the wall stress shear and e is specific energy. g and θ are the gravitational acceleration, and the inclination angle from the vertical, respectively. The subscript k denotes gas phase $(k = g)$ or liquid phase $(k = l)$. The hydraulic diameter D_h is used to calculate the annular and pipe flow areas in different sections of the well. The temporal and spatial coordinates are represented by t and z, respectively. Equations (1)–(3) can be written as follows.

Mass equation

$$\frac{X_k}{c_k^2}\left[\frac{\partial}{\partial t}P_k + v_k\frac{\partial}{\partial z}P_k\right] + \rho_k\frac{\partial}{\partial z}v_k = 0 \tag{4}$$

Momentum equation

$$\frac{\partial}{\partial z}P_k + X_k\rho_k\left[\frac{\partial}{\partial t}v_k + v_k\frac{\partial}{\partial z}v_k\right] = \pm\rho_k g_z\cos\theta \pm \frac{1}{2}\frac{f\rho_k v_k|v_k|}{D_h} \tag{5}$$

Energy conservation equation

$$-\eta_k C_{p,k}\rho_k\left[\frac{\partial}{\partial t}P_k + v_k\frac{\partial}{\partial z}P_k\right] - \frac{\partial}{\partial t}P_k + X_k\rho_k\left[\frac{\partial}{\partial t}v_k + v_k\frac{\partial}{\partial z}v_k\right] +$$
$$C_{p,k}\rho_k\left[\frac{\partial}{\partial t}T_k + v_k\frac{\partial}{\partial z}T_k\right] = \pm\rho_k v_k g_z\cos\theta \tag{6}$$

where X_k is a parameter to take into account the compressibility phase contribution $\left(X_g = 1, X_l = 0,\right)$ η is the Joule–Thomson coefficient which is the most important parameter that affects the temperature prediction (Cazarez et al. 2005), and C_p is the heat capacity at constant pressure. In the obtaining of Eqs. (4)–(6) the next relationships were used:

$$\partial \rho_k = \left.\frac{\partial \rho_k}{\partial P_k}\right|_{T_k} \partial P_k = \frac{1}{c_k^2}\partial P_k \tag{7}$$

$$\partial e_k = \partial \frac{v_k^2}{2} + \partial h_k \tag{8}$$

$$\partial h_k = C_{p,k}\partial T_k - \eta_k C_{p,k}\partial P_k \tag{9}$$

$$\tau_{kw} = \frac{1}{2}f\rho_k v_k |v_k| \tag{10}$$

Equation (7) describes the phase compressibility, where c is the sound velocity in the k phase. Equation (8) represents the specific energy given by the kinetic energy and enthalpy, respectively. The enthalpy definition given in Eq. (9) allows to set the energy equation as function of pressure and temperature. The Newtonian shear stress definition given in Eq. (10) involves the frictional factor f, which is a function of Reynolds number (Re), hydraulic diameter (D_h) and absolute roughness (ε). The friction factor was obtained with the next relation:

$$f = \begin{cases} 16/Re & Re < 2300 \\ \frac{1}{\sqrt{f}} = 1.74 - 2\log\left[2\frac{\varepsilon}{D_h} + \frac{18.7}{\sqrt{f}Re}\right] & Re \geq 2300 \end{cases} \tag{11}$$

3 Two-Phase Flow Model

The two-phase model is based on the Drift-Flux concept, then the relative velocity between the phases is taken into account. The suppositions done for the single-phase model also apply for the two-phase model. The drift flux relations used into the model were proposed by Ishii M. (1977) and Hibiki et al. (2003):

$$v_g = v_m + \frac{\rho_l}{\rho_m}\overline{V_{gj}} \tag{12}$$

$$v_l = v_m - \frac{\alpha_g}{1 - \alpha_g}\frac{\rho_g}{\rho_m}\overline{V_{gj}} \tag{13}$$

$$\overline{V_{gj}} = v_d + (C_o - 1)j \tag{14}$$

$$j = v_m + \frac{\alpha_g(\rho_l - \rho_g)}{\rho_m}\overline{V_{gj}} \tag{15}$$

In Eqs. (12)–(15) the subscript m denotes the mixture property, v_g and v_l are the gas and liquid in situ velocity, respectively; $\overline{V_{gj}}$ is the mean drift velocity, which is a function of the volumetric flux, j, the local drift velocity, v_d, and the distribution parameter, C_o. The constitutive relations proposed by Shi et.al. (2003) are used for the local drift velocity and the distribution parameter calculations. The conservation equations for the two-phase flow model are the Eqs. (4)–(6) with $k = m$ and $X_m = 1$. The mixture density, velocity and the energy coefficients are defined respectively as follows:

$$\rho_m = \alpha_g \rho_g + \left(1 - \alpha_g\right)\rho_l \tag{16}$$

$$v_m = \frac{\alpha_g \rho_g v_g + \left(1 - \alpha_g\right)\rho_l v_l}{\rho_m} \tag{17}$$

$$C_{p,m}\rho_m = \alpha_g C_{p,g}\rho_g + \left(1 - \alpha_g\right)C_{p,l}\rho_l \tag{18}$$

$$\eta_m C_{p,m}\rho_m = \alpha_g \eta_g C_{p,g}\rho_g + \left(1 - \alpha_g\right)\eta_l C_{p,l}\rho_l \tag{19}$$

Other parameters needed for the simulations are given below.

$$\alpha_g = \frac{v_{sg}}{C_o j + v_d} \tag{20}$$

$$v_{sg} = \frac{Q_g}{A_T} = \alpha_g v_g \tag{21}$$

$$v_{sl} = \frac{Q_l}{A_T} = \left(1 - \alpha_g\right)v_l \tag{22}$$

$$j = v_{sg} + v_{sl} \tag{23}$$

where Q is the volumetric flow rate and A_T is the cross section area of the annulus. For the proper calculation of void fraction, an iterative procedure should be done with (20), until the convergence between the superficial velocities from Eqs. (21) and (22) is reached, (23) is calculated with the new superficial velocities.

4 Numerical Solution

The single-phase and two-phase mathematical models were solved using the finite difference technique with an implicit forward scheme (24) for the temporal coordinate and a backward scheme (25) for the spatial coordinate.

$$\frac{\partial}{\partial t}\varphi = \frac{\varphi_z^{t+\Delta t} - \varphi_z^t}{\Delta t} \tag{24}$$

$$\frac{\partial}{\partial z}\varphi = \frac{\varphi_z^{t+\Delta t} - \varphi_{z-\Delta z}^{t+\Delta t}}{\Delta z} \qquad (25)$$

where φ represents pressure, velocity or temperature. The discretized conservation equations solution was reached using the LINPACK package for the simultaneous solution for linear equations (Dongorra et al. 1990).

5 Comparisons and Validations

The single-phase flow model was compared against experimental data (for water and natural gas) from a vertical experimental well reported in the literature by Gaither et al. (1963). Figure 2 shows the comparison between measured and calculated bottomhole pressures (BHP) for gas and liquid in upward flow with different pressures in surface. The points above and below the solid line represent overpredicted and underpredicted values, respectively. This means that the closer points to the solid line have the best fitting to the experimental data. The properties used for the simulations are also given in Fig. 2.

The two-phase flow model was compared with: (1) experimental data (water-nitrogen) from an experimental well composed by a 216.79 mm × 88.9 mm × 60.32 mm annular geometry (Lopes, 1997), and (2) an experimental well with a 159.41 × 88.9 mm annular geometry (Lage et al. 2000). Figure 2 shows the annular pressure profiles and the data used in the simulations.

The two-phase flow model was also validated with oilfield data (DF-nitrogen) reported in the work of Perez-Tellez (2003). Figure 3 shows the pressure profiles of two wells and the data used in the simulations (Fig. 4).

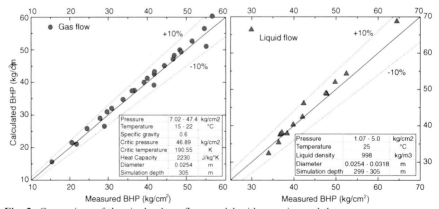

Fig. 2 Comparison of the single-phase flow model with experimental data

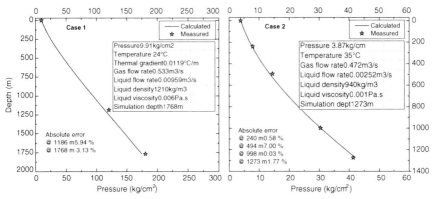

Fig. 3 Comparison of the two-phase flow model with experimental data

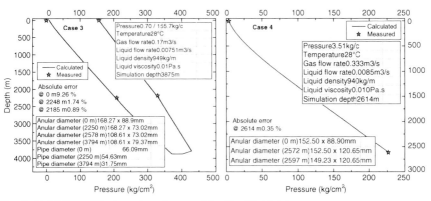

Fig. 4 Validation of the two-phase flow model with oilfield data

6 Conclusions

The drift-flux concept allows obtaining good pressure predictions. In the literature only pressure data are reported because velocity and void fraction profiles are not measured in experiments nor wells. Thus, this work assumed that as long as good pressure predictions are obtained other variables, such as void fraction, will be in good concordance to the physic of the phenomena. This because the equations were coupled solved for pressure, velocity and temperature.

The models have good concordance with experimental and filed data reported in the literature, with absolute errors lower than 10 %.

References

Bird RB, Stewart WE, Lightfoot EN (2002) Transport phenomena, 2nd Edn. Wiley,hoboken, pp 77, 80, 335

Cazarez-Candia O, Vazquez-Cruz MA (2005) Prediction of pressure, temperature and velocity distribution of two-phase flow in oil wells. Petroleum Sci Eng 46:195–208

Dongorra JJ, Bunch JR, Moler CB, Stewart GW (1990) LINPACK user guide. 8th Edition. Soc Ind Appl Math. ISBN 0-89871.172.x

Gaither OD, Winkler HW, Kirkpatrick CV (1963) Single and two-phase fluid flow in small vertical conduits including annular configurations. Paper presented at 37th SPE Annual Fall Meeting, 7-10 October 1962, in Los Angeles, California. Soc Petroleum Eng 441

Hibiki T, Ishii M (2003) One-dimensional drift-flux model and constitutive equations for relative motion between phases in various two-phase flow regimes. Int J Heat Mass Transf 46: 4935–4948

Ishii M (1977) One dimensional drift-flux model and constitutive equations for relative motion between phases in various two-phase flow regimes. ANL-7747 report

Lage ACVM, Time RW (2000) Mechanistic model for upward two-phase flow in annuli. Paper presented at SPE Annual Technical Conference and Exhibition, 1-4 October 2000, Dallas, Texas. Soc Petroleum Eng 63127

Lopes CA (1997) Feasibility study on the reduction of hydrostatic pressure in a deep water riser using a gas-lift method. PhD thesis, Louisiana State University

Perez-Tellez C (2003) Improved bottomhole pressure control for underbalanced drilling operations. PhD thesis, Louisiana State University

Shi H, Holmes JA, Durlofsky LJ, Aziz K, Diaz LR, Alkaya B, Oddie G (2003) Dirft-Flux modelinf of multiphase flow in wellbores. Paper presented at SPE Annual Technical Conference and Exhibition, 5-8 October 2003, in Denver, Colorado. Soc Petroleum Eng 84228

Urbieta A, Perez-Tellez C, Lupo C, Castellanos JM, Ramirez O, Puerto G, Bedoya J (2009) First application of concentric nitrogen injection technique for a managed pressure drilling depleted well in Southern Mexico. Paper presented at IADC/SPE Managed Pressure Drilling and Underbalanced Operations Conference and Exhibition, 12-13 February 2009, San Antonio, Texas. Soc Petroleum Eng 122198

Study of Structural Properties in Complex Fluids by Addition of Surfactants Using DPD Simulation

Estela Mayoral, Eduardo Nahmad-Achar, José Manuel
Martínez-Magadán, Alejandro Ortega and Ismael Soto

Abstract In this work we study the tertiary structure of ionic and surfactant when the pH in the system is modified using electrostatic dissipative particle dynamics simulations (DPD). The dependence with pH and kind of surfactant is presented. Our simulations reproduce the experimental behavior reported in the literature. The scaling for the radius of gyration with the size of the molecule as a function of pH is also obtained.

1 Introduction

Modification and control of structural and interfacial properties between the different components in confined mixed systems e.g. rock/water/oil, by the use of chemical additives, is nowadays an important research area in order to enhanced oil recovery (EOR) retained in the porous rock (Abdel-Wali 1996). EOR is based on the fact that the incorporation of other components into this complex system modifies the collective properties amongst them. The performance of the additives in the system is in strong correlation with their tertiary structure, the characteristics

E. Mayoral (✉)
Instituto Nacional de Investigaciones Nucleares (ININ), Carretera México-Toluca (S/N),
CP 52750 La Marquesa Ocoyoacac, Estado de México, Mexico
e-mail: estela.mayoral@inin.gob.mx

J. M. Martínez-Magadán · A. Ortega · I. Soto
Instituto de Ciencias Nucleares, Universidad Nacional Autónoma de México (UNAM),
Apartado Postal 70-543, 04510 México, D.F., Mexico

E. Nahmad-Achar
Instituto Mexicano del Petróleo (IMP), Eje Central Lázaro Cárdenas S/N,
México, D.F., Mexico

J. Klapp et al. (eds.), *Fluid Dynamics in Physics, Engineering and Environmental
Applications*, Environmental Science and Engineering,
DOI: 10.1007/978-3-642-27723-8_17, © Springer-Verlag Berlin Heidelberg 2013

in the media, and the properties of oil/aqueous interfaces (Hansson and Lindman 1996). In particular, ionic strength, pH, temperature, and pressure play an important role in the behavior of these systems. Understanding how these conditions in the media modify the tertiary structure of the additives is then a fundamental task in order to design new surfactants ad hoc and to develop optimal formulations. The experimental study of the conformation of macromolecules is usually done by dynamic light scattering (DLS), but in many occasions this is complicated due to the different sizes of the molecules involved in multicomponent systems (González-Melchor et al. 2006). Alternatively, numerical simulation offers a viable option to help in the design of new additives that could give a good performance in specific situations, even under extreme conditions that are impossible to handle in laboratory. In these kinds of complex systems, where many particles with different sizes undergo interactions at different time scales, the mesoscopic simulation has prevented to be a good alternative (Español and Warren 1995; Groot and Warren 1997). One of the most important mesoscopic simulation approaches is the dissipative particle dynamics (DPD), introduced by Hoogerbruge and Koelman (1992). In this work we present a study of the radius of gyration for ionic surfactant in different pH conditions. We use electrostatic dissipative particle dynamics simulations to study its dependence with the pH and the kind of polyelectrolyte added. Our simulations reproduce the experimental information reported in the literature (Griffiths et al. 2004). The correspondence between the partial charge on the electrolyte (θ) and the pH is established. Additionally, study the scaling of the radius of gyration with the size of the polymer (N) at different pH conditions. The scaling exponent v at different pH is presented.

2 Methodology

Dissipative Particle Dynamics (DPD) is a *coarse graining* approach which consists of representing a complex molecule (polymer or surfactant) by soft spherical beads joined by springs, interacting through a simple pair-wise potential and thermally equilibrated through hydrodynamics (Español 2002; Español and Warren 1995). The beads follow Newton's equations of motion $\frac{dr_i}{dt} = v_i; \frac{dv_i}{dt} = f_i$, where the force f_i on a bead is constituted by three pair-wise additive components $f_i = \sum_j (f_{ij}^C + f_{ij}^D + f_{ij}^R)$. The sum runs over all nearby particles within a distance R_c. The conservative force is defined as $f_{ij}^C = a_{ij}\omega^c(r_{ij})\hat{r}_{ij}$ for $r_{ij} < r_c$ and zero elsewhere. a_{ij} is a parameter which measures the maximum repulsion between particles i and j, $r_{ij} = r_i - r_j$ and $r_{ij} = r_{ij}|\hat{r}_{ij}|$; the weight function $\omega_c(r_{ij})$ is given by $\omega^c(r_{ij}) = 1 - \frac{r_{ij}}{r_c}$ for $r_{ij} < r_c$, and zero elsewhere. This conservative repulsion force derives from a soft interaction potential and there is no hard-core divergence of this force as in the case of the Lennard-Jones potentials which allows for a more efficient scheme of integration. When we need to introduce a more complex molecule, such as a polymer, we use beads joined by springs, so we also have an extra spring force given by $f_{ij}^s = -Kr_{ij}$ if i is connected with j. The dissipative and random standard DPD forces are given Eq. (1):

$$f_{ij}^D = -\gamma\omega^D(r_{ij})(\hat{r}_{ij} \cdot v_{ij})\hat{r}_{ij} \text{ and } f_{ij}^R = \sigma\omega^R(r_{ij})(\theta_{ij}1/\sqrt{\delta_t})\hat{r}_{ij} \tag{1}$$

where δ_t is the integration time step, $v_{ij} = v_i - v_j$ is the relative velocity, σ is the amplitude of noise and θ_{ij} is a random Gaussian number with zero mean and unit variance. γ and σ are the dissipation and the noise strengths respectively, while $\omega^D(r_{ij})$ and $\omega^R(r_{ij})$ are dimensionless weight functions. These quantities are not independent: they are related by the fluctuation–dissipation theorem (Español and Warren 1995) by $\gamma = \frac{\sigma^2}{2k_BT}$, and therefore $\omega^D(r_{ij}) = [\omega^R(r_{ij})]^2$. This relationship maintains the temperature constant and preserves the total energy in the system. Here k_B is the Boltzmann constant and T is the temperature. The consistency of the results depends principally on how well the essential characteristics for a group of atoms are included into the interaction parameters a_{ij} between the DPD beads. Groot and Warren (1997) established a simple functional form of the conservative repulsion in DPD (a_{ij}) in terms of the Flory–Huggins χ-parameter theory. Since then, the use of solubility parameters for the calculation of χ-parameters in order to get the repulsive parameters for DPD simulations is commonly used. The electrostatic interactions were introduced into DPD via two different methods, by Groot (2003) and by González-Melchor et al. (2006). In both cases the point charge at the center of the DPD particle is replaced by a charge distribution along the particle. Groot (2003) solves the problem by calculating the electrostatic field on a grid. González-Melchor et al. (2006) solve the problem adapting the standard Ewald method to DPD particles.

3 Simulation Details

All simulations were carried out using a DPD electrostatic code. The electrostatic interactions were calculated following González-Melchor et al. (2006). We used dimensionless units denoted with an asterisk; the masses were all equal to 1. According with Groot and Warren (1997), the values for σ and γ were established as 3 and 4.5 respectively, with this $k_BT^* = 1$. The total average density in the system was $\rho^* = 3.0$. The simulation box was cubic with $Lx = Ly = Lz = 8.5$. Periodic boundary conditions were imposed in all directions. The time step used was set to $\Delta t = 0.04$ and we performed 25 blocks of simulations with 10,000 steps each one, and the properties were calculated by averaging over the last 10 blocks. The structure for the molecules of ionic dispersants called was mapped as shown in Figs. 1 and 2. The repulsive interaction parameters were obtained through the solubility parameters for the individual monomers indicated in Fig. 1 and using the Flory–Huggins parameters χ. The parameters a_{ij} were: for equal DPD beads

Fig. 1 Structure and
mapping for PAA

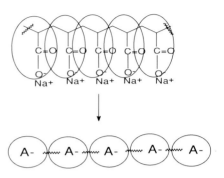

$a_{ii} = 78$, and for different beads $a_{AB} = 78.22$, $a_{BC} = 78.55$, $a_{CA} = 78.25$, $a_{AW} = 182.33$, $a_{BW} = 169.04$, $a_{CW} = 250.13$, where the subscripts correspond to the DPD beads shown in Fig. 1, and the subscript W corresponds to the interaction with water. Monomers in a polymer were joined by Hookean springs with spring constant $K = 100$ and with equilibrium distance $r_{eq} = 0.7$.

4 Results and Discussion

We studied the radius of gyration of an ionic surfactant (PAA) at different values of pH. The structure for this poly-electrolyte is shown in Fig. 1. The PAA, is a polyacrylic dispersant constituted by N monomeric carboxylic units. This is a weak poly-acid with pK_a 5.47. The carboxylic monomeric units could be neutral (if they are protonated) or negatively charged (if they are deprotonated). The degree of ionization depends on the pH of the medium: when we modify the pH some of the monomers stay uncharged (N_0) and the rest become negatively charged ($N^- = N - N_0$). We can express the relationship between the pH and the charge fraction θ as: $pH = \log(\theta/(1-\theta)) + pK_a$ where θ is the ratio between the number of deprotonated (N^-) carboxylic monomeric units and the total number of monomeric units (N). We performed simulations for PAA molecules of different sizes ($N = 12$, 24, 48), and for different values of pH (1, 4.62, 5.25, 5.47, 5.79, 5.94, 6.31, 14); the results for the radius of gyration as a function of pH are shown in Fig. 2.

We can observe that the radius of gyration (Rg) increase with pH, as expected, due to the repulsive intramolecular charge. Setting $Rg = aN^\nu$ with ν the scaling exponent and a the Flory ratio, we obtain the values shown in Table 1 for different charge fractions θ.

Theoretical analysis has shown that, in a good solvent, $\nu = 3/5 = 0.6$ (De Gennes 1979); looking at Table 1 and Fig. 2 we see that the scaling exponent obtained for $\theta = 0.375$ is $\nu = 0.593$. This is the closest value obtained compared with the theoretical value corresponding with the greatest value for $a = 0.556$, indicating that the molecule is completely extended as is expected for a good solvent.

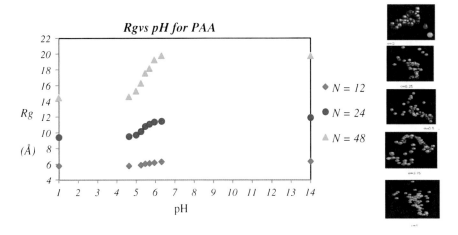

Fig. 2 Rg for PAA as a function of pH and equilibrium configurations at different θ

Table 1 Scaling exponent v for PAA at different θ

θ	0	0.125	0.375	0.5	0.675	0.75	0.875	1
v	0.643	0.631	0.593	0.740	0.745	0.795	0.802	0.779
a	0.449	0.469	0.556	0.403	0.404	0.352	0.436	0.407

Table 2 Simulation and experimental results for the Rg of PAA *NA* not available

PM (gr/mol)	900	1800	3600	20,000	770,000
N	12	24	48	267	10 267
Rg simulation (DPD Units)	0.895	1.454	2.239	6.970	77.550
Rg simulation (Å)	5.7838	9.394	14.464	45.05	500.979
Rg experimental (Å)	NA	NA	NA	50	513

Table 2 shows the estimated results of the radius of gyration, both in DPD units and in Å, for different molecular weights using the scaling approach obtained with the DPD simulations. As well as the experimental data reported. The simulation with $N = 48$ corresponds with a $MW = 3600$ Da. We observed a very good agreement with the experimental results.

5 Conclusions

Electrostatic DPD simulations prove to be very useful in the study of the tertiary structure of polyelectrolyte, by studying the radius of gyration as a function of pH for various molecular lengths. The behavior of the ionic surfactant PAA has been obtained in very good agreement with experimental results. The correct scaling exponent v is also obtained for the ionic surfactant. The quantitative prediction for the radius of gyration using DPD electrostatic simulations and scaling arguments corresponds with the experimental data obtained by DLS experiments.

References

Abdel-Wali AA (1996) Effect of simple polar compounds and salinity on interfacial tension and wettability of rock/oil/brines system. J King Saud Univ Eng Sci 2(8):153–163

De Gennes PG (1979) Scaling concepts in polymer physics. Cornell University Press, Ithaca

Español P (2002) Dissipative particle dynamics. Chall. Mol. Sim. 4:59–77

Español P, Warren P (1995) Statistical mechanics of dissipative particle dynamics. Europhys Lett 30:191–196

González-Melchor M, Mayoral E, Velázquez ME, Alejandre J (2006) Electrostatic interactions in dissipative particle dynamics using the Ewald sums. J Chem Phys 125(1):224107, 1–8

Griffiths PC, Paul A, Khayat Z, Wan KW, King SM, Grillo I, Schweins R, Ferruti P, Franchini J, Duncan R (2004) Understanding the mechanism of action of poly(amidoamine)s as endosomolytic polymers: correlation of physicochemical and biological properties. Biomacromolecules 5(4):1422–1427

Groot D (2003) Electrostatic interactions in dissipative particle dynamics-simulation of polyelectrolytes and anionic surfactants. J Chem Phys 118(1):11265–11277

Groot D, Warren PB (1997) Dissipative particle dynamics: bridging the gap between atomistic and mesoscopic simulation. J Chem Phys 107(11):4423–4435

Hansson P, Lindman B (1996) Surfactant-polymer interactions (Review). Curr Opin Cooloid Interface Sci 1:604

Hoogerbrugge PJ, Koelman JMVA (1992) Simulating microscopic hydrodynamic phenomena with dissipative particle dynamics. Europhys Lett 19:155–160

Study of Slug Flow in Horizontal, Inclined, and Vertical Pipes

Omar C. Benítez-Centeno and Octavio Cazarez-Candia

Abstract In the literature several mathematical models have been reported for the slug flow, which uses the concept of the unit slug in common. This concept requires the longitudes of the liquid slug and the Taylor bubble to be known, for the correct evaluation of the terms: phase-wall and interfacial shear stresses, and virtual mass forces. In this work are presented the results of an experimental study of the upward slug flow (in an acrylic pipe with 6 m of length and 0.01905 m of internal diameter). The working fluids are water and air. They were carried out experiments for several angles of tube inclination from horizontal to vertical. The length of liquid slug and Taylor bubble were measured. Also, using voltage signals of two infrared sensors, was determined: (1) The Taylor bubble velocity by means of the cross correlation and (2) The slug frequency when applying the Fourier transform to these obtained data; this information allows to determine the length of liquid slug. It was observed that the Taylor bubble length decreases when increasing the flow of the liquid so much as the angle of inclination, while the Slug length varies with a similar tendency.

O. C. Benítez-Centeno (✉) · O. Cazarez-Candia
Instituto Tecnológico de Zacatepec, Calzada del Tecnológico
No. 27 Zacatepec, 62780 Morelos, MEX, Mexico
e-mail: omarcbc@gmail.com

O. Cazarez-Candia
e-mail: ocazarez@imp.mx

O. Cazarez-Candia
Instituto Mexicano del Petróleo, Eje Central Lázaro Cárdenas 152 Col San Bartolo
Atepehuacan, 07730 Mexico, D.F., Mexico

J. Klapp et al. (eds.), *Fluid Dynamics in Physics, Engineering and Environmental Applications*, Environmental Science and Engineering,
DOI: 10.1007/978-3-642-27723-8_18, © Springer-Verlag Berlin Heidelberg 2013

1 Introduction

When proposing mathematical models to represent in a realistic way the slug flow, in the literature, diverse positions are presented (mechanistics, slug tracking or slug capturing) where the concept of the unit slug is used, when solving this models numerically the stresses phase-wall and interfacial forces should be calculated, the above mentioned implies to know the lengths of the liquid slug and the Taylor bubble just as it is appreciated in the Eqs. (1) and (2). Where τ is the stress, the sub index k denotes the phase (l liquid or g gas) that has contact with the wall (w), U_g is the phase gas velocity, α_l is the liquid fraction, ρ_l is the density of the liquid, F_{VM} is the virtual mass force, C_{VM} is the virtual mass coefficient, DH is the hydraulic diameter, Db and L_{TB} they are respectively the diameter and the length of the Taylor bubble, L_{LS} is the slug length and finally L_{SU} is the unit slug length.

$$\tau_{kw} = \left[\frac{\tau_{kw}^{(slug)}}{DH_k} \left(\frac{L_{LS}}{L_{SU}} \right) + \frac{\tau_{kw}^{(Taylor\ Bubble)}}{DH_k} \left(\frac{L_{TB}}{L_{SU}} \right) \right] \tag{1}$$

$$F_{VM} = C_{VM}(1 - \alpha_l)\rho_l \left[\frac{\partial U_g}{\partial t} + U_g \frac{\partial U_g}{\partial x} - \frac{\partial U_l}{\partial t} + U_l \frac{\partial U_l}{\partial x} \right] \tag{2}$$

$$C_{VM} = 5 \left[0.66 + 0.34 \left(\frac{1 - \left(\frac{Db}{L_{TB}} \right)}{1 - \left(\frac{Db}{3L_{TB}} \right)} \right) \right] \tag{3}$$

Having appropriate values for L_{LS} y L_{TB}, has an impact to obtain better predictions of pressure drop when mechanistic models are used to simulate liquid–gas two-phase slug flows. For that the next work is presented.

2 Experimental

According to the angle and diameter of the pipe is located inside of the respective flow pattern map, this way is easy to determine the volumetric fluxes used at the experiments of water and air. With a fixed region selected where the slug flow, depending of the inclination angle of the experiment (0°, 3°, 45°, 60° and 90°) and to a superficial gas velocity (0.25–1.0 m/s) and the same values for the superficial liquid velocity were tested, for each value of superficial liquid velocity thirty photographs were taken to measure as a average the unit slug parameters, at the Table 1 are reported this lengths. At the same time on through of an alternating method the slug frequency is measured for the mentioned cases of inclination of the pipe. Starting from the data acquisition coming from the registration of voltage of two infrared sensors what has the capacity to detect the presence of bubbles during a slug flow in a duct, the parameters of the unit slug and the Taylor bubble

Table 1 Measured slug and Taylor bubble lengths (m)

Angle (°)	Slug length (L_{LS})	Taylor bubble length
0	0.264	0.440
3	0.264	0.340
45	0.200	0.180
60	0.195	0.143
90	0.179	0.120

velocity they were determined, this duct with a circular cross section. The experimental equipment is described at detail in the work of Benítez et al. (2011). Table 2 shows the values for the experimental parameters (pressure, water flow rate, air flow rate and temperature).

With the above-mentioned they are proposed slug unities lengths for each angle of inclination in the pipe as a function of their diameter.

The voltage signals data are treated statistically, obtaining the slug frequency (*fs*) by means of Fourier transform; as an example Fig. 1 shows the spectral frequency graph for the 0°case.

Another of the treatments to the acquired data of voltage recorded during each experiment is the application of the cross correlation to obtain the representative time in which the bubbles of Taylor moved among the two sensors located at a distance of separation of 0.908 m, for what the Taylor bubble velocity is the ratio between the distance and the delay time. In Fig. 2 are shown the crossed correlation for the case in which the pipe is prepared horizontally according to the data of the Table 2.

3 Results and Discussion

Well known the slug frequency for each one of the cases of the angular positions of the pipe referred previously and applying the relationship for the slug intermittence (Woods and Hanratty 1996; Cazarez et al. 2012; slug intermittence) the slug length can be calculated by means of the Eq. (4) proposed by Woods et al. 2006. Equation (5) it is calculated the Taylor bubble velocity, C_0 and C_1 they take the values according to (Bendiksen 1984) for angles near to the horizontal, for further angles at 3° and smaller at 90° of agreement with (Kokal 1987) and at the vertical pipe (Issa and Tang 1991):

$$L_{LS} = 1.2\left(\frac{U_{sl}}{fs}\right) \tag{4}$$

$$U_{TB} = C_0 U_M + C_1 \sqrt{gD} \tag{5}$$

Table 2 Experimental data

Parameters	0°	3°	45°	60°	90°
Pressure, [Pa]	138588	109254	111737	142855	138588
Water flow rate, [m³/s]	1.99E–04	9.96E–05	1.88 E–04	1.93E–04	9.44E–05
Air flow rate, [m³/s]	7.13E–05	7.12E–05	6.74E–05	8.58E–05	6.96E–05
Temperature, [°C]	26	27	25	29	30

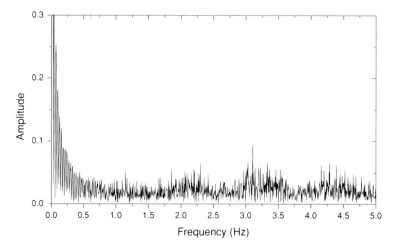

Fig. 1 Domain frequency ($fs = 3.1$ Hz), for the experimental case 0°

Fig. 2 Representative delay time (0.75 s), between the sensor 1 and 2 during a slug flow at 0° case

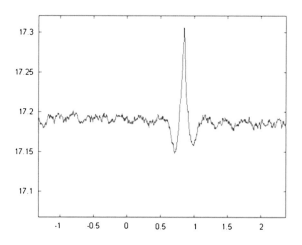

In Figs. 3 and 4 are appreciated the comparison between the calculations and measurements for the slugs lengths and the Taylor bubble lengths respectively, for each one of the cases referred to the experiments of Table 2.

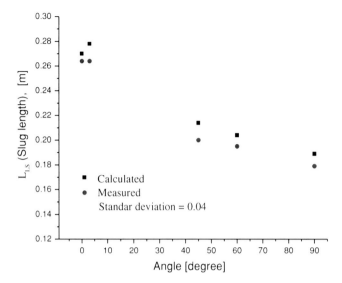

Fig. 3 Slug length calculated and experimental

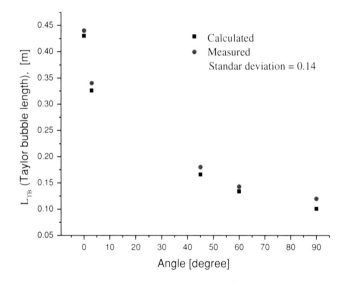

Fig. 4 Taylor bubble length calculated compared with the measured

The behavior tendency for the slug length and to Taylor bubble length it is falling, the percent absolute errors are not bigger to 7 % for the calculated slug lengths to respect at the mensurations and not more than 17 % for the Taylor bubble lengths. Finally in Table 3 the comparisons are shown between the calculations and measurements of the Taylor bubble velocity, the percent absolute errors also reflect a good agreement.

Table 3 Comparison of calculated and measured Taylor bubble velocity (m/s)

Angle (°)	U_{TB} calculated	U_{TB} measured	Abs err %
0	1.229	1.211	1.52
3	0.853	0.825	1.19
45	1.227	1.211	1.31
60	1.324	1.211	9.40
90	0.842	0.841	0.12

4 Conclusions

In this study a methodology is presented to determine the lengths that integrate the unit slug and of the Taylor bubble velocity, the results show agreement among the calculations regarding the mensurations. With the present proposal a tool is suggested to supplement the locks in the solution of numeric models for two-phase slug flow. When the representative longitudes of the unit slug are know, it allows for the numeric models a better approach, this lengths have a direct impact in the terms of closed. In it resides it the contribution of this work.

References

Bendiksen HK (1984) An experimental investigation of the motion of long bubbles in inclined tubes. Int J Multiphase Flow 10(4):467–483

Benitez Centeno OC, Cazarez Candia O, Moya Acosta SL (2011) Experimental study of the slug flow. experimental and theoretical advances in fluid dynamics. Springer, Berlin, vol 3, pp 287–293

Cazarez Candia O, Montoya Hernández DJ, Banwart AC (2012). Simulation of intermittent gas-bubbly oil three phase flow in upward vertical pipe using the two-fluid model. Paper 153472. SPE Latin American and Caribbean Petroleum Engineering Conference, Mexico City, Mexico

Issa RI, Tang ZF (1991) Prediction of pressure drop and holdup in gas-liquid flow in pipes using the two-fluid model, SPE 22534, Unsolicited manuscript

Kokal S (1987) Study of two-phase flow in inclined pipes. PhD dissertation, University of Calgary

Woods BD, Hanratty TJ (1996) Relation of slug stability to sheding rate. Int J Multiphase Flow 22(5):809–828

Woods BD, Fan Z, Hanratty TJ (2006) Frequency and development of slugs in a horizontal pipe at large liquid flows. Int J Multiphase Flow 2(20):902–925

Modeling of Water-Steam Slug Flow in Inclined Pipes Undergoing a Heating Process

P. Mendoza-Maya, O. Cazarez-Candia and S. L. Moya-Acosta

Abstract This work presents a one-dimensional transient mathematical model for the steam-water slug flow. The model is based on the two-fluid modeling technique. This consists in mass and momentum equations for each phase. Total thermal equilibrium was assumed, nucleation is neglected and the difference between steam and water pressures is taken into account. The model was solved using the finite difference technique. The model allows estimating: pressure, steam volume fraction, velocity, mixture temperature, and internal-wall pipe temperature. Different slug frequency correlations were evaluated. It was found that the best prediction of pressure drop was obtained with the Greskovic and Shrier (1971) correlation. The pressure predictions are in agreement with simulations obtained from a commercial simulator.

1 Introduction

Two phase flow in pipes appears in many industrial applications. It appears frequently in steam generators, nuclear engineering, chemical processes, oil production, geothermic, refrigeration, and solar energy by direct steam generation (DSG) in parabolic trough concentrators.

P. Mendoza-Maya (✉) · S. L. Moya-Acosta
Centro Nacional de Investigación y Desarrollo Tecnológico, Prolongación Av. Palmira esq.
Apatzingan, Col Palmira, 62490 Morelos, CU, Mexico
e-mail: pedro_m_maya@yahoo.com.mx

O. Cazarez-Candia
Instituto Mexicano del Petróleo, Eje Central Lázaro Cárdenas Norte 152 col,
07730 DF, San Bartolo Atepehuacan, Mexico
e-mail: ocazarez@imp.mx

J. Klapp et al. (eds.), *Fluid Dynamics in Physics, Engineering and Environmental Applications*, Environmental Science and Engineering,
DOI: 10.1007/978-3-642-27723-8_19, © Springer-Verlag Berlin Heidelberg 2013

Fig. 1 Modes of heat loss
from an evacuated tubular
absorber. (Odeh et al. 1998)

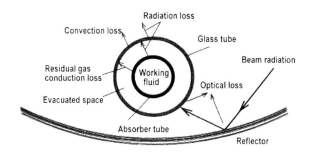

In the direct steam generation process parabolic trough concentrators are used, they consist on a group of big parabolic reflective mirrors that concentrate the sunlight on the focal line where two concentric tubes are placed. The external tube is made of glass. In the annular region vacuum is generated in order to avoid the heat losses by convection. The interior tube is made of steel (absorbent pipe) which heats a flow of water by convection, then, overheated steam is produced (Fig. 1).

The slug flow pattern commonly is present in systems of direct steam generation (Eck et al. 2003; Nattan et al. 2003).

Despite exhaustive experimental and theoretical studies on slug flow, heat and mass transfer phenomena, as well as its hydrodynamic behavior have not been completely understood. Then, in this work a two-fluid model coupled with a heat transfer model is presented. In order to evaluate the effect of different frequency slug correlations on pressure drop.

1.1 Hydrodynamic Model

Slug flow was divided in two sections (Fig. 2): (1) the slug section (liquid flow), and (2) the Taylor bubble region formed by a liquid film at the bottom of the pipe and an elongate bubble placed at the top of the pipe (stratified flow). Due to the arrangement of the phases, the wall-liquid stress ($\tau_{lw} \neq 0$), wall-steam stress ($\tau_{gw} \neq 0$) and interfacial stress ($\tau_i \neq 0$) are present. Due to the pipe heating the mass transfer (Γ_g), the wall-liquid heat flux ($q''_{wl} \neq 0$) and the wall-steam heat flux ($q''_{wg} \neq 0$) are taken into account. The interfacial heat flux ($q''_i = 0$) is neglected because total thermal equilibrium is considered.

In this work we consider that gas and liquid pressures are unequal in the Taylor bubble section (Taitel and Dukler 1976; Ouyang, and Aziz 2002; Cazarez and Espinoza 2008). This consideration was proposed to get good numerical stability.

In Fig. 2, L_{BT} is the bubble length (section considered like stratified flow) and L_S is the slug length (section considered like liquid flow), L_{uslug} is called the slug unit length ($L_{uslug} = L_{BT} + L_S$).

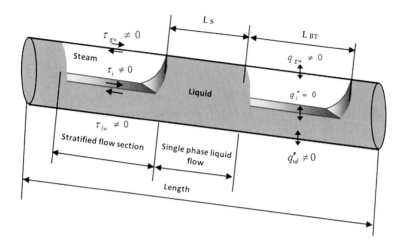

Fig. 2 Slug flow

Liquid and vapour pressures are determined by

$$p_l = p_i + 0.5\rho_l g h_{h,l} \cos\theta \qquad (1)$$

$$p_g = p_i - 0.5\rho_g g \cos\theta (D - h_{h,l}) \qquad (2)$$

Where ρ_g is steam density, ρ_l is liquid density, g is the gravitational constant, and θ is the pipe inclination angle from the horizontal.

The two-fluid model consists in two sets of conservation equations for the balance of mass, momentum and energy for each phase. However considering full thermal equilibrium, the model is based only in the mass and momentum equations for the steam and liquid.

Steam mass conservation:

$$\frac{\alpha_g}{C_g^2 \Omega_{g,s}}\left(\frac{\partial p_i}{\partial t} + v_g \frac{\partial p_i}{\partial z}\right) - \frac{\alpha_g}{C_g^2}\frac{\rho_g g H_{li}}{\Omega_{g,s}}\left(\frac{\partial \alpha_g}{\partial t} + v_g \frac{\partial \alpha_g}{\partial z}\right)$$
$$+ \rho_g\left(\frac{\partial \alpha_g}{\partial t} + v_g \frac{\partial \alpha_g}{\partial z}\right) + \alpha_g \rho_g \frac{\partial v_g}{\partial z} = \left(\frac{L_{BT}}{L_{BT} + L_{SLUG}}\right)\Gamma_g \qquad (3)$$

Liquid mass conservation:

$$\frac{(1 - \alpha_g)}{\Omega_{l,s}C_l^2}\left(\frac{\partial p_i}{\partial t} + v_l \frac{\partial p_i}{\partial z}\right) - \frac{(1 - \alpha_g)}{C_l^2}\frac{g\rho_l H_l}{\Omega_{l,s}}\left(\frac{\partial \alpha_g}{\partial t} + v_l \frac{\partial \alpha_g}{\partial z}\right)$$
$$- \rho_l\left(\frac{\partial \alpha_l}{\partial t} + v_l \frac{\partial \alpha_l}{\partial z}\right) + (1 - \alpha_g)\rho_l \frac{\partial v_l}{\partial z} = \left(\frac{L_{BT}}{L_{BT} + L_{SLUG}}\right)\Gamma_l \qquad (4)$$

Steam momentum equation:

$$\frac{\alpha_g v_g}{C_g^2}\left[\left(\frac{1}{\Omega_{g,s}}\frac{\partial p_i}{\partial t} - \frac{\rho_g g H_{li}}{\Omega_{g,s}}\frac{\partial \alpha_g}{\partial t}\right) + v_g\left(\frac{1}{\Omega_{g,s}}\frac{\partial p_i}{\partial z} - \frac{\rho_g g H_{li}}{\Omega_{g,s}}\frac{\partial \alpha_g}{\partial z}\right)\right]$$

$$+ v_g\rho_g\left(\frac{\partial \alpha_g}{\partial t} + v_g\frac{\partial \alpha_g}{\partial z}\right) + \alpha_g\rho_g\left(\frac{\partial v_g}{\partial t} + 2v_g\frac{\partial v_g}{\partial z}\right)$$

$$+ \alpha_g\left(\frac{1}{\Omega_{g,s}}\frac{\partial p_i}{\partial t} - \frac{\rho_g g H_{li}}{\Omega_{g,s}}\frac{\partial \alpha_g}{\partial t}\right)$$

$$= \left(\frac{L_{BT}}{L_{BT} + L_{SLUG}}\right)\left[v_i^{\Gamma_k}\Gamma_g + F_{wg} + F_i\right] - \left(\alpha_g\rho_g g\right)\cdot sen\theta \tag{5}$$

Liquid momentum equation:

$$\left(\frac{1-\alpha_g}{\Omega_{l,s}}\right)\frac{\partial p_i}{\partial z} + \frac{v_l(1-\alpha_g)}{C_l^2\Omega_{l,s}}\left(\frac{\partial p_i}{\partial t} + v_l\frac{\partial p_i}{\partial z}\right)$$

$$- \left(\frac{(1-\alpha_g)g\rho_l H_l}{\Omega_{l,s}}\right)\left(\frac{\partial \alpha_g}{\partial z}\right) - \left[v_l\rho_l + \left(\frac{v_l(1-\alpha_g)}{C_l^2}\frac{g\rho_l H_l}{\Omega_{l,s}}\right)\right]\left(\frac{\partial \alpha_g}{\partial t} + v_l\frac{\partial \alpha_g}{\partial z}\right)$$

$$+ \rho_l(1-\alpha_g)\left(\frac{\partial v_l}{\partial t} + 2v_l\frac{\partial v_l}{\partial z}\right) = \left(\frac{L_{BT}}{L_{BT} + L_{SLUG}}\right)\left[v_i^{\Gamma_k}\Gamma_l + F_{wl} + F_i\right]$$

$$+ \left(\frac{L_{SLUG}}{L_{BT} + L_{SLUG}}\right)F_{wl} - \left(\alpha_g\rho_g g\right)\cdot sen\theta \tag{6}$$

In Eqs. (3)–(4) Γ_g and Γ_l account for the mass transfer between the phases, α_g is void fraction, C_g and C_l are the velocity of sound in the steam and liquid respectively, v_g is the steam velocity, v_l is the liquid velocity, and $\Omega_{l,s}$ and $\Omega_{g,s}$ are defined by the following equations:

$$\Omega_{g,s} = \left(1 + \frac{g\cos\theta(D - h_{h,l})}{C_g^2}\right) \tag{7}$$

$$\Omega_{l,s} = \left(1 - \cos\theta\frac{gh_{h,l}}{C_l^2}\right) \tag{8}$$

Where D is the internal pipe diameter, ρ_g is the steam density ρ_l is the liquid density and H_{li} is defined by the following equation

$$H_{li} = \frac{D\pi}{4\sqrt{1 - (2\tilde{h}_l - 1)^2}} \tag{9}$$

Where $\tilde{h}_l$ is a dimensionless geometric parameter defined by

$$\tilde{h}_l = \frac{h_l}{D} \tag{10}$$

In Eqs. (5)–(6) $v_i^{\text{T}} \Gamma_g$ is the interfacial momentum due to phase change, τ_{wg} is the steam-wall shear fictional stress, τ_{wl} is the liquid-wall frictional shear stress, τ_i is the interfacial shear stress, S_l and S_g are the liquid and steam perimeters, S_i is the interfacial perimeter, A_l and A_g are the cross-section areas of pipe covered by liquid and steam, respectively.

1.2 Solution of the Mathematical Model

The solution of the mathematical model was carried out using the finite differences technique with an implicit scheme. It was used a first order downstream implicit scheme for spatial derivates and a first order upstream for time derivates. The set of equations in discrete form can be written as:

$$A_j\left(u_j^0\right)u_j^{t+\Delta t} = B_j\left(u_j^0, u_j^t, u_{j-1}^{t+\Delta t}\right) \tag{11}$$

where the superscripts t indicates old time, $t + \Delta t$, indicates new time, 0 represents the dummy variables, and j is the cell where the variables are calculated. Then, the variables with subscripts $j - 1$ and superscripts t are known, therefore these are the inlet variables and the initial conditions. u is a column vector of dependent variables, given by

$$u = \left(p, \alpha_g, v_g, v_l\right)^T \tag{12}$$

The data used for the prediction were: the input mass flow rate $\dot{m} = 0.17\text{kg/s}$, the vapor quality $x = 0.1$, the input pressure $p = 3.75\text{Mpa}$, the wall heat flux along the pipe $q_e'' = 1000\text{W/m}$, inside diameter of pipe $D = 0.025\text{m}$, pipe length $L = 30\text{m}$, and inclination angle $\theta = 10^0$.

1.3 Results and Discussion

The slug frequency (averaged number of slug units passing a given point in the system over unit of time) correlations available in literature affects the predicted Taylor bubble length therefore the pressure drop. The correlations from Mishima and Ishii (1980), Greskovic and Shrier (1971), Gregory and Scott (1969) and Taitel and Dukler (1976), were evaluated. Figure 3 shows the pressure profiles predicted with the present model and a commercial code.

Figure 3 shows that the slug frequency affects the pressure drop, however the correlation from Greskovich and Shrier (1971) allows a good agreement with the pressure drop from a commercial code. Therefore, such correlation is suggested for other predictions.

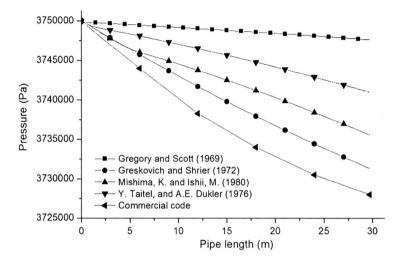

Fig. 3 Pressure drop profiles with different slug frequency correlations

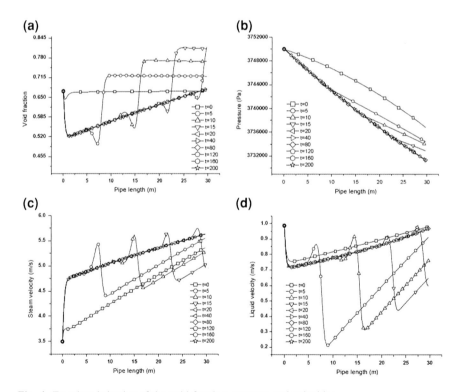

Fig. 4 Transient behavior of the void fraction, pressure and velocities

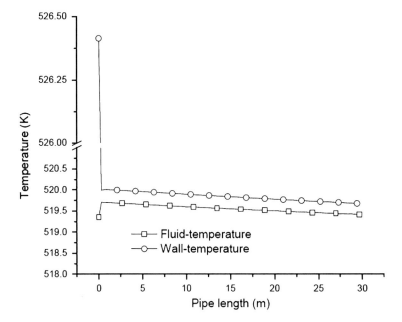

Fig. 5 Fluid-temperature and wall-temperature at steady state conditions

On the other hand, a mesh sensibility analysis was carrying out. It was found that, as the time step is smaller, the predictions are more convergent for larger space step values. Then, the suggested step sizes are $dt = 0.25s$ and $dy = 0.3m$.

Figure 4 shows the transient profiles for the void fraction, liquid velocity, gas velocity and pressure. For the simulation conditions such profiles get the steady state at 40s, and their behaviors are expected. Void fraction and steam velocity increase whereas pressure and liquid velocity decrease along the pipe. These are typical behaviors in a steam–liquid slug flow with steam generation.

Figure 5 shows that the wall pipe temperature is larger than the fluid temperature along the pipe; this behavior is due to the boiling. Temperature profiles diminish because pressure drop occurs and the total thermal equilibrium was considered in the mathematical model.

2 Conclusions

In order to use the two phase flow model technique, the slug flow is divided in a liquid slug region, and a Taylor bubble region. This allowed simulating the steam-water slug flow.

The predictions of pressure drop with different slug frequency correlations have good behavior, but the Greskovich and Shrier (1971) correlation allows obtaining the best predictions regarding to a commercial code. Such correlation is suggested for doing other predictions, however, it is eminent the necessity of compare the predictions against experimental data.

References

Cazarez CO, Espinoza-P.(2008) Numerical Study of Stratified Gas-Liquid Flow

Eck M, Zarza E, Eickhoff M, Rheinlander J, Valenzuela L (2003) Applied research concerning the direct steam generation in parabolic troughs. Sol Energy 74:341–351

Gregory GA, Scott DS (1969) Correlation of liquid slug velocity and frequency in horizontal co-current gas-liquid slug flow. AIChE J 15(6):933–935

Greskovish EJ, Shrier AL (1971) Pressure drop and holdup in horizontal slug flow. AIChE J 17(5):1214–1219

Natan S, Barnea D, Taitel Y (2003) Direct steam generation in parallel pipes. Int J Multiph Flow 29:1669–1683

Mishima K, Ishii M (1980) Theoretical prediction of onset of horizontal slug flow. J Fluids Eng 102:441–444

Ouyang LB, Aziz K (2002) Solution no-uniqueness for separated gas-liquid flow in pipes and wells I Occurrence. Petrol Sci Technol 20:143–171

Odeh S, Morrison GL, Behnia M (1998) Modeling of parabolic trough direct steam generation solar collectors. Sol Energy 62:395–406

Taitel Y, Dukler AE (1976) A model for prediction flow regime transitions in horizontal and near horizontal gas–liquid flow. AIChE J 22:47–55

Crystal-Liquid Transition in Binary Mixtures of Charged Colloidal Particles

Catalina Haro-Pérez, Gualberto Ojeda-Mendoza, Carlos A. Vargas,
Eduardo Basurto-Uribe and Luis F. Rojas-Ochoa

Abstract In this work we present experimental results on the melting of binary colloidal crystals composed by two different sized charged colloidal particles. As the number fraction of the smallest particles is increased, we observe a crystal-liquid transition. The melting line is characterized by studying the structural and dynamic properties of the colloidal suspensions by means of cross-correlation dynamic light scattering. The initial crystal structure consists of a body centered cubic lattice and after reaching a certain concentration of small particles, the system melts. At the melting point the resulting structure measured by static light scattering represents that of a strongly correlated fluid.

C. Haro-Pérez (✉) · C. A. Vargas · E. Basurto-Uribe
Departamento de Ciencias Básicas, Universidad Autónoma Metropolitana-Azcapotzalco,
Av. San Pablo 180, 02200 Mexico, D.F., Mexico
e-mail: cehp@correo.azc.uam.mx

C. A. Vargas
e-mail: cvargas@correo.azc.uam.mx

E. Basurto-Uribe
e-mail: ebasurto@correo.azc.uam.mx

G. Ojeda-Mendoza · L. F. Rojas-Ochoa
Departamento de Física, CINVESTAV-IPN, Av. Instituto Politécnico Nacional 2508,
07360 Mexico, D.F., Mexico
e-mail: gojeda@fis.cinvestav.mx

L. F. Rojas-Ochoa
e-mail: lrojas@fis.cinvestav.mx.

J. Klapp et al. (eds.), *Fluid Dynamics in Physics, Engineering and Environmental Applications*, Environmental Science and Engineering,
DOI: 10.1007/978-3-642-27723-8_20, © Springer-Verlag Berlin Heidelberg 2013

1 Introduction

The crystal-liquid transition is nowadays one of the fundamental phenomena in materials science that attracts a lot of attention. However, despite its enormous implications in industry and biology, it remains far from being fully understood *(Vekilov* 2010*)*. The study of solid–liquid transitions is of broad interest as many crystalline materials with industrial applications and the crystallization of proteins or other biological aggregates behind many human diseases (Charache 1967; Berland 1992) are grown from solutions. To elucidate this scientific issue, colloidal systems have been used lately as model systems due to the experimental advantages they present against their atomic counterparts, e. g. they have larger spatial and time scales. Moreover, charged colloidal systems allow us to tune the interaction potential in order to relate it to their phase diagram. At the moment, there is an increasing interest in this topic and, particularly in the phase diagram of binary colloidal systems (Lorenz 2008) since they can be used to characterize crystallization of metallic alloys (Massalski 1990). The main parameters that determine the final state in binary charged colloidal systems are, apart from total volume fraction, the charge and size asymmetry between the two species and the number fraction of small particles. In our work we will study the melting of charged binary suspensions with high particle size asymmetry, since it is a region where results are scarce (Lorenz 2009).

2 Experimental

The samples under study consist of binary mixtures of charged colloidal particles. The two species of particles are made of polystyrene, and they become charged due to the partial dissociation of surface sulphate groups at deionized aqueous conditions. The two species used in the experiments were purchased from Interfacial Dynamics Corp., Portland, USA, and their main characteristics are listed in Table 1. The particle size and polydispersity were obtained from static and dynamic light scattering.

From the two stock solutions, different binary mixtures of these two particle suspensions are prepared by dilution with ultrapure water as a function of mixing ratios, given by the molar fraction $x_s = n_s/n$, where n_s is the number density of small particles and n is the total number density $n = n_s + n_l$, here n_l is the number density of large particles. The resulting binary mixtures are placed in cylindrical quartz cells with a mixed bed of ionic exchangers in order to deionize the samples. The quartz cells are sealed to avoid contact with air. After a few days, crystallization in some samples could be observed by the naked eye from the iridescence they display under illumination with white light. All the prepared samples have nearly the same large particles volume fraction but different number density of small particles.

Table 1 Main parameters of the two particle suspensions used in our experiments

System	Mean Diameter (nm)	N (surface charged groups per particle)	Stock Volume Fraction %	Polydispersity (%)
YSA	111	3100	8.2	7
YSB	25	170	7.9	17

Table 2 Main parameters of the binary mixtures used in this work

Sample	$\varnothing_{total}$	n (particles/m^3)$\times 10^{18}$	n_s(particles/m^3)	X_s
S1	0.00390	5.45	0	0
S2	0.00374	5.59	4.28×10^{17}	0.08
S3	0.00371	5.64	5.13×10^{17}	0.09
S4	0.00365	5.70	6.84×10^{17}	0.12
S5	0.00358	5.76	8.56×10^{17}	0.15
S6	0.00326	6.07	1.71×10^{18}	0.28
S7	0.00294	6.38	2.57×10^{18}	0.40

The chosen large particles volume fraction was high enough to assure that the scattered signal by the large particles was remarkably larger than that from small ones, in order to study the effect of the presence of small particles on the dynamics and structure of the largest ones. The systems under study have the parameters shown in Table 2.

As we mentioned before, some of the samples present crystallization after deionization as can be seen in sample A, Fig. 1. However for high enough small particle molar fraction, we observe a crystal-liquid transition where the system melts and iridescence disappears as can be observed in sample B. In Fig. 1 we also show for comparison, a highly diluted sample that looks almost transparent, sample C, which was used to characterize the particle size and polydispersity.

In order to study in more detail this transition and its structural and dynamical hallmarks, we performed dynamic and static light scattering experiments. The light scattering experiments were carried out after deionization of the system at a constant temperature of 25 °C for angles from 20° to 150° using a commercial goniometer-based 3D light scattering instrument (LS Instruments AG, Fribourg, Switzerland). The 3D cross-correlation scheme allows correcting the effect of multiple scattering on the scattered light intensity and can be used to successfully characterize non-ergodic turbid suspensions (Haro-Pérez 2011) with transmission values, of the single scattered light, as low as 0.01 (Overbeck 1997; Urban 1998). The light source was a He–Ne laser working at 632.8 nm, and the device is provided with a *multitau* digital correlator (Flex). A complete description of the technique has been given elsewhere (Urban 1998).

Fig. 1 Images of the systems
placed in quartz cells:
a Sample S1 after
deionization. **b** Sample S6
after deionization. **c** Highly
diluted sample used to
measure particle size and
polydispersity

3 Results

3.1 Structure

The measured ensemble static structure factors $S_M(q)$ for all the samples were obtained by following the protocol described in (Haro-Pérez 2011) and the results are shown in Fig. 2. For the sample composed only by large particles at a volume fraction of 0.0039 (S1) we can observe a crystal structure where the analysis of the relative positions of the peaks originated from different crystal planes indicates a BCC (body centered cubic) crystal structure. As we add small particles to the suspensions (systems S2, S3, S4, S5), the BCC lattice is maintained but the peak heights decrease indicating a loss of correlation, until we cross the critical x_s where the system melts (see structure factor of samples S6 and S7). These latter structure factors correspond to a typical strong correlated fluid. We should remark that the appearance of a crystal phase at such low volume fraction is due to the long range electrostatic repulsion among particles.

3.2 Dynamics

To gain further insight into this crystal-liquid transition we investigate the time dependence of the scattered field correlation function, $g^{(1)}(q,t)$. For ergodic systems, $|g^{(1)}(q,t)|^2 = [g^{(2)}(q,t)-1]/\beta$, where β is an experimental constant. In the case of non-ergodic systems, we obtain $g_I(q,t)$ from the ensemble-average following the procedure described in (Haro-Pérez 2011). Dynamics is determined from dynamic light scattering measurements at a wave-vector where the structure does not have a significant influence on dynamics, $S(q) \approx 1$ (Pusey 1978). The corresponding autocorrelation functions are displayed in Fig. 3. As we can see, for the crystal samples ($x_s < 0.28$) the correlation function shows a non-decaying

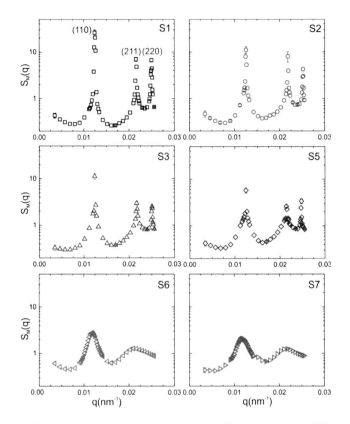

Fig. 2 q-dependence of the measured static structure factor of systems S1 ($x_s = 0$), S2 ($x_s = 0.08$), S3 ($x_s = 0.09$), S5 ($x_s = 0.15$), S6($x_s = 0.28$), S7 ($x_s = 0.40$)

Fig. 3 $|g^{(1)}|^2$ determined at a wave vector where $S(q) \sim 1$ for S1, S2, S3, S4, S5, S6, and S7. To account for the change in q at which we measure $g^{(1)}(q,t)$, the time axis is normalized by q^2

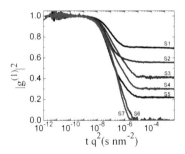

component, and for the fluid samples the correlation function decorrelates completely remarking the ergodic character of the samples. In order to understand the differences between both behaviors, we perform a more detailed analysis by converting the autocorrelation function into mean square displacements *MSD* from $g^1(q,t) = exp(-q^2 MSD/6)$. This can be done in our case since we measured the $g^1(q,t)$ at a wave vector where $S(q) \approx 1$. The mean square displacements obtained

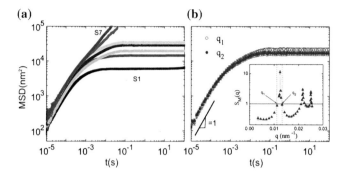

Fig. 4 **a** Mean squared displacement for the systems S1, S2, S3, S4, S5, S6, and S7 (from *bottom to top*). **b** Mean squared displacements for S3 obtained for both *q*-vectors, q_1 and q_2: Inset: wave vectors (*perpendicular dashed lines*) that satisfy S(q) $\approx$ 1

Fig. 5 Localization length in units of the mean interparticle distance, r_{loc}/r_m for all the investigated systems. Solid line: boundary for the melting of a solid (Lindemann criterion)

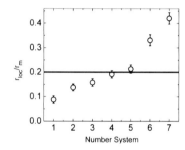

in this way are shown in Fig. 4a. To illustrate the validity of our assumption in extracting the MSD, the Fig. 4b shows the mean squared displacement of sample S3 for two *q*-vectors where $S(q) \approx 1$; and both are almost identical. In all the systems, the MSD's exhibit two time regimes. The initial one is characterized by a linear time-dependence, which represents the diffusive motion of an individual particle in the free volume left by their neighboring particles, also known as *the cage effect*.

Then, the MSD reaches a plateau (for the non ergodic samples), which height is related to the maximum displacement that a particle can experience from its equilibrium position. In this way, a particle localization length can be defined from the plateau height, $l_{loc} = (MSD(t \rightarrow \infty)/3)^{1/2}$ (Wilke 1999). On the other hand, the Lindemann criterion is an empirical rule used to define the melting of a solid (Lindemann 1910).

It can be also interpreted as if the system solidifies when the localization length is less than 20 % of the mean interparticle distance, $l_{loc} < 0.2 \, r_m$ (Löwen 1994).

The so obtained localization lengths are plotted in Fig. 5, where we can observed that the fluid–solid transition is placed between sample S5 and S6, that is for values $0.15 < x_s < 0.28$. We should say that this critical value of x_s, where the system melts, will depend on the sample total volume fraction. Another interesting result is that this dynamic criterion agrees with the transition inferred from structural data, where the static structure factor shows a clear change from a crystal (from system

S1 to S5) to a fluid structure (systems S6 and S7). In a current work, we are characterizing the whole phase diagram as a function of the total volume fraction and the mixing ratios in order to obtain a whole description of the crystal-liquid transition and the influence of the interaction potential on the phase behavior.

4 Conclusions

In this work we have characterized the structural and dynamic properties during melting of charged binary mixtures by means of light scattering. The measured dynamics and structure agree with the observed crystal-liquid transition as the small particle volume fraction of the colloidal mixture is increased. This suggests that the addition of small charged particles to the crystal structure formed by the larger ones screens out the effective interaction among the latter inducing the melting of the crystalline suspension. In order to quantify this effect, further experiments are being carried out.

References

Berland CR, Thurston GM, Kondo M, Broide ML, Pande J, Ogun O, Benedek GB (1992) Solid-liquid phase boundaries of lens protein solution. Proc Natl Acad Sci USA 89:1214–1218

Charache S, Conley CL, Waugh DF, Ugoretz RJ, Spurrell JR (1967) Pathogenesis of hemolytic anemia in homozygous hemoglobin C disease. J Clin Invest 46:1795–1811

Haro-Pérez C, Ojeda-Mendoza G, Rojas-Ochoa LF (2011) Three dimensional cross correlation dynamic light scattering by non-ergodic turbid media. J Chem Phys 134:244902

Lindemann FA (1910) The calculation of molecular vibration frequencies. Phys Z 11:609–612

Lorenz N, Liu J, Palberg T (2008) Phase behavior of binary mixtures of colloidal charged spheres. Colloids Surf A 319:109–115

Lorenz N, Schöpe HJ, Reiber H, Palberg T, Wette P, Klassen I, Holland-Moritz D, Herlach D, Okubo T (2009) Phase behaviour of deionized binary mixtures of charged colloidal spheres. J Phys Condens Matter 21: 464116

Löwen H (1994) Melting, freezing and colloidal suspensions. Phys Rep 237:249–324

Massalski TB, Okamoto H (1990) Binary alloy phase diagrams, 2nd edn. ASM International, Materials Park

Overbeck E, Sinn C, Palberg T (1997) Approaching the limits of multiple scattering decorrelation: 3D light-scattering apparatus utilizing semiconductor lasers. Progr Colloid Polym Sci 104:117–120

Pusey PN (1978) Intensity fluctuation spectroscopy of charged Brownian particles: the coherent scattering function. J Phys A: Math Gen 11:119–135

Vekilov PG (2010) Nucleation. Cryst Growth Des 10:5007–5019

Urban C, Schurtenberger P (1998) Characterization of turbid colloidal suspensions using light scattering techniques combined with cross-correlation methods. J Colloid Interface Sci 207:150–157

Wilke SD, Bosse J (1999) Relaxation of a supercooled low-density Coulomb fluid. Phys Rev E 59:1968–1975

On the Mass Flow Rate of Granular Material in Silos with Lateral Exit Holes

Abraham Medina, G. Juliana Gutiérrez-Paredes, Satyan Chowdary, Anoop Kumar and K. Kesava Rao

Abstract The mass flow rate, $\dot{m}$, associated with the outflow of dry, cohesionless granular material through orifices located in vertical walls of silos was analyzed experimentally in order to reveal the dependence of this quantity on D and w, *i.e.*, the diameter of the orifice and the wall's thickness, respectively. A series of experiments enable us to give a general correlation, which has the form $\dot{m} = \dot{m}_0 - wA$, where A is a constant and $\dot{m}_0$ is the mass flow rate to the minimum allowable wall thickness.

1 Introduction

Since the first studies of Hagen (1852) it is well known that the mean mass flow rate, $\dot{m}$, of the gravity flow of dry granular material emerging from the *bottom exit* of a silo scales essentially as $\dot{m} \sim \rho g^{1/2} D^{5/2}$ where $\dot{m}$ is the mass flow rate (grams/second), ρ is the bulk density, g is the gravity acceleration and D is the diameter of the circular orifice. This result contrasts with that occurring in liquids where the mass flow rate depends on the level of filling above the orifice, moreover, the condition to have a continuous flow of granular matter is that $D > 6d_g$, where d_g is the grain's diameter. After the fundamental study of Hagen was established, many researchers have proved the validity of his law and slight modifications of it have

A. Medina (✉) · G. J. Gutiérrez-Paredes
SEPI ESIME Azcapotzalco, Instituto Politécnico Nacional, Av. de las Granjas 682,
Col. Santa Catarina, 02250 Azcapotzalco, D.F., Mexico
e-mail: amedinao@ipn.mx

S. Chowdary · A. Kumar · K. Kesava Rao
Department of Chemical Engineering, Indian Institute of Science, Bangalore 560012, India

J. Klapp et al. (eds.), *Fluid Dynamics in Physics, Engineering and Environmental
Applications*, Environmental Science and Engineering,
DOI: 10.1007/978-3-642-27723-8_21, © Springer-Verlag Berlin Heidelberg 2013

also been proposed in order to improve the agreement with the experimental data (Beverloo et al. 1961; Mankoc et al. 2007; Ahn et al. 2008; Sheldon and Durian 2010).

Despite the enormous utility of Hagen law few studies have been made to know its validity in the granular outflow when the exit hole is in the vertical wall of a silo (Sheldon and Durian 2010). The aim of this work is to experimentally study the mean mass flow rate when the exit holes are located in the vertical walls of the silos for different orifice sizes, D, and several wall thicknesses, w.

The division of this work is as follows: in the next section we discuss a set of results related to some experiments where the orifice was made on the lateral wall. Then, in Sect. 3, we report new experiments where the influence of D and w has

been studied. There, we also propose a correlation that embraces both changes in D and w and finally, in Sect. 4, we give the main conclusions of this work.

2 Previous Works

To our knowledge, Bragintsev and Koshkovskii (Bagrintsev and Koshkovskii 1977) were the first researchers who studied experimentally the problem of the gravity driven lateral outflow of granular material in silos with vertical walls. They used oval and circular exit holes. Davies and Foye (1991) did experiments in silos with rectangular exit holes and (Kumar Sharma and Kesava Rao 2006) have done experiments only using circular exit holes. (Bagrintsev and Koshkovskii 1977) and Choudary and Kesava Rao (2006) and Kumar Sharma and Kesava Rao (2006) have found that apparently the better correlation among $\dot{m}$ and D was of the form $\dot{m} \sim D^{7/2}$, meanwhile Davies and Foye (1991) and Sheldon and Durian (2010) reported measurements of the mass flow rate in silos with lateral orifices that follows a relation of the type $\dot{m} \sim D^{5/2}$ (in the cases of Bagrintsev and Koshkovskii 1977 and Davies and Foye 1991, D is essentially a hydraulic diameter). Incidentally, Davies and Desai (2008) studied the outflow from vertical slots but no reference to the mass flow rate was reported.

In order to reach a better understanding of the behavior of the mass flow rate in lateral circular exit holes we did a series of experiments where we mainly analyzed the influence of D and the wall thickness, w, on such a quantity. Clearly, the wall thickness will be important for the flow existence because if the silo wall is thick, friction should be a crucial factor that will limit such a flow. Moreover, if $w \rightarrow \infty$ there is no flow. Detailed experiments given in the next section allow quantify these and other facts.

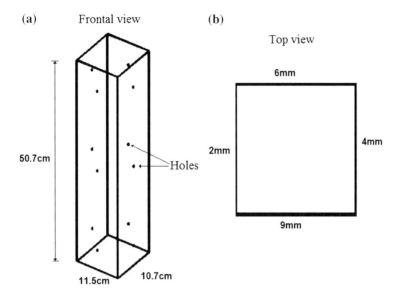

Fig. 1 Square silo with lateral orifices and different thickness wall: **a** schematic view and **b** *top view* where are indicated the different thickness of each wall

3 Experiments

To show the influence of the wall thickness, experiments were made using a squared acrylic box 11 cm inner length, 50.7 cm brimful height, as the one shown in Fig. 1. In such a box each wall has a different thickness (Fig. 1b). More specifically, the wall thicknesses were $w = 0.3, 0.4, 0.6$ and 0.9 cm. During experiments the box was filled up to 50 cm with meshed beach sand of mean diameter $d_g = 0.3$ mm and circular orifices of diameters $D = 0.6, 0.8$ and 1 cm were made on each wall.

We did measurements of the mass flow rate by measuring the weight of the outflow as a function of time. A similar procedure was already reported elsewhere (Ahn et al. 2008). In a first stage we found that the mean mass flow rate is not affected neither by the position of the exit hole respect to the bottom level nor by the brimful height of sand. We measured the quantity $\dot{m}$ for different diameters of orifice $D = 0.6, 0.8$, and 1 cm as a function of w. Such measurements allow to plot data as in Fig. 2 where we show the plot of $\dot{m}$ vs. w for several diameters: each symbol corresponds to a diameter, for instance, symbol ■ indicates data of the mean mass flow rates for all the different wall thicknesses when $D = 1.0$ cm. These symbols are well fitted by the linear correlation (1):

$$\dot{m} = \dot{m}_0 - wA, \tag{1}$$

where A is a constant defined afterwards and $\dot{m}$ is the mass flow rate corresponding to the minimum allowable wall thickness. We determined experimentally that $\dot{m}_0$ obeys the formula

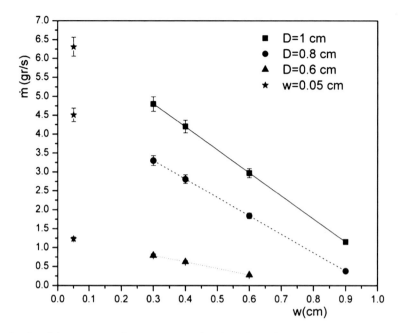

Fig. 2 Plot of the mean mass flow rate, $\dot{m}$, as a function of the thickness wall, w. Each symbol is valid for a given diameter of the exit hole and stars correspond to measurements of $\dot{m}$ when the thickness wall is very thin ($w = 0.05$ cm). The star at the lower part is for $D = 0.6$ cm, star at the middle is for $D = 0.8$ cm and the star at the upper part is for $D = 1.0$ cm. Notice that each trend line tends to its corresponding star when the thickness wall is very small. All these results are well described by the correlations (1) and (2). Values of D and A are given in Table 1. Error bars were included

$$\dot{m}_0 = cpg^1/2D^5/2. \tag{2}$$

In plot of Fig. 2 there are three data represented by stars, actually they correspond to experiments where the wall is very thin, *i.e.*, $w = 0.05$ cm. The upper star corresponds to the value of the mean mass flow rate when the exit hole has a diameter $D = 1$ cm, the star located at the middle part is for the case when $D = 0.8$ cm and finally the star located at the lower part is for an orifice of diameter $D = 0.6$ cm. Thus, if we avail the linear fit and we extend it to values of $w = 0$, it will cross the upper star and also his line will intersect the axis given by $w = 0$.

Therefore, the correlations (1) and (2) establish physically that the mass flow rate diminishes linearly when the thickness wall increases and, when the thickness wall is (ideally) zero, the mass flow rate will obey the Hagen law. Notice also that if we extend the straight line that fits data for $D = 1$ cm, to the right hand side, it will intersect the axis given by $\dot{m} = 0$, consequently this procedure will be useful to predict the critical thickness to arrest the flow. Our experiments allow confirming qualitatively that this condition is true.

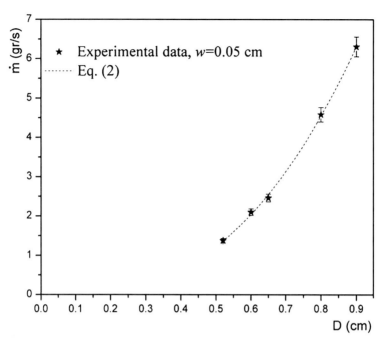

Fig. 3 Plot of the mean mass flow rate, $\dot{m}$, as a function of D for the aluminum-walled can, $w = 0.05$ cm wall thickness. The orifices were made on the vertical wall. The dashed curve is the best fit using Eq. (2) where $\rho = 1.5$ gr/cm^3 and $c = 0.175$. Error bars were included

Table 1 Parameter A as a function of the diameter of the orifice

Diameter the orifice, D (cm) ± 0.05 cm	Parameter A(gr/cm*s) ± 0.06 gr/cm*s
0.6	6.09
0.8	4.86
1.0	1.72

Due to the fact that the infinitely small thickness limit never will be reached, we have made another set of experiments using an aluminum-walled can (10 cm inner radius and 30 cm brimful height), whose nominal thickness wall was $w = 0.05$ cm, and we did circular orifices in the vertical wall with five different diameters. In this case experiments allowed us to assure the validity of the Hagen law to estimate the mass flow rates in silos through the vertical and very slim walls. In Fig. 3 the plot of the mean mass flow rate as a function of D is shown.

As it can be seen in Fig. 3, Eq. (2) fits very well the experimental data where the bulk density is $\rho = 1.5$ gr/cm^3 and the dimensionless constant has the value $c = 0.175$. Similarly, as in the case for $D = 1$ cm, the other cases for $D = 0.8$ cm (●) and $D = 0.6$ cm (▲) obey exactly the same behavior and thus Eqs. (1) and (2) describe very well the dependence of the mass flow rate on D and w.

It is very important to appreciate that there are different critical thicknesses for different diameters of orifices. In Table 1 we give the values of the parameter A of Eq. (1) found experimentally for each diameter.

4 Conclusions

In this work we studied experimentally the problem of the mean mass flow rate of granular material through transversal circular orifices in the vertical walls of silos. Specifically, we characterized the dependence of the mass flow rate as a function of the diameter of the orifice and the wall thickness by using meshed beach sand. Our results show that such a quantity is a decreasing linear function of w and that the effect of the diameter of the exit hole is mainly involved in the basic flow occurring when the wall thickness is very small. In this latest case, the mass flow rate obeys the Hagen law. The proposed experimental correlation can also predict the critical value of w necessary to arrest the granular flow. To our knowledge, this is the first time that this problem has been tackled in the specialized literature and more study is necessary in order to prove that Eq. (1) also embraces more general cases as the existence of non-circular orifices and the outflow of very irregular grains and it also theoretically models this type of flow.

References

Ahn H, Basaranoglu Z, Yilmaz M, Bugutekin A, Zafer Gül M (2008) Experimental investigation of granular flow through an orifice. Powder Tech 186:65–71
Bagrintsev II, Koshkovskii SS (1977) Investigation of the outflow of granular materials through openings in the wall of a vertical cylindrical tube. J Chem Petroleum Eng 6:503–505
Beverloo WA, Leniger HA, van de Velde J (1961) The flow of granular solids through orifices. Chem Eng Sci 15:260–269
Choudary S, Kesava Rao K (2006) Experiments on the discharge of granular materials through vertical and horizontal orifices of a vertical tube. Indian Academy of Science, Project Report
Davies CE, Desai M (2008) Blockage in vertical slots: experimental measurement of minimum slot width for a variety of granular materials. Powder Technol 183:436–440
Davies CE, Foye J (1991) Flow of granular material through vertical slots. Chem Eng Sci 69:369–373
Hagen GH L (1852) Über den Druck und die Bewegung des trocknen Sandes, Bericht über die zur Bekanntmachung geeigneten Verhandlungen der Koniglich Preussischen Akademie der Wissenschaften zu Berlin 35–42
Kumar Sharma A, Kesava Rao K (2006) Experiments on the discharge of granular materials through orifices in the wall of a vertical tube. Indian Academy of Science, Project Report
Mankoc C, Janda A, Arévalo R, Pastor JM, Zuriguel I, Garcimartin A, Maza D (2007) The flow rate of granular materials through an orifice. Granul Matter 9:407–414
Sheldon HG, Durian DJ (2010) Granular discharge and clogging for tilted hoppers. Granul Matter 12:579–585

Traction Force Due to Aqueous Foam Flow Rising in a Vertical Pipe

A. Pérez Terrazo, V. S. Álvarez Salazar, I. Carvajal Mariscal,
F. Sánchez Silva and A. Medina

Abstract This paper discusses in an experimental way the traction force exerted by a flow of aqueous foam through the annular space between a central rod and the inner walls of the pipe. These studies were made to prove if foam behaves as a granular material or as a power-law fluid. Experiments allow concluding that the flow of dry foam is a type of slipping flow where the traction does not depend on the radius of the pipe or on the radius of the rod.

1 Introduction

In this work we are interested in characterizing the overall frictional force (traction), against the surrounding walls, when a foam flow occurs inside a vertical cylindrical pipe. This problem involves the stability of foam in relation to film rupture due to its strong interaction with the solid boundary of the pipe. Thus, in

A. Pérez Terrazo (✉) · V. S. Álvarez Salazar · A. Medina
ESIME Azcapotzalco, Instituto Politécnico Nacional, Av. de las Granjas No. 682,
Col. Sta. Catarina, 02250 Azcapotzalco, D.F., Mexico
e-mail: anterra25@live.com.mx

V. S. Álvarez Salazar
e-mail: catalina-s@hotmail.com

I. Carvajal Mariscal · F. Sánchez Silva
ESIME Zacatenco, Instituto Politécnico Nacional, Av. Instituto Politécnico Nacional S/n,
Unidad Profesional "Adolfo López Mateos" Col. Lindavista, 07738 Gustavo A. Madero,
D.F., Mexico
e-mail: icarvajalm@ipn.com

A. Medina
e-mail: amedinao@ipn.com

J. Klapp et al. (eds.), *Fluid Dynamics in Physics, Engineering and Environmental Applications*, Environmental Science and Engineering,
DOI: 10.1007/978-3-642-27723-8_22, © Springer-Verlag Berlin Heidelberg 2013

267

this case a purely mechanical analysis is tackled and the chemistry of the foaming solutions is left aside.

To our knowledge, the quantification of traction has been absent from studies of the generic foam properties (Weaire and Hutzler 1999). Like granular medium, foam is a disordered system made of a punctuated collection of unit cells, initially grown by gas injection into a quiet liquid. In order to have a monodisperse cell population with an average bubble diameter, d, it is necessary to inject the gas, at a constant rate, and the gas flow rate will be low or medium. Conversely, at high gas flow rates the bubble population will be very disperse, and the cells have a wide range of sizes (Longuet-Higgins et al. 1991; Corchero et al. 2006; Higuera and Medina 2006; Higuera 2005; López-Villa et al. 2011).

Once foam is formed it obeys a set of dynamic phenomena, for example, foam drainage, driven by gravity, through the films or faces of each cell and through the Plateau borders; among each cell there is gas diffusion or coarsening, driven by pressure differences; and film rupture or coalescence occurs driven by film instabilities (Isenberg 1992).

In real fluids, we often observe a bubble rising to the surface, and then floating on the surface. Multiple bubbles interact. If they do not burst, they will stack, forming foam. The water between those stacked bubbles will drain, leaving a micrometer-thin film of liquid between bubbles. The resulting structure is called dry foam.

Our goal in this work is to characterize the overall friction force originated by the simple rise of aqueous foam in a vertical pipe when air is injected. In our experiments we use common detergent foams because they are remarkably stable against film rupture, and are ideal for the study of many properties.

2 The Elements of Foam Structure

Foam is a two-phase system in which gas cells are enclosed by liquid. Simulation of bubbles in wet or dry foam is very challenging since foam can have a complicated liquid/gas interface. Each bubble is surrounded by several faces of thin films, which may meet at junctions, where liquid is clustered. These junctions may move, merge and split. In addition, bubbles or liquid drops can merge and split. Thus, the topology of the interface often changes. This behavior causes difficulties for simulations based on the Lagrangian and Eulerian methods. See Fig. 1.

In this work we will study the formation of foam in a pipe through measurements of the bulk property named traction, which originates when bubbles stack and rise. Traction or drag is a measure of the frictional force due to the pass of a fluid through a pipe. Additional to measurements of traction in a pipe, we also study traction on rods located at the centre of the cylindrical pipe. We believe that the characterization of this force allows us to suggest a model for the rise velocity of dry foam.

Fig. 1 Three-dimensional foam consists of cells whose faces are thin films, meeting in Plateau borders (Weaire and Hutzler 1999)

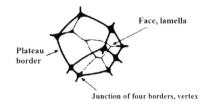

3 Experiments

The physical behavior of the foam depends on its composition, i.e., the volume fraction of gas and liquid in the foam. This feature is known as "quality". In this section we performed experiments where foams have around a 10 % of liquid fraction, i.e., dry foam. Foam was produced by bubbles generated from a single orifice at the lower part of an acrylic cylindrical pipe $r_0 = 23.5$ mm inner radius, 410 mm height and 3 mm wall thickness. Acrylic rods were $r_1 = 12.5$ mm radius, 380 mm height with four rods tipped at different angles: $0°$, $45°$, $60°$ and $70°$. These rods were made so to avoid ascendant normal forces on the lower edge of the rods, which can be added to the shear stress σ_{zr} acting upwards on the face $r = r_0$ and alter the actual measurements of the traction given by the force sensor. See Fig. 2.

Experiments were performed on the transport of air and liquid in at the botton aqueous foam maintained in steady state by a constant gas flux at the bottom. We measured vertical profiles of the upper front of the foam (Fig. 3). The experiments were conducted with distilled water and surfactants in a concentration of 2.5 % (dish soap) diluted with water, remained the same flow of air into the pump for all experiments, thus obtaining the same conditions for all cases.

In steady state the bubbles move upwards with constant speed equal to the measured gas flux, which accounts for all transport of gas. The bubbles also coarsen by gas diffusion at a rate that depends on liquid fraction. The typical behavior of the velocity of the front is given in plot of Fig. 4. Notice that it is a near constant for both cases: with and without the existence of rod.

By the way, in Fig. 5 we show a sketch of the system used to measure the traction force on the rod. It senses the overall frictional force on the lateral area of the rod or the overall force on the inner area of the cylindrical pipe. The force sensor can give data each second.

In Fig. 6 a typical plot of the traction as a function of time is given for traction on the rod and for traction on the inner wall of the cylindrical pipe. Such a plot allows us to conclude that in fact both traction measurements are indistinguishable from one another; the same is maintained for the other rods. Actually, these results are astonishing and understanding their origin is required.

Fig. 2 Rods tipped at different angles respect to the horizontal plane

Fig. 3 Detail of the front of aqueous foam (*upper region*) in a pipe. The hexagonal Plateau borders are appreciable

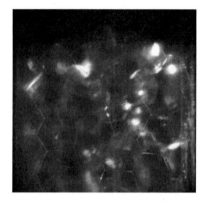

Fig. 4 Typical plot of the instantaneous velocity of the front of the foam as a function of height

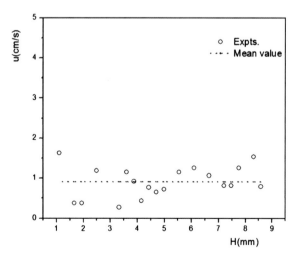

Fig. 5 Schematic of the experimental measurement of the traction force on the concentric rod

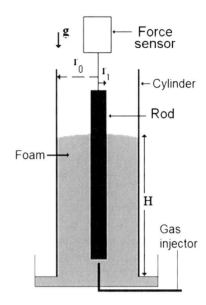

Fig. 6 Traction on the rod and on the inner cylinder's wall

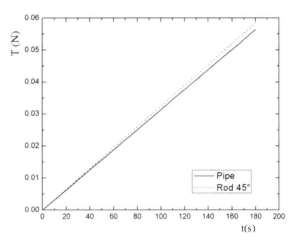

4 A Simple Model

To provide a physical model that embraces all the previous results we propose that the velocity of the foam flow has the form

$$u(r) = u_s + w \ln \frac{r}{R'} \tag{1}$$

where we have taken into account the previous result that the velocity of the front is independent of height. This result was established previously by Feitosa and Durian (2008). In Eq. (1) u_s is the slip velocity on the wall of the pipe ($r = R$) and

w is a velocity. Moreover, we assume that never $r = 0$ because there the velocity diverges. Instead we propose that the minimum value of r is related to the minimum value of the bubble $r = d$. In this case the shear stress on the inner wall of the pipe is

$$\tau_{zr} = \mu \frac{du}{dr} = \mu \frac{w}{R}. \qquad (2)$$

This later result allows finding the traction as

$$T = 2\pi \int_0^H r\tau dz = 2\pi\mu wH. \qquad (3)$$

This result assures that traction depends linearly on H, the level of the front of the foam and it does not depend on the radius of the wall. The same result is valid when we compute the traction on the rod. Both results are in agreement with experiments. We believe that the foundation of this result is that always in a bubble growth inside a pipe with an aqueous solution, the film among the bubble and the wall can be null and thus on the inner wall of the pipe there is a slip velocity (López-Villa et al. 2011).

5 Conclusions

From our experiments of the foam flow in tubes and rods under different configurations we have determined that the traction force does not depend on the radius of the rod or the pipe. Moreover, it is apparent that the proposed velocity also contains a slip velocity that produces the result that traction does not depend on the radius.

References

Corchero G, Medina A, Higuera FJ (2006) Effect of wetting conditions and flow rate on bubble formation at orifices submerged in water. Colloids Surf A 290:41–49
Feitosa K, Durian DJ (2008) Gas and liquid transport in steady-state aqueous foam. Eur Phys J E 26:309–316
Higuera FJ (2005) Injection and coalescence of bubbles in a very viscous liquid. J Fluid Mech 530:369–378
Higuera FJ, Medina A (2006) Injection and coalescence of bubbles in a quiescent inviscid liquid. Eur J Mech B Fluids 25:164–171
Isenberg C (1992) The science of soap films and soap bubbles. Dover, New York
Longuet-Higgins MS Kerman BR Lunde K (1991) The release of air bubbles from an underwater nozzle. J Fluid Mech 230:365–390
López-Villa A, Medina A, Higuera FJ (2011) Bubble growth by injection of gas into viscous liquids in cylindrical and conical tubes. Phys Fluids 23:102102
Weaire D Hutzler S (1999) The physics of foams. Clarendon Press, Oxford

A Discrete Model for Simulating Gas Displacement in Fractured Porous Media

S. Pérez-Morales, A. Méndez-Ancona, M. Ortega-Rocha,
R. Islas-Juárez, R. Herrera-Solís
and G. Domínguez-Zacarías

Abstract A physically based approach to numerical miscible displacement model in a matrix-fracture system was developed. Matrix-fracture configuration was divided in two different domains: fractured media was considered as a free flow channel. Then Navier–Stokes, continuity and convection–diffusion equations were also employed. The matrix is solved with Darcy mass conservation and Diffusion for porous media. The system equation is solved about the scheme of finite differential method and finite volume.

1 Introduction

Miscible displacement refers to a process frequently employed in petroleum industry in which a miscible solvent is injected into oil-bearing rock formations for displacing the oil towards the production well. Papers (Rogerson and Meiburg 1993; Ruith and Meiburg 2000; Zhang and Frigaard 2006; Chao-Ying and Heinz 2004; Wit 2004; Trivedi and Babadagli 2008a, b) focus their research on miscible fluid displacement in porous media, they consider two fluids as incompressible and

S. Pérez-Morales · A. Méndez-Ancona · M. Ortega-Rocha
Región Sur, Exploración y Producción, Instituto Mexicano del Petróleo,
Periférico Carlos Pellicer Cámara, 86209 Villahermosa, TAB, Mexico

R. Islas-Juárez · R. Herrera-Solís
Sede, Exploración y Producción, Eje Central Lázaro Cárdenas Norte 152,
07730 Mexico, D.F., Mexico

G. Domínguez-Zacarías (✉)
Programa de Investigación de Recuperación Hidrocarburos, Eje Central Lázaro Cárdenas
Norte 152, 07730 Mexico, D.F., Mexico
e-mail: gdzacari@imp.mx

J. Klapp et al. (eds.), *Fluid Dynamics in Physics, Engineering and Environmental Applications*, Environmental Science and Engineering,
DOI: 10.1007/978-3-642-27723-8_23, © Springer-Verlag Berlin Heidelberg 2013

gravity effects were neglected. Modeling miscible displacement in naturally fractured reservoirs (NFRs) remains a challenge. In this process, the interaction between the injected fluid and the produced oil is mainly dominated by the matrix-fracture system.

The main procedures to simulate NFRs are a discrete, continuum or hybrid models. Each one simulates the arrangement with very different assumptions and advantages. On continuum models, the system is solved as two superposed domains (Hauge and Aarnes 2009). Fractures are not treated independently and the interaction between fractures and porous media is simulated with transfer functions. On the other hand (Reichenberger et al. 2006) proposed a discrete fracture network model in order to get a natural physical approach instead of applying exchange functions to simulate the interaction between fracture and matrix. Different authors (Hauge and Aarnes 2009; Hoteit and Firoozabadi 2005; Hoteit et al. 2009) agree that a discrete model could be superior to a dual-porosity approach because of the no-use of transfer functions. This could be a relevant improvement over de the dual porosity models as discussed by (Reichenberger et al. 2006). Wu et al. (2001) indicated that a discrete-fracture approximation is "computationally intensive" and implies a deep matrix-fractures spatial and geometrical characterization. A hybrid approach might be the best approximation though; combining the two previous procedures could imply some new ambiguities. Wu et al. (2009) proposed a model of this kind.

Miscible displacement in NFRs has recently gained more research attention. In (1964) Slobod and Howlett experimentally examined the effects of gravity segregation in vertical unconsolidated porous media to determine the effect of density differences of a miscible displacement process; the study becomes crucially concerned with gravity segregation (residence time). Then, in (1969) Thompson and Mungan experimentally studied the displacement rate effect, fracture and matrix properties, on oil recovery efficiency. They concluded that displacement rate is the most important factor in a miscible displacement process. (Zakirov et al. 1991) reported the results of experimental and theoretical investigations of miscible displacement process of compressible fluid within fractured porous media.

Firoozabadi (1995a, b, 2000) used vertical block cores to study the miscible displacement in fractured reservoir, by using a mathematical model of one dimension; they found the gravity number is the most important parameter which affects miscible displacement performance in fractured porous media.

Jamshidnezhad et al. (2003, 2004) presented models for evaluating miscible displacement, comparing a mathematical model against experiment results, they concluded that displacement efficiency depends of solvent injection rate. Experimentally Jamshidnezhad (2004) compared miscible injection against waterflooding obtaining encouraging results.

Pressure, gravity, dispersion/diffusion and capillary drive are main factors of mass transfer between the matrix-fracture systems according to Trivedi and Babadagli (2008a, b), Hoteit and Firoozabadi (2005) modeled diffusion during gas injection in NFRs. In their work concluded that at high pressure, diffusion decreases. And below minimum miscibility pressure (MMP), diffusion may increment recovery. Trivedi and Babadagli (2008a, b, 2006), presented laboratory experiments and numerical simulations to evaluate mass transfer in matrix-fracture configuration; they concluded that Peclet number and mass transfer rate determine recovery along the fracture. And that mass transfer rate is due to diffusion and mixing across non-reactive media. Moctezuma et al. (2010) developed a basic approach to analyze a detailed solution for the interaction matrix-fracture, at fracture scale, in order to identify their major contributions in the oil transfer function between a set of porous media blocks during a miscible displacements. They made the assumption of two separate media in which different equations governing conservation; Darcy flow for the matrix blocks, and Navier–Stokes in the fracture.

The purpose of this paper is to analyze a discrete-fracture approach to model miscible displacement in a fracture-matrix system (Moctezuma 2010). A configuration composed by two sub-independent elements was used: the fracture was taken as a free path orthogonally connected to the matrix. Fractures were characterized. This means that we employed a physically based description of the fluid flow based on an accurate coupled approach where the matrix was solved by Darcy equation and fracture (free-flow path) with Navier–Stokes equation (NSE). Using boundary conditions we gave considerable importance to modeling the interchange between matrix and fracture. The coupling of NSE and Darcy has been extensively studied (Huang et al. 2010; Brush and Thomson 2003; Al-Yaarubi et al. 2005; Kim et al. 2003; Cardenas et al. 2007) for fractured porous media. To our best knowledge miscible displacement in NFRs has not been modeled using this coupled system (NSE-Darcy).

2 Mathematical Models

We consider incompressible flows in 2-D. Fluids are first contact miscible. The fracture-matrix system is based on the Warren and Root (1963) sugar-cube model where fracture is a free straight channel uniformly spaced with a constant thickness of $2d$ and length L. Matrix pore space is a homogenous and isotropic media represented by a parallelepiped with thickness $2a$ and length L orthogonally connected to the fracture. We assume that fracture's diameter is much smaller than its length, this is: a small aspect ratio. The configuration is initially completely saturated with fixed fluid properties. Gravity effects, velocity variations, viscosity and diffusion are neglected. Isothermal conditions are assumed.

The Navier–Stokes (1), continuity (2) and convection diffusion (3) equations describe fluid flow and species conservation in the free medium:

$$\rho(\bar{u} \cdot \nabla)\bar{u} + \nabla p = \mu \nabla^2 \bar{u} \quad \text{on } \Omega_f \tag{1}$$

$$\nabla \cdot \bar{u} = 0 \quad \text{on } \Omega_f \tag{2}$$

Fig. 1 Illustrates the
fracture-matrix configuration.
Let Ω, be a domain in $\mathbb{R}^n$,
n = 2, divided into two
subdomains $\Omega_{i, i} =$ f, m

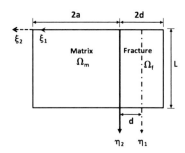

$$\rho(\bar{u} \cdot \nabla)\bar{u} + \nabla p = \mu \nabla^2 \bar{u} \quad \text{on } \Omega_f \qquad (3)$$

where $\bar{u}$ is the velocity vector, ρ is density, μ is viscosity, and p is total pressure in the channel. It is assumed that fracture walls are no-slip boundaries C is solute concentration, t is time, and D is the molecular diffusion coefficient.

Darcy (4), continuity (2) and convection–diffusion (5) equations describe fluid flow in the porous media ($\Omega_{m,}$).

$$\bar{v} = -\frac{k}{\mu} \nabla p \quad \text{on } \Omega_m \qquad (4)$$

$$\frac{\partial C}{\partial t} = \frac{\phi}{\tau} D \nabla^2 C + \bar{u} \cdot \nabla C \quad \text{on } \Omega_m \qquad (5)$$

where $\bar{v}$ is the Darcy velocity vector, k is absolute permeability, μ is viscosity, and p is total pressure in the matrix. C is solute concentration, t is time, ϕ represents effective porosity, τ is tortuosity and D is the molecular diffusion coefficient. Both systems are coupled with boundary conditions and without use transfer function (Fig. 1).

On top of the configuration two coordinate systems were introduced: The first centered in the middle of the fracture, with η–axis from top to bottom and ξ-axis from right to left; the second located on the top right corner of the parallelepiped, which represents the porous media. The η–axis goes from top to bottom and ξ-axis from right to left.

At first, the system is saturated with a fluid; then a displacing fluid is injected. It initiates through the first coordinate system in the middle of the ξ-axis, where fracture ($\xi_1 = 0$) is located. Fluid flow was in y-direction therefore ξ-axis is perpendicular to it. The boundary conditions used to simulate fluid flow from the fracture ($\xi = d$) to the second coordinate system ($\xi_2 = 0$), where matrix is located, are pressures and concentrations. This enables not to compute the flux between the fracture and the matrix. Finally, solute concentrations are used as boundary conditions to model fluid displacement from matrix to fracture.

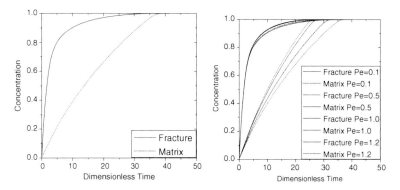

Fig. 2 (*Left*) Concentration profile versus time with dimensionless variables. The concentration profile in the matrix is represented by continued line and fracture by a dashed line. (*Right*) Concentration profile versus time with dimensionless variables. The concentration profile in the fracture is represented by continued line and matrix by a dashed line

3 Numerical Implementation

In this work we used a finite-difference and finite volume technique to discretize the partial differential equations proposed for fracture and matrix. The code was developed in Fortran 90–95 using a 100×100 grid, in both domains with topological effects.

A first numerical result is illustrated in Fig. 2 (left). It can be appreciated that the fracture has a greater solute concentration across time in comparison with the matrix and the solute fully saturates the fracture in a faster way, while the matrix has a slower increase in solute concentration. Figure 2 (right) presents solute concentration change across time for several *Pe* values. It can be seen that solute concentration in the fracture is not very susceptible to a change, although a slight increment in solute concentration occurs with a higher *Pe* value. Matrix is more sensitive to *Pe* variations and as it increases concentration in the porous media gets higher. This *Pe* change might imply that dispersion has a relevant importance within matrix fluid transport. In future research fluid viscosity variation will be taken into account to evaluate its effect over *Pe*.

4 Conclusions

A discrete numerical model to describe miscible displacement in a fracture-matrix system was developed. The model employed the coupling of Navier–Stokes and Darcy equations a physically based approach to solve the fracture-matrix configuration, where the fracture was assumed as free channel and matrix as a porous media. Under the assumptions of: gravity effect neglected, density and viscosity taken as constants, concentration profile advanced more rapidly in the fracture than in the matrix. Dispersion is the more important parameter of fluid transport in the matrix domain. Further research will be done for a model with viscosity and density as concentration functions.

References

Al-Yaarubi AH, Pain CC, Grattoni CA, Zimmerman RW (2005) Navier-Stokes simulations of fluid flow through a rough fracture, in dynamics of fluids and transport in fractured rock. Geophys Monogr Ser162, edited by B. Faybishenko, P. A. Witherspoon, and J Gale pp. 55–64, AGU, Washington

Brush DJ, Thomson NR (2003) Fluid flow in synthetic rough- walled fractures: Navier-Stokes, Stokes, and local cubic law simulations. Water Resour Res 39(4):1085

Cardenas MB, Slottke DT, Ketcham RA, Sharp JM (2007) Navier-Stokes flow and transport simulations using real fractures shows heavy tailing due to eddies. Geophys Res Lett 34:1–6

Chao-Ying J, Heinz H (2004) An experimental study of miscible displacements in porous media with variation of fluid density and viscosity. Transp Porous Med 54:125–144

Firoozabadi A (2000) Recovery Mechanisms in Fractured Reservoir and Field Performance. JCPT 39:13

Hauge VL, Aarnes JE (2009) Modeling of two-phase flow in fractured porous media on unstructured non-uniformly coarsened grids. Transp Porous Med 77(3):373–398

Hoteit H, Firoozabadi a (2005) Multicomponent fluid flow by discontinuous Galerkin and mixed methods in unfractured and fractured media. Water Resour Res 41(11):1–15

Hoteit H, Firoozabadi A, Engineering R (2009) Numerical modeling of diffusion in fractured media for gas-injection and -recycling schemes. SPE J 323–337

Huang Z, Yao J, Li Y, Wang C, Lv X (2010) Numerical calculation of equivalent permeability tensor for fractured vuggy porous media based on homogenization theory. Global Science Press, 1–25. doi: 10.4208/cicp.150709.130410a

Jamshidnezhad M, Montazer Rahmati M (2004) Theoretical and experimental investigations of miscible displacement in fractured porous media. Transp Porous Med 57(1):59–73

Jamshidnezhad M, Montazer-Rahmati M, M and Sajjadian VA (2003) Experimental Study of miscible fluid injection in fractured porous media. J Oleo Sci 52:335–338

Jamshidnezhad M, Montazer Rahmati M, Sajjadian VA (2004) Performance of Iranian naturally fractured cores under miscible displacement experiments. J Jpn Petrol Inst 47:59–63

Kim I, Lindquist WB, Durham WB (2003) Fracture flow simulation using a finite-difference lattice Boltzmann method. Phys Rev E 67:046708

Moctezuma A, Luna E, Serrano E, Domínguez-Zacarías G. One Mathematical model in Fracture-Matrix system for miscible displacement.Manuscript number EGY-D-10-00350 2010. Número de Registro 03-2010_112211414700-01

Reichenberger V, Jakobs H, Bastian P, Helmig R (2006) A mixed-dimensional finite volume method for two-phase flow in fractured porous media. Adv Water Resour 29(7):1020–1036

Rogerson A, Meiburg E (1993) Numerical simulation of miscible displacement processes in porous media flows under gravity. Phys Fluids A 5(11):2644–2660

Ruith M, Meiburg E (2000) Misicible rectilinear displacement with gravity averride. Part I homogeneous porous medium. J Fluid Mech 420:225–257

Slobod RL, Howlett WE (1964) The effects of gravity segregation in studies of miscible displacement in vertical unconsolidated porous media. Soc Petroleum Eng J 1-8

Tan C-T, Firoozabadi A (1995a) Theoretical analysis of miscible displacement in fractured porous media by a one-dimensional model: part i-theory. J Can Petroleum Technol 34:17–27

Tan C-T, Firoozabadi A (1995b) Theoretical analysis of miscible displacement in fractured porous media by a one-dimensional model: Part II—features. J Can Petroleum Technol 34:28–35

Thompson JL, Mungan N (1969) A Laboratory study of gravity drainage in fractured systems under miscible conditions. Soc Petroleum Eng J 2232:247–254

Trivedi J, Babadagli T (2006) Efficiency of miscible displacement in fractured porous media. In: Proceedings of SPE Western Regional/AAPG Pacific Section/GSA Cordilleran Section Joint Meeting. SPE doi: 10.2523/100411-MS

Trivedi JJ, Babadagli T (2008a) Experimental and numerical modeling of the mass transfer between rock matrix and fracture. Chem Eng J 146:194–204

Trivedi JJ, Babadagli T (2008b) Scaling miscible displacement in fractured porous media using dimensionless groups. J Petrol Sci Eng 61:58–56

Warren JE, Root PJ (1963) The Behavior of naturally fractured reservoirs. Soc Petroleum Eng J 426:245–255

Wit De A (2004) Miscible density fingering of chemical fronts in porous media: Nonlinear solutions. Phys Fluid 16(1):163–175

Wu YS, Liu HH, Bodavarsson GS, Zellmer KE (2001) A triple-continuum approach for modeling flow and transport processes in fractured rock. Lawrence Berkley National Laboratory 1-57

Y shu Wu, Qin G (2009) A generalized numerical approach for modeling multiphase flow and transport in fractured porous media introduction. Commun Comput Phys 6(1):85–108

Zakirov SN, Shandrygin AN, Segin TN (1991) Miscible displacement of fluids within fractured porous reservoirs. SPE 22942

Zhang J Frigaard IA (2006) Dispersion effects in the miscible displacement of two fluids in a duct of large aspect. J Fluid Mech 549:225–251

Profile Deformation of a Non Cohesive Granular Material in an Accelerated Box

V. S. Álvarez Salazar, A. Pérez Terrazo, A. Medina
and C. A. Vargas

Abstract When a dry non cohesive granular material confined in a box in such a way that the initial profile of the free surface is a horizontal plane and it is accelerated uniformly, the final equilibrium profile changes to a tilted straight plane whose slope is a function of the magnitude of the acceleration, the magnitude of the gravity acceleration and the friction coefficient of the granular material. Here are presented a simple model, based on the Coulomb's law, that describes correctly such a deformation and some experiments that back the theoretical predictions.

1 Introduction

In his book on Hydrodynamics Daniel Bernoulli (Bernoulli 1968) analyzed, near three centuries ago, the problem of the free surface deformation of the initially horizontal level of quiescent water in a cylindrical vessel when it is accelerated

V. S. Álvarez Salazar (✉) · A. Pérez Terrazo · A. Medina
Instituto Politécnico Nacional, SEPI ESIME UA, Av. de las Granjas No. 282, Col. Sta. Catarina Delegación Azcapotzalco, 02250 Mexico, D.F., Mexico
e-mail: catalina-s@hotmail.com

A. Pérez Terrazo
e-mail: anterra@prodigy.net.mx

A. Medina
e-mail: amedinao@ipn.mx

C. A. Vargas
Departamento de Ciencias Basicas, Universidad Autónoma Metropolitana-Azcapotzalco, Av. San Pablo 180, 02200 Mexico, D.F., Mexico
e-mail: cvargas@correo.azc.uam.mx

J. Klapp et al. (eds.), *Fluid Dynamics in Physics, Engineering and Environmental Applications*, Environmental Science and Engineering,
DOI: 10.1007/978-3-642-27723-8_24, © Springer-Verlag Berlin Heidelberg 2013

uniformly along the horizontal direction. He found that the profile changed to a tilted straight plane whose slope is a function of the gravity acceleration, g, and the acceleration imposed to the vessel during a rigid body translation, a^*. Today, the comprehension of this problem is very important to design safe structures of vehicles carrying liquid cargo and to analyze their driving stability when they are subjected to episodes of sudden breaking or lane change maneuvers (Rumold 2001).

A set of similar problems to the previously alluded occur in vehicles carrying dry cohesionless granular materials (Fleissner et al. 2009) but some important differences appear due to the role played by friction. Thus, the motivation of this study is the determination of how the free surface of a non cohesive granular material in a box changes its shape to final states of equilibrium due to the one-dimensional, horizontal uniform acceleration.

The modeling of this problem can be made through the formulation of a balance of force equation which uses the Coulomb's law; it gives a relation between the tangential (F) and normal (N) forces on a unit volume at the free surface of a non cohesive granular material in sliding contact (Krim 2002). This formula establishes that the equilibrium condition (the slope) is maintained if

$$F \leq \mu N. \tag{1}$$

This formula establishes that the equilibrium condition (the slope) is maintained if the motion occurs if and only if there is equality; the quantity μ is the coefficient of friction of the material. In general terms the start of the distortion of the free surface occurs when the slope of the heap reaches a maximum value, the angle of internal friction φ_c. In this paper the sliding contact condition may be reached under the imposed acceleration, a^*, and the resulting motion equation in this case gives the final state of deformation which, as in the case of accelerated liquids, corresponds to straight slopes if the initial free surface is horizontal and the magnitude of the normalized acceleration $a = a^*/g$ overcomes the friction coefficient $\mu = \tan\varphi_c$.

2 Theory

Here it is supposed that a cohesionless granular material partially fills a box, $2L^*$ length and w^* width, whereas the height of the granular material is H^* respect to the basis ($y^* = 0$). See Fig. 1. The analysis of the deformation of the free surface considers the presence of the gravity acceleration and a one-dimensional, horizontal, uniform acceleration on the box that provokes the initial horizontal profile changes up to a tilted straight plane that does not is deformed anymore (the equilibrium profile has been attained); first the slope of this plane is computed and afterwards the mobilized volume of material and the center of mass are also estimated for different values of the acceleration.

Fig. 1 Schematic of the
accelerated box filled with a
cohesionless granular
material. The initial profile is
a horizontal plane and when
the system is under the
acceleration a^*, the resulting
deformation is an inclined
plane at the angle θ

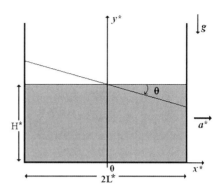

3 The Free Surface Deformation

In Fig. 1 a scheme of the problem is depicted. Initially, the granular material has a
constant level $y^* = H^*$ in the box and suddenly it is uniformly accelerated to the
right hand side with an acceleration a^*, respect to the inertial system fixed to the floor.
Since the point of view of an observer in the (x^*,y^*,z^*) system, fixed to the box, there
exits an acceleration a* in the opposite direction which deforms the free surface as is
sketched in Fig. 1. It is well accepted that when this type of surface deformation
occurs it is possible to build the motion equation using Eq. (1) (Medina et al. 1995;
Shinbrot et al. 2007; Vavrek and Baxter 1994). Thus, for an element of volume of
bulk density ρ, at the free surface making an angle θ respect to the horizontal (Fig. 1),
the motion equation is:

$$\rho(a^* \cos \theta - g \sin \theta) = \mu\rho(a^* \sin \theta + g \cos \theta). \qquad (2)$$

Rearranging terms, scaling the coordinates with the half-length of the box L^*,
i.e., $(x,y,z) = (x^*/L^*, y^*/L^*, z^*/L^*)$, and introducing the dimensionless acceleration
$a = a^*/g$, we obtain the dimensionless differential equation for the slope.

$$\frac{dy}{dx} = \frac{\mu - a}{1 + \mu a}, \qquad (3)$$

Which is valid for any plane $z = const.$, where $0 \le z \le w$ and $w = w^*/L^*$. The
solution of Eq. (3), yields the profile

$$y(x) = \frac{\mu - a}{1 + \mu a} x + H, \qquad (4)$$

where $H = H^*/L^*$, and to determine the constant resulting from the solution of the
differential equation has been used the overall mass conservation of granular
material. Equation (4) allows to notice that there is a critical value of the
dimensionless acceleration, $a = \mu$, below of which the surface is no deformed.
If the acceleration overcomes this threshold value then the initially horizontal

Fig. 2 Dimensionless two
dimensional profiles of the
free surface under uniform
accelerations: $a = 0.6$
(*continuous line*), $a = 0.8$
(◆), $a = 1$ (■), $a = 1.5$
(*dashed line*) and $a = 2$ (◇)

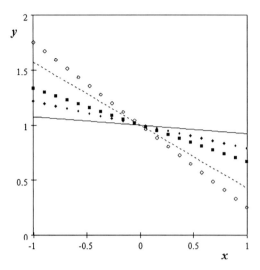

profiles will be deformed to a surface whose profile is a straight plane, with slope
$(\mu - a)(1 + \mu a)$, which always crosses the point $y = H$, the initial height, at the
center of the box.

In Fig. 2 are plotted some nondimensional equilibrium profiles assuming that the
granular material has a friction coefficient $\mu = 0.5$, and that it is contained in a box
with dimensions: length equal to 2 and width w, and initial level of filling $H = 1$.
The dimensionless accelerations were assumed as $a = 0.6, 0.8, 1, 1.5$ and 2. In such
a figure it is clear that the inclination of the resulting profiles increases if a does it.

Note that as in a liquid, if $\mu = 0$, the surface profile given by Eq. (4) transforms in

$$y(x) = -ax + H, \tag{5}$$

which always yields a deformation of the free surface for any value of a, *i.e.*, there
is no a critical value of acceleration as in the case a granular material.

Due to in a non cohesive granular material the number of initial configurations is
practically infinite, whenever the local slope has an angle $\theta < \varphi_c$, which does not
occurs with liquids, it is important to discuss two cases of interest that allows to
notice fundamental aspects of the free surfaces of accelerated systems. First.- when
the initial configuration of the free surface is a symmetric heap, as the one shown in
Fig. 3a, the region where $-L^* \leq x^* \leq 0$ has a positive slope which during accelerated motion also obeys the Eq. (3), but in a different sense. It means that if $a < \mu$,
does not matter how small is a, the free surface in that region will change to a lower
slope. This motion necessarily will change the profile in the zone $0 \leq x^* \leq L^*$ and
thus the complete free surface will be deformed in a complex way.

The study of such a problem is far from the reach of this work because it involves
the coupled motion of both tilted free surfaces. More specifically, this problem
involves rearrangements deep inside the bulk of the system and its treatment requires
extensive numerical computations for the bulk (Aradian et al. 2002).

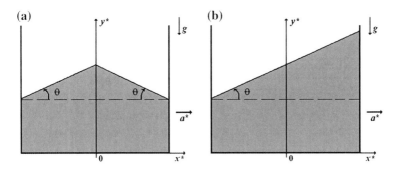

Fig. 3 Schematic of other initial conditions: **a** symmetrical heap **b** Asymmetrical heap. During acceleration the regions with positive slope are immediately deformed

Second. A similar situation occurs when the free surface is a tilted plane making an angle θ respects the horizontal, as in Fig. 3b. When the box is accelerated to the right hand side the granular material in the free surface felt a normal force that tends to decompress the heap and a downhill granular flow occurs on its free surface. Consequently a change in the free surface profile will occur to any value of a. Both of these configurations allow concluding that a knowledge of the initial profile is essential to know the changes of the complete free surface itself; it is also other different situation respect to the classical case of the free surface deformation of a liquid (White 1994).

In next section is estimated the mobilized volume and the new position of the center of mass in the most simple case of free surface deformation, after, in Sect. 5 some experiments with mustard seeds will be presented in order to validate the theoretical results and to show other peculiarities of the free surface deformations of non cohesive granular heaps under uniform accelerations.

4 The Mobilized Volume and the Motion of the Center of Mass

The granular material in a box changes its slope from the horizontal to a tilted slope if $a > \mu$ in accordance with Eq. (4) and, in a first approximation, it is assumed that in such a case the bulk density is the same as in the initial stage (Medina et al. 1995; Shinbrot et al. 2007; Vavrek and Baxter 1994), thus the volume of mobilized material from the right hand side to the left hand side, see Fig. 1, has the dimensionless form

$$V_m = w\left[\frac{a - \mu}{2(1 + \mu a)}\right], \text{if } a > \mu, \tag{6}$$

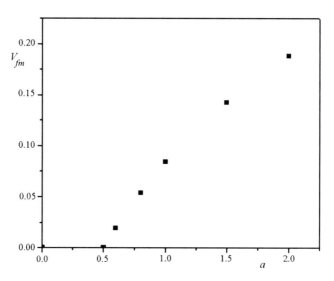

Fig. 4 Fractional volumes as a function of the dimensionless acceleration, a. In this plot have been considered the values of a given in Fig. 2

and the fraction of mobilized volume relative to the total dimensionless volume $V_T = 2wH$, is given by

$$V_{fm} = \frac{V_m}{V_T} = \frac{a - \mu}{4H(1 + \mu a)}, \text{ if } a > \mu. \tag{7}$$

Clearly, the mobilization of granular material changes the original position of the center of mass of the granular material in the box; moreover, this quantity can be of interest to estimate the net torques on tires in vehicles transporting grains (Fleissner et al. 2009). Assuming, once again, that the bulk density is not changed under the deformation of the free surface, it is direct to found that the center of mass can be calculated from the relation

$$r_{cm} = \frac{\int r\,dxdydz}{V_T}, \tag{8}$$

where $\mathbf{r} = (x,y,z)$. Thus, using this latter formula and the Eq. (4) to determine the boundaries that limits the localization of the granular material, is obtained that

$$x_{cm} = \frac{w \int_{-1}^{1} \int_0^{y(x)} x\,dydx}{2Hw} = \frac{\mu - a}{3(a\mu + 1)H}, \text{ if } a > \mu, \tag{9}$$

$$y_{cm} = \frac{w \int_{-1}^{1} \int_0^{y(x)} y\,dydx}{2Hw} = \frac{3H^2(a\mu + 1)^2 + (a - \mu)^2}{6H(a\mu + 1)^2}, \text{ if } a > \mu, \tag{10}$$

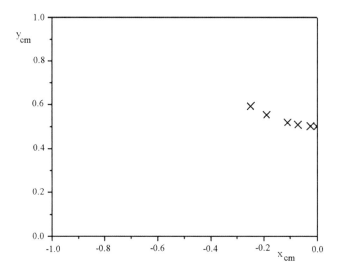

Fig. 5 Position of the center of mass for heaps deformed under the dimensionless accelerations given in Fig. 2. The point in $x = 0$ corresponds to an no deformed heap with a horizontal profile

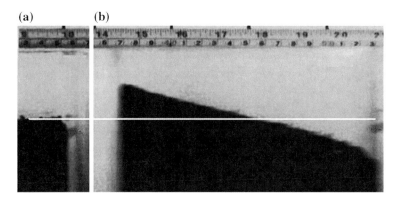

Fig. 6 **a** Partial view of a heap of mustard seeds confined in the box with a profile initially flat and **b** deformed heap of mustard. In this case $a = 1.22$. The white line indicates the position of the initial level (Álvarez 2011)

$$z_{cm} = \frac{w}{2}. \tag{11}$$

It is useful to discuss the behavior of V_{fm} and $\mathbf{r}_{cm}$ in a general context. In Fig. 4 is plotted the fractional volumes V_{fm} assuming the same values of the dimensionless acceleration as aforementioned. From Fig. 4 is easily estimated that the fractional mobilized volume when $a = 2$ is $V_{fm} = 0.1875$, *i.e.*, 18.75 % but in the case of a liquid ($\mu = 0$ in Eq. (7)) this quantity achieves 50 %. This latter result illustrates the important role played by friction which limits the motion of the bulk

(a)   **(b)**

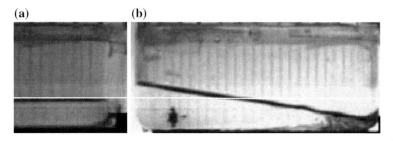

Fig. 7 **a** Partial view of a heap of water confined in the box with a profile initially flat and **b** deformed heap of water. In this case $a = 1.22$. The white line indicates the position of the initial level

Fig. 8 Plot of the accelerated motion of boxes containing mustard seeds: for $a = 1.22$ (■) the initial profile was horizontal and it corresponds to the history of deformation shown in Fig. 6. For $a = 0.62$ (●) there is no deformation of the horizontal free surface and for $a = 0.59$ (▼) the heap in the box is initially symmetric, as shown in Fig. 9, and despite $a < \mu$ in this case occurs an overall deformation

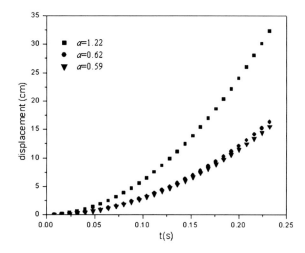

and eventually permit different initial configurations. In Fig. 5 the position of the center of mass, in the plane xy and at $z = w/2$ is plotted under the same conditions: each point indicates the localization of the center of mass when the equilibrium profile for a given acceleration has been achieved.

5 Experiments

It is direct to carry out experiments to validate mainly the theoretical predictions for the surface profiles. In experiments were used mustard seeds which were confined in an acrylic-walled rectangular box, 16.3 cm length and 5 cm width. Mustard seeds have nearly spherical shape and thus their average diameter was $dg = 1.5 \pm 0.1$ mm, density $\rho_g = 1.3$ g/cm^3; the friction coefficient was $\mu = 0.62 \pm 0.01$, and the angle of internal friction took the value $\varphi_c = 32° \pm 0.5°$ (Yamane et al. 1998).

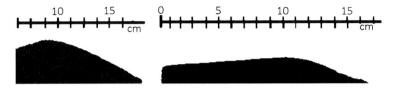

Fig. 9 On the LHS is observed partially an initially symmetric heap of mustard seeds, when the dimensionless acceleration is $a = 0.59$, *i.e.*, this value is lower than that of the friction coefficient, $\mu = 0.62$, there an overall deformation of the heap is observed. It indicates that depending on the initial condition a free surface deformation can be achieved

Fig. 10 A jet can be appreciated when a is large, in this case dimensionless acceleration was close to $a = 1.7$

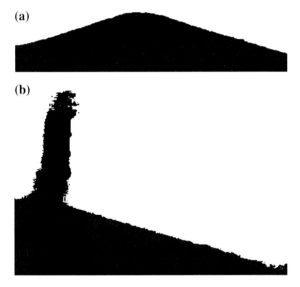

(a)

(b)

Due to the low density of the mustard seeds experiments of one-dimensional, uniform accelerations of different magnitudes were made in the box mounted on a linear air track. In each experiment the box containing the mustard seeds was pulled, near frictionless, by a mass, located at the opposite edge, and falling freely. The motion of the free surface during acceleration events was video-recorded with a high speed camera model Redlake HG-100 K. All video recordings were taken to 125 fps to follow sequentially the free surface deformation and the motion of the box.

In a first experiment the box was filled up to H = 4.8 cm with mustard seeds and on it was imposed a dimensionless acceleration equal to $a = 1.22$, clearly the horizontal free surface will be deformed because $a > \mu$. The determination of the acceleration was made using the formula $s = (a*t^2/2)$ where s is the displacement of a point of the box at a time t. In the experiment the average value of y at each edge of the box was $y \approx 0.245$ (see Figs. 6, 7).

Fig. 11 A jet can be appreciated with water, when *a* is large, in this case dimensionless acceleration was close to $a = 1.7$

Meanwhile using formula (3) it is found that $y = 0.247$, *i.e.*, a relative error of 1 %. In Fig. 8 is shown the plot of the displacement as a function of time, t, to quantify the acceleration of the box. In such a plot there are other values of *a* which correspond to a case where the free surface is horizontal and despite acceleration is $a = 0.62$ there is not appreciated a macroscopic free surface deformation. Finally the case where $a = 0.59$ corresponds to the acceleration of a box with a symmetric heap, as is shown in Fig. 9. There is appreciated that a substantial deformation occurs, even that $a < \mu$. It confirms that during a small acceleration the zone with positive slope can be mobilized because there the normal force plays a decompressing role, instead the compressive one occurring in the zone with negative slope.

For a more high value of acceleration it can be observed a jet as the one shown in Fig. 10 where is noted that the initial condition was a symmetric heap, such structure have appeared when $a = 1.7$. The jet was generated due to the strong uphill surface flow which eventually climbs the wall and as the box is decelerated the jet moves onward and eventually disappears.

Figure 11 we see a jet of water, such structure have appeared when $a = 1.7$. The jet was generated due to the strong uphill surface flow which eventually climbs the wall and as the box is decelerated the jet moves onward and eventually disappears.

6 Conclusions

In this work has been analyzed the problem of the free surface deformation of a non cohesive granular material confined in a box under an uniform, one-dimensional acceleration, *a*. Using a simple model based on the Coulomb's law of friction it has been shown that for an initially horizontal profile there exists a critical value of the dimensionless acceleration that allows to achieve a final, equilibrium profile corresponding to a tilted straight plane. However, if the initial profile has a zone with a positive slope, this condition do not is maintained and it is possible to get a deformation that involves rearrangements deep inside granular material. Experiments, using mustard seeds, back all these results and show also complex evolutions of the bulk material, as the presence of jets when acceleration is strong.

References

Álvarez VS (2011) Fricción en silos y vehículos acelerados con carga granular. Master Science Thesis, Higher School of Electric and Mechanics Engineering, IPN

Aradian A, Raphaël É, de Gennes PG (2002) Surface flows of granular materials: a short introduction to some recent models. C R Physique 3:187–196

Bernoulli D (1968) Hydrodynamics (Dover, New York) translation by Thomas Carmody from hydrodynamica as published by Johann Reinhold Dulseker at Strassburg in 1738

Fleissner FD, Alessandro V, Schiehlen W, Eberhard P (2009) Sloshing cargo in silo vehicles. J Mech Sci Tech 23:968–973

Krim J (2002) Friction at macroscopic and microscopic length scales. Am J Phys 70:890–897

Medina A, Luna E, Alvarado R, Treviño C (1995) Axisymmetrical rotation of a sand heap. Phys Rev E 51:4621–4625

Rumold W (2001) Modeling and simulation of vehicles carrying liquid cargo. Multibody Syst Dyn 5:351–374

Shinbrot T, Duong NH, Hettenbach M, Kwan L (2007) Coexisting static and flowing regions in a centrifuging granular heap. Granul Matter 9:295–307

Vavrek ME, Baxter GW (1994) Surface shape of a spinning bucket of sand. Phys Rev E 50: R3353–3356

White FM (1994) Fluid mechanics, 3rd edn. McGraw-Hill, New York

Yamane K, Nakagawa M, Altobelli SA, Tanaka T, Tsuji Y (1998) Steady particulate flows in a horizontal rotating cylinder. Phy Fluids 10:1419–1427

Part III
Convection, Diffusion and Vortex Dynamics

Experimental and Computational Modeling of Venlo Type Greenhouse

Abraham Rojano, Raquel Salazar, Jorge Flores, Irineo López,
Uwe Schmidt and Abraham Medina

Abstract Currently, experimental data gathered with high frequency and good quality allows us to feed more complex mathematical models by using high speed and large memory computational devices with numerical technique like finite element method. Simple models had been enriched including details of internal flow patterns and temperature profiles. Recent progress in flow modeling by means of computational fluid dynamics (CFD) software facilitates the fast analysis of such scalar and vector fields, solving numerically the transport equations like mass, momentum and heat transfer equations. As a result, this work shows us the mechanical air behavior in terms of velocity and temperature patterns near to the plant benches in a Venlo type greenhouse by using Boussinesq assumptions.

1 Introduction

Each greenhouse has its own physical characteristics given by temperature (T) and velocity (u) distribution as well as its management; however, to reproduce its

A. Rojano (✉) · R. Salazar · J. Flores · I. López
Dirección General Académica, Universidad Autónoma Chapingo, km 38.5 carretera México Texcoco, Chapingo, 56230 Texcoco, Mexico
e-mail: abrojano@hotmail.com

UweSchmidt
Division Biosystems Engineering, Humboldt Universität zu Berlin,
Albrecht-Thaer-Weg 1, 14195 Berlin, Germany

A. Medina
SEPI ESIME-Unidad Azcapotzalco IPN Av, de las Granjas No. 682 Col. Sta. Catarina Azcapotzalco, 02250 Mexico, Mexico
e-mail: amedinao@yahoo.com

J. Klapp et al. (eds.), *Fluid Dynamics in Physics, Engineering and Environmental Applications*, Environmental Science and Engineering,
DOI: 10.1007/978-3-642-27723-8_25, © Springer-Verlag Berlin Heidelberg 2013

intrinsic connection by means of mathematical models begs into some difficulties, even when some assumptions and simplifications are included. First, we need to gather fundamental information at different points in transversal sections of the Venlo type greenhouse. Second, we need to use physical models represented by partial differential equations developed from the conservative principles. Third, we need to use computational procedures in order to simulate the variables involved.

Historically, the modeling of computational fluid dynamics (CFD) to study the ventilation in a greenhouse was conducted several years ago with poor results, as it was reported by Kacira et al. (2004) who compared the numerical methods with the results of wind tunnel. Although, the results agreed poorly with experiment, probably due to the limited capacity of calculation at the time, but somehow, they obtained important information about the flow patterns inside the greenhouse (Mistriotis et al. 1997).

In principle, CFD is applied inside the greenhouse in a discrete number of points corresponding to the space and time domain. The domain is divided into elementary triangular regions, where the conservation equations of mass, momentum and energy are solved. Variables (pressure, velocity, temperature) are obtained in each control element according to an iterative procedure that converges gradually to the solution of nonlinear system of equations (Bournet and Boulard 2010).

2 Background

Since physical principles are written in a mathematical language, the following equations are adapted and represent the mass, momentum and energy conservative principles as vector equations (Anderson 1995):

2.1 Mass Conservation

$$\nabla \bullet u = 0 \tag{1}$$

2.2 Linear Momentum Conservation

$$\rho \frac{Du}{Dt} = -\nabla p + \mu \Delta u + \rho g \tag{2}$$

2.3 Energy Conservation

$$\rho \frac{Dh}{Dt} = \delta \Delta T \tag{3}$$

Subject to appropriate initial and boundary conditions. In these equations, u is the velocity, ρ is the density, t is the time, p is the pressure, μ is the viscosity, g is the gravity, h is the enthalpy, δ is the diffusivity, and T is the temperature. Even though the above equations do not depend on specific coordinates, for feasibility, cartesian coordinates are used to construct the local and global matrices and vectors by using a numerical approximation with finite element method (Majdoubi et al. 2009). Later on, density is calculated by using Eq. 4.

$$\rho = \rho_o(1 - \beta(T - T_o)) \tag{4}$$

β is the coefficient of volume expansion. When the temperature variation is a moderate amount, we have as rule of thumb $\beta < 1/50$.

2.4 Numerical Implementation

Since analytical solutions for this problem are impossible yet, we consider a numeric calculation for the variables represented in the cross section orthogonal to main axis of a Venlo greenhouse. This region defines a sectional area, and boundary condition that encloses the area. The finite element method uses the integral support in a very useful way for the treatment of spatial variables. Thus, using polynomial interpolation it is possible to obtain a system of nonlinear ordinary differential equations. These equations are solved with Newton–Raphson technique and an iterative Gauss–Seidel type method for nonlinear algebraic equations.

3 Experimental Set-Up

In the Department of Biosystems Engineering at the Humboldt University of Berlin, the greenhouse collector system was developed. By cooling surfaces in the roof space of a closed greenhouse, the excess heat energy is dissipated and stored in a water tank. As specific variables in the entire process, temperature and velocity data are obtained and adapted in Microsoft Excel 2003 format for every 15 min.

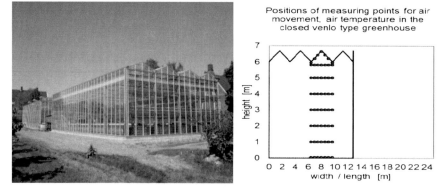

Fig. 1 *Left* Venlo greenhouse. *Right* Diagram of observation points in a bay inside the Venlo greenhouse

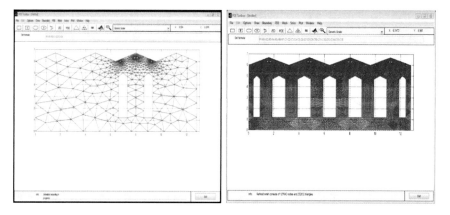

Fig. 2 *Left* Coarse mesh with a single bay system with 1298 triangles and 712 nodes. *Right* A refined mesh with four bays and benches by using 63056 triangles and 32324 nodes

4 Results

With the purposes of illustrate an application of this simulation, we have used data from a Venlo type greenhouse collected in 2010 (see Fig. 1, Left). This greenhouse has upper pipes in order to collect energy, by condensing water. The greenhouse has four bays and o of this is selected as experimental section. First, the information is collected from the bottom to the top (see Fig. 1, Right). In terms of geometry, the domain is split in triangles with different density until we get the most representative model (Fig. 2). Since the first solution obtained from Eq. 3 shows the heat flow trend from the bottom to the top, the next stage is to develop

Table 1 General information

Geometrical dimensions of greenhouse	
Width (m)	12.8
Length (m)	24
Height (m)	6.8
Experimental cabinet	
Width (m)	3.2
Length (m)	24
Height (m)	6.8
Air properties	
Temperature bottom (K)	303
Temperature roof (K)	283
Density at 293 K (kg/m^3)	1.225
Specific heat (J/kg-K)	1006.420
Thermal conductivity (W/m-K)	0.0242
Velocity(m/s)	<0.1
Density a 293 K (kg/m^3)	1.225
Specific heat (J/kg-K)	1006.420
Thermal conductivity (W/m-K)	0.0242
Viscosity (kg/m-s)	1.7894e-5

the connection with the velocity, by using the Boussinesq assumption of Eq. 4 joint to the Eqs. 1, 2, and 3 simultaneously, by using a thermal lattice Boltzmann (LB) technique (Guo et al. 2002). The Table 1 includes more information about the system studied.

5 Conclusions

The mutual collaboration with the project ZINEG (http://www.plantputer.com/) over the years brings about an ambitious goal to understand not only the mechanics of energy flow and air motion but also the optimal management of energy in order to reduce the heat consumption in the plant production. The novel approach of the project lies in the consistent further development of technical systems like measuring and simulating the internal microclimate conditions in the greenhouse with computational fluid dynamics. In this context the Fig. 3 (Left) shows that the precise ascending motion of heat proportional to the temperature differences from bottom to the roof conditions are lower than 10 °C. Even experimental readings by the equipment used are limited in the velocity direction, by using numerical simulations (Fig. 3, Left) allows us to see the air motion as vortices below 0.1 m/s according to experimental measurements. One limitation of this study regards on

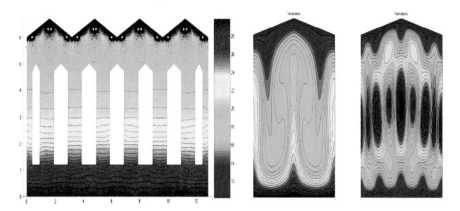

Fig. 3 *Left* Ascending heat flow process by temperature differences. *Right* Air velocity as a result of temperature gradient without benches, with Prandt number equal to 0.71 and Rayleight number equal to 2000

the high amount of energy found in the air humidity as latent heat in the enthalpy which generates not only a strong disturbance in temperature but also in air motion, but this type solution it is coming in the near future.

References

Anderson JD (1995) Computational fluid dynamics: the basics with applications. McGraw Hill, New York

Bournet PE, Boulard T (2010) Effect of ventilator configuration on the distribut ed climate of greenhouses: a review of experimental and CFD studies. Comput Electron Agric 74:195–217

Kacira M, Sase S, Okushima L (2004) Effects of side vents and span numbers on wind-induced natural ventilation of a gothic multi-span greenhouse. JARQ 38(4):227–233

Majdoubi H, Boulard T, Fatnassi H et al (2009) Airflow and microclimate patterns in a one-hectare Canary type greenhouse: an experimental and CFD assisted study. Agric For Meteorol 149:1050–1062

Mistriotis A, Bot G, Picuno P et al (1997) Analysis of the efficiency of greenhouse ventilation using computational fluid dynamics. Agric For Meteorol 85:217–228

Guo Z, Shi B, Zheng Ch (2002) A coupled lattice BGK model for the Boussinesq equations. Int J Numer Meth Fluids 39:325–342

Numerical Simulation of an Open Cavity with Heating in the Bottom Wall

Guillermo E. Ovando-Chacon, Sandy L. Ovando-Chacon, Juan C. Prince-Avelino, Eslí Vázquez-Nava and José A. Ortiz-Martínez

Abstract This work presents the numerical simulation of heat transfer and fluid dynamic in steady state of a two-dimensional Cartesian flow inside a cavity with one inflow and one outflow for Re = 100 and 1,000. The domain of the simulation consists in a rectangular section with two different aspect ratios. Three different lengths of the heater at the bottom wall were analyzed. The governing equations of continuity, momentum and energy for incompressible flow were solved by the finite element method. The velocity fields, isotherms, streamlines and vortex formation were studied. The simulation indicates that is possible to control the dynamic and vortex formation inside the cavity due to the variation of the length of the heater.

G. E. Ovando-Chacon (✉) · J. C. Prince-Avelino · J. A. Ortiz-Martínez
Depto. de Metal Mecánica y Mecatrónica, Instituto Tecnológico de Veracruz, Calzada Miguel A. de Quevedo 2779 Col. Formando Hogar, 91860 Veracruz, Veracruz, Mexico
e-mail: geoc@itver.edu.mx

J. C. Prince-Avelino
e-mail: jcpa@itv.edu.mx

S. L. Ovando-Chacon
Depto. de Química y Bioquímica, Instituto Tecnológico de Tuxtla Gutiérrez, Carretera Panamericana Km. 1080, 29000 Tuxtla Gutierrez, Chiapas, Mexico
e-mail: ovansandy@hotmail.com

E. Vázquez-Nava
Depto. de Ingeniería Química, Universidad Veracruzana, Adolfo Ruiz Cortines s/n, Fracc. Costa Verde, 94294 Boca del Rio, Veracruz, Mexico
e-mail: evazquezn@gmail.com

J. Klapp et al. (eds.), *Fluid Dynamics in Physics, Engineering and Environmental Applications*, Environmental Science and Engineering,
DOI: 10.1007/978-3-642-27723-8_26, © Springer-Verlag Berlin Heidelberg 2013

Fig. 1 Geometry of the
cavity

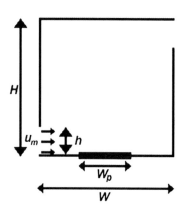

1 Introduction

The numerical simulation of an open cavity with heating in the bottom wall is an
important issue in many technological processes. Madadi and Balaji (2008)
evaluated the optimal location of discrete heat source placed inside a ventilated
cavity. Oztop et al. (2011) numerically examined the steady natural convection in
an open cavity filled with porous media. Radhakrishnan et al. (2007) reported
experimental and numerical investigation of mixed convection from a heat generating element in a ventilated cavity. Zhao et al. (2011) analyzed the characteristics of transition from laminar to chaotic mixed convection in a two-dimensional
multiple ventilated cavity. Najam et al. (2002) presented the simulation of mixed
convection in a T form cavity, heated with constant heat flux and ventilated from
below with a vertical jet. Mamun et al. (2010) studied the effect of a heated hollow
cylinder on mixed convection in a ventilated cavity. Deng et al. (2004) investigated the laminar mixed convection in a two-dimensional displacement ventilated
enclosure with discrete heat and contaminant sources. The main aim of this
numerical investigation is to study the effect of the aspect ratio and the heater
length placed at the bottom wall on the vortex formation and the thermal behavior
in an open cavity.

2 Problem Formulation

This work presents 2D numerical simulations inside an open cavity with two
different aspect ratio, $AR = H/W$, the fluid enters from the left lower side wall and
leaves the cavity through the right upper side wall, see Fig. 1. The heater is
centrally located and three different lengths of the heater, Wp, on the cavities are
investigated. The Reynolds numbers ($Re = U_m h/v$), based on the velocity of the
inlet flow U_m and the height h of the entrance, studied in this investigation were
$Re = 100$ and $Re = 1,000$ for Richardson numbers $Ri = 0.01$ and $Ri = 0.001$.

The governing equations for a non-isothermal incompressible steady state flow are given as:

$$-\frac{1}{Re}\Delta \vec{u} + \vec{u} \cdot \nabla \vec{u} + \nabla p = RiTj \text{ in } \Omega, \tag{1}$$

$$\nabla \cdot \vec{u} = 0 \text{ in } \Omega, \tag{2}$$

$$\frac{1}{Pe}\Delta T + \vec{u} \cdot \nabla T = 0 \text{ in } \Omega, \tag{3}$$

In the above equations $u = (u_1, u_2)$ is the velocity vector, being u_1 y u_2 the transversal and axial velocity components, respectively; v is the kinematic viscosity, p is the pressure, T is the temperature and j is the vertical unitary vector. In the governing equations, the Richardson number, the Reynolds number and the Peclet number are defined as follow:

$$Ri = g\beta h(T_h - T_c)/U_m^2, \quad Re = U_w h/v, \quad Pe = RePe, \tag{2}$$

where g is the gravity, β is the compressibility coefficient, T_h is the hot temperature, T_c is the cold temperature. No slip boundary conditions ($u_1 = u_2 = 0$) were established in all the walls of the cavity and adiabatic walls ($\partial T/\partial n = 0$) were supposed except in the central part of the bottom wall where the heating takes places. The temperature of the inlet flow was fixed to $T = T_c$, meanwhile the isothermal hot plate was fixed to $T = T_h$. The non-dimensional values of this temperature were $T_c = 0$ and $T_h = 1$. The boundary conditions of the inlet flow were $u_1 = U_m$, and $u_2 = 0$. On the outlet flow was imposed $\partial u/\partial n = 0$. The governing equation were solved with the finite element method combined with the operator splitting scheme, see Glowinski (2003). The convergence analysis was done for three different meshes with resolution of 3,600, 14,884 and 19,600 elements. The largest difference of the results between the meshes of 3,721 and 14,884 was 12 %, while the maximum difference of the results between the meshes of 14,884 and 19,600 was 0.92 %. The whole simulations presented in this paper were performed for a cavity with 19,600 elements.

3 Results

Figure 2 shows the velocity field and isotherms, see top panel, for $Re = 100$, $AR = 1$ and $Ri = 0.01$ with three different lengths of the heater plate: $Wp = 0.25$ (left panel), $Wp = 0.50$ (middle panel) and $Wp = 0.75$ (right panel). The bottom panel shows the streamlines indicating the places where the vortices appear. For these cases, it was observed an anti-clockwise, big and strong vortex at the central superior region of the cavity. At the superior left corner a clockwise, weak vortex appears and at the inferior right corner another clockwise vortex emerges from the impingement of the fluid with the corner. The inlet jet emerges horizontally from

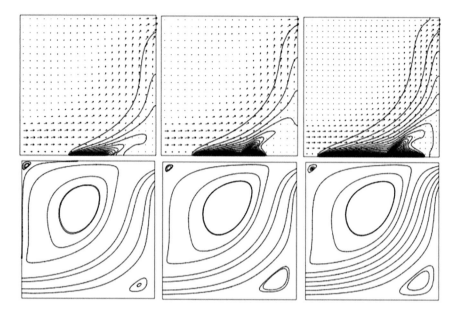

Fig. 2 *Top* velocity fields and isotherms. *Bottom* streamlines. $Re = 100, AR = 1$ and $Ri = 0.01$. *Left* $Wp = 0.25$. *Middle* $Wp = 0.50$. *Right* $Wp = 0.75$

the flow entrances of the cavity, but as moves forwards its horizontal component is decreased and the vertical component is increased. As the length of the heater plate is increased the size of the superior left vortex is reduced while the size of the inferior right vortex is increased due to the removal of heat from the plate. On the other hand, the core of the main vortex tends to move right upward. The heat of the plate fluxes to the right upper part of the cavity due to the main stream of the fluid. As expected high levels of temperature occurs near the heater plate, beyond of this, the contours are distorted and follow the main stream direction toward the outlet of the cavity. The contours tend to concentrate at the left wall of the cavity. As Wp is increased the outlet flow increases its temperature. For $Re = 100$, $AR = 1$ and $Ri = 0.01$ the outlet flow temperature was increased 0.194, 0.195 and 0.393 % for $Wp = 0.25$, 0.50 and 0.75, respectively. For $Re = 100$, $AR = 1$ and $Ri = 0.001$ the outlet flow temperature was increased 0.055, 0.183 and 0.374 % for $Wp = 0.25$, 0.50 and 0.75, respectively.

Increasing Re to 1,000 and reducing $Ri = 0.001$, it can also be observed three vortices along the diagonal opposite to the inlet and outlet of the cavity, see Fig. 3. From the streamlines, it can be seen that the size of the vortices are practically the same for the three different heater lengths while the main vortex core moves as the previous case and for the largest heater length the temperature contours associated with the plate start to penetrate the central part of the cavity. For $Re = 1,000$, $AR = 1$ and $Ri = 0.01$ the outlet flow temperature was increased 0.338, 0.775 and 1.197 % for $Wp = 0.25$, 0.50 and 0.75, respectively. For $Re = 1,000$, $AR = 1$ and

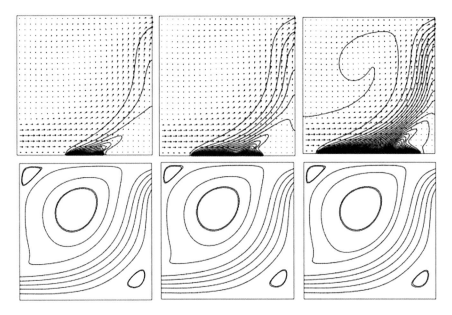

Fig. 3 *Top* velocity fields and isotherms. *Bottom* streamlines. $Re = 1{,}000$, $AR = 1$ and $Ri = 0.001$. *Left* $Wp = 0.25$. *Middle* $Wp = 0.50$. *Right* $Wp = 0.75$

$Ri = 0.001$ the outlet flow temperature was increased 0.315, 0.706 and 1.11 % for $Wp = 0.25$, 0.50 and 0.75, respectively.

When the aspect ratio is increased to 2 the flow behavior becomes more complex than the flow inside the cavity of aspect ratio 1. Figures 4a–c show the velocity field and isotherms, meanwhile Figs. 4a′–c′ show the streamline for $Re = 100$, $AR = 2$ and $Ri = 0.01$ with three different lengths of the heater plate: $Wp = 0.25$ (Fig. 4a–a′), $Wp = 0.50$ (Fig. 4b–b′) and $Wp = 0.75$ (Fig. 4c–c′). For the shortest heater length the streamlines show two vortices. The main anti-clockwise vortex is located at the central superior part of the cavity, while a small clockwise vortex appears in the right lower corner. For $Wp = 0.50$ it can be seen three vortices, the core of the main anti-clockwise vortex moves upward respect to the previous case, the right lower clockwise vortex increases its size while a small, weak, clockwise vortex appears at the left upper corner. For the largest heater length, the streamlines show only two vortices similar to the case of $Wp = 0.25$, however the temperature contours are intensified along the right side wall. For the above three cases the mains vortices are elongated in the vertical direction. For $Wp = 0.50$ and $Wp = 0.75$, the temperature contours associated with the hot plate reach the central part of the cavity. For $Re = 100$, $AR = 2$ and $Ri = 0.01$ the outlet flow temperature was increased 0.243, 0.244 and 0.973 % for $Wp = 0.25$, 0.50 and 0.75, respectively. For $Re = 100$, $AR = 2$ and $Ri = 0.001$ the outlet flow temperature was increased 0.223, 0.228 and 0.928 % for $Wp = 0.25$, 0.50 and 0.75, respectively.

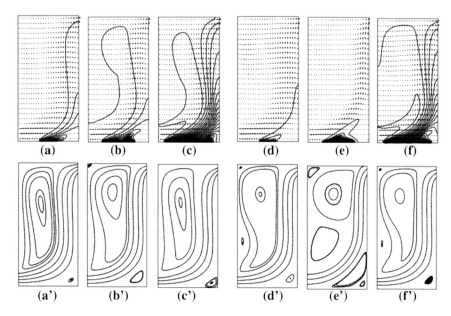

Fig. 4 *Top* velocity fields and isotherms. *Bottom* streamlines. $AR = 2$. $Re = 100$ and $Ri = 0.01$: **a–a'** $Wp = 0.25$. **b–b'** $Wp = 0.50$. **c–c'** $Wp = 0.75$. $Re = 1{,}000$, $Ri = 0.001$: **d–d'** $Wp = 0.25$. **e–e'** $Wp = 0.50$. **f–f'** $Wp = 0.75$

Increasing Re to 1,000 and reducing Ri to 0.001, it can be noted that the number of vortices inside the cavity is four, see Fig. 4d'–f', for the three heater length analyzed. One anti-clockwise vortex is located at the central upper part of the cavity for $Wp = 0.25$ (Fig. 4d'), this vortex is small and weak. Increasing Wp to 0.50 (Fig. 4e') its intensity is increased and for $Wp = 0.75$ (Fig. 4f') this vortex is elongated in the vertical direction. A second clockwise vortex appears at the right lower corner of the cavity and for $Wp = 0.50$ this vortex exhibited the largest size. A third clockwise vortex can be identified at the neighborhood of the left upper corner being the vortex of the case $Wp = 0.50$ the largest one. The fourth clockwise vortex appears near of the lower left wall just above of the entrance of the cavity, for $Wp = 0.25$ this vortex is small and weak, however for $Wp = 0.50$ the size of this vortex is increased and for $Wp = 0.75$ the size of this fourth vortex again decreases as a consequence of the elongation of the central upper vortex. The temperature contours concentrate around the hot plates and deviates toward the left wall. The larger the size of the heater the greater the elongation of contours towards the inside of the cavity is. For $Re = 1{,}000$, $AR = 2$ and $Ri = 0.01$ the outlet flow temperature was increased 0.378, 0.917 and 1.45 % for $Wp = 0.25$, 0.50 and 0.75, respectively. For $Re = 1{,}000$, $AR = 2$ and $Ri = 0.001$ the outlet flow temperature was increased 0.34, 0.076 and 1.36 % for $Wp = 0.25$, 0.50 and 0.75, respectively. The maximum Nusselt number $Nu = 83.7478$ was obtained with the largest heater for $AR = 1$, $Ri = 0.001$ and $Re = 100$, while the minimum

Nusselt number $Nu = 18.6011$ was obtained for $Wp = 0.25$, $AR = 2$, $Ri = 0.001$ and $Re = 1,000$.

4 Conclusions

We have presented in this work results of finite element simulation of the flow inside an open cavity for two different aspect ratios and three different heater lengths located at the bottom wall. The analysis is carried out for the laminar regimen and for the case when the buoyancy effect is outweighed by forced convection. The streamline patterns reveal that for a square cavity three vortices are formed along the diagonal opposite to the inlet and outlet of the cavity. Increasing Re and reducing Ri the heat removed from the heater is transported to the central part of the cavity for $Wp = 0.75$. When the aspect ratio is increased to 2, a more complex behavior takes place in the flow for $Re = 1,000$, observing at less four vortices inside the cavity. The Nu indicates that the heat transfer is enhanced for the case of $Wp = 0.75$ with $AR = 1$, $Ri = 0.001$ and $Re = 100$. So far, this study is limited to $Re = 100$ and $1,000$ and $Ri = 0.01$ and 0.001, the fluid dynamic behavior beyond this value is the subject of ongoing research, but it is useful to understand the vortex formation and thermal behavior of the flow inside an open cavity.

References

Deng QH, Zhou J, Mei Chi, Shen YM (2004) Fluid, heat and contaminant transport structures of laminar double-diffusive mixed convection in a two-dimensional ventilated enclosure. Int J Heat Mass Transf 47:5257–5269

Glowinski R (2003) Numerical methods for fluids, part 3. In: Garlet PG, Lions JL (eds) Handbook of numerical analysis, vol IX. North-Holland, Amsterdam

Madadi RR, Balaji C (2008) Optimization of the location of multiple discrete heat sources in a ventilated cavity using artificial neural networks and micro genetic algorithm. Int J Heat Mass Transf 51:2299–2312

Mamun MAH, Rahman MM, Billah MM, Saidur R (2010) A numerical study on the effect of a heated hollow cylinder on mixed convection in a ventilated cavity. Int Commun Heat Mass Transf 37:1326–1334

Najam M, El Alami M, Hasnaoui M, Amahmid A (2002) Numerical study of mixed convection in a T form cavity submitted to constant heat flux and ventilated from below with a vertical jet. C. R. Mecanique 330:461–467

Oztop HF, Al-Salem K, Varol Y, Pop I (2011) Natural convection heat transfer in a partially opened cavity filled with porous media. Int J Heat Mass Transf 54:2253–2261

Radhakrishnan TV, Verma AK, Balaji C, Venkateshan SP (2007) An experimental and numerical investigation of mixed convection from a heat generating element in a ventilated cavity. Exp Thermal Fluid Sci 32:502–520

Zhao M, Yang M, Lu M, Zhang Yuwen (2011) Evolution to chaotic mixed convection in a multiple ventilated cavity. Int J Therm Sci 50:2462–2472

Natural Convection and Entropy Generation in a Large Aspect Ratio Cavity with Walls of Finite Thickness

D. Pastrana, J. C. Cajas and C. Treviño

Abstract In this work we study the heat transfer and fluid flow process in a vertical cavity of large aspect ratio, $AR = 12$, with walls of finite thickness, heated from two portions localized in the side walls of the cavity near the bottom. The equations of motion are written in non-dimensional form, depending of four non dimensional parameters (the Rayleigh number, the Prandtl number, the ratio of thermal diffusivities of the fluid and the material of the cavity and the non-dimensional width of the walls) and are solved numerically by the use of the SIMPLE algorithm. Calculations are performed for three different values of the Rayleigh number.

1 Introduction

In many of the practical systems where natural convection is present, it is of prime importance to consider the thermal properties of the material that confines the fluid. A deeper knowledge of the combined effects of the heat conduction in the

D. Pastrana (✉) · J. C. Cajas
Facultad de Ciencias, Universidad Nacional Autónoma de México, Yucatán, Mexico
e-mail: dpastrana@ciencias.unam.mx

J. C. Cajas
e-mail: jc.cajas@gmail.com

C. Treviño
Facultad de Ciencias y UMDI Sisal, Universidad Nacional Autónoma de México,
Yucatán, Mexico
e-mail: ctrev@gmail.com

J. Klapp et al. (eds.), *Fluid Dynamics in Physics, Engineering and Environmental Applications*, Environmental Science and Engineering,
DOI: 10.1007/978-3-642-27723-8_27, © Springer-Verlag Berlin Heidelberg 2013

walls of the cavity, and the natural convection in the fluid, can lead to an improvement in the design of thermal exchanging devices.

The conjugated effect of conduction and natural convection has been widely studied, some examples are cited next. Kim and Viskanta (1985) performed a numerical and experimental study of the natural convection in a square cavity made of four conductive walls. They found that in certain configurations of the system, the conductive walls help to partially stabilize the flow and reduce the temperature differences in the cavity, as well as the heat transferred by natural convection. More recently, Liaqat and Baytas (2001), Mobedi (2008) and Zhang et al. (2011) studied the influence of the presence of walls of finite thickness over the natural convection in square cavities with different geometrical configurations. The numerical results reveal that, among other things, the heat transfer increases almost linearly with the ratios of thermal conductivities or diffusivities and decreases if the cavity is inclined. Their results have demonstrated that there is a significant change in the behavior of the resultant flow, in comparison with those obtained in similar systems where the conjugated phenomena is not considered.

In the last years, the heat transfer studies for the design of thermal devices have relied on the second law of thermodynamics and on the minimal entropy production. However, there are few works that consider the entropy generation due to heat conduction and natural convection in enclosures with walls of finite thickness. The works performed by Varol et al. stand out (2008, 2009) where the entropy production due to the conjugated process of heat conduction-natural convection is studied in a square cavity with side walls of different finite thicknesses and in a trapezoidal porous cavity with a solid vertical wall of finite thickness. Among their more important results, the presence of the solid walls affects the temperature and velocity fields inside the cavity, the Bejan number diminishes when the Rayleigh number or the thermal conductivities ratio increases, and the most intense zones of local entropy production are found in the corners of the cavity. They also show that the entropy produced by the viscous effects does not depend on the thickness of the walls, and that the shape of the cavity can be a controlling parameter to decrease the overall entropy production, obtaining an energy save.

The present work is aimed to provide further information on the transient heating process and the entropy production in a large aspect ratio cavity with walls of finite thickness. Which is a system that can be used in many engineering devices.

2 Problem Statement

In this work we study the natural convection and entropy generation in a Boussinesq's fluid, confined by a rectangular cavity with large aspect ratio, $AR = L/H = 12$, and with solid conductive walls of finite thickness, as shown in Fig. 1. L is the interior height of the cavity and H is its interior length. The whole system, conformed by the cavity and the walls, is adiabatically isolated from the

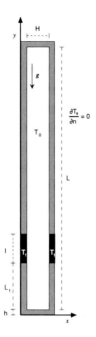

Fig. 1 Schematic representation of the cavity

exterior, except by two portions of height $l = H$ and width h located symmetrically on the side walls of the cavity at a distance L_1 from the lower wall. The two portions are held at constant temperature T_1 which is larger than the initial temperature of the fluid T_0.

Using the length of the cavity H, the diffusion time of temperature H^2/κ_f (κ_f is the thermal diffusivity of the fluid) and the difference between the initial temperature of the fluid and the temperature that induces the buoyancy forces $T_1 - T_0$, the following non-dimensional variables arise,

$$x_i = \frac{x_i^*}{H}, \quad u_i = \frac{u_i^* H}{\kappa_f}, \quad t = \frac{t^* \kappa_f}{H^2}, \quad p = \frac{p^* H^2}{\rho_0 \kappa_f^2}, \quad \theta_f = \frac{T_f - T_0}{T_1 - T_0}, \quad \theta_s = \frac{T_s - T_0}{T_1 - T_0}.$$

$$(1)$$

where x_i and u_i are the non-dimensional ith components of the position and velocity vectors respectively. p, t and θ are used to denote the non-dimensional pressure, time and temperature and the super index $*$ is used to distinguish between dimensional and non-dimensional quantities.

With the use of the non-dimensional variables (1), the governing equations of the system can be written as follows

$$\frac{\partial u}{\partial x} + \frac{\partial v}{\partial y} = 0; \quad \frac{Du}{Dt} = -\frac{\partial P}{\partial x} + Pr\left(\frac{\partial^2 u}{\partial x^2} + \frac{\partial^2 u}{\partial y^2}\right),$$

$$\frac{Dv}{Dt} = -\frac{\partial P}{\partial y} + RaPr\theta_f + Pr\left(\frac{\partial^2 v}{\partial x^2} + \frac{\partial^2 v}{\partial y^2}\right), \qquad (2)$$

$$\frac{D\theta_f}{Dt} = \frac{\partial^2 \theta_f}{\partial x^2} + \frac{\partial^2 \theta_f}{\partial y^2}; \quad \frac{\partial \theta_s}{\partial t} = \alpha\left(\frac{\partial^2 \theta_s}{\partial x^2} + \frac{\partial^2 \theta_s}{\partial y^2}\right),$$

with the following boundary conditions

$$u_i = 0, \ \ i = 1, 2, \ \ \theta_s = \theta_f, \ \ K\frac{\partial \theta_s}{\partial n} = \frac{\partial \theta_f}{\partial n} \quad \text{on the solid-fluid interface;} \quad (3)$$

$$\theta_s = 1, \quad at: y \in \left[\frac{L_1 + h}{H}, \frac{L_1 + h + l}{H}\right], \quad x \in \left[0, \frac{h}{H}\right] \quad \text{and} \qquad (4)$$

$$y \in \left[\frac{L_1 + h}{H}, \frac{L_1 + h + l}{H}\right], \quad x \in \left[\frac{h}{H} + 1, \frac{2h}{H} + 1\right]$$

$$\frac{\partial \theta_s}{\partial n} = 0 \quad \text{on the exterior part of the walls,} \qquad (5)$$

where the usual non-dimensional parameters have been defined, the Prandtl's number $Pr = v/\kappa_f$, with v as the kinematic viscosity; the Rayleigh's number $Ra = \beta g d^3 (T_1 - T_0)/(v\kappa_f)$, β is the volumetric thermal expansion coefficient; the thermal diffusivities ratio $K = k_s/k_f$. The sub indexes f and s distinguish the physical properties of the fluid from those of the solid, and n represents the normal direction to a surface.

2.1 Entropy Production

In the Linear Irreversible Thermodynamics formulation, an explicit expression for the entropy balance is obtained in terms of the velocity and temperature fields. From that expression, the internal entropy production in terms of the non-dimensional variables are

$$\sigma_q^* = \sigma_q \frac{H^2 T_0^2}{k(T_1 - T_0)^2} = \frac{1}{(1 + \varepsilon\theta)^2}\left[\left(\frac{\partial \theta}{\partial x}\right)^2 + \left(\frac{\partial \theta}{\partial y}\right)^2\right]$$

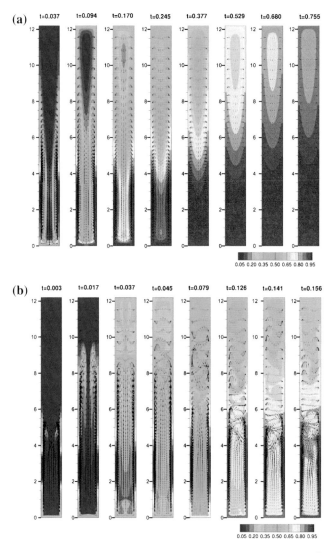

Fig. 2 Temporal evolution of the non-dimensional temperature field: **a** $Ra = 10^4$, **b** $Ra = 10^6$

$$\sigma_v^* = \sigma_v \frac{H^2 T_0^2}{k(T_1 - T_0)^2} = \frac{Ec}{(1 + \varepsilon\theta)} \left[\left(\frac{\partial u}{\partial y} + \frac{\partial v}{\partial x} \right)^2 + 2 \left(\left(\frac{\partial u}{\partial x} \right)^2 + \left(\frac{\partial u}{\partial x} \right)^2 \right) \right]$$

where σ_q represents the entropy production due to heat flux, and σ_v is the corresponding to the viscous effects inside the fluid. Ec is the Eckert number given by $Ec = \nu\kappa/[2c_p(T_1 - T_0)H^2]$ and $\varepsilon = (T_1 - T_0)/T_0$. In the present work $0 < \varepsilon \ll 1$,

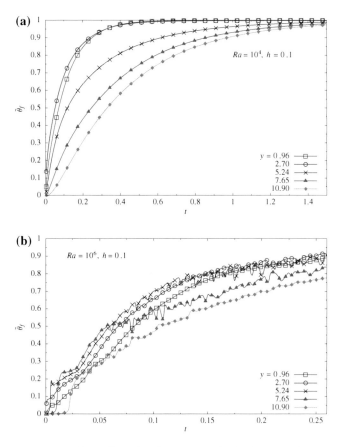

Fig. 3 Transverse averaged non-dimensional temperature in the fluid as a function of time for five different values of the vertical coordinate: **a** $Ra = 10^4$, **b** $Ra = 10^6$

the temperature differences are considered to be small in comparison with the temperature itself.

3 Numerical Method

The governing equations of the system were discretized under the control volumes scheme and solved by means of the SIMPLE algorithm Patankar (1980). Numerical codes in Fortran 90 language parallelized with the standard Open Multi Processing (OpenMP) were developed. The resulting system of algebraic equations are solved with the Tridiagonal Matrix Algorithm (TDMA), a line by line sweep and an iterative method.

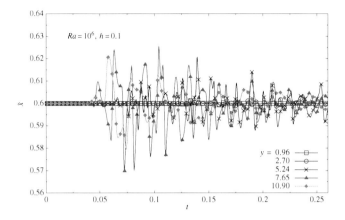

Fig. 4 Thermal centroid for $Ra = 10^6$ as a function of time. For a fixed value of y, positive values of $\tilde{x}(y, t) - 0.6$ represent anticlockwise vortex

Three different meshes, each one with 84 nodes in the horizontal direction and 184 nodes in the vertical one, were used according with the staggered grid scheme, with 10 nodes used for each solid wall. The meshes were generated by the coordinate transformation functions used by Martínez-Suástegui and Treviño (2008). Convergence of the iterative method was declared when the residual of the equations was less than 10^{-10}.

4 Results

Figure 2 shows the evolution of the temperature field for two different values of the Rayleigh number, $Ra = 10^4$ and $Ra = 10^6$. When $Ra = 10^4$, Fig. 2a, two symmetric recirculation regions develop in front of the heat sources whose height increases as the temperature increases on the walls. The shape of the isotherms indicates that the process is dominated by heat diffusion and is symmetric in all the studied time interval. In the case of $Ra = 10^6$, Fig. 2b shows that again two recirculation regions develop in front of the heated portions of the wall. However, convection is stronger than in the previous case, and the conjugated effects of convection and conduction on the walls promote the appearance of zones where thermal energy concentrates, giving place to new recirculation zones which are absent in the cases of smaller Rayleigh number. This leads the system to a unstable vortex configuration and to the symmetry break in the end.

In Fig. 3, the transverse averaged non-dimensional temperature in the fluid is shown for five different values of the vertical coordinate, $\tilde{\theta}_f(y, t) = \int_0^1 \theta f(x, y, t) \, dx$. It can be observed that the thermal signal travels faster to the top of the cavity when the Rayleigh number increases and the overall temperature differences in the fluid decreases.

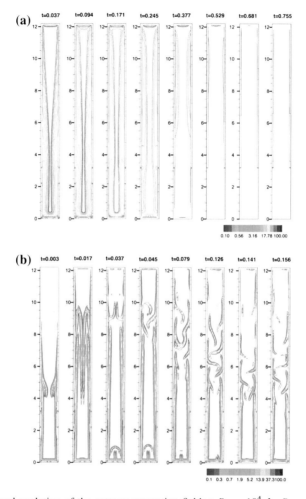

Fig. 5 Temporal evolution of the entropy generation field: **a** $Ra = 10^4$, **b** $Ra = 10^6$

To analyze the vortex dynamics, the first moment of the transverse temperature distribution (thermal centroid) is calculated, $\tilde{x}(y, t) = \int_0^1 x\, \theta_f\, dx / \tilde{\theta}_f(y, t)$, and is shown in the Fig. 4 for $Ra = 10^6$ and five different values of the vertical coordinate. The system loses symmetry in $y \approx 11$ at non-dimensional time $t \approx 0.03$ showing an intense vortex shedding activity with different frequencies.

The temporal evolution of the entropy generation field can be observed in Fig. 5. For the case of $Ra = 10^4$, it can be observed that the larger contributions to the entropy generation are localized in the solid-fluid interface and that the entropy generation is less intense inside the fluid, being practically null along the vertical symmetry axis of the cavity. In the case of the larger value of the Rayleigh number, $Ra = 10^6$, the entropy generation is more intense in general. As a

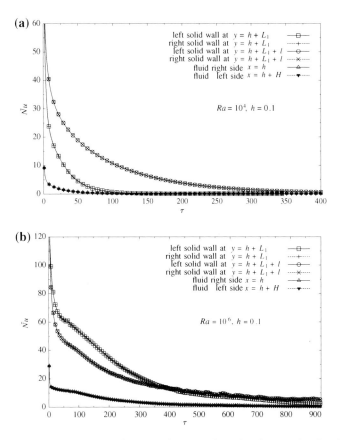

Fig. 6 Average Nusselt number as a function of the non-dimensional convective time in the solid and in the fluid around the heat sources: **a** $Ra = 10^4$, **b** $Ra = 10^6$

consequence of the more intense convective heat flux, new entropy production zones appear inside the fluid and remarkably in mid zone of the base. When symmetry breaks, the entropy production field shows clearly the zones of more intense heat flux. For both cases, the entropy production decreases as time goes by, and the temperature becomes more uniform along the cavity.

Figure 6 shows the average Nusselt number in both, the solid and the fluid around each one of the heat sources as a function of the non-dimensional convective time, $\tau = \sqrt{Ra\,Pr}\,t$

$$\bar{Nu}_f(t) = (-1)^n \int_{h+L_1}^{h+L_1+l} \left.\frac{\partial \theta_f}{\partial x}\right|_{x=x_n} dy \qquad \text{in the fluid,} \qquad (6)$$

where $n = 1, 2$ and makes reference to the heat sources located in the left and right side wall of the cavity respectively.

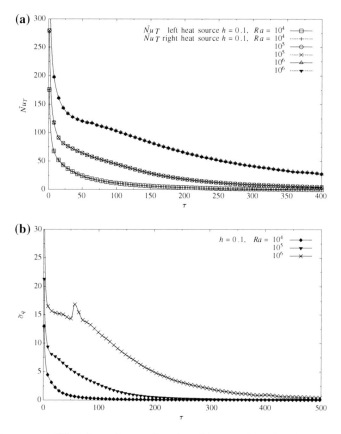

Fig. 7 a Average total Nusselt number as a function of the non-dimensional convective time for $Ra = 10^4$, 10^5 and 10^6. **b** Total entropy production as a function of the non-dimensional convective time for $Ra = 10^4$, 10^5 and 10^6

$$\bar{Nu}_s(t) = K \frac{(-1)^m}{h} \int_0^h \frac{\partial \theta_s}{\partial y}\bigg|_{y=y_m} dx; \quad \bar{Nu}_s(t) = K \frac{(-1)^m}{h} \int_{h+H}^{2h+H} \frac{\partial \theta_s}{\partial y}\bigg|_{y=y_m} dx \quad (7)$$

in the left and right side of the solid respectively. In both expressions $m = 1, 2$ and indicates the bottom or the top of the heat source respectively.

In both cases the heat flux to the solid is larger than the heat flux to the fluid, this is so because of the large value of the thermal conductivities ratio K. When the buoyancy parameter increases, and after an initial period of non-dimensional time, the heat flux that travels across the solid wall in the upward direction decreases in relation to the one that travels across the solid wall in the downward direction. This effect is due to the increment in the speed of convective heat transfer that occurs for high values of the Rayleigh number. When the speed of convective heat transfer in the fluid is comparable to the speed of conductive heat transfer on the solid, the temperature difference in the solid-fluid interface becomes small, thermal energy accumulates in the wall above the heat source and the temperature in this zone tends to be uniform. Thus, the heat flux represented by the Nusselt number decreases.

In Fig. 7a the sum of the three contributions to the average Nusselt number is plotted as a function of the non-dimensional convective time. In this picture, it is clear that as the Rayleigh number increases, the overall heat flux to the system increases as well.

Finally, the Fig. 7b shows the overall entropy production as a function of the non-dimensional convective time τ, and illustrates the direct influence of the increment of the buoyancy forces on the increment of the total entropy production of the system. The curves describe a similar behavior to that followed by the curves of the Nusselt number due to the direct relation that exists between these two quantities. It can be observed as well that when the buoyancy forces are large, the total entropy production increases, however the average temperature in the cavity increases faster in this case. This suggests that a good equilibrium between time of heating and entropy production can be achieved for a practical purpose with little more investigation.

5 Conclusions

Conjugated conduction-natural convection in a rectangular cavity with large aspect ratio and solid conductive walls of finite thickness, was numerically studied with the use of the control volume discretization method and the SIMPLE algorithm. In each case, the isotherms, the entropy production field, the heat flux to the system given by the average Nusselt number and a quantitative information of the vortex dynamics were obtained. Different mechanisms of symmetry break and heat transfer inside the fluid were found depending on the Rayleigh number. In relation to the heat flux and the entropy generation, it was possible to show that an increment in the Rayleigh number provokes an increment in the heat flux over the solid walls and inside the fluid, making a faster process of heat transfer that produces more entropy. Comparing the present results with a previous analysis of the same system but with adiabatic walls, it can be observed that the presence of the solid conductive walls of finite thickness rises the speed of heating and tends to stabilize the process.

References

Kim D, Viskanta R (1985) Effect of wall heat conduction on natural convection heat transfer in a square enclosure. J Heat Transf 107(1):139–146

Liaqat A, Baytas A (2001) Conjugate natural convection in a square enclosure containing volumetric sources. Int J Heat Mass Transf 44:3273–3280

Martínez-Suástegui L, Treviño C (2008) Transient laminar opposing mixed convection in a differentially and asymmetrically heated vertical channel of finite length. Heat Mass Transf 51:5991–6005

Mobedi M (2008) Conjugate natural convection in a square cavity with finite thickness horizontal walls. Int Commun Heat Mass Transf 35(4):503–513

Patankar S (1980) Numerical heat transfer and fluid flow. Hemisphere Publishing Corporation,Washington

Varol Y, Oztop H, Koca A (2008) Entropy generation due to conjugate natural convection in enclosures bounded by vertical solid walls with different thicknesses. Int Commun Heat Mass Transf 35(5):648–656

Varol Y, Oztop H, Pop I (2009) Entropy analysis due to conjugate-buoyant flow in a right-angle trapezoidal enclosure filled with a porous medium bounded by a solid vertical wall. Int J Therm Sci 48(6):1161–1175

Zhang W, Zhang C, Xi G (2011) Conjugate conduction-natural convection in an enclosure with time-periodic sidewall temperature and inclination. Int J Heat Fluid Flow 32(1):52–64

Spectral Analysis of Chaos Transition in a Dynamic System: Application to Backward Facing Step Flow in Mixed Convection

Héctor Barrios-Piña, Stéphane Viazzo, Claude Rey
and Hermilo Ramírez-León

Abstract This work focuses on the study of the transition from steady to chaotic behavior in mixed convection flow over a backward-facing step. Direct numerical simulations are performed in a two-dimensional horizontal channel of expansion ratio $ER = 2$ at step level. The effects of the temperature difference between the heated bottom wall and the inflow temperature are investigated by keeping constant the Richardson number at 1. The covered range of Grashof and Reynolds numbers is respectively $3.31 \times 10^4 \leq Gr \leq 2.72 \times 10^5$ and $182.03 \leq Re \leq 521.34$. The thermodynamic instabilities which cause the onset of unsteady flow are described in detail. A spectral and phase portrait analysis of the temperature time series allows us to observe that the transition from steady to chaotic flow occurs by period-doubling bifurcations.

1 Introduction

The concept "dynamic system" is related to a formal mathematical rule which characterizes the time dependence of a point in a space. The concept includes different types of such rules in mathematics. Any choice tries to measure time and the special properties of the space. This may give an idea of the vastness of the class of objects described by this concept. The study of a dynamic system

H. Barrios-Piña (✉) · H. Ramírez-León
Explotación de Campos en Aguas Profundas, Instituto Mexicano del Petróleo,
Eje Central Lázaro Cárdenas 154, Gustavo A. Madero, 07730 Mexico, D.F., Mexico
e-mail: hbarrios@imp.mx

S. Viazzo · C. Rey
Laboratoire M2P2, UMR 6181 CNRS - Universités d'Aix-Marseille,
Technopôle Château-Gombert, 38 rue F. Joliot-Curie, 13451 Marseille, France

J. Klapp et al. (eds.), *Fluid Dynamics in Physics, Engineering and Environmental Applications*, Environmental Science and Engineering,
DOI: 10.1007/978-3-642-27723-8_28, © Springer-Verlag Berlin Heidelberg 2013

particularly by means of spectral and phase portrait analysis, is one of the ways to characterize the behaviour of a wide variety of applications. In this work, we show how spectral and phase portrait analysis can help us to describe an important phenomenon in Fluid Mechanics: the passage from a steady behaviour to chaos in a given flow.

In this way, we analyse the problem of flow separation due to a sudden change in flow geometry such as backward-facing step. This kind of flow plays an important role in optimizing a wide variety of industrial applications requiring heating or cooling. This problem in laminar and turbulent regime, and in natural, forced and mixed convection has been extensively studied, both experimentally and numerically (Armaly et al. 1983; Barkley et al. 2002; Abu-Mulaweh et al. 1993, 2001, 2002; Hong et al. 1993; Iwai et al. 2000; Abu-Mulaweh 2003; Abe et al. 1994, 1995; Park and Sung 1995; Rhee and Sung 1996; Chen et al. 2006).

Although the effect caused by several parameters on flow and thermal fields over a backward-facing step has received enough attention in the literature, the study of the thermodynamical mechanisms in mixed convection flow which cause the transition to unsteady flow has not been investigated yet. Thus, this transition and the different bifurcations that the flow shows before chaos are investigated here by using a spectral and phase portrait analysis of temperature time series. For this purpose, various direct numerical simulations are carried out in a two-dimensional horizontal channel. In order to disturb the steady state and generate thermal instabilities, the inflow velocity and the temperature difference (between the heated bottom wall of the channel and the colder inflow), are gradually increased up to obtain a chaotic behavior.

2 Governing Equations

The dimensional governing conservation equations, for mass, momentum and energy, in Cartesian coordinates, are used in the following form:

$$\frac{\partial \rho}{\partial t} + \frac{\partial (\rho u_i)}{\partial x_i} = 0, \tag{1}$$

$$\frac{\partial (\rho u_i)}{\partial t} + \frac{\partial (\rho u_i u_j)}{\partial x_j} = -\frac{\partial P}{\partial x_i} + \frac{\partial \tau_{ij}}{\partial x_j} + \rho g \delta_{i2}, \tag{2}$$

$$\frac{\partial T}{\partial t} + \frac{\partial (u_i T)}{\partial x_i} = \left(1 - \frac{R}{c_v}\right) T \frac{\partial u_i}{\partial x_i} - \frac{\gamma}{\rho c_p} \frac{\partial q_i}{\partial x_i}, \tag{3}$$

where $\tau_{ij} = \mu \lfloor \partial u_i / \partial x_j + \partial u_j / \partial x_i - (2/3)\delta_{ij}\partial u_k / \partial x_k \rfloor$. Viscous dissipation in energy equation is ignored. The ideal gas law is used to determine density variations as:

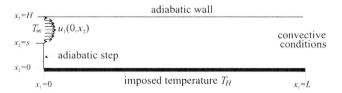

Fig. 1 Flow geometry for the two-dimensional backward-facing step configuration

$$\rho = \frac{P}{RT}, \tag{4}$$

where $R = 287$ J/(kg K) is the ideal gas constant. The thermo-physical properties of the fluid such as dynamic viscosity (μ) and thermal conductivity (κ) are assumed constant.

3 Flow Geometry and Boundary Conditions

In Fig. 1 the computational domain and the geometry of the problem is shown. The expansion ratio, the step height and the channel length are kept constant at $ER = 2$, $s = 4$ cm and $L = 30\,s$, respectively. We denote $\Delta T = T_H - T_\infty$. The Grashof number ($Gr = g\beta\Delta T s^3/v^2$) and Reynolds number ($Re = usv^{-1}$) are function of the step height and the fluid inflow properties.

At the inflow boundary ($x_1 = 0, s \leq x_2 \leq H$), a steady parabolic profile of velocity and a uniform constant temperature $T_\infty = 293$ K are imposed. At the outflow boundary ($x_1 = L, 0 \leq x_2 \leq H$), convective conditions for velocity components and homogeneous Neumann condition for temperature and pressure are imposed. The top wall and the step are assumed to be thermally adiabatic, and the bottom wall is considered at uniform temperature.

4 Numerical Model

The solution method is a fully implicit formulation (Sewall and Tafti 2008). The local density variation is coupled with both temperature and pressure variations, whereas other properties are only coupled to temperature variations. No low Mach number assumption is used in this formulation. The numerical approach is based on a second-order finite difference formulation with a fully staggered arrangement. The coupled set of equations is solved using an iterative predictor–corrector procedure at each time step. The coupling between the mass and the momentum conservation equations is completed through a pressure correction, determined from the following elliptic Helmholtz equation:

$$\frac{\partial^2 \varphi}{\partial x_i \partial x_i} - \frac{\varphi}{\delta t^2 r T^{m+1}} = \frac{\rho^* - \rho^n}{\delta t^2} + \frac{1}{\delta t} \frac{\partial (\rho^* u_i^*)}{\partial x_i}, \tag{5}$$

where the superscripts n and m denote respectively time (or outer) and inner iteration. The symbol (*) means an intermediate or provisional value. The solution of the elliptic Helmholtz equation, Eq. (5), is obtained using a V-cycle Multigrid method with the Strong Implicit Procedure ILU as smoother. The numerical approach is further documented by Barrios (2010).

5 Results and Discussion

Several numerical simulations are carried out for air (Pr $= 0.71$) with a constant Richarson number at Ri $=$ Gr/Re$^2 = 1$ and by imposing the temperature difference ΔT. The Richardson number is kept as constant to maintain in equilibrium natural convection and forced convection. The mean inflow velocity u is obtained from the definition of the Richardson number. In order to investigate the different flow regimes, the initial value of the temperature difference $\Delta T = 3.5$ K, is gradually increased up to $\Delta T = 30$ K. Thus, we cover a range of Grashof and Reynolds numbers of $3.31 \times 10^4 \leq$ Gr $\leq 2.72 \times 10^5$ and $182.03 \leq$ Re ≤ 521.34, respectively. In all cases, numerical simulations use 420×67 grid points, except for the case at $\Delta T = 30$ K, where 510×131 grid points are used.

The velocity fields show a standard pattern which is characterized by the primary recirculation zone downstream of the step and by a secondary recirculation appearing near the corner of the step. This secondary recirculation is located inside the primary recirculation zone with an opposite swirl. In addition, heat is convected upstream by the primary recirculation, and then is accumulated near the step by the secondary recirculating-flow. This physical mechanism of heat accumulation is the cause of the thermodynamical instabilities of the dynamic system. We observe that the flow is steady in the primary recirculation zone for temperature differences up to $\Delta T = 4.35$ K, which corresponds to Gr $= 41,122.18$ and Re $= 202.79$. For higher values of the temperature difference, the flow becomes unsteady. The onset of the instabilities is characterized by thermoconvective vortices which start to appear in the reattachment zone.

In order to quantify the different flow regimes, spectral and phase portrait analyses are performed. Time-histories of temperature at the point $(x_1/s = 8.625, x_2/s = 1)$, which corresponds to the reattachment region, are analyzed.

Among the different possible routes to chaos in a two-dimensional configuration, we observe a route to chaos by period-doubling, associated with sub-harmonic cascades. When the onset of unsteady-state flow occurs, the system exhibits a periodic behavior. In a periodic behavior, only one harmonic peak occurs, which is associated with the fundamental frequency. As the system progresses toward to chaos, more peaks arise, associated with the harmonics, sub-harmonics and their

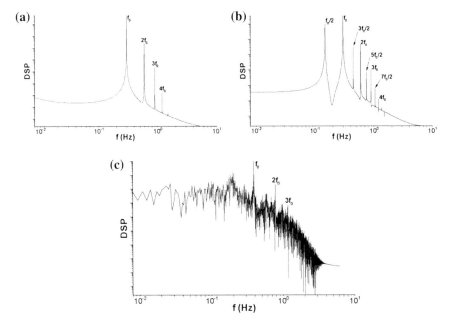

Fig. 2 Power spectrums of **a** the case at $\Delta T = 4.4$ K, Gr = 41,591.33, Re = 203.94. Periodic solution; **b** the case at $\Delta T = 4.45$ K, Gr = 42,060.39, Re = 205.09. Period-2 solution; **c** the case at $\Delta T = 7$ K, Gr = 65,877.90, Re = 256.67. Chaotic behavior

corresponding odd-harmonics of the system. In this particular case, the system exhibits an increase of the fundamental frequency as the temperature difference increases.

Figure 2a shows the power spectrum of the system when exhibiting periodic behavior. The fundamental frequency, $f_0 = 0.2953$ Hz, appears at $\Delta T = 4.4$ K along with four harmonic frequencies, $2f_0$, $3f_0$, $4f_0$, $5f_0$. When the temperature difference is slightly increased at $\Delta T = 4.45$ K, the flow exhibits one bifurcation. The system splits into two frequencies; one which corresponds to the fundamental frequency, and the other which is half the fundamental frequency (see Fig. 2b). The fundamental frequency remains and slightly moves at $f_0 = 0.2969$ Hz. The second frequency is referred to as the sub-harmonic frequency $f_0/2$. The power spectrum in Fig. 2b also exhibits the harmonic frequencies as seen in periodic solution, but now there are odd-harmonic frequencies as well. The odd-harmonic frequencies are found by $3f_0/2$, $5f_0/2$, $7f_0/2$, …

When the temperature difference is increased at $\Delta T = 5.45$ K, the flow exhibits another bifurcation into period quadrupling. In period quadrupling, the system has three stable frequencies. The fundamental frequency of the system moves at $f_0 = 0.3359$ Hz, but now two sub-harmonic frequencies arise at $f_0/2$ and $f_0/4$. Period quadrupling is the final bifurcation observed prior to chaos. The amplitude rang of the period-8 solution is not visible. When $\Delta T = 6$ K, the system starts to exhibit chaotic behavior. In this case, the fundamental frequency remains and

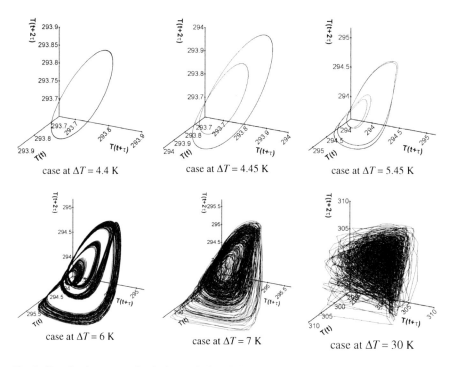

Fig. 3 Pseudo-phase portraits during period-doubling route to chaos

moves at $f_0 = 0.3544$ Hz, and the two sub-harmonic frequencies $f_0/2$ and $f_0/4$ are still visible along whit its odd-harmonics. In addition, many small peaks appear which are associated to the broadening of the frequency spectrum around certain frequencies. It is difficult to distinguish the four distinct peaks for the four frequencies of the system through the time series. This is due to the large number of peaks of different amplitudes.

As the system is more and more chaotic, the power spectrum shows an inverse period-doubling. This process is characterized by the disappearance of the sub-harmonic frequencies before the power spectrum becomes broadband. Figure 2c shows the case when the temperature difference increases at $\Delta T = 7$ K. We observe that the fundamental frequency, $f_0 = 0.3875$ Hz, still remains, and the sub-harmonic frequencies $f_0/2$ and $f_0/4$ are difficult to observe. The power spectrum also reveals the harmonic frequencies $2f_0$, $3f_0$, ... and more broadening around them. Larger temperature differences exhibit a fully developed chaotic behavior.

This period-doubling route to chaos is also visible by a phase portrait analysis. The phase portraits were obtained by using the method of *time delays*. Figure 3 shows the three dimensional pseudo-phase portraits which correspond to the cases when detecting the onset of bifurcations and the onset of chaotic behavior. The location of each point in the 3D pseudo phase space is defined by three data points in the time series of temperature which are separated by a time delay: $T(t)$, $T(t + \tau)$ and $T(t + 2\tau)$, with an arbitrary $\tau = 0.4$ s.

We observe in Fig. 3 the case at $\Delta T = 4.4$ K which exhibits a periodic behavior, the pseudo-phase portrait represents a closed-loop trajectory with a single orbit, as expected for a periodic signal. In the case where the flow is of period-2, at $\Delta T = 4.45$ K, the pseudo-phase portrait forms a closed-loop trajectory composed of two orbits corresponding to the period-doubling. When the bifurcation into period quadrupling occurs at $\Delta T = 5.5$ K, as expected, the pseudo-phase portrait consists of a closed-loop trajectory with four well-defined orbits. When chaotic behavior stars to occur, at $\Delta T = 6$ K approximately, the pseudo-phase portrait reveals the presence of discrete bundles of orbits. This suggests that there is a strange attractor. This behavior still persists at $\Delta T = 7$ K, where the system is more chaotic. When the temperature difference increases at $\Delta T = 30$ K, the pseudo-phase portrait shows strong irregularity and a not defined geometry. Further details and results of this analysis can be found in Barrios (2010).

6 Conclusions

Numerical simulations was performed in mixed convection flow over a backward-facing step in a two-dimensional horizontal channel of expansion ratio $ER = 2$. The Richarson number was kept constant at $Ri = 1$ and the temperature difference was imposed. In order to investigate different flow regimes, the initial value of the temperature difference $\Delta T = 3.5$ K, was gradually increased up to $\Delta T = 30$ K. We have observed that the flow is steady in the primary recirculation zone for temperature differences up to $\Delta T = 4.35$ K, which corresponds to $Gr = 41{,}122.18$ and $Re = 202.79$. For higher values of the temperature difference, the flow becomes periodic prior to chaos. The onset of the instabilities is characterized by thermoconvective vortices which start to appear in the reattachment zone. A spectral and phase portrait analysis allowed us to observe that the transition to chaotic behavior occurs by a period-doubling cascade. The onset of chaotic behavior is observed for temperature differences higher than $\Delta T = 6$ K, which corresponds to $Gr = 56{,}562.16$ and $Re = 237.83$.

References

Abe K, Kondoh T, Nagano Y (1994) A new turbulence model for predicting fluid flow and heat transfer in separating and reattaching flows-I. Flow field calculations. Int J Heat Mass Transf 37(1):139–151

Abe K, Kondoh T, Nagano Y (1995) A new turbulence model for predicting fluid flow and heat transfer in separating and reattaching flows-II. Thermal field calculations. Int J Heat Mass Transf 38(8):1467–1481

Abu-Mulaweh HI, Armaly BF, Chen TS (1993) Measurements of laminar mixed convection in boundary-layer flow over horizontal and inclined backward-facing steps. Int J Heat Mass Transf 36(7):1883–1895

Abu-Mulaweh HI, Armaly BF, Chen TS (2001) Turbulent mixed convection flow over a backward-facing step. Int J Heat Mass Transf 44:2661–2669

Abu-Mulawch HI, Chen TS, Armaly BF (2002) Turbulent mixed convection flow over a backward-facing step-the effect of the step heights. Int J Head Fluid Flow 23:758–765

Abu-Mulaweh HI (2003) A review of research on laminar mixed convection flow over backward- and forward-facing steps. Int J Therm Sci 42:897–909

Armaly BF, Durst F, Pereira JCF, Schonung B (1983) Experimental and theoretical investigation of backward-facing step flow. J Fluid Mech 127:473–496

Barkley D, Gomes MGM, Henderson R (2002) Three-dimensional instability in flow over a bacward-facing step. J Fluid Mech 473:167–190

Barrios H (2010) Développement d'un code de calcul non Boussinesq dédié aux écoulements de gaz chauffé en convection naturelle. PhD Thesis, Universités d'Aix-Marseille, France

Chen YT, Nie JH, Armaly BF, Hsieh HT (2006) Turbulent separated convection flow adjacent to backward-facing step-effects of the step height. Int J Heat Mass Transf 49:3670–3680

Hong B, Armaly BF, Chen TS (1993) Laminar mixed convection in a duct with a backward-facing step: the effects of inclination angle and Prandtl number. Int J Heat Mass Transf 36(12):3059–3067

Iwai H, Nakabe K, Suzuki K, Matsubara K (2000) The effects of duct inclination angle on laminar mixed convective flows over a backward-facing step. Int J Heat Mass Transf 43:473–485

Park TS, Sung HJ (1995) A nonlinear Low-Reynolds-number κ–ε model for turbulent separated and reattaching flows-I. Flow field calculations. Int J Heat Mass Transf 38(14):2657–2666

Rhee GH, Sung HJ (1996) A nonlinear Low-Reynolds-number κ–ε model for turbulent separated and reattaching flows-II. Thermal field calculations. Int J Heat Mass Transf 39(16):3465–3474

Sewall EA, Tafti DK (2008) A time-accurate variable property algorithm for calculating flows with large temperature variations. Comput Fluids 37:51–63

Turbulence Model Validation in Vegetated Flows

R. González-López, H. Ramírez-León, H. Barrios-Piña
and C. Rodríguez-Cuevas

Abstract This work presents the validation of a two-layer mixing-length model for turbulence applied to shallow water flows in presence of vegetation. In order to determine the vertical velocity variations when submerged vegetation exists, a multilayer numerical model-based approach is used. A second-order finite difference method is used for spatial discretization and a semi-implicit lagrangian–eulerian method is used for time discretization. The comparisons to experimental data for validating the turbulence model show good agreement for two cases, a channel with submerged vegetation and other one with both submerged and emergent vegetation.

1 Introduction

Natural river floodplains and adjacent wetlands have vital ecological functions in riverine landscapes. Their vegetation typically comprises a diverse and hetero-geneous combination of herbs, shrubs and trees, which influence sediment, nutrient, and pollutant transport. There are many ways to determine the hydro-dynamics and the shear stress due to the presence of vegetation. Freeman et al. (2000) investigated experimentally the effect of vegetation on flow resistance. Järvelä (2002) analyzed experimentally flow resistance of natural floodplain

R. González-López · H. Ramírez-León · H. Barrios-Piña (✉)
Explotación de Campos en Aguas Profundas, Instituto Mexicano del Petróleo, Eje Central
Lázaro Cárdenas 154, Gustavo A. Madero, C.P. 07730 Mexico, D.F., Mexico
e-mail: hbarrios@imp.mx

C. Rodríguez-Cuevas
Universidad Autónoma de San Luis Potosí, Facultad de Ingeniería, Dr. Manuel Nava # 8,
Zona Universitaria Poniente, C.P. 78290 San Luis Potosí, S. L. P., Mexico

J. Klapp et al. (eds.), *Fluid Dynamics in Physics, Engineering and Environmental Applications*, Environmental Science and Engineering,
DOI: 10.1007/978-3-642-27723-8_29, © Springer-Verlag Berlin Heidelberg 2013

plants, showing that friction factors and vegetal drag coefficients present large variations with the Re number, depth of flow, flow velocity and vegetal characteristics. Logarithmic velocity distributions have been useful for estimating velocity profiles above vegetation layers (Huthoff et al. 2006). Other authors perform hydrologic balances in order to determine flow characteristics (Lal 1998). Some models are based on 2D non-linear diffusion. Other models solve numerically the hydrodynamic equations including vegetation shear stress, in one (Rowinski and Kubrak 2002), two (Arega and Sanders 2004) and three dimensions (Gao et al. 2011).

The present work focuses on the development of a 3D numerical model for vegetated flows using a more sophisticated two-layer mixing-length model for turbulence than the model used by Gao et al. (2011). For this purpose, the two-layer mixing-length model proposed by Stansby (2003) is then employed. Effects due to the variation of vegetation heights on the flow are characterized by a multilayer approach. The present numerical model considers the hydrostatic approximation and solves the shallow-water equations. The formulation of Freeman et al. (2000) was adopted to include the vegetation shear stress into the velocity equations. Two cases of study are treated in order to validate the developed numerical model and to show its capacities, a channel with submerged vegetation and other one with both submerged and emergent vegetation. The results obtained from the present numerical model are compared against experimental measurements finding good agreement.

2 Governing Equations

The system of equations used to describe the velocity fields are the shallow-water equations. A correction term is considered in the velocity equations in order to characterize the shear stress due to the presence of vegetation. After turbulent averaging the governing equations for the velocity components have the following form:

$$
\begin{aligned}
\frac{\partial U}{\partial t} + U\frac{\partial U}{\partial x} + V\frac{\partial U}{\partial y} + W\frac{\partial U}{\partial z} = & -\frac{\rho g}{\rho_0}\frac{\partial \eta}{\partial x} + \frac{\partial}{\partial x}\left(2\nu_E\frac{\partial U}{\partial x}\right) \\
& + \frac{\partial}{\partial y}\left[\nu_E\left(\frac{\partial U}{\partial y} + \frac{\partial V}{\partial x}\right)\right] + \frac{\partial}{\partial z}\left(\nu_E\frac{\partial U}{\partial z}\right) - \frac{1}{\rho_0}\frac{\tau_x^v}{\Delta z},
\end{aligned}
\tag{1}
$$

$$
\begin{aligned}
\frac{\partial V}{\partial t} + U\frac{\partial V}{\partial x} + V\frac{\partial V}{\partial y} + W\frac{\partial V}{\partial z} = & -\frac{\rho g}{\rho_0}\frac{\partial \eta}{\partial y} + \frac{\partial}{\partial y}\left(2\nu_E\frac{\partial V}{\partial y}\right) \\
& + \frac{\partial}{\partial x}\left[\nu_E\left(\frac{\partial U}{\partial y} + \frac{\partial V}{\partial x}\right)\right] + \frac{\partial}{\partial z}\left(\nu_E\frac{\partial V}{\partial z}\right) - \frac{1}{\rho_0}\frac{\tau_y^v}{\Delta z},
\end{aligned}
\tag{2}
$$

$$\frac{\partial W}{\partial z} = -\left(\frac{\partial U}{\partial x} + \frac{\partial V}{\partial y}\right), \tag{3}$$

where U, V and W are the mean velocity components in x, y and z directions, respectively, g is the acceleration due to gravity, ρ is the water density, ρ_0 is a water reference density, η is the free surface elevation, v_E is the kinematic eddy viscosity which consists of turbulent and molecular components, such that $v_E = v_t + v$, Δz is the vegetation layer height, and τ_x^v and τ_y^v are the vegetation shear stresses in x and y directions, respectively. By integrating the continuity Eq. (3), over the water depth and by using a kinematic condition at the free surface, the equation to calculate the free surface elevation is written as:

$$\frac{\partial \eta}{\partial t} = -\frac{\partial}{\partial x}\left(\int_{-h}^{\eta} U dz\right) - \frac{\partial}{\partial y}\left(\int_{-h}^{\eta} V dz\right), \tag{4}$$

where $h(x,y)$ is the water depth.

In Eqs. (1) and (2), the last terms characterize the shear stress due to vegetation, which cause a decrease of the velocity field. These terms vanish in regions without vegetation. In this work, the International System of Units is used.

3 The Turbulence Model

The simple two-layer mixing-length model for turbulence of Stansby (2003) has been implemented in the present numerical model. The characteristic of this model is that the horizontal mixing length is explicitly made a multiple of the vertical mixing length within a general three-dimensional eddy-viscosity formulation. This means that the horizontal mixing length and associated strain rates determine the magnitude of eddy-viscosity which determines vertical mixing. The turbulent viscosity coefficient v_t is then computed through the following mixing-length model:

$$v_t = \left\{ l_h^4 \left[2\left(\frac{\partial U}{\partial x}\right)^2 + 2\left(\frac{\partial V}{\partial y}\right)^2 + \left(\frac{\partial V}{\partial x} + \frac{\partial U}{\partial y}\right)^2 \right] + l_v^4 \left[\left(\frac{\partial U}{\partial z}\right)^2 + \left(\frac{\partial V}{\partial z}\right)^2 \right] \right\}^{1/2} \tag{5}$$

where the vertical length scale $l_v = \kappa(z - z_b)$ for $(z - z_b)/\delta < \lambda/\kappa$ and $l_v = \lambda\delta$ for $\lambda/\kappa < (z - z_b)/\delta < 1$, κ is the von Kármán constant, typically 0.41, $(z - z_b)$ is the distance from the wall, δ is the boundary-layer thickness and λ is a constant, typically 0.09. In the case of shallow-water flows, due to a steady current, the boundary-layer thickness may be assumed to be equal to the water depth h. The horizontal length scale is usually different than the vertical length scale, and the simplest assumption is to assume direct proportionality defined by $l_h = \beta l_v$. The constant β has to be determined experimentally. For parallel flow cases (or near parallel), eddy viscosity reverts to its standard boundary-layer form.

With $l_h = l_v$, it reverts to its correct mathematical three-dimensional form (with negligible vertical velocity).

4 Numerical Model

In order to have good accuracy in the velocity calculation and to consider different vegetation heights, a multilayer model-based approach is used. The model is based on a second-order finite difference formulation. A spatial mesh which consists of rectangular cells with horizontal sizes Δx and Δy, and height Δz, is used. Scalars are located at the middle of cells and velocity components are shifted at the middle of cell sides. The time accuracy is also second-order and the solution method is an adaptation of the semi-implicit eulerian–lagrangian scheme proposed by Casulli and Cheng (1992). This method treats the advection and diffusion terms differently. The solution of the advection terms is given by a lagrangian formulation through the characteristic method and the solution of the diffusion terms is obtained by eulerian formulation through the Adams–Bashforth scheme.

5 Channel with Submerged Vegetation

The validation has been carried out by comparison to the experimental data of Järvelä (2005), who analyzed the behavior of velocity profiles above a vegetated layer for different experiments, three of which were reproduced numerically for comparisons. The numerical simulations were conduced in a channel of 36 long, and 1.1 m wide. The plants covered a 6 m long section at the centre of the channel. The vegetation density was an average of 12,000 stems/m^2. Fifteen meters long sections before and after the vegetation area were covered with 0.1 m thick layer of crushed rock (diameter 16–32 mm), except for the last 2.5 m before the vegetation area. This section was covered with smoother crushed rock (diameter 3–5 mm). Figure 1 shows a diagram of the experimental setup and Table 1 summarizes the initial conditions of the three selected cases, where, Q is the discharge, U_0 is the inflow velocity, $\text{Re} = U_0 h/\nu$, $\text{Fr} = U_0 (gh)^{-1/2}$, S_e is the energy slope, $h_{p,m}$ is the vegetation height and u_2^* is the shear velocity given by $u_2^* = \sqrt{g(h - h_{p,m}) S_e}$.

The velocity profiles above the vegetated layer were compared at three different longitudinal locations for all the cases, at $x = 3.5$, 3.65 and 3.8; $y = 0.4$. For the three numerical simulations, a rectangular elements grid of 144×22 points was used with constant mesh sizes (Δx and Δy). The time step was 0.001s. In order to obtain a good accuracy in the computation of vertical velocity variations, different number of vertical layers with different constant heights were used for each case; 12 layers for the case R4-7, and 14 layers for the cases R4-8, and R4-9. The initial velocity components and the free surface elevation were zero. At the inflow

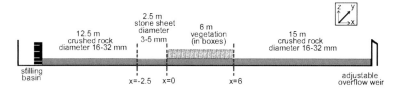

Fig. 1 Experimental setup and definition of the coordinate axes (not in scale) (taken from Järvelä (2005))

Table 1 Experimental setup (taken from Järvelä (2005))

Case	Q (l/s)	h (m)	U_0(m/s)	Re	Fr	Se	u_2^*(m/s)	hp, m(m)	f
R4-7	100	0.4950	0.1836	60,606	0.083	0.0006	0.0416	0.220	0.68
R4-8	100	0.7065	0.1287	60,606	0.049	0.0002	0.027	0.260	0.50
R4-9	143	0.7037	0.1847	86,667	0.070	0.0003	0.0384	0.215	0.44

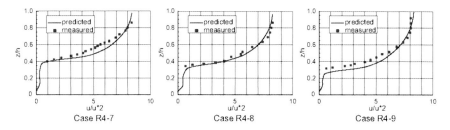

Fig. 2 Comparison of the experimental measurements and the predicted profiles

section, a parabolic U velocity profile was imposed, where the mean velocity is the inflow velocity U_0. In order to consider the thick layer of crushed rock after and before the vegetation zone, the bottom roughness (k_s) was imposed from the corresponding material diameter.

Figure 2 compares the averaged velocity profiles of Järvelä (2005) and the predicted profiles obtained from the present numerical model. In general, the predictions have good agreement with the experimental measurements, yielding the following correlation coefficients: for the case R4-7, R = 0.9827, for the case R4-8, R = 0.9935, and for the case R4-9, R = 0.9947. Järvelä reported that disturbance of the acoustic pulse caused by vegetation elements entering the ADV-sampling volume limited measurements close to vegetation in some test runs. This source of error could be a cause of the differences between predictions and measurements.

6 Channel with Submerged and Emergent Vegetation

The validation for combined submerged and emergent vegetation was carried out by comparison to the experimental data of Liu (2008). In Liu's work, the velocity profiles within a layer of rigid vegetation, represented by dowels, were analyzed

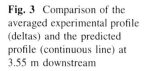

Fig. 3 Comparison of the averaged experimental profile (deltas) and the predicted profile (continuous line) at 3.55 m downstream

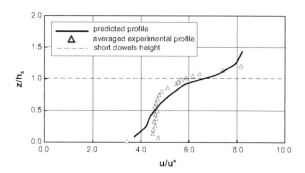

for seven experiments. Four of them were configured with both emergent and submerged vegetation in different spatial configurations. Liu's experiment was conducted in a channel of 4.3 long and 0.3 wide, with 0.1046 m depth, keeping a constant slope of 0.003. The 3 long vegetation test section covered the entire width and was located 1.3 m from the entrance of the flume. The vegetation was simulated by 6.35 diameter (d) acrylic dowels at lengths of 76 for short dowels and 152 mm for large dowels. The discharge was 0.0114 m^3/s. The spacing was determined by non-dimensional parameters, $S_s/d = 8$ and $S_t/d = 16$ for short and tall dowels, respectively in a linear pattern. Instantaneous longitudinal velocities were taken 2.25 m downstream of the beginning of the test section, with a Laser Doppler Velocimeter (LDV). Measurement locations were chosen 4 d downstream and upstream of tall and short dowels, and a single measurement location was selected in the free stream region.

For the numerical simulation, a rectangular elements grid of 86 × 32 points with constant mesh sizes (Δx and Δy) and 9 vertical layers of constant heights were used. The short dowels covered 6 vertical layers in height while the tall dowels covered all the layers. Since the diameter and roughness of short and tall dowels are the same, only a variable density was imposed in the vertical direction to characterize the difference of height between them. The initial velocity components and free surface elevation were zero. The time step was 10^{-5}s. A linear constant velocity profile was imposed at the inflow section.

Figure 3 shows a comparison between the averaged-profile constructed with the measurements and the predicted U-velocity profile. Velocity values were adimensionalized by the shear velocity $u^* = \sqrt{ghS_e}$.

The predictions fit well with the experimental measurements, yielding a correlation coefficient of R = 0.9286. The differences between the measurements and the predictions can be due to local low-scale high-frequency phenomena. In the present numerical model an equivalent blocking area is assumed instead of considering individual obstacles. Consequently, local high-frequency phenomena are out of scale in this numerical simulation.

7 Conclusion

This work presents the validation of a numerical model based on the Navier–Stokes-Reynolds equations which uses a two-layer mixing-length model to characterize turbulence for flows in presence of vegetation.

The present results were validated against the experimental measurements of Järvelä (2005) and Liu (2008). The comparisons between predicted velocity profiles and experimental measurements show a good agreement above and within a vegetated layer. The multilayer approach allows the calculation of the vertical velocity profile from the bottom to the free surface at any location and at any time, which is not possible by using log laws, based analytical models and/or 2D numerical models.

The findings show that the two-layer mixing-length model used here is capable to reproduce the flow behavior in presence of vegetation with good accuracy. The main advantages of using the present turbulence model against others higher order models are that it requires less calibration coefficients and demands lesser CPU-time.

References

Arega F, Sanders BF (2004) Dispersion model for tidal wetlands. J Hydraul Eng 130(8):739–754

Casulli V, Cheng RT (1992) Semi-implicit finite difference methods for three dimentional shallow water flow. Int J Numer Meth Fluids 15:629–648

Freeman GE, Rahmeyer JW, Copeland RR (2000) Determination of resistance due to shrubs and woody vegetation. Final report. Prepared for U.S. army corps of engineers. coastal and hydraulics laboratory. ERDC/CHL TR-00-25

Gao G, Falconer RA, Lin B (2011) Modelling open channel flows with vegetation using a three-dimensional model. J Water Res Prot 3:114–119

Huthoff F, Augustijn DCM, Hulscher SJMH (2006) Depth-averaged flow in presence of submerged cylindrical elements. In: River flow 2006. Taylor & Francis, London, pp 575–582

Järvelä J (2002) Determination of flow resistance of vegetated channel banks and floodplains. In: Bousmar D, Zech Y (eds.) River flow 2002, Swets & Zeitlinger, Lisse, ISBN-90-5809-509-6

Järvelä J (2005) Effect of submerged flexible vegetation on flow structure and resistance. J Hydrol 307:233–241

Lal WAM (1998) Performance comparison of overland flow algorithms. J Hydraul Div ASCE 124(4):342–349

Liu D (2008) Flow through rigid vegetation hydrodynamics. Thesis submitted to the faculty of the Virginia Polytechnic Institute and State University in partial fulfillment of the requirements for the degree of Master of Science in Civil Engineering. Sept, 82 pp

Rowinski PM, Kubrak J (2002) A mixing-length model for predicting vertical velocity distribution in flows through emergent vegetation. Hydrol Sci J 47(6):893–904

Stansby PK (2003) A mixing-length model for shallow turbulent wakes. J Fluid Mech 495:369–384

CFD on Graphic Cards

C. Málaga, J. Becerra, C. Echeverría and F. Mandujano

Abstract In the last decade, the entertainment and graphic design industries demand for higher resolution and more realistic graphics has motivated graphic card manufacturers to develop high performance graphic processing units (GPUs) at low cost. Nowadays GPUs are highly efficient parallel processing units that can be used for general purposes. Such developments open a new alternative within scientific computation, promising high performance parallel computation at everyones reach. Graphic cards seem to be an attractive option, specially for research groups and institutions with limited computational resources and for teaching purposes. Here, an overview of parallel computation on GPUs is presented, as well as the efforts to manipulate and promote this technology within the UNAM, particularly in Fluid Mechanics applications.

1 Introduction

In 2002, commodity graphic cards started to outperform central processing units (CPUs), at least in the theoretical peak performance in single precision computations. By the year 2009, GPUs that could be bought out the shelf had a theoretical peak

C. Málaga (✉) · J. Becerra · F. Mandujano
Facultad de Ciencias, UNAM, Mexico, D.F., Mexico
e-mail: cmi@fciencias.unam.mx

J. Becerra
e-mail: juliansagredo@gmail.com

F. Mandujano
e-mail: frmas@ciencias.unam.mx

C. Echeverría
e-mail: echeverriacarlos@gmail.com

J. Klapp et al. (eds.), *Fluid Dynamics in Physics, Engineering and Environmental Applications*, Environmental Science and Engineering,
DOI: 10.1007/978-3-642-27723-8_30, © Springer-Verlag Berlin Heidelberg 2013

performance of more than $1,300$ single precision GFLOPs (10^9 floating point operations per second), almost ten times more than their multi-core CPU counterpart. For fast double precision computations, more sophisticated and expensive GPUs were needed, and their performance was around five times that of the top CPUs of that time. Parallel computations showed that GPUs could provide the same performance of CPU clusters at a hundredth of its monetary cost in single precision, while in double precision computations, same performance would cost ten times less on specialized GPUs.

As GPUs grew faster and cheaper, the interest to harvest their power for applications others than graphical display originated, around 2006, what is known as GPGPU (General Purpose GPU computation, http://gpgpu.org). GPU manufacturers Nvidia™ and ATI™ provided computer languages that allowed the manipulation of their graphic cards. These languages resemble, and can be considered an extension to the C and C++ languages, there is even a unified language, called OpenCL, also similar to C that allows programing any card and multi-core CPUs to work cooperatively. Fortran and Python versions are also available. Scientific software like Mathematica™, Matlab™ and LabView™ now include the possibility to perform some of their numerical processing on the GPUs. The use of this relatively new technology is spreading and some of the top supercomputers of the last years make use of them (http://www.top500.org).

2 GPU Overview

GPUs are typically composed of hundreds of small processing units called stream processors (SPs). This tiny processors are less powerful than a core on a multi-core CPU but have a high bandwidth memory fetching and dedicate more transistors to parallel computation than CPU cores, normally involved in other duties.

The parallel computation model under which GPUs work is called Single Program Multiple Data (SPMD). This means that one provides a subroutine, called a kernel, that is a single sequence of instructions that can operate on different sets of data independently and parallely to produce some global computation, that could be a time steping of a PDE on a discretized domain. Each independent execution of the subroutine is called a thread and runs on a SP.

Contrary to other parallel strategies, the number of threads in which a given problem can be split is not limited by the number of SPs. In fact, the more threads one can provide the more efficient the GPU computation becomes. The number of threads can exceed in orders of magnitude the available SPs on a card, meaning that each SP will deal with many of these threads. In this way, the same GPU program can run in different cards with no change in the code, adding flexibility and portability.

Additionally, GPUs come with an assortment of different memories that can improve performance depending on the problem and numerical method in hand.

There are memory spaces that can be accessed to read or write data to all SPs on the card. Groups of SPs also share exclusive memory spaces that are much faster than the one that can be addressed globally. There are also memory structures that avoid addressing conflicts when several SPs are looking for the same data. Correct manipulation of these memory levels provide means for fine tunning a given algorithm.

Although GPU programing is heterogeneous, meaning that GPUs and CPUs can cooperate, data transfers between them is extremely time consuming and could deter the overall performance obtained by the GPU working in parallel. Explicit methods profit the most from GPUs as they rarely need extensive serial operations and can be largely parallelized and work entirely on the GPU. In the rest of this exposition, we will comment on some methods for fluid dynamics simulation that we have implemented on GPUs.

3 Methods

Lattice-Boltzmann methods (LBM) for flow simulation are presumably the easiest methods to code including complex boundaries (Guo and Zheng 2002). They compute indirectly approximations to the solutions of Navier-Stokes equations using probability distributions of the BGK-Boltzmann kinetic model equations. There exists a wide variety of such methods that incorporate non-Newtonian effects, multiphase system, thermal effects or even turbulence (Chen and Doolen 1998).

Coding a LBM for three dimensional flows can be at the reach of an undergraduate student, and doing it for GPUs lets you run a system discretized in millions of spacial nodes and watch it run and produce evolving flow structures in real time. Figure 1 shows some examples of LBM computations. In all cases, the flow domain was represented by over a million nodes and the development and evolution of vortical structures, requiring tens of thousands of time steps, were computed in a matter of hours. LBM running on a commodity GPU of over 400 SPs can achieve a speedup of two orders of magnitude if compared with a serial version executed on a CPU (Echeverría 2011).

Thermal versions of LBM are popular to study convection in the Boussinesq approximation (Mandujano and Rechtman 2008). Some of the patterns observed in nature can be reproduced by such methods. Figure 2 shows a 3D example. The implementation of LBM on GPUs provides a valuable tool for the study of fluid dynamics in general and for educational purposes in particular.

When in need of more accurate flow simulations, most researchers look for finite element methods (FEM). Aside from the stability issues of the method when simulating flows at low viscosities, FEM seems to be less efficient on GPUs than finite differences, which fit naturally on GPUs parallel model (Göddeke D et al. 2007). In order to stick to finite differences and avoid stability problems, we chose to implement a high order semi-Lagrangian method (Becerra-Sagredo 2007). This

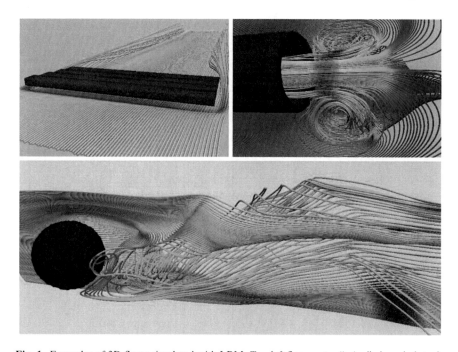

Fig. 1 Examples of 3D flows simulated with LBM. *Top left* flow past a *"wing"* shaped obstacle (3145728 nodes) with *streamlines* showing the wing tip vortex formation. *Top right* oscillatory flow at the exit of a tube at Womersley's number 14.35 (4194304 nodes), streamlines show a vortex ring. *Bottom* flow past a sphere at Reynold's number 240 (1048576 nodes) when the wake becomes unsteady, *streamlines* show vortical structures shed from the sphere

method allows the mesh representing the flow domain to move and deform in a material fashion. Grid points behave as material points for some time, while a high order conservative interpolation scheme reconstructs information back in the original orthogonal grid whenever is needed. Time evolution is computed with a 3-step Runge-Kutta algorithm and the material grid is intepolated back to the original grid periodically to avoid exesive deformation.

This method has superior stability over FEM and is specially suitable for transport equations simulations. On the down side, semi-Lagrangian methods require extensive computations and were considered of little practical interest unless implemented to run in parallel. The advent of GPUs gave us the opportunity. Figure 3 shows solutions to the non-linear shallow water equations obtained with this method.

Being able to solve transport equations with minimal numerical dissipation, opens the option of implementing a vortex method for incompressible flows. Aside from computing the material evolution of vorticity, one has to compute also the stream function by solving a Poisson equation. A multigrid method capable of running entirely on the GPU was implemented to compute the stream function. In order to this, the standard multigrid algorithm had to be modified to avoid the

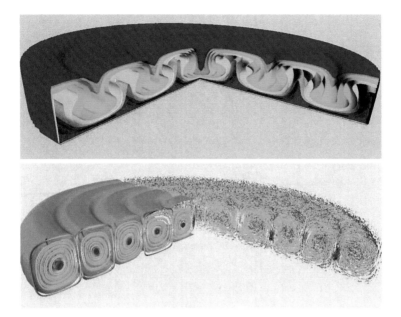

Fig. 2 Thermal-LBM simulation of the convective flow on a cylinder of radius four times its height (1199744 nodes). The cylinder is heated from the bottom and cooled from the top, Rayleigh number is 2×10^4. *Top* isothermal surfaces, a sector has been removed to show the internal structure. *Bottom* the velocity vector field, stream lines and a surface of constant velocity magnitude are shown in different sectors of the cylinder

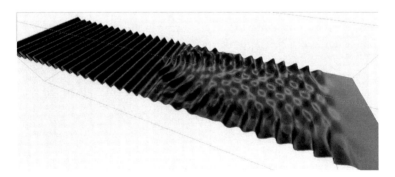

Fig. 3 Non-linear shallow water simulation of a wave tank (65536 nodes). The free surface is shown. A perturbation of the bottom surface at the middle of the channel produces the wave dispersion. A dissipative (viscous) term is included at the end of the channel to avoid reflections

reduction in threads that computation in coarser subgrids would imply during the standard procedure. Figure 4 shows the simulation of a 2D inviscid vortex dipole moving in a box.

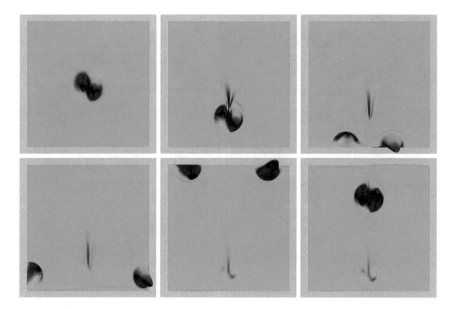

Fig. 4 An inviscid 2D vortex dipole moving in a *square box*. A sequence of different times of the vorticity surface evolution is shown

4 Concluding Remarks

Although new technologies are on their way, for example CPU manufacturers are planning to include SPs on their CPUs, nowadays GPUs represent the cheapest option for parallel computation. They also serve as affordable teaching tools. The School of Science at UNAM is offering a numerical methods course oriented to parallel computation with the aid of GPUs for undergraduate physics students. We believe it is worth the effort of learning new computer languages and understanding new architectures even though faster and more affordable computer technologies may soon appear, specially when preparing future scientists.

Acknowledgments This work was partially funded by UNAM-PAPIIT-IN118608.

References

Chen S, Doolen DD (1998) Lattice Boltzmann method for fluid flows. Annu Rev Fluid Mech 30:329–364

Echeverría C (2011) Simulación numérica de flujos en ductos en procesadores gráficos de alto rendimiento. BSc. Thesis, UNAM, 001–00323-E1-2011

Göddeke D, Strzodka R, Mohd-Yusof J, McCormick P, Buijssen SHM, Grajewski M, Turek S (2007) Exploring weak scalability for FEM calculations on a GPU-enhanced cluster. Parallel Comput 33:685–699

Guo Z, Zheng C (2002) An extrapolation method for boundary conditions in lattice Boltzmann method. Phys Fluids 14(6):2007–2010
Mandujano F, Rechtman R (2008) Thermal levitation. J Fluid Mech 606:105–114

Vortex in the Wakes of Airplanes

David Flores-García, Tiburcio Fernández-Roque
and Jorge Hernández-Tamayo

Abstract This paper describes the disturbances caused by wing trailing vortices and their effect on aircraft flying into them. We briefly discuss the physical characteristics of the wing trailing vortices together with their control and alleviation and emphasis is done about the importance that it has in order to diminish the problems generated by the increase in air traffic and in the fly safety.

1 Introduction

The lifting surfaces of an aircraft generate vorticity, which then rolls up into two dominant counter-rotating trailing vortices. The roll-up process is usually completed within a distance equivalent to several wingspans. An inadvertent encounter with a leading aircraft's wake vortex can be dangerous and should be prevented. The hazard to aircraft is primarily due to rolling moments, which can exceed the roll capability of the encountering aircraft. The increase in air traffic is currently outpacing the development of new airport runways. This is leading with massive air traffic congestion, resulting in costly delays and cancellations. So it is very important to investigate new technologies that will allow increased airport capacity maintaining the standards requirements for safety.

The wake produced by an aircraft may be separated into three categories:

(a) The slipstream effects from propellers (or jet engines).

D. Flores-García (✉) · T. Fernández-Roque · J. Hernández-Tamayo
Escuela Superior de Ingeniería Mecánica y Eléctrica, Unidad Ticoman, Instituto Politécnico Nacional, 07340 Mexico, D.F., Mexico
e-mail: jdbp09_david72@hotmail.com

J. Klapp et al. (eds.), *Fluid Dynamics in Physics, Engineering and Environmental Applications*, Environmental Science and Engineering,
DOI: 10.1007/978-3-642-27723-8_31, © Springer-Verlag Berlin Heidelberg 2013

Fig. 1 Wing trailing vortices
(Proctor and Switzer 2000)

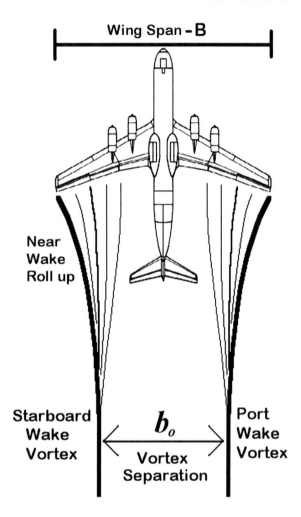

(b) The turbulence associated with drag due to fuselage and others airplane
     components.
(c) The wing trailing vortices or wake turbulence.

When the effect of this wake on following aircraft is considered, it has been
shown that (a) and (b) are negligible compared with (c) (Kerr and Dee 1960). This
paper, therefore, deals exclusively with the disturbances caused by trailing vortices,
and their effect on aircraft flying into them.

Trailing vortices are an unavoidable product of finite-span lifting wings.
Differences in pressure between the upper and lower surfaces of the wings produce
swirling vortices that trail the airplane as it flies through the air. These vortices can
persist for several minutes and translate many kilometers behind the generating
airplane (Proctor and Switzer 2000; Cottet et al. 2000) (Fig. 1). Under most
circumstances, the vortices propagate downward away from the flight path or they

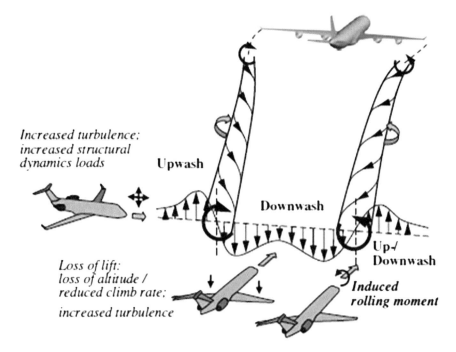

Fig. 2 Effects of the wing trailing vortices (Breitsamter 2011)

are carried away by crosswinds (Fig. 2). In practice, viscous and turbulent diffusion as well as atmospheric turbulence and vortex related instabilities result in the decay of strength of wing tip vortices; however, they have been observed to persist tens of kilometres downstream of large aircraft (Faddy 2005).

The risk exists because of the sheer size of these vortices and the high swirl velocities contained in them. For example, swirl velocities of the order of ± 18.3 m/s may be found in the vortices shed by a C5-A Galaxy 2.4 km astern or 30 s after the aircraft's passage (Bera 1993).

To avoid vortex encounters, minimum separation distances are imposed for airplanes on approach to the same airport under instrument flight rules. This separation also applies in all phases of flight. For example, the Civil Aviation Authority (CAA) of New Zealand requires that two airplanes in cruise must be at least 7.4 between two heavy aircraft, and 11.1 km between a heavy aircraft and a light aircraft. These distances apply when one aircraft is operating directly behind (within 0.93 km laterally) another, or is crossing behind, at the same level and up to 0.305 km below (CAA 2008). This allows time for the vortices to move out of the flight path. The imposed separation distances are key elements affecting airport capacity.

The danger is particularly severe during landings and take-offs for two reasons. First, the extension of the flaps of the leading aircraft may create trailing vortices that are even stronger than the wing tip vortices. Second, the proximity of the

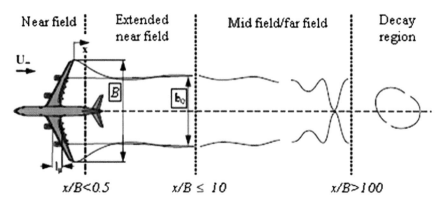

Fig. 3 Stages of wake vortex lifespan (Breitsamter 2011)

following aircraft to the ground means that a small perturbation in its trajectory may be disastrous (Orlandi 1998).

Commercial airplanes with flaps extended produce multiple trailing vortices that remain distinct for some distance behind the airplane. In the simplest case, the wing produces tip vortices at the wingtips and flap vortices at the outboard edge of the inboard flaps. These two pairs of Co-rotating vortices and a pair of counter-rotating vortices from the horizontal tail form the basic flaps-down vortex system. Details of the airplane configuration determine how far behind the airplane that the multiple vortex pairs remain as distinct vortices.

2 Physical Characteristics of the Wing Trailing Vortices

Typically, a vortex develops a circular motion around a core region. The core size can vary in size from only a few centimeters in diameter to a meter or more, depending on the type of aircraft. From larger aircraft, the speed of the air inside this core can be up to 100 m/s. The core is surrounded by an outer region of the vortex, as large as 30 m in diameter, with air moving at speeds that decrease as the distance from the core increases. Wake vortices can persist for 3 min, or longer, in certain conditions (CAA 2008).

Considering the downstream development, a vortex wake can be divided into four regions (Fig. 3): (i) The near field, $x/B \leq 0.5$, which is characterized by the formation of highly concentrated vortices shed at all surface discontinuities. (ii) The extended near field, $0.5 < x/B \leq 10$, where the wake roll-up process takes place and the merging of dominant vortices (e.g. shed at flap edge, wing tip, etc.) occurs, leads gradually to two counter-rotating vortices. (iii) The mid and far field, $10 < x/B \leq 100$, where the wake is descending in the atmosphere and linear instabilities emerge. (iv) The dispersion region, $x/B > \sim 100$, where fully developed instabilities cause a strong interaction between the two vortices until they collapse (Breitsamter 2011) (Fig. 3).

Proctor and Switzer (2000) give the next equations for diverse wing trailing vortices variables:

(a) The separation of the wake vortex pair immediately following roll up, is defined from conventional assumptions for an elliptically loaded wing as:

$$b_0 = \frac{\pi B}{4} \tag{1}$$

(b) The circulation is:

$$\Gamma_0 = \pm \frac{4Mg}{\pi B \rho U_\infty} \tag{2}$$

where M is the mass of the aircraft, ρ is air density, g is acceleration due to gravity, and U_∞ is air speed. The wake vortex system initially sinks at $w_0 = \Gamma_0/2\pi b_0$ due to the mutual interaction of the vortices.

Breitsamter (2011) gives the equation for the associated time scale t_0, which define the time interval at which the vortex pair moves downward by a distance equal to the vortex spacing b_0.

$$t_0 = \frac{b_0}{w_0} \tag{3}$$

The timescale t_0 is a measure of the "wake vortex age" depending strongly on the load factor $s = b_0/B : t_0 \sim s^3$.

The greatest hazard from wake turbulence is induced roll and yaw. The wake vortex impact on a follower aircraft is typically quantified by the induced rolling moment (Breitsamter 2011):

$$C_{l,f} = \frac{C_{La,f} \Lambda_f}{4} \frac{U_\infty}{U_f} \int_{-1}^{+1} \frac{\overline{w}(\eta)}{U_f} \frac{l_f}{b_f} \eta d\eta; \eta = \frac{2y}{b_f} \tag{4}$$

This rolling moment coefficient depends on the wake vortex induced vertical velocity ($\overline{w}/U_\infty$), the velocity ratio of leading and following aircrafts (U_∞/U_f) and the lift slope, $C_{La,f}$, aspect ratio, Λ_f and the inverse relative span, (l_f/b_f) of the follower aircraft.

The strength of the vortex is governed by the weight, speed, and shape of the wing of the generating aircraft. The vortex characteristics of any given aircraft can also be changed by extension of flaps or other wing configuring devices. However, as the basic factor is the weight, the vortex strength increases proportionately with increase in aircraft operating weight. Peak vortex tangential speeds up to almost 91.5 m/s have been recorded. The strongest vortices are produced by heavy aircraft flying slowly in a clean configuration at high angles of attack. Flight tests have shown that the vortices from larger (transport category) aircraft sink at a rate of several 10 m/s, slowing their descent and diminishing in strength with time and distance behind the generating aircraft.

3 Control and Alleviation of Wing Trailing Vortices

The control and alleviation of wakes is a subject of paramount importance in aircraft industry and remains a challenging problem of strong economical interest (Revelly et al. 2006). Depending on the particular application, wake control can have various goals and can be achieved either by passive or active strategies. Passive control mostly operates through shape optimization and often results in the addition of appendices to the surface of the obstacle. Active control implies that one is ready to impart energy on the flow by means of actuators on the surface of the obstacle, keeping in mind that this energy must be included in the global energy budget to conclude on the efficiency of the particular control strategy (Cottet and Poncet 2004).

The design of schemes to accelerate the decay of trailing vortices has been the topics of several theoretical, experimental and numerical investigations. Fundamental ideas for the reduction of the wake vortex hazard are aimed to affect the spatial vorticity distribution in the near field, thus alleviating the induced rolling moment (low vorticity vortex), or to enhance inherent instability mechanisms of the vortex systems, in order to enforce an accelerated decay (quickly decaying vortex) (Crouch et al. 2001).

A variety of passive means in the sense of additional or changed configuration elements, like wing fins, flap edge elements, spoiler elements, vortex plates, etc., were and are tested. Further, special wake vortex topologies are regarded to trigger inherent instabilities. In addition, active solutions like oscillating flaps and/or spoilers are studied (Heyes and Smith 2004). The fact has to be stressed that for wake vortex reduction the suggested configurationally means will not have any unfavorable effects on the flight characteristics and performance, i.e. in context of certification, the flight envelope must remain unchanged. Furthermore, with all these considerations, the question is to be answered, to what extent the modifications made at the aircraft, i.e. in the near field, are still effective in the far field to rearrange the vorticity distribution for wake vortex alleviation and/or to enhance the wake vortex decay.

4 Conclusions

Wing tip trailing vortices or wake turbulence affects aircraft of all sizes; wake turbulence incidents are not confined to operations involving heavier aircraft there are incidents involving all aircraft types. Because of the spacing requirements for airplanes approach, trailing vortices play a role in determining the capacity of commercial airports.

Because the wake-turbulence spacing is often larger than the spacing required by other factors such as radar resolution or runway occupancy, under some conditions it adds to the congestion and delays in the air transportation system. The safety hazards associated with trailing-vortex upsets and the impact of vortex separations on airport capacity have motivated numerous studies aimed at wake-vortex upset prevention.

The most effective means of upset prevention would be remove the threat by alleviating or destroying the vortices. Short of this, the next best solution would be to enable the following aircraft to tolerate the vortex encounter. If the vortex threat can not be removed, or tolerated, then it must be avoided.

References

Bera RK (1993) Wing tip vortices of cyclonic proportions. Curr Sci 64(3):69–70 Indian Current Science Association

Breitsamter C (2011) Wake vortex characteristics of transport aircraft. Prog Aerosp Sci 47:89–134

Civil Aviation Authority (CAA) (2008) Wake turbulence, New Zealand Government

Cottet GH, Poncet P (2004) Simulation and control of three-dimensional wakes. Comput Fluids 33:97–713 Elsevier Ltd

Cottet G-H, Sbalzarini I, Muller S, Koumoutsakos P (2000) Optimization of trailing vortex destruction by evolution strategies, proceedings of the summer program, Center for Turbulence Research

Crouch JD, Miller GD, Spalart PR (2001) Active-control system for breakup of airplane trailing vortices. AIAA J 39(12):2374–2381

Faddy JM (2005) Flow structure in a model of aircraft trailing vortices. Doctor of philosophy thesis. California Institute of Technology, Pasadena, California

Heyes L, Smith DAR (2004) Spatial perturbation of a wing-tip vortex using pulsed span-wise jets. Experiments in fluids, vol. 37, Springer, New York

Kerr TH, Dee F (1960) A flight investigation into the persistence of trailing vortices behind large aircraft. A.R.C. Technical Report No. 489. Ministry of Aviation Aeronautical Research Council, London

Orlandi P, Carnevale GF, Lele SK, Shariff K (1998) DNS study of stability of trailing vortices. Proceedings of the summer program, Center for Turbulence Research

Proctor FH, Switzer GF (2000) Numerical simulation of aircraft trailing vortices, 9th conference on aviation, range and aerospace meteorology, 11–15 Sept 2000, Orlando Florida American Meteorology Society

Revelly A, Iaccarino G, Wu X (2006) Advanced RANS modeling of wingtip vortex flows, Proceedings of the summer program, Center for Turbulence Research

The Activity of *La Bufadora*, A Natural Marine Spout in Northwestern Mexico

Oscar Velasco Fuentes

Abstract *La Bufadora* is a natural marine spout characterized by frequent eruptions of sea water. It is located about 20 km to the southwest of Ensenada, on a cliff of basaltic andesite where the essential elements for the occurrence of this type of spouts are present; namely: a littoral cave with a thin opening, a sea level which is always close to the opening's tip point and surface waves that vary from mild to strong all year round. We analyzed the activity of La Bufadora, under various conditions of surface waves and ocean tide, by monitoring the recurrence time of the eruptions (T). It was found that T typically lies in the range 13–17 s, which is also the dominant period of the surface waves, and that more and longer periods of inactivity appear as the tide ebbs.

1 Introduction

Marine spouts—variously known as blowholes, spouting horns, marine geysers, etc.—are the beautiful result of a chance combination of faulted rocks, sea level at just the right height and powerful waves (see, e.g., Bell 2007). Their working is simple: when a wave crest arrives it compresses the air inside the cave; when a wave trough arrives the air rushes out of the cave producing a jet of spray and a thunderous sound.

Blowholes are not uncommon but, since most are small, they usually remain unnoticed. Large ones, in contrast, become true tourist attractions, as it has been the case with the blowholes of Halona (Hawaii, U.S.A.), Kiama (New South

O. Velasco Fuentes (✉)
Departamento de Oceanografía Física, CICESE, Ensenada, Mexico
e-mail: ovelasco@cicese.mx

J. Klapp et al. (eds.), *Fluid Dynamics in Physics, Engineering and Environmental Applications*, Environmental Science and Engineering,
DOI: 10.1007/978-3-642-27723-8_32, © Springer-Verlag Berlin Heidelberg 2013

Fig. 1 *Left* La Bufadora is located 20 km southwest of Ensenada, Baja California. *Right* A view of the mouth of the cave that gives rise to La Bufadora

Wales, Australia) and Ensenada (Baja California, México). The latter, known as *La Bufadora* (from the Spanish verb *bufar*, signifying *to roar* or *to snort*), is located in Papalote Bay, a small inlet of triangular shape on the Pacific coast of northwestern México (see Fig. 1). The cave finds itself on a cliff of basaltic andesite that dates back to the early Cretaceous period, some 120 million years before present. The axis of the bay, oriented in southwest direction, has a length of about 825 m; the mouth, where the water depth reaches 50 m, is about 1,000 m wide. The tidal range in this area may reach 270 cm during spring tides, but the average difference between high and low water is only 160 cm.

Papalote Bay is exposed to persistent swells during the whole year. There are, however, no actual measurements there and only a few short time series in and around the Port of Ensenada, 20 km north of La Bufadora, in the neighbouring Todos Santos Bay (Fig. 1). The longest series available was taken just outside the port facilities from August 1986 to July 1989. The significant wave height (denoted by H_s and defined as the mean wave height, trough to crest, of the highest third of the waves) was observed to have a marked seasonality. The highest waves were found in winter, when H_s was between 60 and 90 cm during 50 % of the time and occasionally reached 240 cm; the lowest waves were found in summer, when H_s was between 50 and 60 cm during 50 % of the time and occasionally reached 120 cm (Martínez Díaz de León 1993). The peak period (denoted by T_p and defined as the inverse of the frequency at which the wave energy spectrum reaches its maximum) showed no seasonality. Over 90 % of the time T_p was in the range

11–19 s, whereas 55 % of the time it was in the range 14–16 s (Hernández Hernández 2004).

The purpose of this paper is to present a simple method to analyze the activity of La Bufadora, and to characterize this under various conditions of surface waves and ocean tide. Although there are numerous studies of the temporal behavior of intermittent eruptive phenomena such as volcanoes and geysers (see, e.g., Simkin 1993; Rojstaczer et al. 2003), this is, to the best of our knowledge, the first one devoted to a marine spout.

2 Data and Observations

We scheduled our field trips so that we could observe the activity of La Bufadora when the tide was around the mean sea level (MSL), above the mean higher high-water (MHHW), or below the mean lower low-water (MLLW). To ascertain this, we used the tide forecasting program Mar (González 2011), which uses historic hourly data of sea level obtained by the tide stations of CICESE and Mexico's National University (UNAM). The differences between predicted and measured values at Ensenada's tide station are typically less than ±10 cm.

Similarly, we got the properties of the surface waves from Surfline (2011). This private company uses the wave model Wavewatch III of the U.S. National Oceanic and Atmospheric Agency (Tolman 2002), together with its own data and algorithms, in order to forecast wave and surf conditions along the coast. Thus we were able to monitor the peak period T_p, the significant height H_s, and direction of the swell predicted to arrive at Papalote Bay at the time of observation.

Our measuring technique was simple: at each scheduled time we focused a video camera on the northwestern wall of the cliff and recorded footage during one hour, the standard length of digital video (DV) tapes. The video footage was sampled every 0.2 s, then each frame was transformed into a black and white image and its mean intensity (I) was computed as the average of the grey levels (0–256) of all pixels (480 × 640). We found this to be an efficient diagnostic of the occurrence of eruptions because the white color of the mixture of water, foam and spray ejected by La Bufadora increases the intensity I of the images by 50–100 units, depending on the weather conditions and the strength of the eruption. Furthermore, by defining a suitable cut-off intensity, we were able to automate the computation of the time elapsed between successive eruptions, hereafter referred to as recurrence time T (see Fig. 2). Note that I reaches a lower than average value after each eruption because we shot in automatic exposure mode.

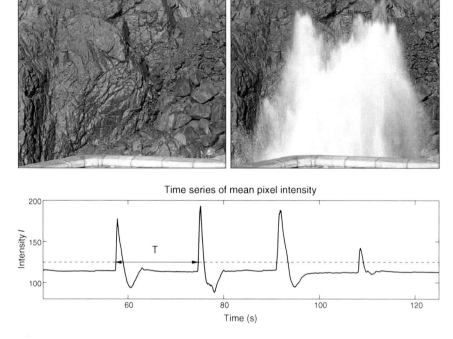

Fig. 2 *Top left* A view of the west wall of the cliff under normal conditions; the mean pixel intensity (*I*) of this image is 113. *Top right* A view of the west wall during an eruption of La Bufadora; now $I = 171$. *Bottom* The eruptions of La Bufadora can be easily identified in a time series of *I*

3 Results

We made the first measurements on 8 March 2005, when the tidal range was largest even though it was two days before the new moon. The first recording session started at 7:07 h, shortly before the tide reached high water (193 cm above MSL). Figure 3 shows the one-hour time series of the intensity *I* and the corresponding Welch estimate of the power spectral density. The vertical line indicates the peak spectral value at a frequency of 0.0714 Hz, or a period of $T = 14$ s, which coincides with the period of the southwest swell. Figure 4a shows a return map of T_{n+1} versus T_n: as expected, there is a clear clustering of points along the lines $T_n = 14$ and $T_{n+1} = 14$ s.

The second recording session started at 13:26 h, about half an hour before the tide reached low water (40 cm below MSL). Figure 4b shows that there is a slight shift of the cluster of points towards 15 s. The most important change, however, is the total number of eruptions: only 73 in one hour, compared with the 117 observed in the previous session. Since, according to Surfline, the characteristics of the surface waves remained fairly constant in the six hours elapsed between these two series the changes in the activity of La Bufadora should be attributed to the

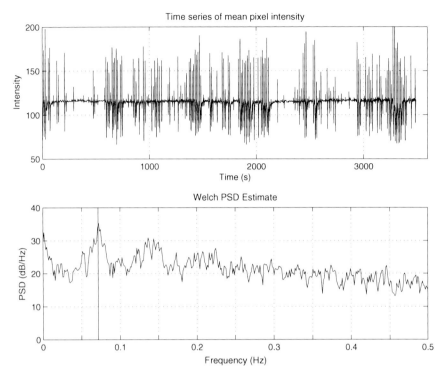

Fig. 3 *Top* Time series of *I* showing the activity of La Bufadora on 8 March 2005 during the hour succeeding the high tide (7:08 am). *Bottom* The corresponding estimation of the power spectral density (PSD). The *vertical line* corresponds to the frequency of the most important component of the incoming swell

change in sea level (233 cm). A similar behavior was observed on 11 December 2011 when the tidal range was 200 cm and the dominant swell had a period of 17 s all day long. The recurrence time therefore clustered around 17 s during both high and low water, but in the former case we observed 157 eruptions in one hour, whereas in the latter we observed only 60.

Observations with the tide around MSL were scheduled on different days to single out the effect of the surface waves. The most regular activity was observed on 3 May 2005. The measurements started at 16:26 h, a few minutes after the tide reached MSL. We observed 225 eruptions in one hour with a recurrence time mostly in the range 14–15 s (see Fig. 4). On this day the dominant swell was coming from the southwest and had a period of 14 s. The most irregular activity was observed on 12 May 2005. Measurements started at 15:40 h, and during the whole hour of observation the tide remained within 2 cm of MSL. We observed 86 eruptions in one hour with no predominant recurrence period (see Fig. 4). The mean and the standard deviation of *T* were, respectively, 37.5 and 31.3 s. This day was characterized by two swells coming from the southwest, one with a period of 12 s and the other with a period of 16 s.

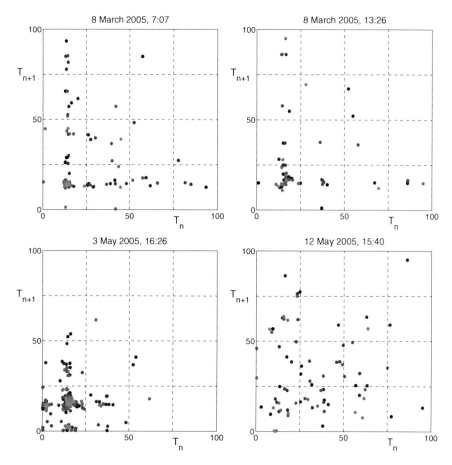

Fig. 4 Return map of the recurrence time: T_n is the time elapsed between the nth eruption and the following one. *Top* Similar surface waves but different sea level; *Bottom* Similar sea level but different surface waves

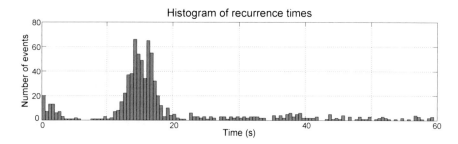

Fig. 5 Histogram of the recurrence time T for the complete set of observations. The figure shows only $T < 60$ s (88 % of the observed events)

The histogram with the whole set of data (Fig. 5) shows that the recurrence time T lies between 11 and 19 s in 60 % of the events. Furthermore, there are two peaks in this range: 23.5 % of the events have $T = 16 \pm 1$ s, whereas 22.6 % of them have $T = 14 \pm 1$ s. There is also a significant number of events (8 %) in the range 0–4 s. These are most probably the result of the sloshing of water in the complex geometry of the cliff where the cave is located.

4 Conclusions

Although our observations are limited in number and they were made only during spring, late autumn, and late winter, we can derive some general conclusions about the activity of La Bufadora.

The interval between eruptions, or recurrence time T, is in general fairly constant and its value matches the peak period (T_p) of the dominant swell, confirming that La Bufadora is simply driven by the ocean surface waves. There is a remarkable agreement between the most frequent recurrence times observed in La Bufadora in 2005 and 2011 (13–17 s) and the most frequent peak periods of the surface waves measured at the Port of Ensenada during 1986–1989 (11–19 s, see Hernández Hernández 2004). This result suggests that the wave climate has remained rather constant for the last quarter of a century.

At the current sea level La Bufadora is active during the whole tidal range, but the eruptions produced at the extremes of this range exhibit important differences, both quantitative and qualitative. The former are deduced from the measurements presented in Sect. 3, the latter from direct observations made during the recording sessions. Succinctly stated, the characteristics of the eruptions are as follows. When the sea level is high the eruptions consist of a jet of water, they are numerous and usually occur in long trains of fairly periodic eruptions separated by lapses of inactivity of tens of seconds to a couple of minutes. When the sea level is low the eruptions consist of a cloud of spray, they are scarce and occur in isolated trains of a few eruptions separated by periods of inactivity of a few minutes.

The best time to witness a good performance of La Bufadora is during the high water of a winter day; the less favourable time is during the low water of a summer day. Anyway, it will be a very rare occasion when it takes more than ten minutes to observe a good train of eruptions.

Acknowledgments I am grateful to Francisco Ocampo and Manuel Figueroa for suggestions and comments on an earlier version of this paper.

References

Bell FG (2007) Engineering geology. Elsevier, London

González JI (2011) Predicción de Mareas en México. (http://oceanografia.cicese.mx/predmar)

Hernández Hernández RR (2004) Optimización del sistema de bombeo por energía de oleaje para el puerto de Ensenada. Tesis de Maestría, CICESE

Martínez Díaz de León A (1993) Probability distribution of wave height in Todos Santos Bay, B.C. Mexico. Ciencias Marinas 19:203–218

Rojstaczer S, Galloway DL, Ingebritsen SE, Rubin DM (2003) Variability in geyser eruptive timing and its causes: Yellowstone National Park. Geophys Res Lett 30:1953

Simkin T (1993) Terrestrial volcanism in space and time. Annu Rev Earth Planet Sci 21:427–452

Surfline (2011) Online surf forecasting and reports. (http://www.surfline.com)

Tolman HL (2002) Distributed-memory concepts in the wave model WAVEWATCH III. Parallel Comput 28:35–52

A CNC Machine for Stationary Drop Deposition and Coalescence in Liquid–Liquid Systems

F. Peña-Polo, L. Trujillo, J. Klapp and L. Di G. Sigalotti

Abstract The controlled deposition of a dispensed liquid drop onto the surface layer of another liquid is a process that is widely applied in the industry. In most applications, the deposition of stationary drops requires a micrometric translational approach to the surface. Here we describe a computer controlled apparatus that has been constructed to perform precision translation of pendant drops and deposition at liquid and solid surfaces. Different settings of the experimental setup can be easily implemented in the laboratory for use in a variety of other applications, including surface tension measurements and wire bonding in microelectronics. Some experimental tests of partial drop coalescence with a miscible liquid are presented which validate the reliability of the apparatus.

F. Peña-Polo (✉) · L. Trujillo · L. D. G. Sigalotti
Centro de Física, Instituto Venezolano de Investigaciones Científicas, IVIC,
Apartado Postal 20632, Caracas 1020-A, Venezuela
e-mail: franklin.pena@gmail.com

L. Trujillo
The Abdus Salam International Centre for Theoretical Physics, ICTP, Trieste, Italy
e-mail: leonardo.trujillo@gmail.com

J. Klapp
Instituto Nacional de Investigaciones Nucleares, ININ, Km. 36.5 Carretera México-Toluca,
52750 La Marquesa, Estado de México, Mexico
e-mail: jaime.klapp@hotmail.com

J. Klapp
Departamento de Matemáticas, Cinvestav del I.P.N., 07360 México, D.F., Mexico

L. D. G. Sigalotti
e-mail: leonardo.sigalotti@gmail.com

J. Klapp et al. (eds.), *Fluid Dynamics in Physics, Engineering and Environmental Applications*, Environmental Science and Engineering,
DOI: 10.1007/978-3-642-27723-8_33, © Springer-Verlag Berlin Heidelberg 2013

1 Introduction

The controlled deposition and coalescence of micrometer-sized drops in liquid-liquid systems play an important role in many industrial processes such as oil separation and petroleum refineries (Rommel et al. 1993). With recent advances in biotechnology and nanotechnology, the formation and deposition of drops has begun to become important on increasingly smaller scales. For instance, drops ranging in size from hundreds of micrometers down to tens of micrometers are finding applications in protein crystallization (Zheng et al. 2004) and microcultures for cellular studies (Martin et al. 2003). In general, the industrial applications span across the whole range of engineering disciplines, from the distribution of agrochemicals to the fabrication of microlenses, and should be intended as examples of the potential as well as of the multidisciplinarity of this topic.

In most applications, a drop is formed at the tip of a syringe and its controlled deposition with negligible velocity on the free surface of another liquid requires a micron-scale translational approach of the drop to the surface. In particular, for small drops in the microliter range, stationary deposition demands the use of a motorized syringe to achieve a reasonably good precision at contact. Computerized operation of the process of drop deposition eliminates human errors in approaching the point of contact with the surface and removes excessive vibrations induced to the drop by hand manipulation of the syringe. A computer numerical control (CNC) apparatus has been constructed to perform this task, with commands encoded in a workstation computer to enhance reproducibility in laboratory experiments of drop deposition (Peña-Polo et al. 2010). The mobile tools of the apparatus, i.e., a charged-coupled device (CCD) camera, which is employed to capture static and dynamic drop images, and the electronic syringe are driven by standard unipolar stepper motors, which can be adjusted to produce discrete linear displacements of 5.6 μm and 150–200 ms of duration. A drawback of CNC machines is that they often lack any form of sensory capability to detect problems with the machining process, leaving to the operator the choice of manually abort the process. In addition, they assume that stepper motors are perfectly accurate and never mis-step. To overcome part of these problems, a new version of the apparatus has been equipped with optical encoder strips, which work as absolute position sensors to monitor tool position and with torque sensors on the drive system to detect abnormal strain when both the CCD camera and/or the syringe are just moving.

A parallel port (PP) interfacing is implemented to control the apparatus, with a bus speed of about 1.3 MHz so that reading and writing of the data is done every 1 μs. The data and images are stored in a computer for later analysis or easy transfer to other software applications. Once the drop is formed at the capillary tip, the front panel also offers the possibility of running an image processing software for digitization of the drop profile and measurement of other drop characteristics as its projected surface area, its shape, its surface tension, and its dynamics during micrometric translation and coalescence with the other liquid.

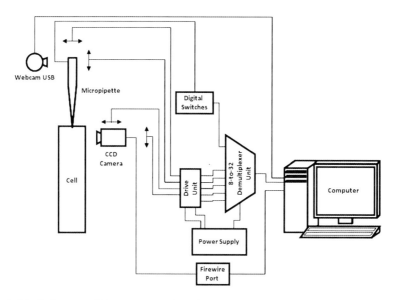

Fig. 1 Block diagram of the apparatus. The *left–right* and *up–down arrows* indicate horizontal and vertical displacements of the syringe and recording CCD camera, respectively

2 Brief Description of the Apparatus

The apparatus consists of a Plexiglas® vessel filled with a liquid solvent and mounted on a rigid metallic support and a metallic tower provided with an internal elevator car coupled to telescopic rails to allow positioning of the syringe and movement of the CCD camera for drop visualization. Pendant drops were formed at plastic capillary tips of diameters ranging from 0.3 to 1.2 mm, using a BRAND Transferpette® electronic as the drop delivery device. The mobile tools are driven by standard stepper motors, which divide a full rotation into 48 discrete translational steps. The resolution of the motor driving the micrometric descent of the syringe is calibrated to 96 steps per revolution, implying discrete vertical displacements of the capillary tip of ~ 5.6 µm per step. When the duration of this stepping is calibrated to time intervals longer than 150 ms, the induced drop vibrations are strongly attenuated (Peña-Polo et al. 2010). The apparatus is controlled through a PP interfacing, which allows the input of up to 9 bits or the output of 12 bits at any one given time, thus requiring minimum external circuitry. The circuit is an 8-to-32 data line demultiplexer circuit so that there is enough capacity to connect up to eight motors, with five wires each, to the same PP. The drive circuit works with four integrated circuits for bus interfacing, each one being an 8-bit transparent latch with three-state outputs. Data transfer is then controlled in 32 independent outputs, in groups of eight outputs per activated latch. In addition, three ULN2803 integrated circuits are used to run the stepper motors with each circuit being an eight-line driver. A block diagram of the apparatus is shown in Fig. 1.

The control software has been developed in the laboratory and is based on the National Instruments LABVIEW^{TM} programming environment on a Microsoft© Windows® platform. The front panel on the screen of the computer allows to control the speed and sense of rotation of the stepper motors, handling and positioning of the electronic syringe as well as monitoring its function state and position. This latter task, which was not implemented in the previous version of the apparatus, is accomplished by optical encoder strips (Klaassen 2002). For each stepper motor, the drive circuit incorporates a L298 integrated monolithic circuit, which is directly connected to the motor in a bipolar configuration. In this way, any slipping of the stepper motor is immediately detected and the missed step corrected. This technique permits high precision motion and monitoring tool position in a much safer manner. To detect abnormal strain during the displacement of the mobile components, a reaction torque sensor has been also bolted to the rotating shaft of each stepper motor and to a fixed plate which then connects to a clutch. In addition, some small mechanical modifications have been designed and implemented to enhance the rigidity of the mobile assembly and reduce the occurrence of induced vibrations on the pendant drop. New tests of deposition on liquid and solid surfaces with different drop volumes have shown that these improvements have notably increased the reproducibility for the same experimental setup, implying a high degree of reliability for laboratory experiments and engineering applications.

The front panel on the screen of the computer includes several menu selection keys for controlling the speed and sense of rotation of the stepper motors, the handling and position of the syringe as well as its functioning state. It also offers the possibility of running an image processing software to visualize the drop dynamics on-line, perform edge detection, calculate the drop shape parameter, measure the drop surface tension, its radius of curvature at the tip, the coordinates of the center of mass, and other properties including its projected surface area.

3 Some Experimental Tests and Discussion

Calibration of the apparatus has shown that stationary deposition of drops on liquid and solid surfaces is achieved accurately using translational steppings with lengths of 5.6–11.2 μm and durations δt between 150 and 200 ms (Peña-Polo et al. 2010). Several experiments conducted for distilled water drops have shown that the image analysis software reproduces the surface tension of water for drops larger than about 2–3 μl (Sigalotti et al. 2011). The results compare quite well with recent accurate surface tension measurements for pendant micro-drops (Lin et al. 1996; Yeow et al. 2008). Other tests with methylene blue drops have also yielded accurate surface tension measurements.

Among the several applications, the apparatus was originally designed to study the inflow (partial coalescence) of a stationary pendant drop that is brought into contact with a quiescent flat surface of a liquid with which it is miscible. Although

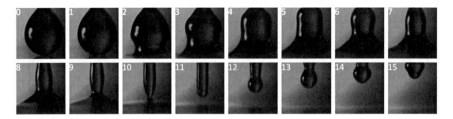

Fig. 2 Side-view images of the contact and partial coalescence of a methylene *blue drop* of 8 μl with a reservoir of pure distilled water. The numbers on the left top of each frame gives the time in milliseconds.

the first investigation of this problem was reported more than a century ago (Thomson and Newall 1885), it has only received considerable attention over the past two decades (Arecchi et al. 1989, 1996; Shankar and Kumar 1995; Residori et al. 2000; Thoroddsen and Takehara 2000; Blanchette and Bigioni 2006). As a further experimental test, here we shall describe the morphology of the partial coalescence of stationary pendant drops above and below the surface of the host liquid. In Fig. 2 we report a sequence of photos showing the details after contact of a methylene blue drop of volume $V = 8$ μl with a surface layer of pure distilled water. The square frames fully contain the region occupied by the drop. The images have a resolution of 144×144 pixels and the event was recorded with a video acquisition rate of 1,000 fps. The evolution bears a strong similarity with recently reported experiments of the partial coalescence of an ethanol drop beginning at rest on a reservoir of the same liquid (Blanchette and Bigioni 2006). At contact with the surface a bridge is formed between the drop and the host liquid. This gives rise to capillary waves which rise up the drop surface, deforming the drop volume into a bell shape (3 and 4 ms) and then into a stretching cylindrical column (5–9 ms) before pinching off at about 10 ms.

In Fig. 3 we show side-view images of the process of coalescence as viewed inside the pool of distilled water. This time the drop consists of a water/glycerin solution (80 % distilled water, 20 % glycerin) of volume $V = 10$ μl, doped with a small amount of sodium fluorescein (10^{-8} moles/l) for visualization. After the first 2 or 3 ms during drop penetration, a large surface-parallel acceleration occurs very near the free surface, causing flux of positive vorticity into the drop liquid (Rood 1994; Dooley et al. 1997). This circulatory motion soon develops into a single-branched vortex spiral sheet (van Kuik 2004). As the bulk of the drop descends, the spiraling which carries with it an increasing volume of dyed liquid, rapidly organizes itself into a circular tube of vorticity, much like a ring torus. After about 14 ms, the lower cap of the falling drop deforms into a bulge, reminiscent of the transient laminar jet which carried part of the drop liquid before pinching-off occurred above the surface. After about 20 ms, a second vortex ring forms at the circular top of the bulge owing to the roll-up of the liquid coming from the impacting jet. The new vortex strengthens and expands in radius, while the rear one stretches due to the straining field of the former and merges with it by 30 ms.

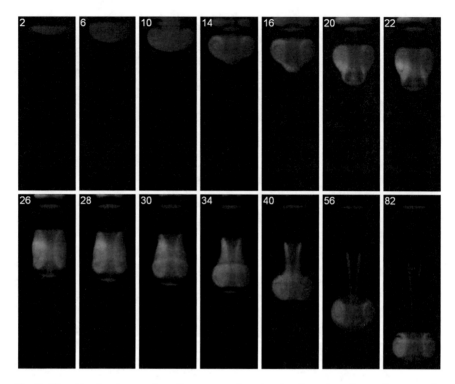

Fig. 3 The *side-view images* show the process of vortex-ring formation after coalescence of a pendant drop of volume 10 µl composed of a water/glycerin solution with a reservoir of pure distilled water. The numbers on the *left top* of each frame gives the time in milliseconds

Only the lower part of the weakened vortical fluid is sucked by the pursuing vortex, while the upper part rapidly diffuses into a well-defined conical trailing wake, which stretches and elongates for the remainder of the evolution, as shown in the last image of Fig. 3 after 82 ms.

In passing, we note that our setup can be used to perform drop deposition and splashing on solid surfaces without significant modifications. Applications of contact dispensing drops are also important in a variety of other processes, including micromachined fountain-pen techniques (Moldovan et al. 2006), biofluid dispensing applications (Ben-Tzvi et al. 2007), electrowetting-assisted drop deposition (Leïchlé et al. 2007), and drop deposition on hydrophobic surfaces (Qian et al. 2009). Other uses may involve wire bonding in microelectronic applications (Prasad 2004; Harman 2010).

Acknowledgments F. P.-P. acknowledges the organizer of the XVII Annual Meeting of the Fluid Dynamics Division (XVII-DDF) of the Mexican Physical Society, with special mention to Anne Cros. This work has been partially supported by CONACyT-EDOMEX-2011-C01-165873 project.

References

Arecchi FT, Buah-Bassuah PK, Francini F, Pérez-Garcia C, Quercioli F (1989) An experimental investigation of the break-up of a liquid drop falling in a miscible fluid. Europhys Lett 9(4):333–338

Arecchi FT, Buah-Bassuah PK, Francini F, Residori S (1996) Fragmentation of a drop as it falls in a lighter miscible fluid. Phys Rev E 54(1):424–429

Ben-Tzvi P, Ben Mrad R, Goldenberg AA (2007) A conceptual design and FE analysis of a piezoceramic actuated dispensing system for microdrops generation in microarray applications. Mechatronics 17:1–13

Blanchette F, Bigioni TP (2006) Partial coalescence of drops at liquid interfaces. Nat Phys 2:254–257

Dooley BS, Warncke AE, Gharib M, Tryggvason G (1997) Vortex ring generation due to the coalescence of a water drop at a free surface. Exp Fluids 22:369–374

Harman G (2010) Wire bonding in microelectronics, 3rd edn. McGraw-Hill Professional, New York

Klaassen KB (2002) Electronic measurement and instrumentation. Cambridge University Press, Cambridge

Le T, Tanguy L, Nicu L (2007) Electrowetting-assisted drop deposition for controlled spotting. Appl Phys Lett 91:224102-1

Lin S-Y, Wang W-J, Lin L-W, Chen L-J (1996) Systematic effects of bubble volume on the surface tension measured by pendant bubble profile. Colloids Surf A Physicochem Eng Aspects 114:31–39

Martin K, Henkel T, Baier V, Grodrian A, Schön T, Roth M, Köhler JM, Metze J (2003) Generation of larger numbers of separated microbial populations by cultivation in segmented-flow microdevices. Lab Chip 3:202–207

Moldovan N, Kim K-H, Espinosa HD (2006) Design and fabrication of a novel microfluidic nanoprobe. J Microelectromech Syst 15:204–213

Peña-Polo F, Trujillo L, Di Sigalotti LG (2010) A computer-controlled apparatus for micrometric drop deposition at liquid surfaces. Rev Sci Instrum 81:055107

Prasad SK (2004) Advanced wirebond interconnection technology. Kluwer Academic Publishers, Dordrecht

Qian B, Loureiro M, Gagnon DA, Tripathi A, Breuer KS (2009) Micron-scale droplet deposition on a hydrophobic surface using a retreating syringe. Phys Rev Lett 102(16):164502-1

Residori S, Pampaloni E, Buah-Bassuah PK, Arecchi FT (2000) Surface tension effects in the zero gravity inflow of a drop into a fluid. Eur Phys J B 15:331–334

Rommel W, Blass E, Meon W (1993) Plate separators for dispersed liquid-liquid systems: hydrodynamics coalescence model. Chem Eng Sci 48:159–168

Rood EP (1994) Interpreting vortex interactions with a free surface. J Fluid Eng 116:91–94

Shankar PN, Kumar M (1995) Vortex rings generated by drops just coalescing with a pool. Phys Fluids 7(4):737–746

Sigalotti L Di G, Peña-Polo F, Trujillo L (2012) An image analysis procedure for measuring the surface tension of pendant micro-drops. J Comput Methods Sci Eng (in press)

Thomson JJ, Newall HF (1885) On the formation of vortex rings by drops falling into liquids, and some allied phenomena. Proc R Soc Lond 39:417–436

Thoroddsen ST, Takehara K (2000) The coalescence cascade of a drop. Phys Fluids 12(6):1265–1267

van Kuik GAM (2004) The flow induced by Prandtl's self-similar vortex sheet spirals at infinite distance from the spiral kernel. Eur J Mech B/Fluids 23:607–616

Yeow YL, Pepperell CJ, Sabturani FM, Leong Y-K (2008) Obtaining surface tension from pendant drop volume and radius of curvature at the apex. Colloids Surf A Physicochem Eng Aspects 315:136–146

Zheng B, Tice JD, Ismagilov RF (2004) Formation of droplets of alternating composition in microfluidic channels and applications to indexing of concentrations in droplet-based assays. Anal Chem 76(17):4977–4982

Part IV
Meteorology and Pollution

Numerical Study of Wind Field Adjustment with Radial Basis Functions

Rafael Reséndiz, L. Héctor Juárez, Pedro González-Casanova,
Daniel A. Cervantes and Christian Gout

Abstract A collocation method based on radial basis functions (RBF) is introduced for wind field adjustment in meteorology. The numerical solutions are
shown to be more accurate than those obtained with the finite element method
(FEM), and they also require much less computational effort. A detailed analysis
shows how inconsistent boundary conditions may affect numerical solutions.

1 Introduction

The variational method proposed by Sasaki (1958) to recover a vector wind field $\mathbf{u}$
from an observed wind field $\mathbf{u}^0$, in a given bounded domain Ω, uses the continuity
equation $\nabla \cdot \mathbf{u} = 0$. The method is based on the minimization of the Lagrangian

R. Reséndiz (✉) · L. Héctor Juárez
Departamento de Matemáticas, UAM-I, Av. San Rafael Atlixco 186, Col. Vicentina,
09340 Mexico, D.F., Mexico
e-mail: rafael.resendiz@gmail.com

L. Héctor Juárez
e-mail: hect@xanum.uam.mx

P. González-Casanova · D. A. Cervantes
Instituto de Matemáticas, UNAM, Circuito Exterior CU, 04510 Mexico, D.F., Mexico
e-mail: casanovapg@gmail.com

D. A. Cervantes
e-mail: dcchivela@gmail.com

C. Gout
INSA Rouen, LMI, Av. de l'Université, BP 08, 76801 St Etienne du Rouvray cedex, France
e-mail: christian.gout@insa-rouen.fr

J. Klapp et al. (eds.), *Fluid Dynamics in Physics, Engineering and Environmental
Applications*, Environmental Science and Engineering,
DOI: 10.1007/978-3-642-27723-8_34, © Springer-Verlag Berlin Heidelberg 2013

$$L(\mathbf{u}, \lambda) = \frac{1}{2} \int_{\Omega} \{S(\mathbf{u} - \mathbf{u}^0) \cdot (\mathbf{u} - \mathbf{u}^0) + \lambda[\nabla \cdot \mathbf{u}]\} \, dV, \tag{1}$$

where λ is a Lagrange multiplier and S is a diagonal matrix with weighting parameters $S_{ii} = \alpha_{ii}^2 > 0$, $i = 1, 2, 3$, called Gaussian precision moduli, which are related to the scales of the respective components of the velocity field. The initial wind field $\mathbf{u}^0$ is a horizontal field, because meteorological stations usually do not measure the vertical component. The Euler–Lagrange equations of (1) are:

$$\mathbf{u} = \mathbf{u}^0 + S^{-1}\nabla\lambda, \tag{2}$$

The unknown $\mathbf{u}$ is obtained from (2), after the multiplier λ is computed from the elliptic equation $-\nabla \cdot (S^{-1}\nabla\lambda) = \nabla \cdot \mathbf{u}^0$. To close this equation, two types of boundary conditions are commonly used: $\lambda = 0$, for open or "flow through" boundaries (like truncated boundaries), and $\partial\lambda/\partial\mathbf{n} = 0$,, for closed or "no flow through" boundaries (like the surface terrain). Many authors recommend, these boundary conditions (Kitada et al. 1983; Ratto 1996; Sherman 1978).

However, they are physically and mathematically inconsistent and may degrade numerical solutions, sometimes by several orders of magnitude (Flores et al. 2010). There have been several sophisticated developments in the numerical simulations of this model as, for instance, the application of multigrid methods (Wang et al. 2005), the application of genetic algorithms to estimate parameters (Montero et al. 2005), and modelling (Ferragut et al. 2010), but it seems that the analysis of boundary conditions on truncated boundaries, and its relation with the parameters α_{ii}^2, has not attracted the attention of the meteorologist. In this work we do a detailed analysis of boundary conditions using a meshfree method based on RBF (Kansa 1990). We also show that this method is computationally cheaper and produces more accurate solutions than the FEM.

2 Formulation of the Problem and the RBF Collocation Method

The boundary $\partial\Omega$ of the domain is decomposed as $\Gamma_N \cup \Gamma_D$, where Γ_N is the part of the boundary associated to the surface terrain (topography), and Γ_D is the rest of the boundary (vertical and top boundaries). We consider that the unknown adjusted wind field $\mathbf{u}$ belongs to the following normed closed space

$$\mathbf{V} = \{\mathbf{v} \in \mathbf{L}_2(\Omega) \ : \ \nabla \cdot \mathbf{v} \in L_2(\Omega), \nabla \cdot \mathbf{v} = 0, \mathbf{v} \cdot \mathbf{n} = 0 \ on \ \Gamma_N\}, \tag{3}$$

equipped with the norm $\| \cdot \|_{S,\Omega}$ associated to the inner product defined by

$$\int_\Omega \dagger S\mathbf{u} \cdot \mathbf{v} \ d\mathbf{x},$$

where

$$\mathbf{v} \cdot \mathbf{w} = \Sigma_1^d \dagger v_i \ w_i$$

is the usual scalar product in $\mathbb{R}^d$, and $d = 2$ or 3 is the space dimension. Given $\mathbf{u}^0$, we may find $\mathbf{u} \in \mathbf{V}$ as the projection of $\mathbf{u}^0$ onto $\mathbf{V}$, i.e. we may find the the minimum of the convex quadratic functional $J : \mathbf{V} \to \mathbb{R}$, defined by

$$J(\mathbf{v}) = \frac{1}{2} \| \mathbf{v} - \mathbf{u}^0 \|^2_{S,\Omega} = \frac{1}{2} \int_\Omega \dagger S(\mathbf{v} - \mathbf{u}^0) \cdot (\mathbf{v} - \mathbf{u}^0) \ d\mathbf{x}, \text{ for all } \mathbf{v} \in \mathbf{V} \quad (4)$$

The **unique solution** $\mathbf{u}$ must satisfy the necessary (and in this case, sufficient) first order conditions

$$\frac{\partial}{\partial \varepsilon} J(\mathbf{u} + \varepsilon \ \mathbf{v})|_{\varepsilon=0} = \int_\Omega \dagger S(\mathbf{u} - \mathbf{u}^0) \cdot \mathbf{v} \ d\mathbf{x} = 0, \quad \forall \mathbf{v} \in \mathbf{V}. \quad (5)$$

This relation implies that $S(\mathbf{u} - \mathbf{u}^0)$ *must belong to the orthogonal complement of $\mathbf{V}$ in* $\mathbf{L}_2(\Omega)$, which is $\mathbf{V}^\perp = \{\nabla q : q \in H^1(\Omega), q = 0 \text{ on } \Gamma_D\}$, (Flores et al. 2010). Therefore, $S(\mathbf{u} - \mathbf{u}^0) = \nabla\lambda$, with $\lambda = 0$ on Γ_D. With this properties, we obtain the following **saddle-point problem** for $\mathbf{u}$ and λ (left), as well as its correspondent **elliptic problem** for λ (right).

$$S\mathbf{u} - \nabla\lambda = S\mathbf{u}^0, \nabla \cdot \mathbf{u} = 0 \text{ in } \Omega, \quad -\nabla \cdot (S^{-1}\nabla\lambda) = \nabla \cdot \mathbf{u}^0 \text{ in } \Omega, \quad (6)$$

$$\lambda = 0 \text{ on } \Gamma_D, \qquad\qquad \lambda = 0 \text{ on } \Gamma_D, \quad (7)$$

$$\mathbf{u} \cdot \hat{\mathbf{n}} = 0 \text{ on } \Gamma_N. \qquad - S^{-1}\nabla\lambda \cdot \hat{\mathbf{n}} = \mathbf{u}^0 \cdot \hat{\mathbf{n}} \text{ on } \Gamma_N. \quad (8)$$

The *elliptic problem* is obtained eliminating $\mathbf{u}$ from the *saddle-point problem*. *Collocation method based on radial basis functions* The main idea (Kansa 1990) is to introduce a set of collocation points $\{\mathbf{x}_i\}_{i=1}^n$, with n_i points in the interior of Ω, n_D points on Γ_D, and n_N points on Γ_N (see (7)), so that $n = n_i + n_D + n_N$, and approximate λ by a linear combination of the form

$$\lambda_h(\mathbf{x}) = \sum_{j=1}^n \omega_j \phi(\|\mathbf{x} - \mathbf{x}_j\|), \quad (9)$$

where $\phi(r)$ is a base radial basis function defined for the non-negative numbers $r = \|\mathbf{x} - \mathbf{x}_j\|$. The unknown coefficients $\{\omega_i\}_{i=1}^n$ are computed from the linear system of equations obtained when we substitute $\lambda_h(\mathbf{x}_i)$ in Eq. (6) for each interior node $\mathbf{x}_i \in \Omega$, and in Eq. (7) for each boundary node $\mathbf{x}_i \in \Gamma_D$, and in Eq. (8) for each boundary node $\mathbf{x}_i \in \Gamma_N$. Concerning the election of the radial basis function, there

Table 1 Comparison of accuracy with different boundary conditions for the RBF method, and with the numerical solution obtained with the FEM (Flores et al. 2010)

Case	1 (Benchmark) $c = 10.8$	2 (FEM)	2 (RBF) $c = 5.1$	3 (FEM)	3 (RBF) $c = 11.3$
e_r	5.9×10^{-5}	1.9×10^{-2}	5.5×10^{-2}	4.0×10^{-4}	7.1×10^{-4}
mdiv	2.0×10^{-7}	4.1×10^{-2}	-3.8×10^{-5}	1.8×10^{-2}	2.7×10^{-6}

are several possibilities (Buhmann 2003): multiquadric $\phi(r) = \sqrt{r^2 + c^2}$, inverse multiquadric $\phi(r) = \left(\sqrt{r^2 + c^2}\right)^{-1}$, thin plate spline $\phi(r) = (cr)^2 \ln(cr)$, Gaussian$(r) = \exp(-cr^2)$, among many others. There are also compactly supported RBF (Wendland 2005). In this work we will always consider **multiquadric** RBF. Observe that we must choose the *shape paramenter c* before discretization. After the coefficients ω_j are computed, the recovered wind field is obtained from (2):

$$\mathbf{u}_h(\mathbf{x}) = \mathbf{u}^0(\mathbf{x}) + \sum_{j=1}^{n} \uparrow \omega_j \, S^{-1} \nabla \phi(\| \mathbf{x} - \mathbf{x}_j \|). \tag{10}$$

3 Influence of the Boundary Conditions

We shall consider the following 2-D synthetic divergence-free wind field: $\mathbf{u}(x, z) = (x, -z)$, with initial field $\mathbf{u}^0(x, z) = (x, 0)$ in $\Omega = (1, 2) \times (0, 1)$. Given the diagonal matrix S, with coefficients $S_{11} = \alpha_1^2$ and $S_{33} = \alpha_3^2$, we easily obtain the exact multiplier $\lambda(x, z) = 0.5\alpha_3^2(1 - z^2)$. We will consider the solution of (6) and (8) with different boundary conditions on Γ_D besides (7). In particular, we will consider the following cases:

Case 1 Exact Dirichlet boundary conditions (benchmark, Fig. 1).

Case 2 Homogeneous Dirichlet boundary conditions, given by (7).

Case 3 $\lambda = 0$ *on* Γ_T, and $-S^{-1}\nabla\lambda \cdot \hat{\mathbf{n}} = 0$ *on* Γ_V.

In case 3, $\Gamma_T = $ top boundary, $\Gamma_V = $ vertical boundaries, so that $\Gamma_D = \Gamma_T \cup \Gamma_V$, and the boundary conditions are motivated by the following argument: the horizontal component of the wind field is known, thus $u \cdot \hat{n} = u^0 \cdot \hat{n}$ must hold on Γ_V, and from (2) we obtain $-S^{-1}\nabla\lambda \cdot \hat{n} = 0$ on Γ_V. Concerning the numerical calculation, we set $\alpha_1 = 1$ and $\alpha_3 = 0.001$, and we apply Kansa's method with $n = 36$ collocation points in a regular distribution ($n_i = 16$, $n_D = 14$, $n_N = 6$). Knowing the exact solution, it is possible to compute the **relative error**, and **mean divergence**, respectively defined as

$$e_r = \frac{\| \mathbf{u} - \mathbf{u}_h \|_2}{\| \mathbf{u}_2 \|}, \quad \text{and} \quad mdiv = mean_{\{\mathbf{x}_i\}}\{\nabla \cdot \mathbf{u}_h(\mathbf{x}_i)\}. \tag{11}$$

In the formula for *mdiv*, $\{\mathbf{x}_i\}$ denotes the set of interior nodes. For completeness, we compare these results with those obtained by the FEM (Table 1).

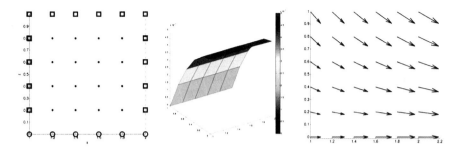

Fig. 1 *Case 1 Left* Regular distribution of collocation points in Ω with Γ_N (*circles*) and Γ_D (*squares*) with h = 0.2. *Middle* Multiplier λ_h. *Right* Wind field $\mathbf{u}_h$

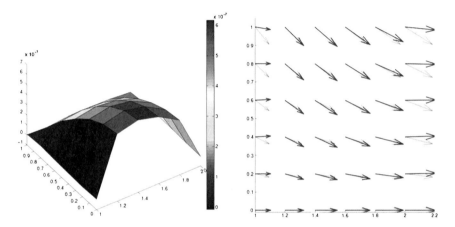

Fig. 2 *Case 2 Left* Multiplier λ_h. *Right* $\mathbf{u}_h$ in *bold line* and $\mathbf{u}$ in *thin line*

As expected, the benchmark (case1) yields the more accurate solution. The influence of poor boundary conditions on the vertical truncated boundaries (case 2) is clearly observed in Fig. 2. The error propagates from Γ_D to the interior of the domain and degrades the numerical solution. On vertical boundaries, when the boundary condition $\lambda = 0$ (case 2) is replaced by $S^{-1}\nabla\lambda \cdot \hat{\mathbf{n}} = 0$ (case 3), the relative error improves by two orders of magnitude, and the mean divergence improves by one order of magnitude. Compared with the FEM, the RBF collocation method reduces the mean divergence for at least three orders of magnitude, although the obtained relative error is about the same order.

4 Ghost Nodes, Influence of Matrix S, and 3-D Experiments

We may improve the accuracy of numerical solutions by the introduction of "ghost nodes" around the truncated boundary. The idea is to obtain an extended domain Ω_E, mainly beyond truncated boundaries, and impose boundary conditions on the

Table 2 Comparison of numerical results obtained with ghost nodes and the results obtained in case 3

Case	3 (RBF), $c = 11.3$	4 (RBF), $c = 3.1$	4 (RBF), $c = 4.0$
α_3	0.001	0.001	1
e_r	7.1×10^{-4}	6.4×10^{-3}	3.1×10^{-4}
mdiv	2.7×10^{-6}	6.7×10^{-8}	-6.3×10^{-8}

We also compare the numerical results with those obtained when $S = I$ ($\alpha_3 = 1$)

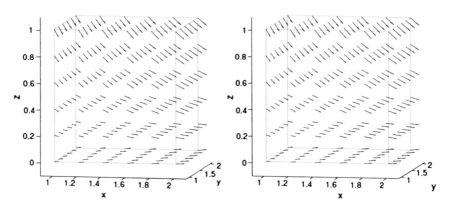

Fig. 3 *Case 4* Three-dimensional solution with ghost nodes on Γ_D. *Left* Solution for $\alpha_1 = \alpha_2 = 1$, and $\alpha_3 = 0.001$. *Right* solution when $S = I$. Field **u** in *thin line* and $\mathbf{u}_h$ in *bold line*

boundary of the extended part of the domain. After finding the numerical solution with some numerical schemes, we discard the solution on the ghost nodes $(\Omega_E \backslash \Omega)$, and we only keep the solution values in Ω.

Case 4 A layer of ghost nodes is introduced beyond the vertical boundaries of $\Omega = (1, 2) \times (0, 1)$, to get the extended domain $\Omega_E = (0.8, 2.2) \times (0, 1)$. Kansa's method is applied with the boundary conditions as in case 3.

Table 2 summarizes the numerical results when we solve the problem considering matrix S as before ($\alpha_1 = 1$, $\alpha_3 = 0.001$), and when $S = I$ ($\alpha_1 = 1$ and $\alpha_3 = 1$). We also compare these results with the best solution obtained before (case 3). The introduction of ghost nodes reduces the mean divergence, and the introduction of the identity matrix reduces the relative error.

Three-dimensional numerical examples One of the main features of Kansa's method is that it is a dimension-independent algorithm. Besides, the programming process is much simpler than other discretization methods, making this method attractive to solve 3D elliptic problems. Now we consider the synthetic exact wind field $\mathbf{u}(x, y, z) = (x, y, -2z)$ defined in $\Omega = (1, 2) \times (1, 2) \times (0, 1)$, with initial vector field $\mathbf{u}^0(x, y, z) = (x, y, 0)$, and exact multiplier $\lambda(x, y, z) = \alpha_3^2(1 - z^2)$. We compute the numerical solution, given by (10), with $n = 216$ collocation points, arranged in a regular grid, with $n_i = 64$ interior points, $n_T = n_N = 36$ points on the top and bottom boundaries, and $n_V = 80$ points around the vertical boundaries. This time we consider four cases:

Table 3 Comparison of the numerical solutions for the 3D problem

Results with matrix $S = diag(1.0, 1.0, 0.000001)$			
Case	1 ($c = 11.4$)	2 ($c = 4.6$)	3 ($c = 5.3$)	4 ($c = 10.6$)
e_r	2.1×10^{-4}	7.3×10^{-3}	3.2×10^{-2}	5.9×10^{-3}
mdiv	-1.5×10^{-6}	1.0×10^{-6}	9.7×10^{-5}	-2.1×10^{-6}
Results with matrix $S = I$			
Case	1 ($c = 10.1$)	2 ($c = 11.3$)	3 ($c = 6.3$)	4 ($c = 10.5$)
e_r	2.1×10^{-5}	1.9×10^{-4}	4.4×10^{-4}	7.3×10^{-5}
mdiv	8.4×10^{-7}	-5.5×10^{-6}	-1.6×10^{-6}	3.1×10^{-6}

Case 1 Exact Dirichlet boundary conditions (benchmark).

Case 2 $\lambda = 0$ *on* Γ_T, and $-S^{-1}\nabla\lambda \cdot \hat{\mathbf{n}} = 0$ *on* Γ_V.

Case 3 A layer of ghost nodes beyond Γ_V, and b.c. as in case 2.

Case 4 A layer of ghost nodes beyond $\Gamma_D = \Gamma_T \cup \Gamma_V$, and b.c. as in case 2.

The problem is solved on each case with two different diagonal matrices, as shown in Fig. 3 and in Table 3, where we summarize the numerical results. Concerning realistic numerical solutions, it is observed that the best strategy is given by case 4. Once again, solutions with $S = I$ are more accurate, since the relative error is reduced by one or two orders of magnitude for each case.

5 Conclusions

Collocation methods based on radial basis functions are a simple reliable alternative for the reconstruction of 2D and 3D wind fields, it is a truly mesh free method, dimensional independent, easier to implement and cheaper than other well-known discretization schemes. For instance, to get about the same accuracy for the 2D problem, the triangular mesh for the FEM includes 6,561 nodes, 5,241 of them in the interior, while the RBF uses only 36 nodes, 16 of them in the interior.

The true velocity wind field is recovered with excellent accuracy. Our numerical study indicates that the appropriate boundary conditions for this problem are $\lambda = 0$ *on* Γ_T, $-S^{-1}\nabla\lambda \cdot \hat{\mathbf{n}} = 0$ *on* Γ_V, and $-S^{-1}\nabla\lambda \cdot \hat{\mathbf{n}} = \mathbf{u}^0 \cdot \hat{\mathbf{n}}$ *on* Γ_N. The introduction of ghost nodes produces an additional reduction of mean divergence and the introduction of the identity matrix, $S = I$, reduces the relative error and simplifies both the model and the algorithm. Also, it indicates that the dependence of solutions from matrix S, obtained by other authors, is actually related to inconsistent boundary conditions, which degrade numerical solutions by the introduction of high gradients on truncated boundaries.

Finally, we want to mention that these methodologies can be extended and applied to experimental fluid dynamics and computer vision. In particular, the reconstruction of velocity fields from experimental data, obtained by means of the PIV technique, is an important issue, (Adrian 2005). Its relation with computer vision is established by optical flow estimation (Ruhnau and Schnorr 2007).

References

Adrian RJ (2005) Twenty years of particle image velocimetry. Exp Fluids 39(2):159–169

Buhmann MD (2003) Radial basis functions: theory and implementations. Cambridge University Press, Cambridge

Ferragut L, Montenegro R, Montero G, Rodríguez E, Asensio ML, Escobar JM (2010) Comparison between 2.5-D and 3-D realistic models for wind field adjustment. J Wind Eng Ind Aerodyn 98(10–11):548–558

Flores CF, Juárez LH, Nuñez MA, Sandoval ML (2010) Algorithms for vector field generation in mass consistent models. J Numer Methods PDE 26(4):826–842

Kansa EJ (1990) Multiquadrics a scattered data approximation scheme with applications to computational fluid dynamics ii: solutions to parabolic, hyperbolic and elliptic partial differential equations. Comput Math Appl 19(8/9):147–161

Kitada T, Kaki A, Ueda H, Peters LK (1983) Estimation of the vertical air motion from limited horizontal wind data–a numerical experiment. Atmos Environ 17(11):2181–2192

Montero G, Rodríguez E, Montenegro R, Escobar JM, González-Yuste JM (2005) Genetic algorithms for an improved parameter estimation with local refinement of tetrahedral meshes in a wind model. Adv Eng Softw 36:3–10

Ratto CF (1996) An overview of mass-consistent models. In: Lalas DP, Ratto CF (eds) Modeling of atmosphere flow fields. World Scientific Publications, Singapore, pp 379–400

Ruhnau P, Schnörr C (2007) Optical stokes flow estimation: an imaging-based control approach. Exp Fluids 42(1):61–78

Sasaki Y (1958) An objective analysis based on the variational method. J Met Soc Jpn 36:77–88

Sherman CA (1978) A mass-consistent model for wind fields over complex terrain. J Appl Meteorol 17(3):312–319

Wang Y, Williamson C, Garvey D, Chang S, Cogan J (2005) Application of a multigrid method to a mass-consistent diagnostic wind model. J Appl Meteorol 44(7):1078–1089

Wendland H (2005) Scattered data approximation. Cambridge University Press, Cambridge

Dispersion of Air Pollutants in the Guadalajara Metropolitan Zone

Hermes Ulises Ramírez-Sánchez
and Mario Enrique García-Guadalupe

Abstract The objective of this work is to identify the main factors influencing the dispersion of air pollution in the Guadalajara Metropolitan Zone (ZMG). To this end, we analyzed the behavior of atmospheric pollutants, the speed of the winds and the presence of thermal inversions (IT) over the past 10 years. The results showed that during the early hours of the day, the dominant factors are the influence of IT and the frequency of calmer winds (winds between 0 and 5 km/h), which do not allow the dispersal of pollutants in the area. Past the noon (12:00 h) the solar radiation increases the temperature of the bottom layer of the IT, reaching the equilibrium temperature that breaking the IT, thus starting the dispersion of pollutants. This process occurs in addition to the increase of the wind speed in the afternoon, generating horizontal dispersion to the outside of the ZMG, possibly affecting the Toluquilla Valley. In conclusion, the dominant factors in the dispersion of air pollutants in the ZMG are thermal inversions and the wind speed.

1 Introduction

Different works have reported that weather has a great influence on the accumulation and dispersion of air pollutants from the sources of emissions. The natural environment (geography and topography) and urbanization (urban infrastructure, transport and industry), affect the air quality in large cities where high concentrations of pollutants prevail. These problems become more important because the

H. U. Ramírez-Sánchez (✉) · M. E. García-Guadalupe
Astronomy and Meteorology, Institute—University of Guadalajara,
Av. Vallarta 2802 Col. Arcos Vallarta, 44130 Guadalajara, JAL, Mexico
e-mail: ramirez@astro.iam.udg.mx

J. Klapp et al. (eds.), *Fluid Dynamics in Physics, Engineering and Environmental Applications*, Environmental Science and Engineering,
DOI: 10.1007/978-3-642-27723-8_35, © Springer-Verlag Berlin Heidelberg 2013

projected growth of the world's urban population increases both the sources of pollution and the number of exposed persons. Mobile sources are the primary cause of air pollution in large cities (Molina-Molina 2005). Céspedes (2005) reported that the meteorological factors that influence the concentration of pollutants in the air are the transport and dispersion, turbulence and limitations on the vertical dispersion for the thermal inversion effect (IT).

In Mexico, the work on environmental degradation dates back to 1980. Mexico City has a high-density of population, vehicles and industry resulting in accelerated technological development; therefore, it is not surprising that the most notorious environmental deterioration in this megacity is the air pollution (Molina-Molina 2005). In this city and in Los Angeles, the mountains trap pollutants long enough so that they suffer chemical mutations. In this case, there must be implemented stringent measures to prevent the accumulation of unacceptable levels of harmful contaminants. The objective of this work is to identify the main factors influencing the dispersion of air pollution in the ZMG, Jalisco, Mexico, during the period 2001–2010.

2 Materials and Methods

The ZMG has a population of 4.5 million inhabitants and a fleet of vehicles from 2 million in daily circulation. Thus, its air quality problem relates to the urban and technological development (Ramírez et al. 2009). The sources of air pollutants are transport (73.5), industry (1.1), services (4.2), and the vegetation and soil (21.2 %) (SEMARNAP/GEJ/SS 1997). Other factors are the deterioration of green areas by changes of land use and forest fires in urban forests; in particular, in the forest "La Primavera". The ZMG is located at the center of the State of Jalisco (Fig. 1a). It covers the municipalities of Guadalajara, Tlaquepaque, Tonalá and Zapopan. In reference to its topography, it is situated in the basin of the Rio Grande Santiago Valley, the Atemajac Valley and the Tonalá plain; between the mountains of the Sierra Madre Occidental and the Neovolcanic belt. Its relief is partially a natural physical barrier to the movement of the wind to impede the dispersion of pollutants from the source of emission to outside the area. It is subject to the influence of anticyclone systems generated both in the Gulf of Mexico and in the Pacific Ocean, causing great atmospheric stability by preventing vertical mixing of the air. In addition, become of its latitude it receives abundant solar radiation making its atmosphere highly photoreactive.

The Ministry of the Environment and Sustainable Development for the Government of Jalisco (SEMADES/GEJ) operates an automatic network for atmospheric monitoring (RAMA) with eight stations distributed in the ZMG (Fig. 1b), which measure atmospheric pollutants: carbon monoxide (CO), nitrogen dioxide (NO_2), ozone (O_3), sulfur dioxide (SO_2) and particles smaller than 10 µm (PM_{10}) and same meteorological variables. The analysis of IT was determined by atmospheric radioprobes and virtual probes obtained from the page www.ready.noaa.gov/ready/amet.html. A database of winds was built for the period 2001–2010, reporting the

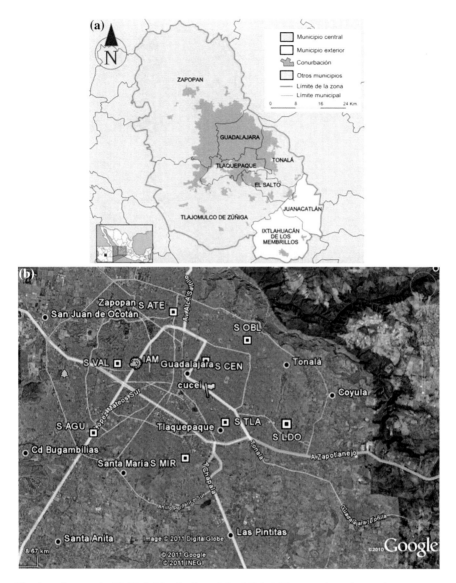

Fig. 1 **a** Case study: ZMG. **b** Distribution of the SEMADES network stations 2001–2010

wind daily, monthly and yearly from which the winds means, maximums and minimums were calculated. Also, with information on concentrations of atmospheric pollutants databases were built, obtaining mean and maximum trends for the year, monthly and day the behavior in the period for each pollutant. The analysis of IT was performed for the period 2003–2007.

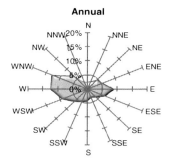

Fig. 2 Dominant winds 2001–2010

3 Results

3.1 Behavior of the Wind in the ZMG: Direction

The dominant wind comes from the West-Northwest, West and West-Southwest 13.17, 12.72 and 9.43 % respectively; followed by the winds from the East-Northeast, East and East-Southeast with 5.62, 9.06 and 6.53 %. In all cases, their speeds are from 6 to 19 km/h and, temporarily reach from 20 to 38 km/h. The directions North-Northeast, North, and North-Northeast occur with 2.94, 2.66, and 2.97 % frequency respectively. The South-Southwest, South and Southeast represent 4.96, 4.39 and 3.43 % of the total frequency. Northeast (4.76), Southeast (4.31), Southwest (6.30) and Northwest (6.75) complete 100 %. The winter-spring period has 24.08, with a circulation from Western winds with directions West-Southwest, West and West-Northwest; while the summer-autumn period has 16.2, with a circulation of Eastern winds with directions North-East, East-Northeast, East, East-Southeast and Southeast. The winds of the North and South share 8.33, having little significance in local circulation (Fig. 2).

3.2 Behavior of the Wind in the ZMG: Speed

The winds were marked by periods of calm (Beaufort scale 0–1), with an average frequency of 38.57 %. This demonstrates the great potential of accumulation of pollutants in the area due to the lack of ventilation that favors dispersion. The months in which the frequency is dominated by calm winds occurs are October to January, with 38.56, 46.04, 43.92 and 41.81 % respectively. The months between June and August have high frequencies for calm winds; however, the presence of precipitations clean the atmosphere. The characteristic period with decreased calm winds occurs between February and June. Frequencies were classified in daily intervals: night (00:01–06:00), morning (06:01–12:00), afternoon (12:01–18:00)

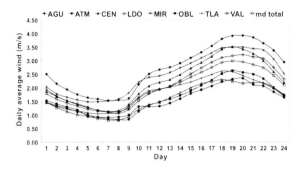

Fig. 3 Daily average wind (2001–2010)

and evening (18:01–24:00), observing periods of significant calm winds in the night and in the morning with more than 50 %; in the afternoon and evening there is greater flow of the wind (Fig. 3).

3.3 Behavior of Atmospheric Pollutants in the ZMG

3.3.1 PM$_{10}$ (NOM-025-SSA1-1993—modified DOF-26/09/2005 (120 µg/m^3))

The highest daily average concentrations occurred to the South-southeast of the ZMG (above the EPA standard: 50 µg/m^3 and peaks above the NOM). Maximums were from 7 to 11 and 19 to 23 h (Fig. 4). These maximums are the result of major periods of stability of the local atmosphere and the emission of pollutants both mobile and fixed in the above ranges. In terms of the monthly performance 2001–2010, the maximum concentrations were above the NOM and EPA throughout the period (140–500 µg/m^3). Average maximums ranged from 43 to 188 µg/m^3 and averages 20–95 µg/m^3. The annual performance of the period showed concentrations above the NOM and the EPA. The maximums corresponded to 2004 (499.8), 2003 (412.4), 2001 (403.3) and 2007 (347.9 µg/m^3). A significant decrease was observed in the year 2010 (137.7 µg/m^3) but it was still above standard. The average maximum ranged from 46 to 115 µg/m^3 and the average of 20–45 µg/m^3. PM$_{10}$ is identified as the most significant pollutant primarily in the South and Southeast of the ZMG.

3.3.2 O$_3$ (NOM-020-SSA1-1993—DOF-23/12/1994-modified 30/10/2002 (0.11 ppm))

The average daily behavior shows maximum concentrations between 12 to 19 h (Fig. 5), given the location of the ZMG (latitude 20°N), solar radiation is a determining factor in the generation of this secondary pollutant. The minimum occurred at

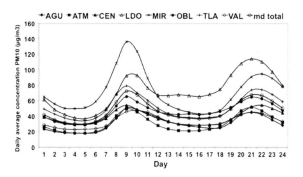

Fig. 4 Daily average behavior of PM₁₀

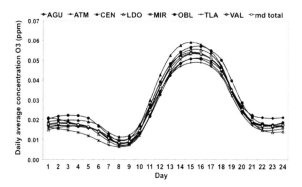

Fig. 5 Daily average behavior of O₃

night and in the morning. The O_3 is located within the EPA standard (0.09 ppm) and NOM in the decade analyzed. In relation to the monthly performance, virtually the entire analyzed period exceeded EPA and NOM standards at extreme levels. The maximum concentrations ranged from 0.0950 to 0.2870 ppm. The average maximum varied from 0.0364 to 0.1110 ppm; while the monthly averages ranged from 0.0171 to 0.0520 ppm. The highest peaks were observed in the intervals of drought and summer, characterized by increased solar radiation in the area. The maximum concentrations per year were located in 2005 (0.2290), 2010 (0.2280), 2009 (0.1770) and 2008 (0.1890 ppm), above EPA and NOM standards. The average maximum ranged from 0.0512 to 0.0764 ppm and averages from 0.0244 to 0.0338 ppm.

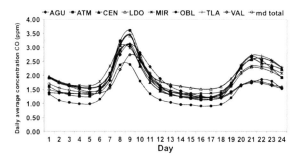

Fig. 6 Daily average behavior of CO

This pollutant has a greater influence towards the Centre and Southeast of the ZMG, in such a way that it presents itself as a risk factor for the population, in particular, when the atmosphere is highly photoreactive.

3.3.3 CO (NOM-021-SSA1-1993—DOF-23/12/1994 (11 ppm))

The daily average CO featured a peak of maximum concentrations; between 7 and 11 a.m. (Fig. 6). The most affected areas were located to the North, Central and Southeast of the ZMG. This pollutant remains within the EPA (9 ppm) and NOM regulations in the daily average. As for the monthly performance, the maximum concentrations exceeded the NOM by 50 and the EPA by 25 %. The maximums of CO ranged from 4.50 to 53.50 ppm. These concentrations are risk factors for the population. The events were recorded in dry periods (March-June). In the annual performance, the maximum concentrations occurred in 2001 (53.50) and 2002 (47.20); above the EPA and NOM standards. The average maximum ranged from 3.0327 to 4.9303 ppm and averages from 1.2839 to 2.3065 ppm. The difference between extreme maximums mean, maximums mean and means, demonstrates the presence of peaks over short intervals.

3.3.4 NO$_2$ (NOM-023-SSA1-1993—DOF-23/12/1994 (0.21 ppm))

The daily behavior showed two peaks of maximum concentrations; the first, from 7 to 12 and the second between 20 and 24 h (Fig. 7). The most affected areas were located to the North, Centre and South. The average concentrations were observed to be EPA (0.05) and NOM (0.21 ppm) regulations. In the monthly performance,

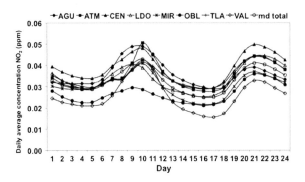

Fig. 7 Daily average behavior of NO₂

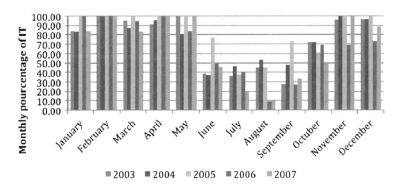

Fig. 8 Monthly % of IT

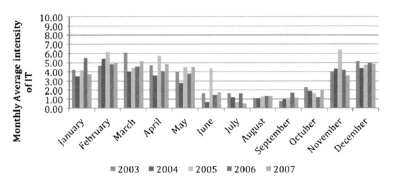

Fig. 9 Monthly average intensity of IT (°C)

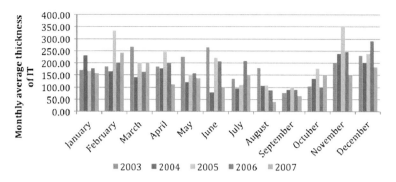

Fig. 10 Monthly average thickness of IT (m)

extreme values ranged from 0.0630 to 0.5230 ppm; these represent risk factors for the population. The average maximum varies from 0.0290 to 0.0951 ppm and monthly averages from 0.0155 to 0.0513 ppm. In the annual performance, the maximum concentrations occurred in 2001 (0.5210), 2002 (0.4260), 2003 (0.5230) and 2004 (0.4190 ppm), with higher values than the standards (EPA and NOM). The average maximum ranged from 0.0420 to 0.0644 ppm and averages from 0.0219 to 0.0399 ppm.

3.3.5 SO_2 (NOM-022-SSA1-1993—DOF-23/12/1994 (0.13 ppm)

The daily average concentrations showed a peak between 8 and 12 h. The higher values were observed in the Central, South and South-Southeast of ZMG, staying within the NOM (0.13) and EPA (0.03 ppm) regulations. In the monthly performance, the maximum ranged from 0.020 to 0.5240 ppm. The average maximum ranged from 0.0024 to 0.0321 ppm and the monthly averages from 0.0015 to 0.0134 ppm. The most affected areas are located in Central, southern and Southeast of the ZMG. The extreme maximums for the annual behavior were recorded in 2001 (0.2960), 2002 (0.5340), 2004 (0.4120) and 2010 (0.1780 ppm), with levels higher than the NOM and EPA regulations. The average maximum ranged from 0.0076 to 0.0205 ppm and averages between 0.0037 and 0.0112 ppm.

3.4 Behavior of Thermal Inversions in the ZMG

In the period 2003–2007, moderate to strong ITs were the most frequent and occurred in the months of drought (November to May). February is the month, which has the greatest frequency of IT. Those of lower intensity were recurrent in the months of June to October, during the rainy season (Fig. 8). Most frequent ITs of average intensity were recorded during the months of drought (November to

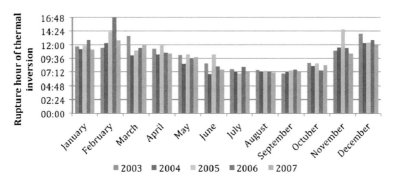

Fig. 11 Rupture hour of IT

May) with gradients above 4 °C (Fig. 9). The average thicknesses of the IT ranged between 50 and 350 m with maximums in the months of November, December and February (Fig. 10). Hours of breaking varied between 7 and 15 h. The maximums were presented during the months of November to May (Fig. 11).

4 Discussions

Atmospheric concentration of pollutants showed variable behavior, however, PM_{10} showed the higher concentrations followed by O_3, NO_2, CO and SO_2. The most affected areas were the South-Southeast, Southeast and East-southeast of the ZMG where a significant percentage of days exceed the NOM. The months of April to June showed high concentrations of O_3 and CO; whereas, December to March revealed acute concentration of CO, SO_2, PM_{10} and NO_2. An implicit factor which makes this happen is the frequent occurrence of low temperatures that accentuate the duration of the IT; in addition to the low humidity of the environment, which is an unfavorable situation for the dispersion of air pollutants. SEMARNAP/SS/GEJ (1997) conducted a study on the wind in Guadalajara in the period 1985–1990. They reported a prevailing wind from the West with 15.5 of the total frequency, followed by easterly winds (7.5); both with speeds of 5–20 km/h and, on a temporary basis from 21 to 35 km/h. The periods of calm winds showed a frequency of 44.3 %. Winds from the North and South share 5 % of the total frequency, representing little significant impact on the local circulation. The findings of this study differ from those of SEMARNAP/SS/GEJ, this may be due to multiple factors or be an indication of a change of wind patterns over the past 20 years; a period in which the area has experienced an exponential urban growth.

5 Conclusions

This work revealed, periods of calm with a frequency of 38.57 %, this results in the great potential of pollutant concentration in the area due to lack of ventilation that is conducive to dispersal to great distance. The most significant periods of stability occurs in the night and morning with more than 50 %, while in the afternoon and evening, there is an increase in the flow of the winds. In addition, it is most likely that this behavior is influenced by the orographic conditions of the ZMG and urbanization, as reported by Glynn and Heinke (1999); Barry and Chorley (1999). High levels of PM_{10} are the result of the influence of large periods of stability of the local atmosphere and polluting emissions from mobile and fixed sources, it is considered as the most significant contaminant, and it may represent a risk factor for the population. The O_3 has greater influence to the Centre and Southeast of the area; being a risk factor for the population, in particular, when the atmosphere is highly photoreactive. The highest peaks of O_3 can be seen at intervals of drought and in summer, characterized by increased solar radiation in the area and greater chemical reaction with primary pollutants. The problems of air pollution by NO_2 are also significant. This pollutant is key to the generation of secondary pollutants (O_3) and, therefore, a pollutant of potential risk in the ZMG. In relation to SO_2, the difference in the extreme maximum concentrations compared with the maximum average and average, demonstrate that the extremes occur in short intervals and pollution generated by this pollutant is not yet a major problem in the ZMG.

In the period 2003–2007 the moderate and strong ITs were the most frequent and observed in the months of November to May. February is the month observed to have the greatest frequency of IT. In short, the dominant factors in the dispersion of air pollutants in the ZMG are thermal inversions and the wind speed.

References

Barry RG, Chorley RJ (1999) Atmósfera, tiempo y clima. Omega S.A., Barcelona, España. 441 páginas. ISBN: 84-282-1182-5

Céspedes R (2005) Impacto de la contaminación atmosférica en la salud. Proyecto Aire Limpio: Evaluación del efecto de la contaminación atmosférica en la salud de la población del Municipio Cochabamba–Bolivia, pp 4–25

Glynn HJ, y Heinke GW (1999) Ingeniería ambiental. Pearson Educación-Prentice Hall, México, pp 216–217). ISBN 970-17-0266-2. 778 páginas

Molina LT, Molina MJ (2005) La calidad del aire en la megaciudad de México. Un enfoque integral. Fondo de Cultura Económica, 2005. ISBN: 968-16-7580-0. 463 páginas. México

Ramírez H, Andrade M, Bejaran R, García M, Wallo A, Pompa A, De la Torre O (2009) The spatial–temporal distribution of the atmospheric polluting agents during the period 2000–2005 in the Urban Area of Guadalajara, Jalisco, Mexico. J Hazard Mater 165(1–3):1128–1141

Modeling of Funnel and Gate Systems for Remediation of Contaminated Sediment

Fei Yan and Danny D. Reible

Abstract Capping is typically used to control contaminant release from the underlying sediments. While conventional capping doesn't necessarily provide the removal of contaminants, incorporating a "funnel and gate" reactive barrier with capping has the potential to treat contaminants or limit contaminant migration. The purpose of this study was to develop a model of funnel and gate systems for remediation of contaminated sediment. Numerical modeling of vertical two dimensional water flow and solute transport was built in COMSOL MULTI-PHYSICS 3.4. The model was employed to evaluate the performance of the funnel and gate system, i.e. residence time, removal efficiency, and breakthrough curve. Two types of gates, reactive and adsorptive gates, were evaluated for the remediation of phenanthrene contaminated sediment. The simulated results showed that the performance of the reactive gate depended on Damkohler number at the gate, and the adsorptive gate could effectively slow contaminate migration into water body, and decrease the maximum concentration at the gate. This model could potentially serve as a design tool of funnel and gate systems for a range of typical sediment capping conditions.

F. Yan
Department of Civil, Architectural, Environmental Engineering,
The University of Texas at Austin, 1 University Station C1786
Austin, TX 78712-0273, USA
e-mail: feiyan@austin.utexas.edu

D. D. Reible (✉)
Department of Civil, Architectural, Environmental Engineering, Bettie Margaret Smith
Chair of Environmental Health Engineering, The University of Texas at Austin,
1 University Station C1786 Austin, TX 78712-0273, USA
e-mail: reible@mail.utexas.edu

J. Klapp et al. (eds.), *Fluid Dynamics in Physics, Engineering and Environmental Applications*, Environmental Science and Engineering,
DOI: 10.1007/978-3-642-27723-8_36, © Springer-Verlag Berlin Heidelberg 2013

1 Introduction

Contaminated sediment has become a major concern at many sites throughout the United States and the world. A variety of organic and inorganic contaminants such as chlorinated solvents, aromatic hydrocarbons and heavy metals have been found in contaminated sediment and pose a risk to ecology and human health. An effective remediation method is needed for the management of contaminated sediment. Because of its high cost and limited effectiveness (Palermo et al. 1998; Reible et al. 2003), dredging and ex-situ disposal is sometimes not an efficient method for the treatment of contaminated sediment.

An alternative for the management of contaminated sediment is in situ capping—the placement of clean material (usually sand) on the sediment to isolate the contaminants into the overlying water (Palermo et al. 1998; Wang et al. 1991). In situ capping can be a relatively cost-effective and noninvasive approach compared to dredging but may not provide adequate containment under high groundwater upwelling conditions. By maintaining proper hydraulic control, groundwater seepage carrying contaminants can be funneled to a relative small area, or "gate", where physical, chemical, or biological processes promote the degradation or removal of contaminants (McMurtry and Elton 1985; Starr and Cherry 1994). By placing a funnel and gate reactive barrier at the sediment water interface, it provides opportunities for successful remediation of contaminated sediments subjected to groundwater upwelling (Fig. 1).

The design of funnel and gate systems generally requires bench scale treatability study, site investigation and computer modeling. Various models have been developed to simulate groundwater flow, contaminant fate and transport, and coupled geochemical and physical processes of funnel and gate systems for groundwater remediation (Bilbrey and Shafer 2001; Gupta and Fox 1999; Hatfield et al. 1996; Hudak 2004; Klammler et al. 2010; Painter 2004; Sedivy et al. 1999; Smyth et al. 1997). Though these models were successful in predicting hydraulic performance of funnel and gate systems in groundwater aquifers, contaminant fate and transport were not a primary focus in these studies. Further, there has been no successful application of funnel and gate models to problems of contaminated sediment. The objective of this study was to develop a funnel and gate model for sediments that describes both water flow and solute transport and could form the basis for more complete models that incorporate explicitly modeling of redox conditions and microbial degradation.

In this study, a general numerical model for water flow and solute transport in contaminated sediment and funnel/gate systems was presented. In particular, groundwater flow that discharges into sediment due to seepage or tidal fluctuations was considered to fully describe the system. Two types of gates, reactive and adsorptive gates, were evaluated for their performance to mitigate contaminant migration. Contaminant fate and transport in flow-through systems were investigated for multiple scenarios by varying reaction rates and adsorption coefficients. The model may be used to simulate contaminant behavior in natural systems or to predict contaminant removal by funnel and gate systems. In addition, the model could also be potentially used to evaluate the ability of hydraulic control by funnel

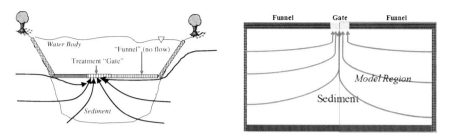

Fig. 1 A funnel and gate for the remediation of contaminated sediment

and gate systems. The model can serve as a design tool for funnel and gate systems for a range of typical sediment capping conditions.

2 Methods

A typical funnel and gate system for the remediation of contaminated sediment is shown in Fig. 1. On top of the contaminated sediment is a capping layer, with a large portion of "funnel" and a small portion of "gate". The engineered funnel has a low hydraulic conductivity, which guides the groundwater flow through the gate. The gate zone, filled with reactive materials and/or adsorptive materials, has a high hydraulic conductivity compared with the funnel. Above the funnel and gate is usually the water column, and the overlying water level may fluctuate significantly under tidal conditions. On the left and right borders of the sediment, pore water in the sediment usually interacts with groundwater flows. The direction of groundwater flow depends on the water level above the sediment and groundwater level. The layer underlies the sediment is often a low permeable sedimentary rock bed, which serves as a confining layer for contaminants migrating from upper sediment.

Steady state groundwater flow in the sediment and funnel/gate systems was simulated on vertical plane. The governing equation is as follows:

$$\nabla[K\nabla H] = 0. \tag{1}$$

Where K is hydraulic conductivity and H is hydraulic head.

Solute transport was simulated within the same domain as a transient problem. The governing equation of solute transport is as follows:

$$\left(1 + \frac{\rho_b K_d}{\theta}\right)\frac{\partial c}{\partial t} = \nabla\left[D\nabla c - \frac{v}{\theta}c\right] - \lambda c. \tag{2}$$

Where ρ_b is bulk density of the porous medium, K_d is partition coefficient between concentration in solid phase to concentration in liquid phase, D is hydrodynamic dispersion coefficient, v is Darcy velocity, θ is porosity, λ is first order reaction rate and c is solute concentration (mass per liquid volume).

The contaminated sediment domain was 15 m wide by 3 m deep, with a 0.15 m thick capping layer (a funnel and gate system) on the top. The gate was at the center of the capping layer, and gate width was set as one-tenth of the total capping width, so its width was 1.5 m. As only the right half of the system was simulated, the width of sediment and capping layer was halved in the simulated model.

The model described a homogeneous and isotropic sediment with a hydraulic conductivity K_{sed} of $3.53 * 10^{-5}$ m/s (10 ft/d). The gate was assumed to have the same hydraulic conductivity as the sediment in the basic scenario. The funnel was set as a low permeable media with a hydraulic conductivity of 10^{-9} m/s, which was a reasonable value for conventional soil-bentonite cutoff walls—a typical material of subsurface vertical barrier (Yeo et al. 2005). Sensitivity analyses were performed by varying hydraulic conductivity of gate (K_{gate}) over five orders of magnitude, resulting different K_{gate}/K_{sed} values.

In the groundwater flow model, the bottom and the left side of the model domain were specified as no flow boundaries, whereas the top and the right side of the model domain were specified as constant head boundaries. The head difference between the right and top boundary (ΔH) was set as 1 m.

In the solute transport model, phenanthrene was selected as a model compound due to its occurrence in sediment and toxicity to ecology. Because phenanthrene is a recalcitrant and persistent substance in a reduced sediment environment (i.e. away from the sediment–water interface), it was assumed that there was no degradation of phenanthrene in the sediment. In in-active gate scenario, the funnel and gate zones were assumed to have the same parameters as the sediment, i.e. no degradation and the same partition coefficient.

Two types of gates, reactive and adsorptive gates, were investigated for multiple scenarios. For the reactive gate, reactive materials are usually placed in the gate zone to remove the contaminants. Different first order reaction rates in gate zone were evaluated in the simulation as follows:

(1) $\lambda = 0/d$ (in-active gate scenario); (2) $\lambda = 1/d$; (3) $\lambda = 10/d$; (4) $\lambda = 100/d$; (5) $\lambda = 1000/d$.

An adsorptive gate is amended with a sorbent, which will prolong contaminant isolation by sequestering contaminants and retarding their transport from the sediment into the water body. Adding a sorbent layer to a sand cap has been successfully applied to isolate contaminants (Palermo et al. 1998; Murphy et al. 2006; Talbert et al. 2001). In the simulation of the adsorptive gate, sorbent with different partition coefficients K_d in gate zone were evaluated for the following scenarios:

(1) $K_{d\_gate} = K_{d\_sed}$ (in-active gate scenario); (2) $K_{d\_gate} = 10 * K_{d\_sed}$; (3) $K_{d\_gate} = 100 * K_{d\_sed}$; (4) $K_{d\_gate} = 1000 * K_{d\_sed}$

No flux conditions were assumed at the left, right and bottom boundaries, whereas at the top boundary, no dispersive flux condition was assumed.

The sediment was assumed to have a homogeneous initial concentration of unity ($c_0 = 1$ mg/L), and funnel/gate were assumed to be devoid of contaminant initially ($c_0 = 0$).

Simulation of groundwater flow and solute transport was conducted using Earth Science Model of COMSOL MULTIPHYSICS 3.4.

3 Results and Discussion

3.1 Groundwater Flow Model

Figure 2 shows the simulation results in the two-dimensional domain for the basic scenario. Groundwater flow was forced to go via the gate zone due to the low hydraulic conductivity of the funnel zone. Results shows that the model was successful in predicting the flow pattern of funnel and gate systems.

The gate zone and its vicinity had the largest flow velocity among the entire model region. It was caused by the placement of the funnel that had a significantly different hydraulic conductivity from the surrounding sediment/gate. Particularly, the maximum velocity was found at the joint point where funnel, gate and sediment meet.

As shown in Fig. 3, the vertical velocity had a significant variation along the horizontal direction in gate zone. The vertical velocity at the center of the gate ($x = 0$) was the smallest, and it increased towards the funnel zone. The difference in flow rate across the gate zone resulted in a difference in residence time, suggesting the level of contaminant removal varied across the gate zone. This is important for gate design and construction of funnel and gate systems because flow rate can affect the rates of change to gate permeability and reactivity over time. Significant variation in flow rate in the gate may lead to replenishment of the reactive material in some portions of the gate before others.

In the construction of funnel and gate systems, the materials used in the gate zones sometimes have different hydraulic conductivity. Table 1 illustrates the effect of K_{gate}/K_{sed} on the velocity in the gate zone. The harmonic mean was used instead of arithmetic mean because the former could be used to calculate mean residence time (Selim and Ma 1998). A decrease of hydraulic conductivity of gate zone caused a lower flow rate. However, an increase of K_{gate}/K_{sed} didn't cause an increase of flow rate. The harmonic means of vertical velocity v_y were 98 and 97 % of the basic scenario for $K_{gate}/K_{sed} = 10$ and $K_{gate}/K_{sed} = 100$, respectively (Table 1). Profiles of v_y at the entrance and exit of gate zone are shown in Fig. 3. If K_{gate}/K_{sed} increased, v_y at the edge of the gate increased but at the center of the gate decreased. These results indicate that increasing hydraulic conductivity of the gate didn't increase average flow rate in the gate zone, but increased lateral variation. The implication of these results is that a larger K_{gate}/K_{sed} should always be avoided if a uniform flow field is one of the goals in gate design.

3.2 Solute Transport Model

3.2.1 Reactive Gate

The solute transport model simulated 50 years to capture concentration changes in funnel/gate and sediment for the reactive gate. First order reaction rates over four orders of magnitude were specified to gate zone in different scenarios. The coupled

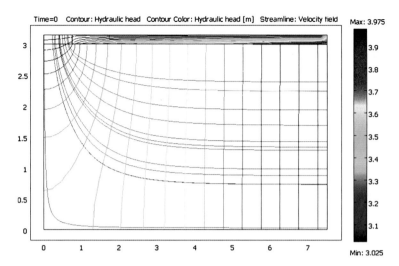

Fig. 2 Contour of hydraulic head, and streamline (velocity field) at steady state of funnel/gate and sediment

Fig. 3 Vertical velocity at the entrance and exit of the gate for $K_{gate}/K_{sed} = 1$ and $K_{gate}/K_{sed} = 100$

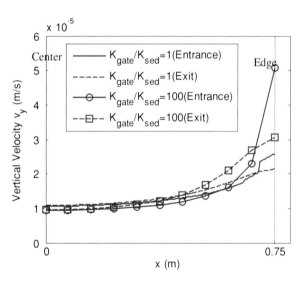

reaction and transport processes in gate zone were characterized by a dimensionless number, Damkohler number (Da), which is defined as follows (Rosner 2000):

$$Da = \theta\lambda l / U_y. \tag{3}$$

Where θ is porosity, λ is first order reaction rate ($\lambda = 0, 1, 10, 100, 1000/d$), l is the thickness of the gate, i.e. 0.15 m, and U_y is the harmonic mean of vertical velocity in gate zone, i.e. $1.34 * 10^{-5}$ m/s.

Table 1 Sensitivity analysis of K_{gate}/K_{sed} on harmonic mean of vertical velocity in gate zone

K_{gate}/K_{sed}	Harmonic mean of v_y in gate zone	Ratio of harmonic mean of v_y to that of the basic scenario
0.01	2.03×10^{-6}	0.15
0.1	9.11×10^{-6}	0.68
1	1.34×10^{-5}	1
10	1.31×10^{-5}	0.98
100	1.30×10^{-5}	0.97

Fig. 4 Breakthrough curves of the funnel/gate for different Damkohler numbers. Values were average concentrations at the exit of the gate

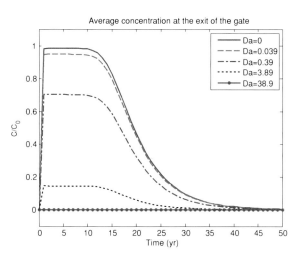

The Damkohler number describes the effect of reaction ($\theta\lambda$) relative to that of convection (U_y/l). At high Damkohler numbers (Da $\gg$ 1), reaction times are much faster than transport times for a given length scale.

In this study, all the solute transport simulations were performed under the basic scenario of groundwater flow, so the velocity didn't vary, and Da was only dependent on λ. Figure 4 shows breakthrough curves of the funnel/gate for different Damkohler numbers. Initially concentrations at the exit of the gate were zero, and then it increased to a steady state concentration and kept at this level for about 10 years. The levels of this steady state concentration were inversely related to Da. Concentrations dropped to 0.1 % of the initial concentration if Da was as large as 38.9.

3.2.2 Adsorptive Gate

The solute transport of the adsorptive gate was simulated with several different partition coefficients K_{d_gate} for a period of up to 500 years. To quantify the difference in adsorption capacity between gate zone and sediment, a parameter R was defined as the ratio of partition coefficient in the gate zone (K_{d_gate}) to partition coefficient in sediment (K_{d_sed}). As shown in Fig. 5, breakthrough of contaminant occurred as soon as simulation started for the in-active gate scenario. The

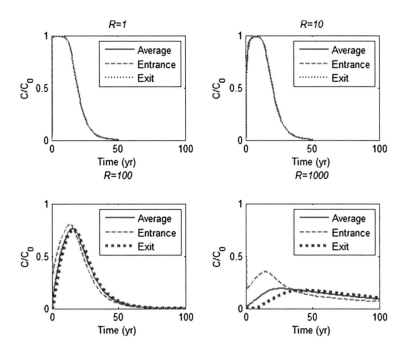

Fig. 5 Concentrations at the entrance and exit of the gate, and mean concentrations in the gate zone for different scenarios

time of peak concentration was delayed by a larger partition coefficient in the gate zone (K_{d_gate}), and the spreading or "tailing" of breakthrough curves was also enhanced by a larger K_{d_gate}. Further, for in-active gate ($R = 1$) and $R = 10$ cases, there was almost no difference between concentrations at the exit and the entrance, but for $R = 100$ and $R = 1000$, peak concentrations at the exit arrived later, and the magnitude was smaller compared to that at the entrance. Analysis of mass balance shows that a larger value of R caused a slower transport rate. The time when 90 % of the mass was transported out of the system was 21.7, 22.8, 38.4 and 235.8 years for $R = 1$, 10, 100 and 1000, respectively.

4 Conclusions

A numerical model of funnel and gate systems for remediation of contaminated sediment was developed in this study. Two dimensional groundwater flow and solute transport simulations were carried out for multiple scenarios.

The groundwater flow model was used to simulate hydraulic head, velocity field, etc. at steady state. Velocity field around gate zone shows that the maximum velocity in the model region was at the joint of gate, funnel and sediment. Significant lateral variation in vertical velocity occurs across gate zone, which may cause difference in durability and longevity of gate materials at different locations. Sensitivity analysis reveals that decreasing hydraulic conductivity of the gate (K_{gate}) decreased flow rate in gate zone, and reduced the lateral variation in vertical velocity. Increasing hydraulic conductivity in the gate increased flow rate at the edge and decrease flow rate at the center in gate zone, resulting in more significant lateral variation. Further, increasing K_{gate} resulted in a small drop of harmonic mean of vertical velocity.

Transient solute transport simulations were conducted to evaluate the performance of reactive and adsorptive gates. Removal efficiency of the reactive gate was dependent on Damkohler number in gate zone. An adsorptive gate with a larger adsorption capacity could effectively slow contaminate migration into water column, and decrease concentrations at the gate.

This study demonstrated that the model presented here could potentially serve as a powerful tool to predict groundwater flow and contaminant transport of funnel and gate systems for remediation of contaminated sediment.

Acknowledgments The project described was supported by Award No.: 1R01ES016154-01 from the National Institute of Environmental Health Sciences, National Institute of Health. The content is solely the responsibility of the authors and does not necessarily represent the official views of the NIEHS or the NIH.

References

Bilbrey LC, Shafer JM (2001) Funnel-and-gate performance in a moderately heterogeneous flow domain. Ground Water Monit Rem 21(3):144–151

Gupta N, Fox TC (1999) Hydrogeologic modeling for permeable reactive barriers. J Hazard Mater 68(1–2):19–39

Hatfield K, Burris DR, Wolfe NL (1996) Analytical model for heterogeneous reactions in mixed porous media. J Environ Eng-Asce 122(8):676–684

Hudak PF (2004) Effects of funnel and gate geometry on capture of contaminated groundwater. Bull Environ Contam Toxicol 72(3):557–563

Klammler H, Hatfield K, Kacimov A (2010) Analytical solutions for flow fields near drain-and-gate reactive barriers. Ground Water 48(3):427–437

McMurtry DC, Elton RO (1985) New approach to in situ treatment of contaminated ground waters. Environ Prog 4:168–170

Murphy P, Marquette A, Reible D, Lowry GV (2006) Predicting the performance of activated carbon-, coke-, and soil-amended thin layer sediment caps. J Environ Eng-Asce 132(7):787–794

Painter BDM (2004) Reactive barriers: hydraulic performance and design enhancements. Ground Water 42(4):609–617

Palermo M, Maynord S, Miller J, Reible D (1998) Guidance for in situ subaqueous capping of contaminated sediments. Great Lakes National Program Office, Chicago

Reible D et al (2003) Comparison of the long-term risks of removal and in situ management of contaminated sediments in the Fox River. Soil Sediment Contam 12(3):325–344

Rosner DE (2000) Transport processes in chemically reacting flow systems. General Publishing Company, Toronto, p 436

Sedivy RA, Shafer JM, Bilbrey LC (1999) Design screening tools for passive funnel and gate systems. Ground Water Monit Rem 19(1):125–133

Selim HM, Ma L (1998) Physical nonequillibrium in soils: modeling and application. Ann Arbor Press, Chelsea, p 8

Smyth DJ, Shikaze SG, Cherry JA (1997) Hydraulic performance of permeable barriers for in situ treatment of contaminated groundwater. Land Contam Reclamation 5(3):131–137

Starr RC, Cherry JA (1994) In situ remediation of contaminated ground-water—the funnel-and-gate system. Ground Water 32(3):465–476

Talbert B, Thibodeaux LJ, Valsaraj KT (2001) Effectiveness of very thin soil layers in chemical release from bed sediment. Environ Prog 20(2):103–107

Wang XQ, Thibodeaux LJ, Valsaraj KT, Reible DD (1991) Efficiency of capping contaminated bed sediments insitu.1. laboratory-scale experiments on diffusion adsorption in the capping layer. Environ Sci Technol 25(9):1578–1584

Yeo SS, Shackelford CD, Evans JC (2005) Membrane behavior of model soil-bentonite backfills. J Geotech Geoenviron Eng 131(4):418–429

Analysis of Transport Parameters for a Cr(VI) Contaminated Aquifer in México

Lázaro Raymundo Reyes-Gutiérrez, Ramiro Rodríguez-Castillo, Elizabeth Teresita Romero-Guzmán and José Alfredo Ramos-Leal

Abstract In the Buenavista area of Leon City, Mexico, Cr(VI) groundwater contamination was detected, originating from an industrial landfill with chromium compounds. A 2D vertical simulation model was established for the Buenavista study area. Laboratory and field data were incorporated into a finite element groundwater flow model and a solute transport model to analyze the transport parameters in the Buenavista shallow aquifer. A sensitivity analysis was performed to obtain values representative of the transport parameters (hydraulic conductivity [K], longitudinal, horizontal and vertical transverse dispersivities [α_L, α_{TV}], distribution coefficient [K_d], the initial concentration [C_o] and pumping rates [Q]). This analysis allowed a good calibration of the model. The incorporation of the resulting set of parameters in the finite element model enabled the reproduction the observed contaminant plume in Buenavista close to 95 % match. The values obtained were $\alpha_L = 50.0$ m, $\alpha_{TV} = 2.5$ m, $K_d = 0.007$ mL/g, $C_o = 160$ mg/L and $Q = 100$ m^3/d. The sensitivity analysis indicated that the dispersion of the Cr(VI) plume is most sensitive to variations in hydraulic conductivity, the

L. R. Reyes-Gutiérrez (✉) · E. T. Romero-Guzmán
Instituto Nacional de Investigaciones Nucleares ININ, Carretera México-Toluca S/N,
52045 Mexico, C.P., Mexico
e-mail: raymundo.reyes@inin.gob.mxraregu@gmail.com

E. T. Romero-Guzmán
e-mail: elizabeth.romero@inin.gob.mx

R. Rodríguez-Castillo
Instituto de Geofísica, Universidad Nacional Autónoma de México, UNAM, Mexico,
Mexico
e-mail: ramiro@geofisica.unam.mx

J. A. Ramos-Leal
Instituto Potosino de Investigación Científica y Tecnológica A.C., IPICyT,
Camino a la Presa San José 2055, 78216, San Luis Potosí, C.P., Mexico
e-mail: jalfredo@ipicyt.edu.mx

J. Klapp et al. (eds.), *Fluid Dynamics in Physics, Engineering and Environmental Applications*, Environmental Science and Engineering,
DOI: 10.1007/978-3-642-27723-8_37, © Springer-Verlag Berlin Heidelberg 2013

distribution coefficient, longitudinal and transverse dispersivity and pumping rates. In addition to the sensitivity analysis, it was observed that Q strongly affects the plume geometry.

1 Introduction

Chromium leaching from an industrial waste landfill affected the groundwater quality of a shallow aquifer located close to the site, in the Buenavista area of León Valley, in central Mexico. These wastes, containing Cr(VI), originated at the Química Central (QC) industrial facility, which utilized chromium in industrial processes. The concentration of Cr(VI) reached more than 50 mg/L in the La Hulera well, which serves as part of their water supply. Following the discovery of chromium contamination in the groundwater, a proposed remediation scheme was analyzed based on groundwater pump-and-treat through a simulation–optimization approach, where the simulation model would allow for the definition of different choices for the location of wells and pumping rates (Reyes-Gutiérrez 1998). Figure 1 shows the study area and the simulation domain for the field of existing wells adjacent to QC, as well as a network of six multi-level piezometers at a depth of 30 m installed by the Geophysics Institute of UNAM (Rodríguez et al. 1991) in late 1990. Vertical and periodic monitoring, conducted at three different depths every 15 days during 3 years, permitted the definition of the plume geometry of Cr(VI) in the affected area. The plume is located in a sandy aquifer with a relatively fast groundwater flow of $v = 0.66$ m/d. The resulting plume geometry allowed for the estimation of preliminary values of the dispersion coefficient using breakthrough curves analysis (Reyes-Gutiérrez 1998). The longitudinal and transverse dispersivities were determined using the Jiao method (1993). Armienta (1992) performed batch tests of undisturbed cores of the piezometers in order to obtain the distribution coefficient [K_d] of the aquifer material.

1.1 Description of the Buenavista area, León Guanajuato Valley

Figure 2 shows a shallow aquifer and a semi-confining unit that form the aquifer system: the shallow formation is composed of fine-medium sands and fine gravel with an overlay of a discontinuous clay lens that has very low hydraulic conductivity ($K = 1 \times 10^{-6}$ m/s). A clay package underlies the shallow aquifer. Castelán and Villegas (1995) provide a detailed description of the local geology. The regional flow has a northeast–southwest direction. Together with these features, the flow is partially modified by the extraction rates in the local area.

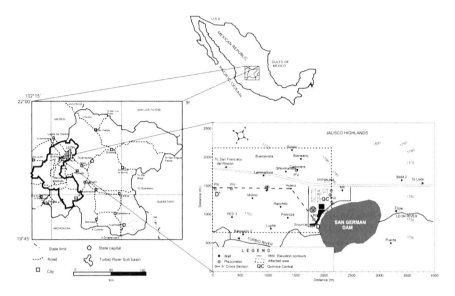

Fig. 1 The study area. Buenavista Guanajuato

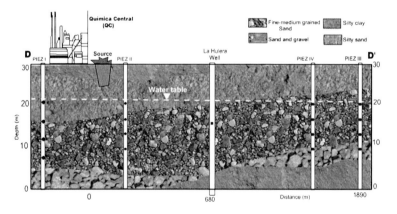

Fig. 2 Geological profile, piezometers I, II, III, IV and La Hulera well

1.2 Contaminant Source

A chromium industrial waste landfill was built in 1982, with a surface area of 3,750 m^2 and a capacity of 38,400 tons. It was located in a clay unit with low hydraulic conductivity and designed to a depth of 6 m; nevertheless, its capacity was breaking the container.

A sandy interface was found at said depth, allowing for the leaching of Cr(VI) into the water table. The exposure to residues was suspended in late 1993, at which time they were removed and deposited into a temporary container. Although the residues were deposited inside impermeable polyethylene bags (caliber 1000) they exceeded the landfill capacity and, as a result, the cover was broken. These conditions allowed for chromium leaching and, in early 1985, contamination of groundwater with chromium was detected, affecting the shallow aquifer. The chromium entrance into the aquifer was delayed for a time after the start of the deposition process. This delay was estimated as follows: considering that the material that underlies the source in the landfill is sand, and with an effective porosity of 0.35, a flow velocity of 0.3 m/d was estimated by means of the values of K_v obtained by an infiltration test. The depth of water level is 2.5 m below the source. Thereby, the mean transit time for leached chromium from the source to the aquifer is 8.3 days. The vertical hydraulic gradient under the landfill is a constant value near 1.0, and it was assumed that the transit time in 1985 was similar. In addition, the horizontal velocity along the aquifer was estimated at 0.66 m/d (Reyes-Gutiérrez 1998). Therefore, the arrival time of Cr(VI) to La Hulera well in 1985, located 680 m from the source, was 2.7 years. In addition to this information, the entrance of chromium into the aquifer started in 1982.

2 Experimental

A numerical model was integrated based on the Istok code (Istok 1989). The SUTRA (Voss 1984) model was also used (Reyes-Gutiérrez 2007), obtaining similar results. The following assumptions were made during the development of the numerical model: (1) the flow is in steady-state; (2) the movement of the contaminant is in the section plane, (3) the porous medium is homogeneous and isotropic in comparison to the dispersivity, and; (4) the adsorbed and soluble phases of Cr(VI) are related through equilibrium sorption isotherms.

2.1 Governing Equations

The governing equations for the two-dimensional simulation model of the Buenavista area can be written in vector form. For the groundwater flow in saturated porous media, the equation can be written as:

$$S_s \frac{\partial h}{\partial t} - \nabla_{xy} \cdot (\mathbf{K} b \cdot \nabla_{xy} h) = \pm Qj(x, y, t) \tag{1}$$

and for the solute transport, the equation describing advective–dispersive transport in a uniform flow field, with effect of chemical reactions, can be written as:

$$\frac{\partial C}{\partial t} = \nabla_{xy} \cdot (\mathbf{D} \cdot \nabla_{xy} C) - \mathbf{V} \cdot \nabla_{xy} C \pm \Gamma_j(x, y, t) \tag{2}$$

The two main equations are linked through Darcy's law:

$$\mathbf{V} = -\frac{\mathbf{K}}{n} \cdot \nabla h \tag{3}$$

where S_s is the specific storage coefficient, h is the hydraulic head, K is the hydraulic conductivity tensor, b is the aquifer saturated thickness, Q_j is the source/sink term for flow (pumpage and recharge), C is the solute concentration, D is the directionally dependent dispersion tensor, V is the vector velocity field, Γ_j is a source-sink term of solute that represents the rate at which the dissolved species is removed from the solution (solute mass per unit volume of solution per unit time), and n is the aquifer effective porosity. The hydrodynamic dispersion tensor is assumed to be of Fickian form, dominated by advection (Bear 1979):

$$D_{ij} = \alpha_T |v| \delta_{ij} + (\alpha_L - \alpha_T) \frac{v_i v_j}{|v|} \tag{4}$$

where δ_{ij} is the Kronecker delta, α_L and α_T are the longitudinal and transverse dispersivities, respectively, and $|v|$ is the magnitude of V. Following assumption 4, we used the contaminant transport K_d model (Langmuir 1997):

$$C_s = K_d C; \quad \text{and} \quad \frac{\partial C_s}{\partial C} = K_d$$

where C_s is the adsorbed concentration in the solid phase and K_d is the distribution coefficient. This model assumes that the interaction between the solid phase and the groundwater is reversible and that under isothermal conditions the concentration in solution is a function of dispersion and general form of the 2D transport equation is:

$$\nabla_{xy} \cdot (\mathbf{D} \cdot \nabla_{xy} C) - \mathbf{V} \cdot \nabla_{xy} C = \left[1 + \frac{\rho_b K_d}{n} \right] \frac{\partial C}{\partial t} \tag{6}$$

where $R = [1 + (\rho_b/n)K_d]$ is the retardation factor and ρ_b is the bulk density. The boundary conditions for Eqs. (1) and (6) generally are of three types: (a) specified head and concentration along a boundary (Dirichlet condition); (b) specified concentration gradient and discharge flux or dispersive flux across a boundary (Neumann condition), and; (c) specified head and concentration gradient, or total flux, both along a boundary and across that boundary, yielding a combination of (a) and (b) (Cauchy condition). Darcy's law and contaminant transport are simulated with the Galerkin finite element method, subject to appropriate initial and boundary conditions. The transport equation is derived from the principle of mass conservation. Linear triangle and linear rectangle elements are used to facilitate the discretization of the area.

2.2 Stability Criteria

The groundwater flow and mass-transport models provide stable and accurate results when they are employed with proper spatial and temporal discretization. In general, spatial discretization requires a fine mesh in areas where accurate results are required or where parameters vary greatly over short distances; the spatial discretization should also be consistent with the dispersivity parameters. Daus et al. (1985) recommends grid spacing, in which the grid Peclet number (ratio of spatial discretization and dispersion length) and the Courant number (ratio of advective distance during a time-step to spatial discretization) should match the following constraints in the section plane:

$$Pe_x = \frac{\Delta x}{\alpha_L} \leq 2, \; Pe_y = \frac{\Delta y}{\alpha_{TV}} \leq 2 \tag{7}$$

and

$$C_x = \frac{V_x \Delta t}{\Delta x} \leq 1, \; C_z = \frac{V_y \Delta t}{\Delta y} \leq 1 \tag{8}$$

where Pe_x, and Pe_y are the Peclet numbers in the x and y directions, respectively; C_x, and C_y are the Courant numbers; Δx is the grid spacing in the direction of flow and Δy is the grid spacing in the direction perpendicular to the flow; α_L, and α_{TV} are the longitudinal and vertical transverse dispersivities, respectively, and; Δt is the time-step. The numerical calculations were done with the finite element mesh shown in Fig. 3a. In order to define the transport model, the transport parameters used are a horizontal groundwater velocity of magnitude $v_x = 0.66$ m/d, $\Delta x = 100$ m, $\Delta y = 3$ m, $\alpha_L = 50$ m, $\alpha_{TV} = 2.5$ m, with a solution time $\Delta t = 1$ day. This discretization yields local grid Peclet numbers $Pe_x = 2.0$ and $Pe_y = 1.2$, and Courant numbers $C_x = 0.0066$ and $C_y = 0.22$. These parameters correspond to equally dominant dispersion and advection taking place along longitudinal and transverse directions in a velocity field that is uniform in the longitudinal direction.

2.3 Conceptual Model of the Buenavista Aquifer

The conceptual model represents the conditions along cross-section D-D' of Fig. 1, hydrogeological conditions Fig. 2 and it is based on the assumption that the flow field is found in steady-state. The aquifer is permeable and is assumed to have homogeneous properties and variable thickness. Figure 3 describes the physical system, together with the finite element mesh consisting of 128 nodes and 190 elements that were used for the computations. For the numerical model, constant-head conditions were specified along the east and west edges of the model domain, the top was simulated as a specified-flow boundary and the bottom of the aquifer was approximated as a no-flow boundary. The flow rate (or recharge) specified

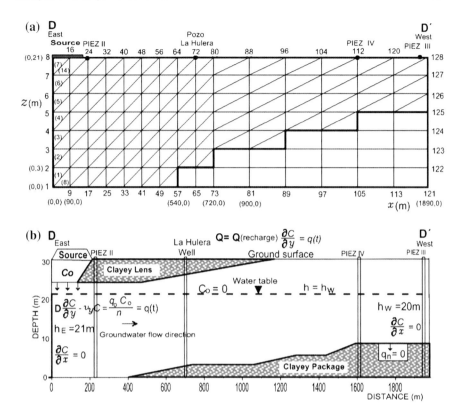

Fig. 3 **a** Finite element mesh **b** Boundary conditions along cross-section D-D'

along the water table is 25 cm/year, the hydraulic conductivity is 100 m/d vertically and 50 m/d horizontally, and the porosity is 0.35. A constant point source is situated in a shallow unsaturated aquifer where Cr(VI) is introduced instantaneously at the concentration of 90 mg/L; only the saturated zone was considered for convenience. A fully-penetrating well (La Hulera) was represented by a linear sink in the flow domain. These values were used in the model calibration runs and matched the water table calculated with field observation.

2.4 Model Characteristics

The flow model was calibrated by first adjusting the distribution of hydraulic conductivity until a satisfactory match with observed head data was achieved. Because the aquifer has not been subject to significant pumping since 1985, it is assumed to be in a steady-state condition. Therefore, the hydraulic conductivity determined by the flow model calibration was used to generate a velocity field. The aquifer is heterogeneous, and the distribution of the aquifer hydraulic conductivity was taken from a total of six measurements of hydraulic conductivity available for the whole field site. Horizontal and vertical hydraulic conductivities were determined through the iterative calibration procedure. The values for the hydraulic conductivity vary from 5.0 m/d to 200 m/d. The base case for this simulation, with $K_x = 100$ m/d, resulted in a residual sum of squares of $C_2 = 2.49$ m^2, and it was observed in this analysis that of 100 realizations, only four simulations had better calibrations (the realizations $R = 6$, $C_2 = 8.75$, $K_x = 71.87$ m/d; $R = 69$, $C_2 = 3.71$, $K_x = 113.36$ m/d; $R = 90$, $C_2 = 6.93$, $K_x = 75.28$ m/d; and $R = 92$, $C_2 = 6.60$, $K_x = 75.77$ m/d). Figure 4 shows the final calibrated horizontal and vertical hydraulic conductivities (K_x and K_y) for layer 1, for which uniform horizontal and vertical hydraulic conductivity values of 100 and 50 m/d, respectively, were taken for the calibration.

The model was calibrated using a 1-day measurement of piezometric heads at available groundwater piezometers and wells. The simulated heads are shown in Fig. 4a. A comparison of the observed and calibrated heads is presented in Fig. 4b, which shows good agreement. Mean deviation between observed and simulated heads is 0.095 m.

3 Results and Discussion

3.1 The Input Model

3.2 Sensitivity Analysis

To analyze the Cr(VI) movement pattern, a sensitivity analysis was done which permitted knowing which parameters have a greater effect on the spreading of chromium in the Buenavista shallow aquifer. The parameters evaluated were: the hydraulic conductivity, K_x, K_y; the distribution coefficient, K_d; the longitudinal and transverse dispersivities; the initial concentration C_o, and; the pumping rate, Q. A wide range of values for each parameter was tested to estimate the most suitable values in order to obtain a more representative parameter set for the Buenavista aquifer. Sensitivity analysis is a measure of the effect of change in one factor on another factor (McElwee 1982, 1987; Wagner and Gorelick 1986, 1987; Hill 1998; Zheng and Bennett 2002).

The dispersivity variation effect was analyzed taking values from the best match with the analytical solution of $\alpha_L = 50.0$ m and $\alpha_{TV} = 2.5$ m. The chromium was assumed to be non-reactive. Molecular diffusion in the aquifer was assumed a negligible contributor to the solute spreading at the field scale, such that hydrodynamic dispersion was solely related to mechanical dispersion, which was computed as a function of the specified dispersivity of the medium and the velocity of the flow field. The initial chromium concentrations in the aquifer were assumed to be zero and the source concentration was assumed to be 90 mg/L. The values

(a)

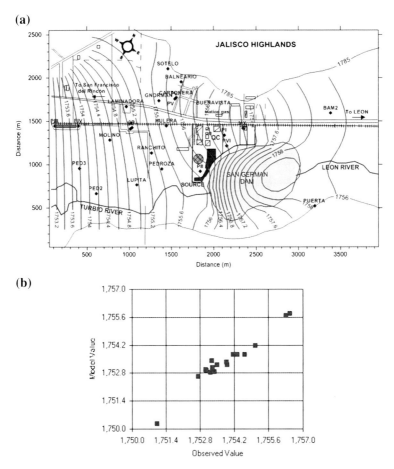

(b)

Fig. 4 **a** Map of steady-state hydraulic heads and velocities field of the calibrated model, **b** Scatter plot of simulated and measured hydraulic heads for layer 1 after calibration

$K_x = 100$ m/d and $K_y = 50$ m/d were chosen for the hydraulic conductivity components of the calibration. Thereby, a base case was obtained and the solute transport model was run for a stress period of 10 years, from 1982 (start of landfill operation) to 1992. Every parameter (K, K_d, α_L, α_{TV}, Q and C_o) was analyzed in cross-section. Figure 5 shows that the calculated plume for the base case is similar in geometry and concentration distribution to that observed in 1992.

3.3 Plume Evolution in Cross-Section

The chromium plume evolution is analyzed using a transverse and longitudinal dispersion essentially equal to zero, using $\alpha_L = 0$ and $K_d = 0$, so as to analyze the

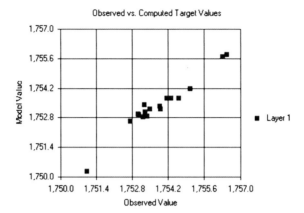

Fig. 5 Cr(VI) [mg/L] vertical distribution: (**a**) observed plume in October 1992, along cross-section D-D' and (**b**) base-case configuration

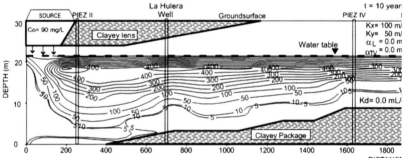

Fig. 6 Evolution of the plume for $t = 10$ years for the case of the advective transport of a conservative ion, $K_d = 0$, and Cauchy boundary condition in water table

importance of the advective process in the investigated domain and to obtain an idea of the magnitude of the dispersive process involved in transport. Figure 6 shows that the chromium plume is forced to move in the preferential direction of the flow.

3.4 Longitudinal Dispersivity Variation

Generally, longitudinal dispersivities are assigned much larger values than vertical dispersivity (Domenico and Schwartz 1998). Consequently, it is crucial to quantify the extent to which longitudinal dispersion contributes to the spreading of the contaminant. In this case, the vertical transverse dispersivity remained constant, at $\alpha_{TV} = 5.0$ m, and the longitudinal component was varied from 25 to 100 m. It was expected that when increasing α_L, the longitudinal propagation of the plume would

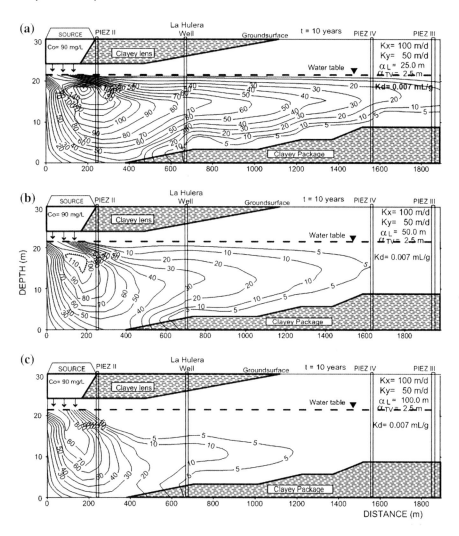

Fig. 7 Sensitivity with regard to the longitudinal dispersivity with $\alpha_{TV} = 2.5$ m:
(a) $\alpha_L = 25.0$ m, (b) $\alpha_L = 50.0$ m and (c) $\alpha_L = 100.0$ m

increase; however, this did not happen. This effect may be explained by the type of Cauchy boundary condition applied under the source, which supplies mass to the system proportional to the recharge and to the concentration of the recharge flow. The effect of this boundary type causes the concentration to diminish when the value of α_L is increased. The contrary case would be obtained using a Dirichlet boundary condition, indicating that a concentration similar to that of the source would introduce greater mass into the system, causing a longer plume. Figure 7b shows, however, that the best result was obtained using a Cauchy boundary

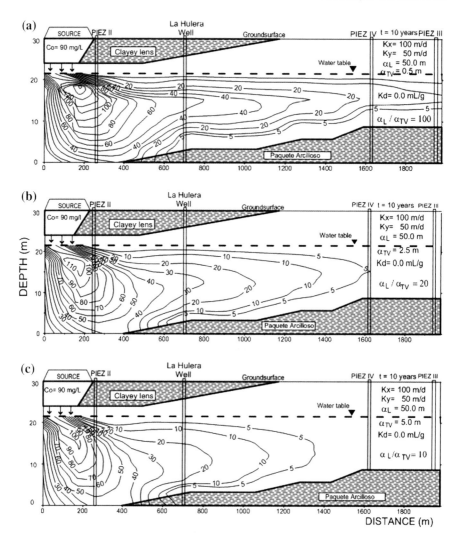

Fig. 8 Sensitivity with regard to the vertical transverse dispersivity with $\alpha_L = 50.0$ m: (a) $\alpha_{TV} = 0.5$ m, (b) $\alpha_{TV} = 2.5$ m and (c) $\alpha_{TV} = 5.0$ m

condition and the elected value of α_L was 50 m, since the simulated plume morphology is consistent with that of the observed plume.

3.5 Vertical Transverse Dispersivity

Sensitivity analyses of the vertical transverse dispersivity were performed for five different values of α_{TV} (0.05, 0.25, 0.5, 2.5 and 5.0 m). The longitudinal

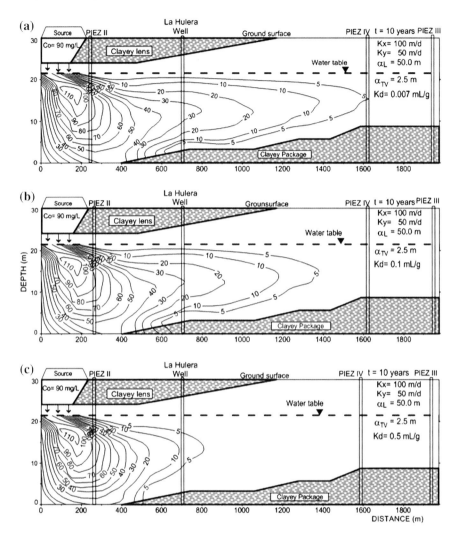

Fig. 9 Effect of varying the distribution coefficient for the Cr(VI) concentrations distribution: (a) $K_d = 0.007$ mL/g, (b) $K_d = 0.1$ mL/g and (c) $K_d = 0.5$ mL/g

component remained fixed at $\alpha_L = 50.0$ m, and a distribution coefficient of $K_d = 0.007$ mL/g was used; in this way, the plume geometry is well-defined longitudinally and transversally. When the plume advances horizontally, α_{TV} decreases the concentration. When increasing the value of α_{TV}, the plume propagation also increases toward the base of the aquifer, shortening the front of the plume. The plume geometry is very sensitive to the value of α_{TV}. With lower values of α_{TV}, the plume front increases in the longitudinal direction along the water table and diminishes toward the base of the aquifer. Figure 8b shows the best value obtained with $\alpha_{TV} = 2.5$ m.

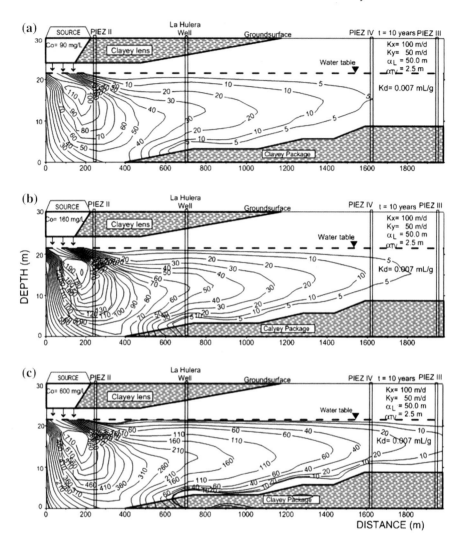

Fig. 10 Sensitivity of the plume with regard to the initial concentration, *Co*: (**a**) *Co* = 90 mg/L, (**b**) *Co* = 160 mg/L and (**c**) *Co* = 600 mg/L

3.6 Retardation Coefficient Variations

The rate at which the Cr(VI) travels through the aqueous phase is given by v_c, the groundwater flow rate is v_w, and the ratio of v_w/v_c yields the retardation factor R, which is given in Eq. (6). The distribution coefficient was varied from $K_d = 0.007$ to 0.1 and to 0.5 mL/g (sand, silty-sand, and silt–clay, respectively). These values correspond to retardation coefficients of R = 1.0, 2.0 and 5.0, respectively, and as

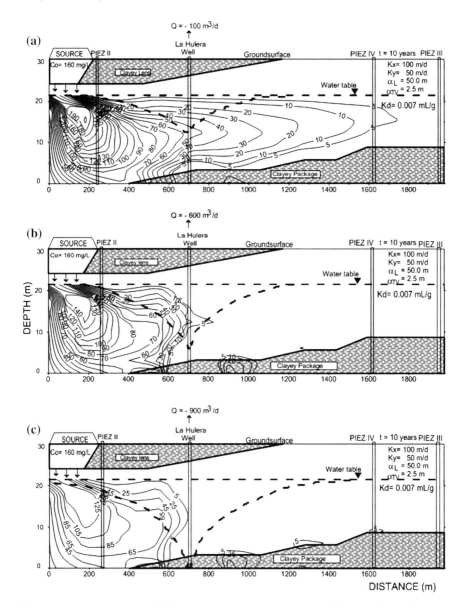

Fig. 11 Pumping effect on Cr(VI) concentration produced by La Hulera well for 10 years: (a) Q = 100 m³/d, (b) Q = 600 m³/d and (c) 900 m³/d

expected, increasing R generated a retarded plume. Figure 9a shows that the value of $K_d = 0.007$ mL/g was the best value and reflected the low capacity of adsorption of the chromium in the shallow aquifer.

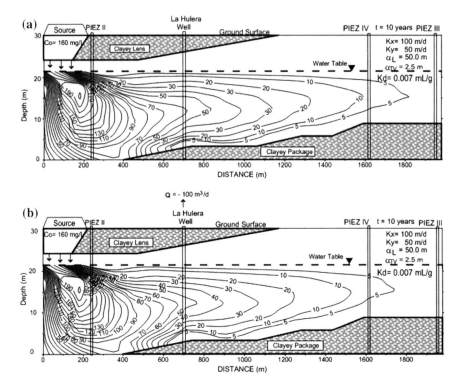

Fig. 12 Simulated plume for best match with the observed plume (Fig. **7a**) in November of 1992: a) steady-state case and **b**) unsteady-state case. Concentration isolines in mg/L

3.7 Variations in the Initial Concentration of Chromium, C_o

The different values chosen for the initial concentration of chromium were $C_o = 90$, 120, 160, 200 and 600 mg/L. All these cases were analyzed using the base case run, where $\alpha_L = 50.0$ m, $\alpha_{TV} = 2.5$ m and $K_d = 0.007$ mL/g. Figure 10 shows that the best match to the dates from 1992 corresponds to $C_o = 160$ mg/L.

3.8 Pumping Rate Variation

The unsteady-state case was analyzed in the cross-section plane. The pumping rate was varied using the rates of $Q = 100$, 200, 600 and 900 m³/d. When the extraction increases, the gradient of concentration of the plume decreases longitudinally and transversally. The pumping rate acts as a sink (hydraulic barrier) for the mass flow, restricting the plume advance. Figure 11 shows the pumping rate variation. The value $Q = 100$ m³/d was selected. The La Hulera and La Cartonera wells were operating with this value during the source operation, given that they operate 9 h per day, with an approximate pumping rate of 125 m^3/d. The plume propagation in the flow domain is related to the operation of La Hulera well.

The plume is retarded in the discharge area to the west of the source. Also, a good correlation was obtained between the numeric simulations (Fig. 12a) and the plume of Cr(VI) observed in 1992 (Fig. 12b).

These results justify the implementation of a remediation schedule for pump and treat to retard the advance of the chromium plume in the affected area. In addition, periodic observations of the piezometers were interrupted in 1992 and restarted in late 2001. The observed plume was configured with the concentrations in the piezometers, corresponding to four different depths. The comparison of the results indicates that the numeric solution tends to increase the dispersion over time due to the continuous input of Cr(VI).

The hydrogeologic parameters that control the 2D flow field are relevant to model the transport of Cr(VI). These include local stratigraphy, aquifer geometry, hydraulic conductivity and the flow boundary conditions. Dispersion, taking into account the plume scale, also plays a very important role in the movement of the plume. The set of sensitivity simulations retained the values $\alpha_L = 50.0$ m and $\alpha_{TV} = 2.5$ m for the dispersivities, and a good reproduction of the vertical concentration distribution was obtained for the 10-year simulation period, from 1982 to 1992.

The set of sensitivity simulations retained the values $\alpha_L = 50.0$ m and $\alpha_{TV} = 2.5$ m for the dispersivities, and a good reproduction of the vertical concentration distribution was obtained for the 10-year simulation period, from 1982 to 1992. These values are in the same range as those reported by Gelhar et al. (1992). It is possible that the aquifer material, composed of sands, caused these relatively high values, with the layer of sand contributing to high flow velocity when the flow can be considered steady. Contrasts were analyzed between the longitudinal and transverse dispersivities to three orders of magnitude to observe the effect produced by the dispersivity on the migration of the Cr(VI) plume along the preferential flow direction.

4 Conclusions

A two-dimensional simulation model was built based on the modified Istok code with finite elements. A sensitivity analysis of the transport parameters was done to determine the best dataset for the Buenavista shallow aquifer in León Guanajuato, Mexico. A group of transport parameters was obtained for the best match to the simulation of the Cr(VI) plume in the Buenavista area: $\alpha_L = 50.0$ m, $\alpha_{TV} = 2.5$ m, $K_d = 0.007$ mL/g, $C_o = 160$ mg/L and $Q = 100$ m^3/d. Only one of the parameters analyzed, C_o, varied from the initial values. This suggests that the procedure to obtain the other parameters was appropriate. A good correlation was obtained between the numeric simulations and the Cr(VI) plume observed. The plume was modeled in the vertical plane, in which case the match was acceptable. In general, the vertical transverse plume propagation is not negligible, even with the influence of the local abstraction regime; the transverse migration retarded the advance of the plume front in comparison with the pure advective advance, and high values of transverse dispersivity provoked a deformed plume with a high retardation of the plume front.

The numerical dispersion control is important in the advective–dispersive transport simulations. It is possible to control the longitudinal numerical dispersion with the Peclet and Courant criteria which, for this case, were values of $P_e = 2.0$ and $C = 0.22$, respectively. The numerical dispersion diminished when aligning the finite elements of the mesh with the main flow direction, taking the water table as reference. Pumping plays a very important role in the fast plume movement. By increasing the extraction rate in the La Hulera well, the advancement of the plume toward the discharge area will be constrained, limiting the extension of the contamination in the shallow aquifer and reducing the concentration of Cr(VI) in the flow field.

The implementation of a two-dimensional model can be considered successful for this case because the sandy aquifer formations are relatively homogeneous. If there were aquifer heterogeneity, a two-dimensional model would probably not be the most suitable one. Incorporating the estimated parameters obtained by the piezometer network into field and laboratory data justifies the use of a 2D model.

Acknowledgments The authors wish to thank Chemical Central for its open doors policy. To Miss. Ellen Sue Weiss by technical help in English review of manuscript and we wish to thank her.

References

Armienta MA (1992) Contribución al estudio de los mecanismos de transporte del cromo en el acuífero de León Gto. Ph. D. Thesis, Mexico, UNAM, pp 213

Bear J (1979) Hydraulics of groundwater. McGraw-Hill, New York

Castelán RA, Villegas CI (1995) Control estratigráfico de la dispersión de compuestos de cromo en la zona de Buenavista, Edo. de Guanajuato. Bachelor Thesis in Geology. ESIA-IPN, Mexico, p 70

Daus T, Frind EO, Sudicky EA (1985) Comparative error analysis in finite element formulations of the advection-dispersion equation. Adv Water Resour 8(2):86–95

Domenico PA, Schwartz FW (1998) Physical and chemical hydrogeology. John Wiley and Sons, New York

Gelhar LW, Welty C, Rehfeldt KR (1992) Critical review of data on field-scale dispersion in aquifers. Water Resour Res 28(7):1955–1974

Hill MC (1998) Methods and guidelines for effective model calibration, with application to: UCODE, a computer code for universal inverse modeling, and MODFLOWP, a computer code for inverse modeling with MODFLOW. U.S. Geological Survey Water-Resources Investigations Report 98–4005, p 90

Istok J (1989) Groundwater modeling by the finite element method. [Water resources monograph 13]. American Geophysical Union, Washington

Jiao JJ (1993) Data analysis methods for determining two-dimensional dispersive parameters. Groundwater 13(1):57–62

Langmuir D (1997) Aqueous Environmental Geochemistry. Prentice-Hall Inc, Englewood Cliffs

McElwee CD (1982) Sensitivity analysis and the ground-water inverse problem. Gound Water 20(6):723–735

McElwee CD (1987) Sensitivity analysis of groundwater models. In: Bear J, Corapcioglu MY (eds) Advances in transport phenomena in porous media, NATO advanced study institute series, pp 751–817

Reyes-Gutiérrez LR (1998) Parámetros que controlan la dispersión de compuestos de Cr en un acuífero de conductividad hidráulica variable. Master Thesis (Geophysics). UNAM, México D. F. p 130

Reyes-Gutiérrez LR (2007) Análisis de un sistema de remediación acuífera mediante bombeo y tratamiento en Buenavista, Guanajuato. Ph. Degree Thesis Geophysics). UNAM, México D. F. p 192

Rodríguez CR, Armienta MA, Villanueva S, Díaz P, González MT (1991) Estudio hidrogeoquímico y modelación matemática del acuífero del Río Turbio para definir acciones encaminadas a proteger de contaminantes la fuente de abastecimiento de la Cd. de León, Gto. Technical Report il. IGF UNAM, CNA SARH. , p 140, June 1991

Voss IC (1984) A finite-element simulation model for saturated-unsaturated. Fluid-density-dependent groundwater flow with energy transport or chemically-reactive single-species solute transport. U.S. Geological Survey, Water-Resources Investigations Report 84-4369, p 391

Wagner BJ, Gorelick SM (1986) A statistical methodology for estimating transport parameters: theory and applications to a one-dimensional advective-dispersive systems. Water Resour Res 22(8):303–1315

Wagner BJ, Gorelick SM (1987) Optimal groundwater quality management under parameter uncertainty. Water Resour Res 23(7):1162–1174

Zheng C, Bennett GD (2002) Applied contaminant transport modeling, 2nd edn. Jhon Wiley and Sons Inc, New York 621

Numerical Simulation of Dispersion and Sorption of Se(IV) Through Packed Columns with Non-Living Biomass: Experimental and Numerical Results

Carlos E. Alvarado-Rodríguez, Jaime Klapp-Escribano, Elizabeth T. Romero-Guzmán, Zayre I. González-Acevedo and Ricardo Duarte-Pérez

Abstract A continuous fixed bed study was carried out using the non-living biomass *Lemna minor* as a biosorbent for the removal of Se(IV) from an aqueous solution. A 3D numerical model was constructed for solving the *Navier–Stokes–Brinkman* and mass transport equations using the finite element technique with 161,764 tetrahedral elements. Experimental and numerical results were obtained and compared to validate the model, obtaining correlation factors of up to $R^2 = 0.95$. From the sensibility analysis, the parameters for the Thomas model best match were obtained, with a value of $k = 2.95 \times 10^{-3}$ m/s, $K_f = 0.1201$ L/kg, $\alpha_L = 0.5$ m and $\alpha_{TH} = \alpha_{TV} = 0.005$ m was possible to predict the breakthrough curves of sorption of Se(IV) in packed column with non-living biomass of *Lemna minor* in aqueous solution. Using the numeric model, seven fixed bed columns with different

C. E. Alvarado-Rodríguez · J. Klapp-Escribano (✉) · E. T. Romero-Guzmán · Z. I. González-Acevedo · R. Duarte-Pérez
Instituto Nacional de Investigaciones Nucleares, Carretera México-Toluca S/N, La Marquesa, 52750 Ocoyoacac, Estado de México, Mexico
e-mail: jaime.klapp@inin.gob.mx; q_l_o@hotmail.com

E. T. Romero-Guzmán
e-mail: elizabeth.romero@inin.gob.mx

Z. I. González-Acevedo
e-mail: zayre.gonzalez@inin.gob.mx

R. Duarte-Pérez
e-mail: ricardo.duarte@inin.gob.mx

C. E. Alvarado-Rodríguez
División de Ciencias Naturales y Exactas, Universidad de Guanajuato, Campus Guanajuato Noria Alta S/N, Guanajuato, Guanajuato, Mexico

J. Klapp-Escribano
Departamento de Matemáticas, Cinvestav del I.P.N., 07360 Mexico, D.F., Mexico

J. Klapp et al. (eds.), *Fluid Dynamics in Physics, Engineering and Environmental Applications*, Environmental Science and Engineering, DOI: 10.1007/978-3-642-27723-8_38, © Springer-Verlag Berlin Heidelberg 2013

dimensions were simulated, and from the simulation results the removal of 15 % of Se(IV) was obtained.

1 Introduction

For many years, selenium has been a largely unrecognized pollutant, particularly in developing nations, and has been overshadowed by issues involving contaminants such as industrial chemicals, heavy metals, pesticides, and air pollutants just to name a few. A continuous intake greater than 8 mg of selenium per day can produce harmful health effects. The EPA limits the amount of selenium allowed in drinking water supplies to 50 parts total selenium per billion parts of water (50 ppb) (Lemly 2004; OMS 2003). In Mexico, there have been reports of selenium contamination in water, as is the case of Guanajuato, which has been observed due to leaching of mine tailings (IEE 2001). In the state of Puebla, the denim laundries have polluted the waters leaving residues of selenium (Hurtado and Gardea 2007).

Dispersion plays a critical role in numerous processes and practical applications, including contaminant transport in groundwater, filtration, etc. Hydrodynamic dispersion in a porous medium occurs as a consequence of two different processes: (i) molecular diffusion, which is originated from the random molecular motion of solute molecules, and (ii) mechanical dispersion, which is caused by non-uniform velocities and the flow path distribution. Both processes cannot be separated in a flow regime (Nützman et al. 2002).

Mathematical models are useful for understanding fixed-bed column dynamics and assisting in the design and optimization by reducing the amount of time-consuming and repetitive experiments (Chu and Hashim 2007). As the mathematical model is based in the experiment characteristics, the numerical results can be used to verify the results of the column experiments and ensure the experiment design or propose a change in the experiment.

The biosorption of Se(IV) by the *Lemna minor* biomass was simulated using experimental data. In this work the simulation was used as a tool for representing the Se(IV) sorption in a natural porous sorbent such as the *Lemna minor* biomass, comparing the experimental and simulated breakthrough curves. The breakthrough curves can be calculated using different equations such as those of Thomas, Wolborska, Clark (Aksu and Gönen 2004) and Ogata and Banks (1961). For this work we developed a 3D numerical model on the software COMSOL Multiphysics 3.5a that can calculate the Se(IV) transport as a time function and obtain the breakthrough curves by solving the fluid flow dynamic equations.

Table 1 Characteristics of the columns and packing biomass (Rodríguez 2011)

Column	Length (cm)	Diameter (cm)	Lemna minor (g)	Bed height (cm)
1	22	2.0	6.9	19
2	15	1.5	1.5	11
3	10	1.0	0.5	8

2 The Experimental Setup

The biomass was washed with 0.01 M HCl solution to remove impurities of wastewater and then it was dried. The dry biomass was milled and meshed to obtain homogeneous particles of 2 mm in size. With this biomass, glass columns of different dimensions were packed manually, maintaining the mass amount in every column, according to Table 1.

The sorption of Se(IV) by the biomass packed columns was performed by passing 300 mL of 0.02 mg/L Se(IV) solution using a peristaltic pump at a flow rate of 2 mL/min. At the outflow, 30 aliquot fractions of 10 mL were taken. Subsamples of these fractions were diluted with 6 M HCl and treated in a microwave oven, to reduce Se(VI) to Se(IV) according to the methodology proposed by Brunori et al. (1998). Finally, total selenium was determined with a hydride generator coupled with Atomic Absorption Spectrometry. The experimental results of columns with different scales were compared with the numerical results to ensure that the model can be used to different scales. The K_p value (constant in the linear isotherm) was calculated from the experimental results as it is indicated below. The parameters used in the model were obtained by column Se(IV) sorption experiments. These methods are commonly used for calculating K_d values from experimental data.

3 The Mathematical Model

In the literature we can find numerous models, mostly 1D, that are used to correlate the dynamics obtained from numerical calculations to experimental data, and to predict the breakthrough curves for sorption in fixed bed columns. We obtain the free flow in the unpacked zones by solving the 3D *Navier–Stokes–Brinkman* and mass transport equations. The *Navier–Stokes* (1) and continuity (2) Eqs. can be written in the form:

$$\rho \frac{\partial v}{\partial t} + (\mathbf{v} \cdot \nabla)\mathbf{v} = -\nabla p + \mu \nabla^2 \mathbf{v} + \rho \mathbf{g} \tag{1}$$

$$\frac{\partial \rho}{\partial t} = -(\nabla \cdot \rho v) \tag{2}$$

where ρ is the density, v is the Darcian velocity, t is the time, p is the pressure, μ is the viscosity and g is the gravity.

For the flow through the biomass (porous system), the *Brinkman* (3) and the modified continuity (4) Eqs. were used to calculate the velocity field (Bird et al. 2003):

$$-\nabla p - \frac{\mu}{k} v_0 + \mu \nabla^2 v_0 + \rho \mathbf{g} = 0 \tag{3}$$

$$\varepsilon \frac{\partial p}{\partial t} = -(\nabla \cdot \rho 2 v_0) \tag{4}$$

where k is the permeability, v_o is the average lineal velocity in porous media, v/ε and ε is the porosity. The value of porosity was determined in laboratory.

The Eq. 5 gives the mass transport through the media, nevertheless it is necessary obtain a second equation that estimate the relation between the solution concentration and the Se(IV) sorbed amount. A linear isotherm has an advantage to describe the sorption with only one parameter. Se(IV) transport and its sorption through the biomass are calculated by solving the transport Eq. 5 and a lineal isotherm Eq. 6 respectively:

$$\varepsilon \frac{\partial c}{\partial t} + \rho_b \frac{\partial c_p}{\partial c} \frac{\partial c}{\partial t} + \nabla \cdot [-\varepsilon D_L \nabla c + vc] = 0 \tag{5}$$

$$c_p = K_f c \tag{6}$$

where c is the solute concentration, ρ_b is the porous medium density, c_p is the concentration of the solute sorbed in the biomass (the mass amount of solute sorbed per unit of biomass), D_L is the hydrodynamic dispersion tensor and K_f are constants of lineal isotherms.

Figure 1 shows the complete domain divided in three zones, inlet, outlet and packed zone. The packed zone boundaries are the lower and upper boundary of the in and out zones respectively. The complete domain was divided in 161,764 tetrahedral elements, using the element finite method to solve the equations and the UMFPACK, that is a set of routines for solving unsymmetric sparse linear systems to solve the lineal system.

Equations 1 and 3 were solved with the following boundary conditions: The upper limit is a Dirichlet condition ($v = -v_o n$). For the lower limit was use the pressure condition $p = p_0$, where $p_0 = 0$. The Eq. 5 was solved with the following boundary conditions: the upper and lower limits are Cuachy conditions, $v_n = v * C_{in}$ (inward flux) and $-v_n = v * C_{out}$ (outward flux), respectively. The other condition in the walls of the column is a Neuman condition $v_n = 0$. The solution of the model was obtained with the COMSOL Multiphysics software version 3.5a. All parameters: permeability coefficient (k), dispersivity coefficient values of the longitudinal dispersivity (α_L), horizontal (α_{TH}) and vertical (α_{TH}) transversal dispersivity were taken from the literature for a porous media similar to it, Table 2.

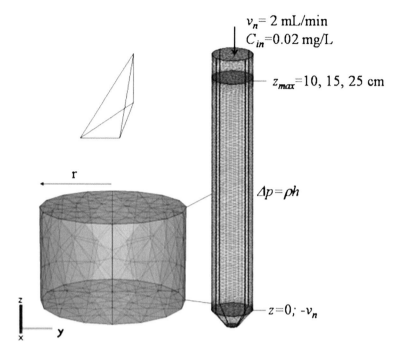

$v_n = 2$ mL/min
$C_{in} = 0.02$ mg/L

$z_{max} = 10, 15, 25$ cm

$\Delta p = \rho h$

$z = 0; -v_n$

r

Fig. 1 Initial and boundary conditions to solve equations (1), (3) and (5) with tetrahedral elements

Table 2 Model parameters ([a]Bear 1972; [b]Marín 2010)

Symbol	Parameters	Value
ρ	Density of Lemna minor[b]	109 kg/m^3
εn	Porosity[b]	15 %
k	Permeability[b]	1.4×10^{-3} m/s
K_f	Constant of lineal isotherm	0.1201 L/Kg
α_L	Longitudinal dispersivity[a]	0.5 m
α_{TH}	Horizontal transversal dispersivity[a]	0.005 m
α_{TV}	Vertical transversal dispersivity[a]	0.005 m

4 Results and Discussion

Columns with the dimensions shown in Table 1 were simulated with the parameters in Table 2 and the results compared with the experimental results. The numerical simulations were performed for the columns of Table 3 with a feed flow of 50, 250, 500 L/day according to a consumption of 5 L/day by person for 10, 50 and 100 persons. The results of a sensitivity analysis show that the transport of Se(IV) is controlled by the parameters k, K_f and the dispersivity coefficients. Thereafter, the Se(IV) sorption percent was calculated using the numerical results.

Table 3 Dimensions of simulated columns

Column	Diameter (cm)	Length (cm)
A	18	55
B	16	70
C	12	52
D	10	74
E	14	53
F	12	72
G	16	150

The velocity field for column 1 in Table 1 obtained by *Navier–Stokes* and *Brinkman* equations are showed in Fig. 2.

The velocity field for column 1 in Table 1 obtained by *Navier–Stokes* and *Brinkman* equations are showed in Fig. 2.

Velocity field in porous media for column 1 in Table 1 obtained by *Brinkman* equations and the *Reynolds* number in the same column are showed in Fig. 3.

According to the results, the highest velocity is in the center of the column. Similar results are presented by Sigalotti et al. (2003) for flows at low Reynolds numbers.

The numerical results were dependent with the mesh and it was calibrate comparing the numerical and experimental results of the velocity. The best correlation obtained was $R^2 = 0.98$. With an initial selenium concentration of 0.02 mg/L in Fig. 1 $C_{in} = 0.02$ mg/L, the dispersion and sorption of Se(IV) are showed in Fig. 4: (a) contains the dispersion of Se(IV) through the column 1 in Table 1 for 15, 50, 80, and 160 min respectively and (b) contains the sorption percent of Se(IV) through the column 1 in Table 1 for 15, 50, 80, and 160 min respectively.

When the sorption percent is 100 %, the column is saturated and the inlet concentration is the same to the outlet concentration because the biomass does not sorb more. From these results, the outlet concentration was taken and comparing with the experimental results.

In order to validate the computer model developed for this study, a comparison of the model output was made against an analytical solution. Thomas model (Thomas 1944; Han et al. 2007) presents a solution to the one dimensional advection–dispersion equation assuming a continuous source and initial conditions of no tracer within the column, Eq. 7.

$$\frac{C}{C_0} = \frac{1}{1 + exp\left(\frac{k_{Th}q_oX}{Q_{Th}} - k_{Th}C_0t\right)} \tag{7}$$

where k_{Th} is Thomas constant related with the sorption velocity, Q_{Th} is the volumetric flow, t is the saturation time, q_o is the maximum sorption capacity, C_o is the initial concentration and X is the biomass quantity. The time duration of each simulation was at least the length of the applicable laboratory experiment, and

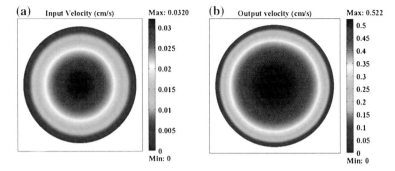

Fig. 2 **a** Inlet velocity field and **b** outlet velocity field by solution of *Navier–Stokes* equations

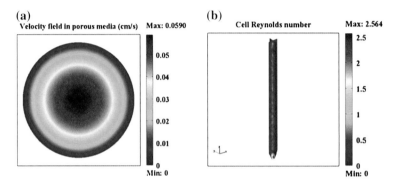

Fig. 3 **a** Velocity field in porous media by solution of *Brinkman* equation. **b** Cell *Reynolds* number in column 1 in Table 1

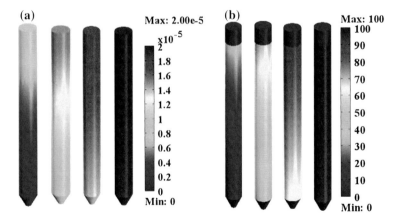

Fig. 4 **a** Dispersion of Se(IV) through the column 1 in Table 1 for 15, 50, 80, and 160 min respectively, **b** Sorption percent of Se(IV) through the column 1 in Table 1 for 15, 50, 80, and 160 min respectively

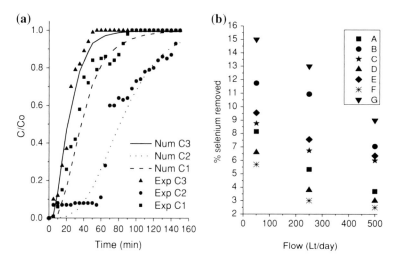

Fig. 5 **a** Comparison of experimental and numerical breakthrough curves of columns in Table 1 packing with Lemna minor, column 1 (■), column 2 (●), column 3 (▲). **b** Percent retention or adsorbed of Se(IV) for columns in Table 3

ranged from approximately 180 min to approximately 4,560 min. The agreement between the analytical and numerical models was generally good, with small differences between modeled values at the crest of the solute input. These differences are believed to be due to incipient oscillation in the concentration estimation of the numerical model, possibly due to truncation error in the dispersion calculation.

In the Fig. 5a are showed the results of breakthrough curves obtained by numerical results and experimental data for columns 1, 2 and 3 of the Table 1.

The correlation factor (R^2) between experimental data and numerical results in Fig. 5a are 0.977, 0.951 and 0.983 for column 1, 2 and 3 respectively. Similar correlations were obtained by Aksu and Gönen (2004) and Hasan et al. (2010) using different models. According to the match good correlations, the model can be used to obtain the breakthrough curves of Se(IV) sorption systems with different dimensions and porous media. Percent removals of Se(IV) in columns in Table 3 with different dimensions are presented in Fig. 5b.

Columns in Table 3 were simulated and their breakthrough curves were used to calculate the removal percent in each one with the equation (8), (9), (10), and (11). A 34, 51 and 32 percentages were obtained for 1, 2 and 3 column respectively, Table 1.

$$q_{total} = QA = Q \int_{t=0}^{t=t_{total}} C_{ad}dt \qquad (8)$$

$$C_{ad} = C_o - C \qquad (9)$$

$$m_{total} = \frac{C_o Q t_{total}}{1000} \qquad (10)$$

$$\% \ removal = \frac{q_{total}}{m_{total}} * 100 \qquad (11)$$

where q_{total} is the rate Se(IV) sorbed and the amount of packed biomass, Q is the inlet flow, A represent the total amount of retention of Se(IV), t_{total} is the time of the break point, Co is the initial concentration, C is the solution concentration, m_{total} is the total selenium fed to the column, *removal* % is the percent Se(IV) removal in the column.

Columns with more height and diameter obtained more percent removal. This can be attributed to the fact that when there was an increase in bed height, the axial dispersion got decreased in the mass transfer and as a result the diffusion of the metal ions into the sorbent got increased. Thus, the Se(IV) got enough time to be sorbed by the biomass. Similar results were reported by Hasan (2010).

5 Conclusions

A 3D model was configured to represent the real system and obtain a breakthrough curves from simulations of the dispersion and sorption of Se(IV) in aqueous solutions.

The model represent the Se(IV) sorption system with correlations up to $R^2 = 0.97$. Based on the above, the model can predict breakthrough curves obtaining acceptable results. According to the results, the removal percent of Se(IV) increases when the inlet velocity decreases and the removal percent decreases when the inlet velocity increases. With a value of $K_f = 0.1201$ L/kg, $\alpha_L = 0.5$ m y $\alpha_{TH} = \alpha_{TV} = 0.005$ m is possible to predict the breakthrough curves of Se(IV) sorption in a packed column with non living biomass of *Lemna minor*. The highest percent retention of Se(IV) achieved is 15 % with a column of 16 cm and 150 cm of radio and length respectively.

The transport of Se(IV) in aqueous solution is controlled by the parameters k, K_f and the dispersivity coefficients principally.

From the sensibility analysis, the parameters for the Thomas model best match were obtained. The permeability values were between 1×10^{-2} and 1×10^{-4} m/s, with a $k = 2.95 \times 10^{-3}$ m/s for this system, the rest of the parameters had the initial values.

Acknowledgments The authors wish to thank the financial support provided by the Consejo Nacional de Ciencia y Tecnología (CONACyT) and to Dra. María Teresa Olguín Gutiérrez from ININ. We appreciate very much the comments and suggestions of a reviewer, which help to improve the manuscript. This work has been partially supported by CONACyT-EDOMEX-2011-C01-165873 project.

References

Aksu Z, Gönen F (2004) Biosorption of phenol by immobilized activated sludge in a continuous packed bed: prediction of breakthrough curves. Proccess Biochem 39:599–613

Bear J (1972) Dynamics of fluids in porous media, Elsevier, Scientific Publishing Co, Chennai

Bird RB, Stewart WE, Lightfoot EN (2003) Fenómenos de transporte. Barcelona, 4–29

Brunori C, De la Calle MB, Morabito R (1998) Optimization of the reduction of Se(VI) to Se(IV) in a microwave oven. Fresenius J Anal Chem 360:26–30

Chu KH, Hashim MA (2007) Copper biosorption on immobilized seaweed biomass: column breakthrough characteristics. J Environ Sci 19:928–932

Han R, Wang Y, Zou W, Wang Y, Shi J (2007) Comparison of linear and nonlinear analysis in estimating the Thomas model parameters for methylene blue adsorption onto natural zeolite in fixed-bed column. J Hazard Mater 145:331–335

Hurtado JR, Gardea TJ (2007) Evaluación de la exposición a selenio en los altos de Jalisco, México. Salud Pública de México, 49(4)

Hasan SH, Ranjan D, Talat M (2010) Agro-industrial waste 'wheat bran' for the biosorptive remediation of selenium through continuous up-flow fixed-bed column. J Hazard Mater 181:1134–1142

IEE (2001) Instituto de Ecología del Estado de Guanajuato. Informe ambiental del Estado de Guanajuato 2001:33

Marín AMJ (2010) Remoción de Arsénico en solución empleando biomasas no vivas de malezas acuáticas. Universidad Autónoma del Estado de México, Tesis de maestría

Nützman G, Maciejewski S, Joswig K (2002) Estimation of water saturation dependence of dispersion in unsaturated porous media: experiments and modelling analysis. Adv Water Resour 25:565–576

Lemly A Dennis (2004) Aquatic selenium pollution is a global environmental safety issue. Ecotoxicol Environ Saf 59:44–56

Ogata A, Banks RB (1961) A solution of the differential equation of longitudinal dispersion in porous media. U.S. Geological Surveys Professional Papers, 411-A

OMS (2003) Selenium in drinking-water. Documento de referencia para la elaboración de las Guías de la OMS para la calidad del agua potable. Ginebra (Suiza), Organización Mundial de la Salud (WHO/SDE/WSH/03.04/13)

Rodríguez M C E (2011) Optimización de condiciones dinámicas para sorción y desorción de selenio en solución empleando biomasas no vivas de malezas acuáticas. Tesina de licenciatura, ININ-ITT Toluca Edo. De México

Sigalotti LDG, Klapp J, Sira E, Melean Y, Hasmy A (2003) SPH simulations of time-dependent Poiseuille flow at low Reynolds numbers. J Comput Phys 191:622–638

Thomas HC (1944) Heterogeneous ion exchange in a flowing system. J Am Chem Soc 66:1664–1666

Tridimensional Analysis of Migration of ^{226}Ra Through a Saturated Porous Media

N. Perez-Quezadas, E. Mayoral, J. Klapp, E. de la Cruz
and R. González

Abstract Transport simulation of radioactive contaminants in fluids through saturated porous media provides a powerful tool to prevent the risk they could represent to the environment. This kind of waste is stored in containers. These containers are collocated in subterranean places for being confined. Proper election of those places requires the appropriate classification according to the properties of the site. This work is presenting the 3D numerical simulation of migration of ^{226}Ra through a tridimensional saturated porous media. The numerical solution is obtained solving the transport equations using Darcy's approximation for flux in saturated media by finite element method (FEM). The governing equations for this model are described as well as the numerical tool employed. The results are analyzed at different simulation time showing the behavior of the system.

1 Introduction

Due to the necessity to develop processes involving radioactive materials, it is important to implement safety measures for the treatment of wastes generated by the nuclear industry. The radioactive wastes are dangerous since they might cause

N. Perez-Quezadas · E. Mayoral (✉) · J. Klapp · E. de la Cruz
Instituto Nacional de Investigaciones Nucleares, La Marquesa,
Ocoyoacac 52750, Estado de México, Mexico
e-mail: estela.mayoral@inin.gob.mx

R. González
Facultad de Ciencias, Universidad Autónoma del Estado de México,
El Cerrillo Piedras Blancas 50200, Estado de México, Mexico

J. Klapp
Departamento de Matemáticas, Cinvestav del I.P.N, México 07360, Mexico

J. Klapp et al. (eds.), *Fluid Dynamics in Physics, Engineering and Environmental Applications*, Environmental Science and Engineering,
DOI: 10.1007/978-3-642-27723-8_39, © Springer-Verlag Berlin Heidelberg 2013

negative impact in the environment and in human beings. The best way to manage
these wastes is by isolation until they decay to a stable element. They are confined
in containers and later on these containers are grounded into the subsurface to keep
them away from any other activity. The sites are chosen to be stables, which means
free of seismic activities, among others characteristics. In order to select an adequate site to deposit this material, the site is required to be studied in detail, as well
as a risk analysis is needed. Nowadays the study, prevention, and prediction of
migration of harmful contaminants through subsurface is an important matter since
it is necessary to predict and avoid the ecological impact. Contamination of sites
by radioactive materials comes mostly, from unsuitable storage of this waste.
These kinds of studies, as the current presented in this work, are necessaries.

2 Mathematical Modeling

Modeling fluid flow through porous media requires a set of equations that describe
the phenomena. It will be necessary to make use of Darcy's law and its corresponding Darcy's equation, as well as transport equation which gave us the evolution of the contaminant plume through time. Solving these equations one can
obtain the representation of the phenomena through time. It will be described in
the next subsections.

2.1 Transport Equations for a Saturated Porous Media

The governing equation for saturated porous media is (Bear 1979):

$$\Theta \frac{\partial C}{\partial t} + \rho \frac{\partial C_p}{\partial C} \frac{\partial C}{\partial t} + \nabla \cdot [-\Theta D \nabla C] = -\boldsymbol{u} \cdot \nabla C + R_L + R_P + S_c \qquad (1)$$

Where Θ is the pore volume fraction, $\boldsymbol{u}$ is the velocity field with x, y, and z as
its components, C represents the solute concentration in the liquid, C_p the sorbed
solid particles, D is the dispersion coefficient. Also R_L is the reaction in liquid, R_P
is the reaction in solid and S_c is the source term. The first and second terms in the
left hand side contains the retardation factor RF, in some cases this coefficient is
known, it describes how sorption slows the solute velocity, and it can be written as:

$$RF = 1 + \frac{\rho}{\theta} \frac{\partial C_p}{\partial C} \qquad (2)$$

If the contaminant moves at the average linear velocity of the fluid then RF = 1.
For RF > 1, the contaminant velocity is smaller than the fluid velocity owing to
residence time on solids. The third term in the left hand side involving gradients
has to do with mechanical dispersion and molecular diffusion.

The first term in the right hand side involving the velocity u represents the advective flux which means the flux associated with the mean flow velocity. Flow velocities in porous media are very low. In ground water flow is essential to know the fluid movement through interstices in a porous media. This is possible with Darcy's law. Applying Darcy's law requires the gradient in hydraulic potential driving fluid in the porous media (Fetter 1994). Considering two points in the domain, the velocity can be calculated knowing pressure and elevation potential in both points, by:

$$u = -\frac{\kappa}{\eta}(\nabla H) \tag{3}$$

In this equation, u is the Darcy's velocity or specific discharge vector [m/s]; κ is the permeability of the porous media [m^2]; η is the dynamic viscosity of the fluid [Pa·s]; and H is the hydraulic head. Considering diffusion as a small quantity it can be neglected, therefore dispersion is the only term considered and it can be calculated by the expression:

$$Dij = \alpha_T|V|\delta_{ij} + (\alpha_L - \alpha_T)\frac{ViVj}{|V|}, \tag{4}$$

Where α_L and α_T are transversal and longitudinal dispersion components respectively, δ_{ij} is the Kronecker's delta, $|V|$ is the velocity vector module V, where V is the effective velocity and represents the average displacement of water molecules. Its components can be calculated by Darcy's law which considers flux through solid matrix goes from higher levels to lower levels besides this flux depends on pressure and depth. As the flux goes down it losses energy due to friction, as a consequence the hydraulic pressure gradient appears which can be expressed as a linear relation where the flux velocity Q is directly proportional to energy losses and K, and inversely proportional to the length a accomplish by the fluid L. The coefficient K depends on the soil properties and the hydrodynamics properties of the fluid, and it is called permeability or hydraulic conductivity.

3 Methodology

The problem is written as a system of partial differential equations with boundary conditions. The first step is to solve Darcy's equation to obtain the velocity field. Later transport equation needs to be solved. As it was mentioned previously, solute transport is a time-dependent equation; considering $RF = 1$ in expression (1) it could be described by the equation:

$$\Theta\frac{\partial C}{\partial t} + \nabla \cdot [-\Theta D\nabla C + uC] - \lambda C = 0 \tag{5}$$

where Θ is the porosity, D the hydrodynamic dispersion tensor [m^2/s], C the dissolved concentration [kg/m^3], u the Darcy's speed [m/d], t the time [s], and we

Fig. 1 Domain distribution.
Six levels from *top*
representing a heterogeneous
media

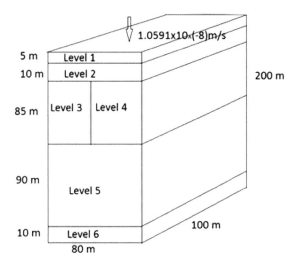

have introduce the decay constant $\lambda = \ln 2/t_{1/2}$ [s^{-1}], where $t_{1/2}$ is the half-life time and describe the radioactive decay rate. The disintegration rate is proportional to the number of nuclei present in the sample: $dN/dt = -\lambda N$, where N is the number of nuclei radioactive, for the ^{226}Ra $\rightarrow$ ^{222}Rn decay ($t_{1/2} = 1599$ years). In this work is assumed that ^{226}Ra decays directly to the stable isotope ^{206}Pb, because all half-life timescales from ^{222}Rn to ^{206}Ti are much shorter than for ^{226}Ra (De la Cruz 2011a). The numerical tool used to obtain the solution was the finite element method which is implemented in the software COMSOL Multiphysics 3.5. This method is powerful machinery to get the numerical solution of this kind of problems (De la Cruz 2011b). It consists in discretizing the equation to be solved in space and time. In two dimensions the space is discretized by triangles; in three dimensions the space is discretized by tetrahedral. Moreover, time is divided in time steps for which the solution is found. It is needed to set up boundary conditions to be coupled in the model. These conditions can be Dirichlet, Neumann or Robin. At the same time some other experimental parameters and relations are needed to describe completely the real system such as porosity, dispersion coefficients, viscosity, and permeability, among others.

4 Solving de Model

In this section the particular system solved is described along with the boundary conditions, parameters and constitutive relations.

The first case analyzed is a 3D domain which has a rectangular area of 80 × 100, and is 200 m depth from the surface. This domain is a heterogeneous

Table 1 Characteristics of each level in the porous media

Levels	Level 1	Level 2	Level 3	Level 4	Level 1 5	Level 6
Porosity	0.3	0.25	0.35	0.25	0.30	0.20
Permeability K	10^{-10}m^2	10^{-11}m^2	10^{-7}m^2	10^{-11}m^2	10^{-10}m^2	10^{-13}m^2

Fig. 2 Velocity Field on face XZ

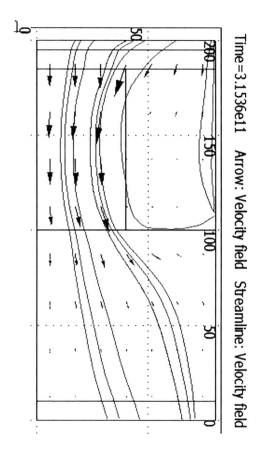

Time=3.1536e11   Arrow: Velocity field   Streamline: Velocity field

system with six different layers as Fig. 1 shows. A normal flux of water is considered on the top surface due to the rain. The amount of this flux is 1.0591×10^{-8} m/s. On this face, there is an area where the contaminant ^{226}Ra is resting at t = 0 and it is drained into the volume during the process. This substance is radium which has a half-life of 1,599 years and decay into radon gas. There is a zero flux boundary condition through the rest of the boundaries.

The varying parameters are porosity and saturated permeability along the domain as is described on Table 1.

After solve Darcy's equation, the velocity field is obtained, it can be seen in Fig. 2, on XZ plane, where the arrows are describing the velocity field direction and the streamlines are describing the trajectory of the field.

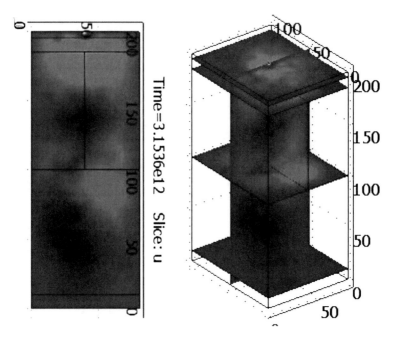

Fig. 3 *Left* Concentration distributions on face XZ. *Right* Concentration distribution in a 3D domain

Fig. 4 *Left* Concentration on the second horizontal plane (*top-down*) at three different time steps (*Right*) Concentration on the fourth horizontal plane (*top-down*)

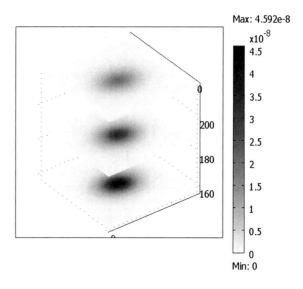

The Fig. 2 includes the length in meters. The simulation time is $t_f = 3.1536 \times 10^{11}$ in seconds. It is important to see the trajectory of the flux since it will define the path of the contaminant inside the domain.

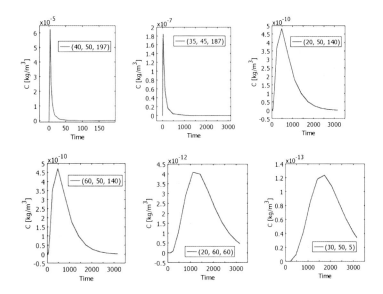

Fig. 5 Concentration at points (40,50,197), (35,45,187), (20,50,140), (60,50,140), (20,60,60), (30,50,5) during the first hour

Figure 3 (left) shows the distribution of the contaminant along the domain in XZ plane and in the right side shows the 3D version. It is clear the concentration tends to flow in the same way velocity lines do. The simulation time is $t_f = 3.1536 \times 10^{12}$ s. It can be seen the source of contaminant on the top surface. As well as, each horizontal and vertical plane separates the change of properties in the media.

In the left of Fig. 4 we can see the concentration on the second horizontal plane (top-down) at three different times, $t_1 = 157.68$, $t_2 = 186.0624$, $t_3 = 217.5984$ s. Also, the right of Fig. 4 shows the concentration on the fourth horizontal plane (top-down) at the same time. In both plots the first plane generated is the lowest plane and the last one is the top plane. Comparing both planes, we can observe that that the concentration goes from higher to lower value in second plane and the opposite occurs the fourt plane.

The next (Fig. 5) describe the concentration at different points in the domain along a short period of time. After this time the concentration runs along values close to zero. The six points analized are: (40,50,197), (35,45,187), (20,50,140), (60,50,140), (20,60,60), (30,50,5).

5 Conclusions

This work presented the 3D simulation of ^{226}Ra through a saturated porous media. The results presented are useful to know the transport and decay of ^{226}Ra through the subsurface. The most important point of this work is the simulation which is developed in a 3D domain. The simulation gives the evolution of the concentration along thousands of years. And it is clear that the 200 meters depth considered, are long enough to see its evolution. This specific behavior of the flow obtained numerically is the result of physical and chemical properties of the soil layers as well as the pressure distribution along the system and the particular properties of the contaminant and its interaction with the soil. The concentration is analyzed in specific points. It is important to emphasize that 3D simulation of this kind of problems is useful to adopt preventive measures and to take decisions for environmental impact and remediation.

References

Bear J (1979) Dynamics of fluids in porous media. Dover, New York

De la Cruz E, Gonzalez R, Klapp J, Longoria LC, Mayoral E (2011a) Migration and decay of ^{226}Ra in a saturated porous media in experimental and theoretical advances in fluid dynamics, series environmental sciences and engineering series. Springer, New York, pp 341–348

De la Cruz E, Klapp J GR, Longoria LC, Duarte R ME (2011b) Numerical simulation of ^{226}Ra migration and decay in a saturate porous medium. Rev Int Contam Ambie 27(3):215–221

Fetter CW (1994) Applied hydrogeology. Prentice, New Jersey

Evaluation of a Temporary Repository of Radioactive Waste

Roberto González-Galán, Eduardo de la Cruz-Sánchez,
Jaime Klapp-Escribano, Estela Mayoral-Villa,
Nora Pérez-Quezadas and Salvador Galindo Uribarri

Abstract The confinement of radioactive waste (radionuclides) in underground installations has to take into account its influence on the environment, the solvent action and the waste groundwater drag. This work evaluates by computer simulations how radionuclides migrate through the subsurface of a typical site in Mexico. The simulations show preferential routes that the contaminant plume follows over time. Results indicate that the radionuclides flow is highly irregular and it is influenced by failures in the area and its interactions in the fluid–solid matrix. The obtained concentration of the radionuclide is as expected.

R. González-Galán (✉)
Facultad de Ciencias, Universidad Autónoma del Estado de México,
El Cerrillo, Piedras Blancas C.P, 50200 Estado de Mexico, Mexico
e-mail: rgonzalez470@yahoo.com.mx

E. de la Cruz-Sánchez · J. Klapp-Escribano · E. Mayoral-Villa · S. G. Uribarri
Instituto Nacional de Investigaciones Nucleares, Carretera México-Toluca s/n,
La Marquesa, Ocoyoacac C.P, 52750 Estado de México, Mexico
e-mail: jaime.klapp@inin.gob.mx

N. Pérez-Quezadas
Instituto de Geofísica, Universidad Nacional Autónoma de México,
Mexico D.F., 04510 Mexico

J. Klapp-Escribano
Departamento de Matemáticas, CINVESTAV-IPN, Mexico D.F., 07360, Mexico

J. Klapp et al. (eds.), *Fluid Dynamics in Physics, Engineering and Environmental Applications*, Environmental Science and Engineering,
DOI: 10.1007/978-3-642-27723-8_40, © Springer-Verlag Berlin Heidelberg 2013

1 Introduction

Computer simulation tools are useful in evaluating and selecting confinement radionuclide sites. These tools are: (1) a mathematical model that describes the main phenomena that affects the process, and (2) a computational model that solves the equations of the mathematical model. In recent decades there has been a growing need to perform this type of study to define appropriate strategies for radionuclide disposal.

For oil industry applications, Coats and Smith (1964) have proposed a mechanism for the fluid flow through a virtually immobile solid matrix, Corapcioglu and Baehr (1987) studied the pollution of a porous media by petroleum products. The works of Neretnieks (1980), Pollock (1986) and Walton (1994) focused its attention on the subsurface transport of radionuclides and the hydrodynamic behavior of the barriers on confinement sites.

In the present will deal with radionuclides transport calculations. For this purpose we first solve the equations that describe the water flow that is seeping through the ground. A classical approach for the solution of this problem (Bear 1979), consists of considering the porous medium as a continuum and the point-to-point spatial variations of hydrogeological characteristics as an average over a representative element of the volume (REV). The applicability of this approach has been studied by Long et al. (1982) and Schwartz (1988). Water hydrodynamic is calculated from a conservation equation based on Darcy's law. On the other hand, transport equation considers that the variation on radionuclides concentration is due to advection, molecular diffusion, and mechanical dispersion. This equation also takes into account that the radionuclides dissolved in water or adsorbed by the solid matrix may decay. This work presents the mathematical formulation of the model for the simulation of the transport of radionuclides through the ground, and reports the results for a specific site.

2 The Mathematical Model

For this work, the ground is considered as the interaction of two continuums (media): a mobile medium (the fluid) and a fixed medium (the porous matrix). The fixed medium serves as the frame through which the mobile medium moves. The dynamics of the fluid is governed by the continuity equation and using Darcy's law can be written in the form

$$\frac{\partial}{\partial t}(\rho_f \theta_s) + \nabla \cdot \rho_f \left[\frac{k}{\eta} \left(\nabla p + \rho_f g \nabla y \right) \right] = \rho_f Q_s, \tag{1}$$

where ρ_f is the density of the fluid (kg/m^3), θ_s the fraction of the rock or soil that is occupied by the fluid, p the pressure (Pa), y the vertical coordinate (m), k the permeability of the porous medium (m^2), η the viscosity of fluid (Pa•s), ∇ the spatial gradient operator, g the gravity acceleration (m/s^2), and Q_s represents the sinks and/or sources of the fluid in the simulation domain. In Eq. (1), the term $\mathbf{u} = \nabla p + \rho_f g \nabla y$ is called "Darcy's velocity" and corresponds to the transport equation advective term. The water that seeps into the subsoil may contain a significant amount of dissolved substances which are called solutes. In the case of underground installations for radionuclide disposal, this is a factor to be taken into account. The governing equation for a saturated porous medium with water, with adsorption by the solid matrix and radioactive decay is

$$\theta_s \frac{\partial C_i}{\partial t} + \rho_b \frac{\partial C_{pi}}{\partial C_i} \frac{\partial C_i}{\partial t} + \nabla \cdot [\theta_s \mathbf{D}_L \nabla C_i + \mathbf{u} C_i] - \theta_s \frac{\ln 2}{\lambda_i} C_i - \rho_b \frac{\ln 2}{\lambda_{pi}} \left(\frac{\partial C_{pi}}{\partial C_i} \right) C_i = f,$$

(2)

where the $\theta_s C_i$ term represents the liquid radionuclides concentration (mass per liquid volume) of the $i - th$ species considered, C_{pi} is the adsorbed radionuclides concentration per unit mass in the solid matrix, ρ_b is the "bulk" density of the porous medium, $\mathbf{D}_L$ represents the hydrodynamic dispersion tensor, the $\mathbf{u}$ vector is the velocity of the fluid as given in a previous paragraph, and the $\theta_s(\ln 2/\lambda_i)$ term corresponds to the radioactive decay constant in the liquid phase; similarly $\rho_b(\ln 2/\lambda_i)(\partial C_i/\partial C_i)$ is the radioactive decay constant in the solid phase, and the term f is the solute source.

The Eq. (2) establishes that the time variation of the concentration of the radionuclides is mainly due to the following processes: (1) the advection $\mathbf{j}_A = \mathbf{u} C_i$ is the process through which the dissolved radionuclide are dragged by the moving groundwater, (2) the molecular diffusion $j_d = -\theta_s D_d \nabla C_i$ due to the concentration gradients and determined by Fick's law, where D_d is the molecular dispersion coefficient, (3) the mechanical dispersion, $\mathbf{j}_m = -\theta_s \mathbf{D}_m \nabla C_i$, which acts diluting and reducing the concentration of the solute dissolved in groundwater.

The components of the mechanical dispersion tensor, $\mathbf{D}_m$, are defined by the velocity of the fluid and two types of scattering: the longitudinal dispersion that is the scattering that occurs along the fluid streamlines, and the transversal dispersion that is the dilution that occurs normal to the pathway of the fluid flow (Bear 1979). In Eq. (2) the diffusion and the mechanical dispersion are combined in a single term composed by the mechanical dispersion tensor and the diffusion coefficient to form the hydrodynamics tensor $\theta_s \mathbf{D}_L = ID_d + \theta_s \mathbf{D}_m$. On other hand, the interaction of the radionuclides with the porous matrix is established by the adsorption of dissolved radionuclides by the granules that make up the solid matrix. The proportion of the absorbed to the dissolved radionuclides concentration is determined by the relationship between phases: $k_d C_{pi} = C_i$, where k_d is the proportionality constant or isotherm of the sorption that governs how the radionuclide is absorbed by the solid phase.

The terms $-\theta_s(\ln 2/\lambda_i)$ and $-\rho_b(\ln 2/\lambda_i)(\partial C_{pi}/\partial C_i)$ corresponds to the radioactive decay of each species in the fluid phase, and in the solid phase (absorbed), respectively. λ_i is the decay constant of the specie i. In this work we studied the case of a radioactive chain, where the disintegration of a radionuclide, so-called

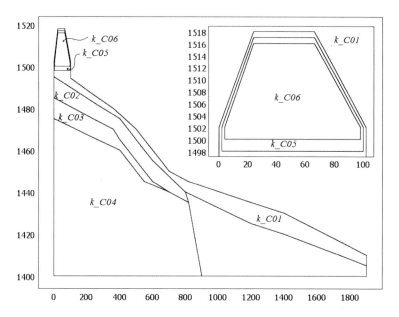

Fig. 1 The subsurface strata in the computational domain. The *horizontal* (x) and *vertical* (y) axis are given in meters

"father", produces elements that are also unstable and will disintegrate. The decay chain is: $^{230}Th \xrightarrow{\lambda_{Th}} {}^{226}Ra \xrightarrow{\lambda_{Ra}} {}^{206}Pb$, where λ_{Th} and λ_{Ra} are the decay constants of ^{226}Ra and ^{230}Th, respectively. In reality, the decay chain does not directly transmutes ^{226}Ra to the stable element ^{206}Pb, as this process undergoes a series of decay steps, however, in this work we have a secular equilibrium process Choppin et al. (2002), as the decay time of ^{230}Th and the ^{226}Ra are much larger than the decay time of the ^{226}Ra intermediate decay products, and so that these are neglected.

3 Simulations

Our site under study has a desert climate with an annual rainfall of 300 mm and an average temperature of 18 °C (Rojas 1996).

Figure 1 shows a cross-section model of the subsoil layers. In our case six layers are considered. Layer *K_C04* is the most superficial and is composed of a material so-called "caliche". The *K_C02* layer is composed of conglomerated sand, and below both of them we find the conglomerate layer *K_C03*. The deepest layer, the *K_C04* is composed of volcanic stone and limestone on a sandy matrix. The radioactive wastes were stacked on a clay bed of 1 m thickness forming a truncated pyramid. The pyramid is covered with a layer of the same clay used in

Table 1 Hydro-geological parameters of the substrates

Stratum	K (cm^2)	θ	k_d (m^2/Kg)
K_C01	1×10^{-5}	0.249	6.96×10^{-2}
K_C02	1×10^{-4}	0.2	1.06×10^{-2}
K_C03	1×10^{-3}	0.3	1.06×10^{-2}
K_C04	7.4×10^{-8} (cm^2/s)	0.15	6.96×10^{-2}
K_C05	7.4×10^{-8} (cm^2/s)	0.15	3.83×10^{-2}
K_C06	7.2×10^{-6} (cm^2/s)	0.1	3.06×10^{-2}

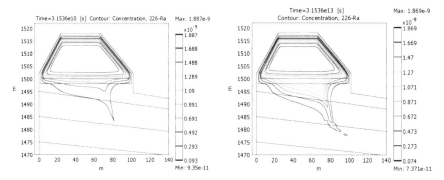

Fig. 2 Plume contaminant concentration for the times 1,000 years (*left panel*) and 1 million years (*right panel*)

the bed. The whole arrangement is covered with soil of the region. In Table 1 we summarize the hydrogeological parameters of the sub-soils, where K represents the permeability for strata K_C01–K_C03, and the hydraulic conductivity for strata K_C04–K_C06. θ is the porosity, and finally, k_d is the adsorption isotherm.

In our model, water enters through the upper boundaries and seeps to the deepest layers. When the water arrives to the K_C04 stratum, the radionuclide is dissolved and dragged to deeper strata which initially are non-contaminated. Figure 2 shows the time evolution of the ^{226}Ra plume. Graphs shown in Fig. 3 were obtained for the position x = 65 m, y = 1,495 m; in both sides, the dashed line shows the situation when the barrier has a failure; solid line corresponds to the non-failure event. In the left side, we show the ^{226}Ra concentration with time. The right side shows the ^{230}Th–^{226}Ra concentration ratio. The model was solved using the finite element method, with the COMSOL Multiphysics software with a spatial mesh that consist of 73,144 elements and inputs exponentially increasing of 10 years from 10 to 100 years, 100 years increments in the period from 100 to 1,000 years, and 1,000 years increments in the period of 1,000 to 10,000 years, etc. until 1 million years.

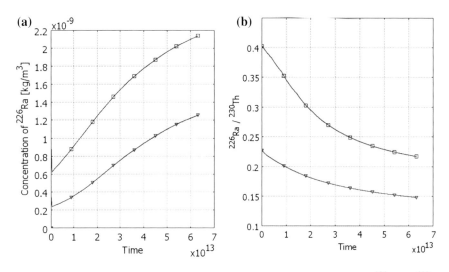

Fig. 3 **a** Concentration at x = 65 m, y = 1,485 m along time; **b** relationship between ^{226}Ra and ^{230}Th

4 Discussion and Conclusions

In Fig. 2 the color bar indicates the magnitude of the ^{226}Ra concentration; the time is given in seconds and the graphs correspond to 10,000 and 1 million years. The maximum concentration is of the order of 10^{-6} gr/m^2 for final times of the simulations; the values obtained for times <1,000 years were about 0.5×10^{-9} gr/m^2 (see Fig. 3) which is consistent with Rojas (1996). Our graphs show that the contaminant plume has its source in region near the locations x = 10 m, y = 1,500 m and x = 100 m, y = 1,500 m, thereafter it moves towards the regions 60 m < x < 75 m; the above results suggest a local change in pressures and velocities that conveys the plume to that area. The shape of the plume is strongly influenced by the site lithology. We obtained a Peclet number that indicates that the flux is mainly advective. The values shown in Fig. 3 indicate that the transit time of radionuclide through the solid matrix is consistent with the values reported by Rojas (1996) and Luo et al. (2000). Concentration values shown in dashed line are significantly higher than the values for the solid line. In Fig. 3b we show the relationship between the ^{230}Th and ^{226}Ra radionuclide, the ratio is higher in the dashed line than the solid line, which is an indication that the mobility of ^{226}Ra is higher in the model including a failure.

The concentration of ^{226}Ra obtained for times <1,000 years are below the picograms, so that the model suggests that this type of installations are an effective alternative for radionuclide disposal, however, in case of failures in the engineered barrier, concentrations observed were 5 times higher; this indicates that it is essential a careful design of the facility and that it must have a permanent monitoring program. The computer simulation we have just shown is a powerful tool for evaluating and recognizing potential sites that can serve as nuclear waste disposal facilities. In addition it serves to forecast the impact of these installations, and even potential disasters.

Acknowledgment J. Klapp thank ABACUS, CONACyT grant EDOMEX-2011-C01-165873.

References

Bear J (1979) Dynamics of fluids in porous media. Dover Publications Inc, New York, p 764

Coats KH, Smith BD (1964) Dead-end pore volume and dispersion in porous media. J Soc Petrol Eng 4:73–84

Corapcioglu MY, Baehr AL (1987) A compositional multiphase model for groundwater contamination by petroleum products 1, Theoretical considerations. Water Resour Res 23(1):191–200

Choppin GR, Liljenzin JL, Rydberg J (2002) Radiochemistry and nuclear chemistry, Butterworth-Heinemann, ISBN 0750674636, 9780750674638, p 709

Long JCS, Remer JS, Wilson CR, Witherspoon PA (1982) Porous media equivalent for networks of discontinuous fractures. Water Resources Res 18(3):645–658

Luo S, Ku TL, Roback R, Murrell M, McLing TL (2000) In-situ radionuclide transport and preferential groundwater flows at INEEL (Idaho): decay-series disequilibrium studies. Geochim Cosmochim Acta 64:867–881

Neretnieks I (1980) Diffusion in the rock matrix: an important factor in radionuclide retardation? J Geophys Res 85(B8):4379–4397

Pollock DW (1986) Simulation of fluid flow and energy transport processes associated with high-level radioactive waste disposal in unsaturated alluvium. Water Resour Res 22(5):765–775

Rojas Martínez VP (1996) Un enfoque biológico sobre la migración del 226-Ra en los estratos someros subyacentes en el depósito de estériles en Peña Blanca Chihuahua. Tesis de Licenciatura FES- Zaragoza, UNAM

Schwartz F, Smith L (1988) A continuum approach for modelling mass transport in fractured media. Water Resour Res 24(S):1360–1372

Walton JC (1994) Influence of evaporation on waste package environment and radionuclide release from a tuff repository. Water Resour Res 30:3479–3487

Part V
General Fluid Dynamics

Friction Coefficient in Plastic Pipelines

E. A. Padilla, O. Begovich and A. Pizano-Moreno

Abstract Currently, plastic pipes are beginning to be more used because of its low cost, easy installation, flexibility, durability, low weight, etc. However, this kind of pipes may have a friction coefficient sensitive to variations of flow and temperature, which can affect in the design of an analytic leak diagnosis system. The variations of friction in a plastic pipeline can be caused by the following: the flow is not in a fully developed regime. In the same way, temperature affects the viscosity of water which affects the value of the Reynolds number and then, the value of the friction coefficient. This work illustrates the sensitivity of the friction coefficient due to changes in flow and temperature with experiments performed in a plastic pipeline prototype located at the Centro de Investigación y de Estudios Avanzados del Instituto Politécnico Nacional (CINVESTAV-Guadalajara).

1 Introduction

Water is an essential natural resource for humanity therefore the waste of this vital liquid must be avoided. A way to do that, in pipelines, is by implementing monitoring systems by using algorithms for leak detection and isolation (LDI).

E. A. Padilla (✉) · O. Begovich · A. Pizano-Moreno
Centro de Investigación y de Estudios Avanzados del Instituto Politécnico Nacional,
Av. Del Bosque 1145, 45019 Guadalajara, Jalisco, Mexico
e-mail: epadilla@gdl.cinvestav.mx

O. Begovich
e-mail: obegovi@gdl.cinvestav.mx

A. Pizano-Moreno
e-mail: apizano@gdl.cinvestav.mx

J. Klapp et al. (eds.), *Fluid Dynamics in Physics, Engineering and Environmental Applications*, Environmental Science and Engineering,
DOI: 10.1007/978-3-642-27723-8_41, © Springer-Verlag Berlin Heidelberg 2013

Fig. 1 Moody chart

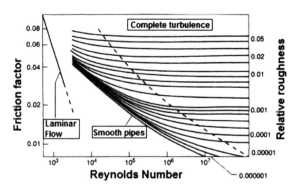

However, due to several phenomena present in plastic pipelines, robust LDI algorithms are still in via of research, since in this kind of pipes a flow regime that assures a constant friction factor is difficult to obtain.

The variations in the friction coefficient in plastic pipes can be caused by: a) A small relative roughness, so that is difficult to reach a complete turbulence zone where the value of the friction factor is almost constant; b) The effect of temperature changes on density and viscosity, these variations affect the flow regime and friction coefficient.

This paper is organized as follows: In Sect. 2, we talk about variation of the friction coefficient due to a non-fully developed flow regime. In Sect. 3 we discuss the variation of friction caused by the action of temperature. Section 4 describes the need of friction compensation for models used to the design of LDI algorithms. Section 5 shows some real-time experimental results. Finally, in Sect. 6 the conclusions are stated.

2 Friction Coefficient Variations Due to Flow Changes

In the Moody chart, in Fig. 1, it can be seen that for pipes with a relative roughness less than 1×10^{-3}, such as plastic, it is difficult to reach a complete turbulent flow, where the friction coefficient can be considered constant. Then, in this kind of pipes, it is possible to obtain a more appropriate value for the friction coefficient by using the *Swamee-Jain* Eq. (1). In this expression, the friction factor is a function of the Reynolds number (2):

$$f_r(z,t) = \frac{0.25}{\left[\log_{10} \left(\frac{\varepsilon}{3.7D} + \frac{5.74}{Re^{0.9}} \right) \right]^2} \tag{1}$$

$$Re = \frac{Q(z,t)D}{A\upsilon} \tag{2}$$

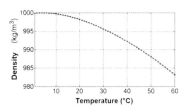 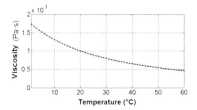

Fig. 2 Density and viscosity of water with respect temperature where $T \in (0^{\circ}\,C, 60\,^{\circ}C)$

where ε is the roughness of the pipe, D is the diameter of the pipe and Re is the Reynolds number, A is the cross-sectional area of the pipe and v is the fluid kinematic viscosity. Notice that the Reynolds number is a linear function of the flow.

3 Variations of the Friction Coefficient Caused by Action of Temperature

When a pipe is exposed to temperature changes, density and viscosity also change their values, as is illustrated in Fig. 2. Such variations impact on both, the flow regime and the friction coefficient. It can be seen from (2) that the viscosity affects the Reynolds number. Since the friction value depends on the Reynolds number, a variation occurs on the friction coefficient when the Reynolds number changes.

4 Friction Compensation for Models Used to the Design of LDI Algorithms

Leak detection algorithms in pipelines are analytical methods used to determine the occurrence of a leak (leak detection) and its intensity and location (leak isolation). The design of these systems is based on a mathematical model that emulates the dynamics of water in a pipe (Begovich and Pizano-Moreno 2008; Verde 2005; Besançon et al. 2007; Benkherouf and Allidina 1998). Thus, this model must be able to take into account variations caused by temperature and by changes in friction if it is desired that such model predicts the dynamics of a liquid in a more exact way.

In literature which deals with analytic LDI algorithms, the friction coefficient is considered constant. However, if the fluid dynamics is affected by a friction coefficient sensitive to variations of flow and temperature, then the model used will not follow the dynamics of the pipe properly, this fact leads to a bad LDI algorithm. One way to compensate errors caused by variations in the friction is by measuring the flow and constantly recalculating the friction factor by using the *Swamee-Jain equation* as showed Navarro et al. (2011).

Table 1 Parameters of the pipeline prototype

Parameter	Symbol	Value	Units
Length between sensor	L_T	68.54	m
Internal diameter	D	0.0635	m
Cross-section area	A	0.0031669	m^2
Wall thickness	τ	0.01270	m
Roughness	ε	7×10^{-6}	M

5 Experimental Results

In this section we present some real-time experiments in order to show the sensitivity of the friction coefficient due to changes in flow and temperature. These experiments were performed by using a pilot plastic pipeline. A brief description of this prototype is presented below and the parameters of the pipeline are given in Table 1.

The pipeline prototype is equipped with two pressure head sensors and two water flow sensors at the ends of the pipeline; a pump capable of delivering a maximum pressure head of 19.5 m which pushes the water through the pipe. The pipes of this prototype are of a material known as Polypropylene Co-polymer Random (PCR). More details about this prototype can be found in (Begovich and Pizano-Moreno 2008).

5.1 Experiment 1. Friction Coefficient with Flow Variations

This experiment was performed when the temperature did not vary significantly around of 15 °C. Initial conditions for this experiment and parameters such as density and viscosity are given in Table 2. First it is attempted to maintain a steady flow for a while and then a change in the operating point (o.p.) was induced at 200 s. In Fig. 3 can be seen that the Reynolds number changes at the same time that the flow changes. Also, as it was expected, it can be seen that the friction coefficient is affected by the change of flow.

5.2 Experiment 2. Variation of the Friction Caused by the Action of Temperature

For this experiment, the inlet pressure head is set to 18 m. In this case, the pressure head and the flow are constant. Under the previous scenario, the experiment was made when the temperature increased. Parameters and initial conditions are given in Table 3. The friction coefficient calculated by using the *Swamee-Jain equation*

Table 2 Pipeline parameters at 15 °C and initial conditions for the experiment 1

Parameter	Symbol	Value	Units
Average flow rate	Qss	4.46448×10^{-3}	m^3/s
Density	ρ	9.99102×10^2	Kg/m^3
Viscosity	υ	1.11375×10^{-3}	Pa-s
Relative roughness	ε/D	1.10×10^{-4}	–
Reynolds number	Re	7.8622×10^4	–
Friction	fr	1.9363×10^{-2}	–

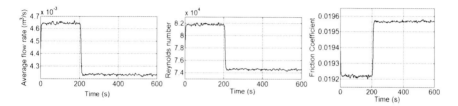

Fig. 3 Average flow rate with respect to Reynolds number and influence of flow on friction

Table 3 Initial conditions for the experiment 2

Parameter	Symbol	Value	Units
Pressure head at inlet	Hin	18	m
Average flow rate	Qss	8.7×10^{-3}	m^3/s
Relative roughness	ε/D	1.10×10^{-4}	–
Friction	Fr	1.991×10^{-2}	–

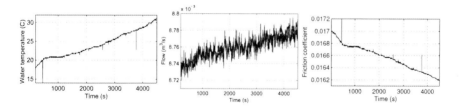

Fig. 4 Influence of temperature on the flow and friction factor

is shown in Fig. 4. It can be seen that the value of friction decreases as the temperature increases. It is also shown that the flow slowly increases; this happens due to the friction decreases and more flow is able to circulate through the pipeline.

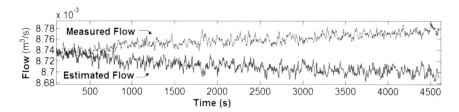

Fig. 5 Estimated flow from a standard model without temperature compensation vs. flow measures

5.3 Need to Get Better Models

In this experiment the model in Verde (2005), without temperature compensation, is used to estimate the flow in our pipeline in the presence of a gradual increase in temperature. In Fig. 5 it is shown the estimated flow and the measured flow. As it can be seen, the estimation of the model is incorrect. Then it is a challenge to find a more exact model.

6 Conclusions

A friction coefficient sensitive to flow variations and temperature can difficult the design systems for leak diagnosis based in models. To avoid this situation, it is necessary that the models used to design LDI algorithms take into account the friction sensitivity to flow variations and temperature changes.

Acknowledgments This work was supported by the project FOMIX 2009-05-125679 CONACYT-Gobierno del Estado de Jalisco.

References

Begovich O, Pizano-Moreno A (2008) Application of a leak detection algorithm in a water pipeline prototype: difficulties and solutions. In: Proceedings 5th international conference on electrical engineering, computing science and automatic control CCE 2008, pp 26–30
Benkherouf A, Allidina AY (1988) Leak detection and location in gas pipelines. IEE Proc 135:142–148
Besançon G, Georges D, Begovich O, Verde C, Aldana C (2007) Direct observer design for leak detection and estimation in pipelines. European control conference, pp 5666–5670
Navarro A, Begovich O, Besançon G, Dulhoste JF (2011) Real-time leak isolation based on state estimation in a plastic pipeline. IEEE multi-conference on system and control denver, CO, USA, pp 28–30 September 2011
Verde C (2005) Accommodation of multi-leak location in a pipeline. Control Eng Pract 13:1071–1078

Super Free Fall in Concentric Pipes

C. Treviño, S. Peralta, Carlos A. Vargas and A. Medina

Abstract In the present work we have analyzed experimentally the characteristics of the free surface of a liquid column starting off of the rest in a vertical container (concentric tubes). We have two tubes of different areas $A_1 < A_2$ interconnected where a great transition due to this difference of areas is registered. To small times for which the viscous effects are despised, the free surface in the superior tube reaches acceleration greater than gravity (g).

1 Introduction

Nowadays there's no more information about that configuration (concentric pipes), but there are some experiments about super free fall which were made in a chain by Calking and March (1989). They solve the falling chain in both for

C. Treviño (✉)
Unidad Interdisciplinaria Sisal UNAM, Yucatán, Mexico
e-mail: ctrev@servidor.unam.mx

S. Peralta · A. Medina
Instituto Politécnico Nacional Escuela Superior de Ingeniería Mecánica y Eléctrica, Sección de Estudios de Posgrado e Investigación, Av. de las granjas Col, Santa Catarina, Delegación Azcapotzalco, Mexico
e-mail: peraltasalomon@hotmail.com

C. A. Vargas
Universidad Autónoma Metropolitana Azcapotzalco, Av. San Pablo 180 Col. San Pedro Xalpan, Azcapotzalco, D.F., Mexico
e-mail: cvargas@correo.azc.uam.mx

A. Medina
e-mail: amedinao@ipn.mx

J. Klapp et al. (eds.), *Fluid Dynamics in Physics, Engineering and Environmental Applications*, Environmental Science and Engineering,
DOI: 10.1007/978-3-642-27723-8_42, © Springer-Verlag Berlin Heidelberg 2013

Fig. 1 Scheme for
concentric tubes, used with
the experiment. Z-axis is
towards at the superior part
and opposite to the action of
the gravity

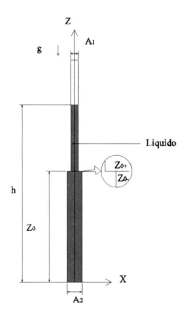

experimentally and numerically and found that the tip of the chain falls much
faster than acceleration of g. And in experimental form in a conical tube were
made by Villermaux and Pomeau (2010). For our case of interest only Daniel
Bernoulli (1738) has made such kind of experiments and he sealed in the bottom
with a balloon. In our case we follow the free surface of liquid and found that the
free surface reaches an acceleration greater than g.

2 Theory

When an ideal liquid is in a vertical, cylindrical pipe and suddenly the low part of
the cylinder is opened, the gravity will accelerate the liquid. The main flow is one
dimensional and the conservation equations of mass and momentum are, respectively (Fig. 1).

This is a tube with area A_1, connected at $z = z_0$, with another bigger on area
A_2, it is clear that $A_2 > A_1$ and height $h > z_0$, filled with liquid at time $t = 0$. If
the inferior cover takes off suddenly $z = 0$, considering a one-dimensional flow,
the equations of movement take the form.

$$\left(\frac{\partial u}{\partial t}\right) + u\left(\frac{\partial u}{\partial z}\right) = -\left(\frac{1}{\rho}\right)\left(\frac{dp}{dz}\right) - g. \tag{1}$$

The equation of conservation mass is $A_1 u_1(t) = A_2 u_2(t)$ solving for $u_2(t)$ we
have: $u_2(t) = \alpha u_1(t)$, where $\alpha = A_1/A_2 < 1$. If the pressure is $p = 0$ in $z = 0$ and
$z = h$.

Fig. 2 Graphic of a dimensional position for the free surface of liquid, for different values of $\beta = ((Z_o)/(h_o))$. With $\alpha = A_1/A_2 = 0.17892$

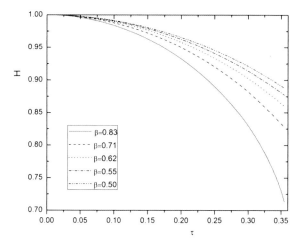

Fig. 3 Graphic which shows velocity vs time for the free surface, with $\beta = ((Z_o)/(h_o))$ and $\alpha = A_1/A_2 = 0.17892$

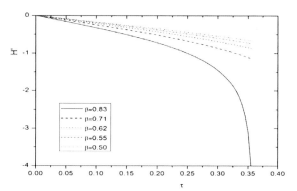

$$\int_{z0+}^{h} \frac{\partial u}{\partial t} dz + \int_{z0+}^{h} u \frac{\partial u}{\partial z} dz = \frac{1}{\rho} \int_{z0+}^{h} u \frac{dp}{dz} dz - \int_{z0+}^{h} g dz, \tag{2}$$

Integrating the equation (Villermaux and Pomeau 2010) from z_{0+} until h this equation takes the form:

$$h''(h - Z_{0+}) = \left(\frac{1}{\rho}\right) pZ_{0+} - g(h - Z_{0+}), \tag{3}$$

Integrating the equation (Villermaux and Pomeau 2010) but now from 0 until z_{0-} the equation takes the form:

$$\int_{0}^{z0-} \frac{\partial u}{\partial t} dz + \int_{0}^{z0-} u \frac{\partial u}{\partial z} dz = \frac{1}{\rho} \int_{0}^{z0-} u \frac{dp}{dz} dz - \int_{0}^{z0-} g dz, \tag{4}$$

$$\alpha h'' Z_{0-} = -\left(\frac{1}{\rho}\right) pZ_{0-} - gz_{o-}, \tag{5}$$

but we now that: $h'' = (\partial u_1)/(\partial t) = (\partial^2 h_1)/(\partial t^2), u_1 = (\partial h)/(\partial t) = h'$ and also: $(\partial u_2)/(\partial t) = \alpha(\partial u_1)/(\partial t) = \alpha h''$ from this equations and in the interface: $z = z_0 = z_{0+}$ and joining the Eqs. (3) and (5)

$$\left(\frac{1}{\rho}\right)(p_{zo+} - p_{zo+}) = h''[h - (1 - \alpha)z_0] + gh, \tag{6}$$

The conditions at the jump (are considering without losses) the Eq. (1)

$$\left(\frac{1}{\rho}\right)(p_{zo+} - p_{zo+}) = u_1^2(1 - \alpha^2) = -\left(\frac{1}{2}\right)(h')^2(1 - \alpha^2) \tag{7}$$

The Eqs. (6) and (7), the equation for h(t) takes the form:

$$h''[h - (1 - \alpha)Z_0] + (1/2)(h')^2(1 - \alpha^2) + gh = 0 \tag{8}$$

With the initial conditions h (0) = h_o and h' (0) = 0. At the start point then:

$$h''(0) = -\left(\frac{gh}{h - (1 - \alpha)z^0}\right) \tag{9}$$

Which is bigger than g, if $\alpha < 1$. This equation gave us the complete history of the free surface (acceleration) during the whole experiment but of course, this solution is just in physics variables, for generalizing the solution we have to dimensionless the Eq. (8), for making the results general. We have to introduce our dimensionless parameters:

$$H = \left(\frac{h}{h^0}\right), \tau = \left(\frac{t}{\sqrt{\left(\frac{h^0}{g}\right)}}\right), \beta = \left(\frac{Z^0}{h^0}\right)$$

Then we have

$$\left(\frac{d^2 h}{dt^2}\right)[h - (1 - \alpha)Z^0] + \left(\frac{1}{2}\right)\left(\left(\frac{dh}{dt}\right)\right)^2(1 - \alpha^2) + gh = 0 \tag{10}$$

Introducing parameters

$$\left(\frac{h^0}{\left(\frac{h^0}{g}\right)}\right)\left(\frac{d^2 h}{dt^2}\right)[h^0 H - (1 - \alpha)h^0\beta] + \left(\frac{1}{2}\right)\left[\left(\frac{h^{02}}{\left(\frac{h^0}{g}\right)}\right)\left[\left(\frac{dH}{d\tau}\right)\right]^2\right](1 - \alpha^2)$$
$$+ gh^0 H$$
$$= 0 \tag{11}$$

Factorizing and eliminating equals terms

Fig. 4 Graphic that shows the relationship between time vs acceleration. In which it is shown the evolution of the free surface of fluid with the values of $\beta = (Z_o)/(h_o)$ and $\alpha = A_1/A_2 = 0.17892$

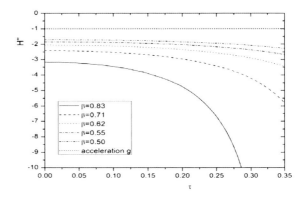

Fig. 5 Relationship between kinetic energy a potential energy in which can be observe the conversion of potential energy in kinetic for different β

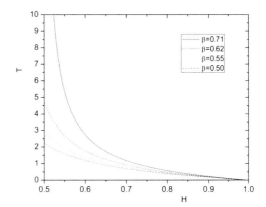

$$h^0 g \left(\frac{d^2 h}{dt^2}\right)[H - (1-\alpha)\beta] + h^0 g \left(\frac{1}{2}\right)\left[\left(\frac{dH}{d\tau}\right)\right]^2 (1-\alpha^2) + gh^0 H = 0 \quad (12)$$

Dividing all by $h_o g$ we have

$$\left(\frac{d^2 h}{dt^2}\right)[H - (1-\alpha)\beta] + \left(\frac{1}{2}\right)\left[\left(\frac{dH}{d\tau}\right)\right]^2 (1-\alpha^2) + H = 0 \quad (13)$$

Finally we have

$$\ddot{H}[H - (1-\alpha)\beta] + \left(\frac{1}{2}\right)[\dot{H}]^2 (1-\alpha^2) + H = 0 \quad (14)$$

Differential Eq. (14) is a non-linear, second order equation which only can be solved numerically. The results of solving the equation are shown as follows:

Graphic in which is shown the position vs time so that $\beta = 0.8333$ and $\alpha = A_1/A_2 = 0.17892$ is faster than the other values for beta (shown in the graphic). Position and time are both without dimensions, which can be seen in the evolution of the free surface Fig. 2.

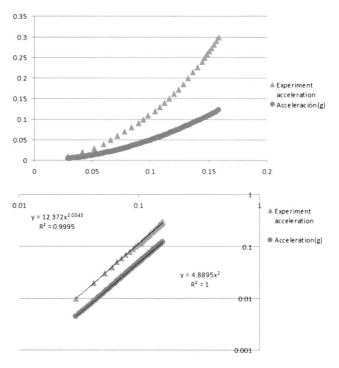

Fig. 6 Superior graph data in normal scale, inferior graphic made in scale log–log scale in which we can observe the difference between the free fall and the super free fall

Fig. 7 Experiments in which the profile of free surface could be observed

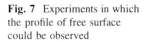

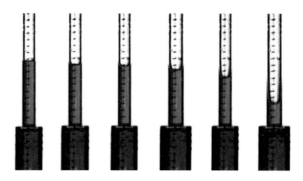

Graphic of evolution of position vs time a dimensionless in which is shown that the velocity of free surface is much faster than the other values of β and the development of it Figs. 3 and 4.

Fig. 8 Graphic in which we have different heights of the free surface for all the experiments for concentric tubes, with different values of $\beta = 0.8333$, $\beta = 0.714$, $\beta = 0.625$ and $\alpha = 0.17892$

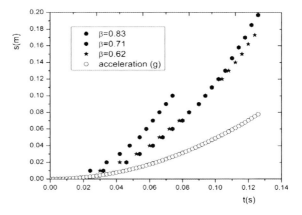

Fig. 9 Graph of sequence for experiment results versus analytical with $\alpha = 0.1789$ and $\beta = 0.8333$ can be observed

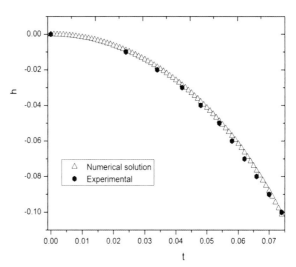

3 Another Point of View

If we dimensionless the Eq. (13) where $T = (1/2)\,((dH)/(d\tau))^2$ which is the kinetic energy dimensionless and H is the potential energy dimensionless, for $\alpha = 1$ (free fall) it results $T = 1 - H$. The results are shown below Fig. 5.

4 Development of Experiment

We have two tubes joined in a concentric form, whose dimensions are: for tube 1; $L = 0.5$ m, $\varphi = 0.0435$ m and tube 2; $L = 0.5$ m, $\varphi = 0.0565$ m joined in a concentric form. The experiment was made by sealing the inferior part of the test tube with help of a pneumatic circuit, and opening it in a sudden form. It helped us managing the closing and the opening and gave us a very good control while acquiring data to analyze and verify the numerical results vs experimental results.

4.1 Experimental Results

After making several experiments we found that:

This graphic was obtained from the experiment with data $\beta = 0.83333$. and $\alpha = A_1/A_2 = 0.17892$ (Fig. 6).

Concentric pipes in which the results have been obtained (Figs. 7 and 8)

In graph, β is the three comparison of different that were used in the experiment and it can be appreciated that $\beta = 0.83$ is the one that has a greater displacement in the time. In the low part the free fall can be observed.

4.2 Comparison of Experimental and Numerical Results

After making several experiments with different heights of filling and several runs for the solution of the second order nonlinear equation we find the following data. In solving the Eq. (14) in physical variables it is compared with our experiment and it is shown in Fig. 9.

5 Conclusions

- The super free fall happens in concentric liquid systems.
- The inviscid theory describes the experiments well with flow of low viscosity such as the water or ethanol.
- Protuberances in free surface are not observed.

References

Bernoulli D (1738) Hydrodynamica. Argentorati (Strassburg). Johann Reinhold Dulsecker, Germany

Calkin MG, March RH (1989) The dynamics of a falling chain: I. Am J Phys 57:154–157

de Sousa CA, Rodrigues VH (2004) Mass redistribution in variable mass systems. Eur J Phys 25:41–49

Goldstein (1980) Classical mechanics, 2nd edn. Addison-Wesley, Reading, MA

Hairer E, Wanner G (1996) Solving ordinary differential equations II:stiff and differential algebraic problems, 2nd edn. Springer-Verlag, Berlin

Hamel G (1949) Theoretische mechanik grundlehren der mathematischenwissenschaften, Bd LVII. Springer, Berlin, Aufgabe 100, pp 643–645

Hirata K, Craik ADD (2003) Nonlinear oscillations in three-armed tubes. Eur J Mech B–Fluids 22:3–26

Lamb H (1923) Dynamics, 2nd edn. Cambridge University Press, Cambridge, p 149 Chap.VII, Example XII.5

Love EH (1921) Theoretical Mechanics, 3rd edn. Cambridge University Press, Cambridge, p 261

McMillen T, Goriely A (2003) Whip waves. Physica D 184:192–225

Newton I (1687) Philosophiae Naturalis Principia Mathematica. London

Paterson AR (1983) A First course in fluid dynamics. Cambridge University Press, Cambridge, MA

Pedro MR, Jung S (2010) How cats lap: water uptake by felis catus. Science 30:1231

Penney WG, Price AT (1952) Finite periodic stationary gravity waves in a perfect liquid. Phil Trans R Soc A 244:254–284

Rosenberg M (1977) Analytic dynamics of discrete systems. Plenum, NewYork, pp 332–334

Schagerl M, Steindl A, Steiner W, Troger H (1997) On the paradox of the free falling folded chain. Acta Mech 125:155–168

Sommerfeld A (1952) Lectures on theoretical physics academic, vol I. New York, Mechanics, pp 28–29, and Problem I.7 on p 241, and solution on p 257

Taylor GI (1953) An experimental study of standing waves. Proc R Soc A CCXVIII:44–59

Tomaszewski W, Pieranski P, Geminard J-C (2006) Am J Phys 74:776

Villermaux E, Pomeau Y (2010) Super free fall. J Fluid Mech 642:147

Wong CW Yasui K Falling chains arXiv.org:physics 0508005

Super Accelerated Flow in Diverging Conical Pipes

A. Torres, F. J. Higuera and A. Medina

Abstract The motion of the upper free surface of a liquid column released from the rest in a vertical conical container is analyzed theoretically and experimentally. An inviscid one dimensional model for a weakly increasing cross section in vertical tubes describes how the recently reported super free fall of liquids occurs in liquids of very low viscosity. Experiments show that the inviscid one dimensional model developed here agrees with theoretical results.

1 Introduction

Consider a chain of length L initially attached at both ends to an horizontal support. As one end is suddenly released, the chain begins to fall in the gravitational field. Moreover, if the initial distance between the ends of the chain is very closed to L, that is, when the chain is initially stretched to its maximum length, the vertical motion of the chain tip becomes identical to the motion of a freely falling body (Tomaszewski et al. 2006). However, when the horizontal separation Δl between the ends of the chain is shorter than L, that is, the chain is tightly folded, the falling chain tip will

A. Torres (✉) · F. J. Higuera · A. Medina
Instituto Politécnico Nacional SEPI ESIME Azcapotzalco,
Av. de las Granjas 682, Colonias Santa Catarina 02250,
Azcapotzalco D.F., Mexico
e-mail: higherintellect@hotmail.com

F. J. Higuera
e-mail: fhiguera@aero.upm.es

A. Medina
e-mail: amedinao@ipn.mx

J. Klapp et al. (eds.), *Fluid Dynamics in Physics, Engineering and Environmental Applications*, Environmental Science and Engineering,
DOI: 10.1007/978-3-642-27723-8_43, © Springer-Verlag Berlin Heidelberg 2013

attains an acceleration that is larger than the gravity acceleration, g^*, i.e., there is a super free fall. All of these results have been confirmed both in experiments and numerical simulations (Tomaszewski et al. 2006; Calkin et al. 1989).

For this work the problem of the free fall of a mass of low viscosity liquid in a vertical, weakly expanding (conical) pipe (Villaermaux et al. 2010) is revisited. In this case, it has recently been found that the motion of the liquid free surface has an effective acceleration that overcomes the acceleration due to the gravity (Villaermaux et al. 2010). They have argued that this motion in a conical pipe super accelerates downward due to a force originated by a positive pressure gradient at the upper interface. Thus, the pressure force added to the pure gravitational body force induces this type of motion.

The theoretical model here developed, also based on the slender slope approximation, allows showing that this condition is unnecessary. Instead, it appears that the relative levels of filling are crucial to get different values of supper free fall in this geometry.

The division of this work is as follows, in the next section the formulation of the governing equations for the one dimensional motion of an inviscid liquid in a vertical cone is given. After that, in Sect. 3 the numerical results concerning to the supper accelerated motion of the free surface for several initial levels of fillings are analyzed. Section 4 is dedicated to the experimental analysis. And finally, in Sect. 5 the conclusions and perspectives of this work are presented; it cannot be considered as homogeneous fluidization.

2 Theory

When an ideal liquid is in a vertical, cylindrical pipe and suddenly the lower part of the cylinder is opened, the gravity will accelerate the liquid. The main flow is one dimensional and the conservation equations of mass and momentum are, respectively:

$$\frac{\partial u}{\partial z} = 0, \tag{1}$$

$$\rho \frac{\partial u}{\partial t} = -\frac{\partial p}{\partial z} + \rho g. \tag{2}$$

In the previous equations u is the downward velocity, z is the vertical velocity pointing downward, p is the pressure, ρ is the liquid density and t is the time. The solution of Eq. (1) allows finding that

$$u = u(t).$$

Now consider the same liquid in a vertical, conical pipe where the initial level of the free surface measured from its apex is $H_2(0)$ and the position of the bottom exit is H_1 (Fig. 1). When the bottom exit is suddenly opened the liquid falls due the gravity action. The continuity and momentum equations, in the slender slope

Fig. 1 Three dimensional
projection of a vertical
conical pipe filled with an
inviscid liquid. In the figure
the coordinate system and the
positions of the *upper free
surface* ($z = H_2 0$) and the
lower free surface ($z = H_1$)
which appears when the pipe
is suddenly opened are shown

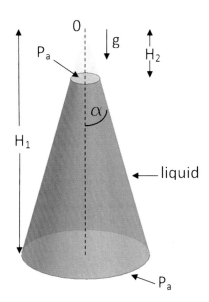

approximation, i.e., when the angle of aperture α is small, which is valid for a
smoothly expanding tube, are now:

$$\frac{1}{z^2}\frac{\partial}{\partial z}(z^2 u) = 0, \tag{3}$$

$$\rho\left(\frac{\partial u}{\partial t} + u\frac{\partial u}{\partial z}\right) = -\frac{\partial p}{\partial z} + \rho g. \tag{4}$$

In this case the velocity can be obtained from Eq. (3), in the form $z^2 u = A(t)$
and therefore

$$u = \frac{A(t)}{z^2} \tag{5}$$

Using this last result in Eq. (4), it is obtained that

$$\rho\left(\frac{\partial A}{\partial t} + 2\frac{A^2}{z^3}\right) = -\frac{\partial p}{\partial z} + \rho g. \tag{6}$$

If Eq. (6) is integrated from H_2 to H_1 the resulting height-averaged momentum
equation is:

$$\frac{dA}{dt}\left(\frac{H_1 - H_2}{H_1 H_2}\right) - \frac{A^2}{2}\left(\frac{H_1^4 - H_2^4}{H_1^4 H_2^4}\right) = g(H_1 - H_2). \tag{7}$$

By the way, the liquid velocity is given also by

$$u = \frac{dH_2}{dt},$$

And using the later equation in Eq. (5) it is easy to find that

$$\frac{dH_2}{dt} = \frac{A(t)}{H_2^2}, \tag{8}$$

or

$$\frac{dH_2^3}{dt} = 3A(t). \tag{9}$$

A second derivative of the previous equation gives

$$\frac{d^2 H_2^3}{dt^2} = 3\frac{A(t)}{dt}. \tag{10}$$

The substitution of Eq. (10) into Eq. (7) yields

$$\frac{1}{3}\left(\frac{d^2 H_2^3}{dt^2}\right)\left(\frac{H_1 - H_2}{H_1 H_2}\right) - \frac{1}{18}\left(\frac{dH_2^3}{dt}\right)^2\left(\frac{H_1^4 - H_2^4}{H_1^4 H_2^4}\right) = g(H_1 - H_2). \tag{11}$$

Rearranging terms and using the identity

$$\frac{d^2 H_2^3}{dt^2} = 3H_2^2\frac{d^2 H_2}{dt^2} + 6H_2\left(\frac{dH_2}{dt}\right)^2,$$

In Eq. (11) it is finally found that

$$\frac{d^2 H_2}{dt^2} = \frac{H_1}{H_2}g + \frac{1}{2}\left(\frac{dH_2}{dt}\right)^2\left\{\frac{1}{H_1} + \frac{H_2}{H_1^2} + \frac{H_2^2}{H_1^3} - \frac{3}{H_2}\right\}. \tag{12}$$

In this later equation it should be noted that the term on the left side is the acceleration of the upper free surface of the liquid, H_2. On the right side the term $(H_1/H_2)g$ appears which can be much larger than g if $H_2 \ll H_1$. The second term has acceleration units but involves the square of velocity of the free surface. At this point it is convenient to take this equation to a non-dimensional form in order to generalize the problem to any length L of a conical tube.

Considering these non-dimensional variables and substituting in the last equation

$$\xi = \frac{H_2}{H_1} \quad y \quad \tau = \frac{t}{\sqrt[2]{\frac{H_1}{g}}}$$

A new expression in terms of ξ and τ is obtained.

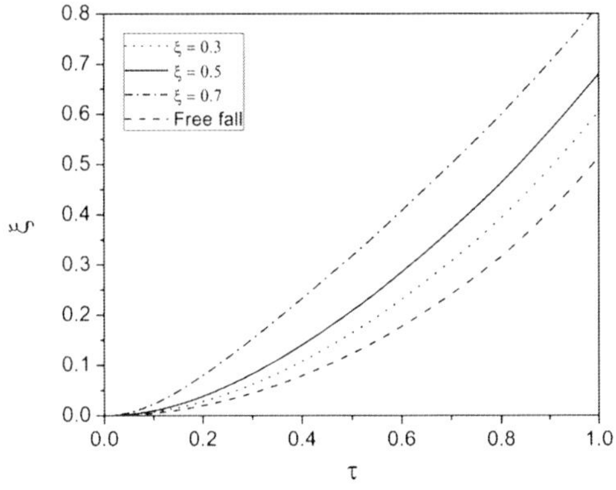

Fig. 2 Plots showing the
evolution on the position of
the *upper free surface*, $\xi -$
$\xi(0)$ as function of time τ. It
is assumed that initially, the
upper free surface was at
position $\xi(0)$. Three different
initial positions and the free
fall case are assumed, as it
can be seen in the box in the
plot

Fig. 3 Instantaneous
velocities of the upper free
surfaces, $\xi' = d\xi/d\tau$, as
function of time. The
different curves correspond to
initial positions, $\xi(0)$ of
Fig. 1

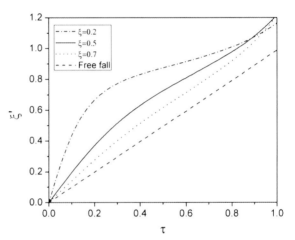

$$\frac{d^2\xi}{dT^2} = \frac{1}{\xi} + \frac{1}{2}\left[\frac{d\xi}{dT}\right]^2\left[1 + \xi + \xi^2 - \frac{3}{\xi}\right].\tag{13}$$

3 Numerical Solutions

For the numerical solution it was assumed that ξ goes from zero to a maximum
value of 1. The solution of Eq. (13) obeys an initial value problem and its solution
was obtained by using a Runge–Kutta method of fourth-order. In order to study the
effect of the level of filling ξ, for overall motion of the liquid, three different values
of this quantity were chosen: $\xi(0) = 0.2$, $\xi(0) = 0.5$, $\xi(0) = 0.7$. The first value
corresponds to a high level of filling. Whereas $\xi = 0.7$ indicates that the level of
filling is close to the bottom exit; note that the same intervals of time were used,
this criteria helps to observe the behaviour of the free surface at the large time of

Fig. 4 Plots of the non-
dimensional instantaneous
accelerations of the upper
free surfaces including the
free fall case, as functions of
time. Plots correspond to
cases given in Fig. 2

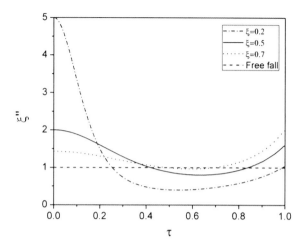

empty. As a result, in Fig. 2 is plotted the spatial evolution of the free surfaces, $\xi - \xi(0)$, as a function of time for three different values of $\xi(0)$. There, the free fall case is also plotted. It has been noted three cases the rate of change of the position of the free surface as a time function is faster than that corresponding to the free fall. It is confirmed through the Figs. 3 and 4 where the corresponding instantaneous velocities and the accelerations of the free surfaces are plotted.

From Figs. 3 and 5 it is possible to conclude that, for $\xi(0) = 0.2$, $\xi(0) = 0.50$ and $\xi(0) = 0.70$ and any value of ξ that the analytical behavior plotted here does not correspond to an Uniformly Accelerated Motion (U.A.M.), in Fig. 4 are given the plots of the non dimensional accelerations, and it shows that the accelerations for the three level of filling begin the super accelerated movement and this acceleration decreases under the gravity acceleration is constant (Free fall case) and tends to the value of gravity (1 for the non dimensional case) just at the end of the cone.

Note that in Fig. 4 when the lines of the different accelerations intersect for a second time the line of the Free fall case, the free surface ξ is just at the bottom exit of the cone. In Fig. 6 the values of the different accelerations (equally spaced in length) have been plotted on its corresponding initial values of ξ. Note as it is predicted in Fig. 4 when the different lines of acceleration achieve value of 1 for a second time the, the free surface of each value is located exactly at the bottom end of the conical tube falling to the same value of g.

In other words in a first stage the motion begins fast, it has a high acceleration, after that, the acceleration of this one decreases to a value located under the gravity and finally the acceleration rises again achieving the value of g at exactly at the bottom end of the conical pipe.

Fig. 5 Velocities ξ' as a function of position ξ. Note that the initial position ξ was changed to a new friendly value $1 - \xi$ in order to understand the behavior of the free surface from this interesting perspective in a better way

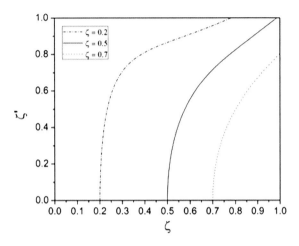

Fig. 6 Plots of the non dimensional accelerations ξ'' as a function of position ξ. As it has been made for Fig. 3 the value of ξ were changed for a new one $1 - \xi$

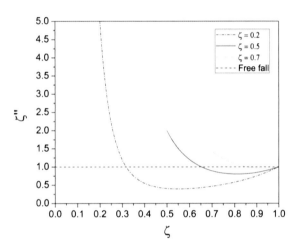

4 Experiments

In this part, an experiment of the super accelerated system it is discussed. As it has been said the main idea of the experiment is easy, first an inviscid liquid was confined into a tube with a divergent geometry, and then the bottom exit of it one is opened abruptly, finally we proceeded to observe the behavior of the free surface falling due to gravity force only. For this experiment two conical tubes with specific geometries consigned in Table 1, have been used both tubes have been chosen with a smoothly expanding cross section (slender slope theory must be completed), moreover we used two different liquids with very low viscosities, water (mass density $\rho = 10^3$ kg/m^3, surface tension $\sigma = 70 \times 10^{-3}$ N/m,

Table 1 Geometry of the glass conical tubes used in the experiments

	Length (cm)	Upper diameter (cm)	Bottom diameter (cm)
Tube 1	23	2.46	4.81
Tube 2	36	2.85	5.77

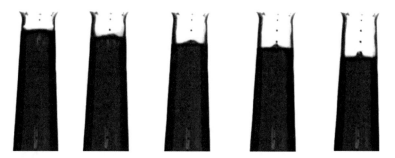

Fig. 7 Sequence of pictures showing an early time of the experiment. A nipple at the middle part of the free surface was observed

Fig. 8 Plots showing the trajectory of the free surface compared with the numerical solution scaled for the same values

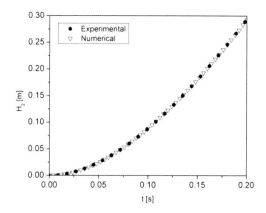

viscosity $v = 10^{-6}$ m^2/s) and ethanol (mass density $\rho = 810$ kg/m^3, surface tension $\sigma = 22 \times 10^{-3}$ N/m, viscosity $v = 1.52 \times 10^{-6}$ m^2/s).

For the plots in Fig. 8 the tube 2 (Fig. 7) was filled with water to an initial position of $H_2 = 0.16$ m; this experiment was compared with the numerical calculations at the same H_2.

5 Conclusions

It is apparent that the theoretical model here developed predicts the super free fall in the motion of the upper interface of a liquid in a conical container when it falls only during the action of the gravity field. Moreover, it has been shown that the effective acceleration depends strongly on the level of filling, $H_2(0)$ and on the length of the container, H_1. Experiments gave here show that the theoretical predictions agree with the experimental ones. Experiments by other authors (Villaermaux et al. 2010) have shown that during the free fall the formation of nipples at the middle part of the free surface was observed. The simple model developed here does not allow quantifying the properties of such structure, but their increase appear when there is super accelerated flow at the beginning of the flow.

References

Calkin MG, March RH (1989) The dynamics of a free falling chain: I. Am J Phys 57(2):154

Tomaszewski W, Pieranski P, Géminard J-C (2006) The motion of a freely falling chain tip. Am J Phys 74:776

Villaermaux E, Pomeau Y (2010) Super free fall. J Fluid Mech 642:147

Analysis of the Blasius' Formula and the Navier–Stokes Fractional Equation

J. R. Mercado, E. P. Guido, A. J. Sánchez-Sesma, M. Íñiguez and A. González

Abstract The objective of this paper is to find the relationship between the Blasius formula for friction factor and the Navier–Stokes Fractional equation. The renormalization that produces changes of scale of the boundary layer equations contains the essential hypothesis of the thinness of said layer, a characteristic which appears in all important outcomes, such as friction force and drag coefficient and gives rise to a multi-fractal description. A generalization of experimental results for the Blasius friction factor that is interpreted and generalized as a multi-fractal is obtained. Applying Hadamard functionals the friction factor is described as a fractional derivative whose order depends on the *spatial occupancy index*. Results are applied to the interactions between currents and boundaries (in rivers, deserts and hurricanes).

1 Introduction

In Mercado et al. (2010) and Mercado et al. (2009) the Navier–Stokes Fractional equations (NSFE) are presented. The fundamental approach is based on the claim that the viscous stresses produce a momentum flow that is described through a fractional approach of Darcy's law and that the divergence of the flow coincides with the temporary change of momentum, according to Newton's law. The NSFE is simplified through its application to boundary layer equations, as well as its fractional character, which is achieved considering its relatively thin thickness,

J. R. Mercado (✉) · E. P. Guido · A. J. Sánchez-Sesma · M. Íñiguez · A. González
Instituto Mexicano de Tecnología del Agua, IMTA, Paseo Cuauhnáhuac 8532, Jiutepec,
Mor 62550, Colonia Progreso CP, Mexico
e-mail: rmercado@tlaloc.imta.mx

J. Klapp et al. (eds.), *Fluid Dynamics in Physics, Engineering and Environmental Applications*, Environmental Science and Engineering,
DOI: 10.1007/978-3-642-27723-8_44, © Springer-Verlag Berlin Heidelberg 2013

implies that the main speed is set in the downstream direction, with a relatively greater vertical speed gradient compared to the longitudinal one, which leads to the speed to uphold the no-slip condition at the boundary of the channel and conversely, with mild pressure gradients in the vertical direction compared to the strong pressure gradient in the longitudinal (Landau and Lifshitz 1987). This renormalization and the application of the chain rule for fractional derivative produce an indexed Reynolds number (IR). If the downstream velocity potential and the potential for vertical transverse velocity are introduced, the momentum equation is expressed in terms of speed potential. It is obtained that stress show a power law of the IR.

The friction force that is obtained by the integrating of friction stresses is expressed in a dimensionless form by introducing the friction factor. The expression for the Darcy-Weisbach friction factor has the form of a power law of the IR whose power value is called the Blasius exponent which depends on the *spatial occupancy index* (SOI). When stress is reintroduced in calculating the friction force it is observed that it can be described as a fractional derivative in terms of a Hadamard functional whose order also depends on the SOI. We obtain a relationship between the order of the derivative and the Blasius exponent which is described by an increasing function, so that the higher the SOI greater the exponent of the indexed Reynolds number will be, which is achieved with laminar motion; on the contrary, the greater the turbulence the lower SOI and the lower the Blasius exponent.

2 Navier–Stokes Fractional Equation

The fluid motion is described from the Eulerian viewpoint considering a volume of fluid limited by a boundary surface, with its momentum per unit volume given by $\rho\mathbf{v}$. Because of its importance the internal friction interaction is considered first. The fractional gradient is expressed by $\nabla_M^\beta \rho\mathbf{v}$, where ρ is the mass density, $\mathbf{v}$ velocity, β is the SOI, and M the mixing scale for the different spatial directions. The diffusivity of momentum is the $\alpha-$ kinematic viscosity ν_α, so Darcy's flow momentum is $\mathbf{q}_D$. The rate of change of momentum per unit of time is the negative divergence or convergence, of Darcy's flow, and it is achieved choosing M such that the flow is proportional to the negative fractional Laplacian, (Zhang 2011), as shown in (1), with $\alpha = 1 + \beta$,

$$\mathbf{q}_D = -\nu_\alpha \nabla_M^\beta \rho\mathbf{v}, \quad \frac{d}{dt}\rho\mathbf{v} = -\nabla \cdot \left(-\nu_\alpha \nabla_M^\beta \rho\mathbf{v}\right) = -\nu_\alpha(-\Delta)^{\alpha/2}\rho\mathbf{v} \quad (1)$$

After that, taking into consideration the contribution of pressure variations to fluid momentum variations through the pressure gradient forces, in such a way that the viscous stress friction and the hydrostatic pressure constitute the tensor $\mathbf{T} = \nu_\alpha \nabla_M^\beta \rho\mathbf{v} - p\mathbf{I}$, which gives way to the law of deformation. Then an external potential force by unit volume is introduced, of the type $-\nabla \rho \phi$ Afterwards, incompressibility is taken into account. The derivative material that makes up the local and advective variations is made explicit. However, the objectivity requirement demands the invariance under coordinate changes, so the advective contribution must be modified, and the vorticity term arises. Finally, the contribution to the inertial force of the vorticity term is written on the right side of the equation as $\mathbf{v} \times rot\mathbf{v}$, and may be thought of as originating from an external force that energizes the evolution of the velocity field through its vorticity, entering in contradiction to the viscous force, while the other term is interpreted as a restriction along the current lines containing the Bernoulli equation. The coefficient v_α can be compared to Boussinesq turbulent viscosity,

$$\frac{\partial}{\partial t}\mathbf{v} = -v_\alpha(-\Delta)^{\alpha/2}\mathbf{v} + \mathbf{v} \times rot\mathbf{v} - \nabla\left(\frac{1}{2}(\mathbf{v} \cdot \mathbf{v}) + \frac{p}{\rho} + \phi\right) \tag{2}$$

As previously discussed, the boundary layer equations are obtained from the NSFE by simplifications that are induced from the premise of a relatively thin thickness. Now, the bi-dimensional boundary layer equations are considered in the stationary or permanent version, along with the conservation of mass, as a null divergence, as shown in Eq. (3).

$$u\partial_x u + v\partial_y u = v_\alpha \partial_y^\alpha u, \quad \partial_x u + \partial_y v = 0 \tag{3}$$

Given a potential form for main downstream speeds, represented through the potential of the velocity $\psi(u,v)$, $u = \partial_y\psi$, $v = -\partial_x\psi$ the under-potential function $g(\xi)$ appears as a solution for the differential equation of fractional Blasius equation (Mercado 2010). Stress is calculated through $\tau_{xy} = \mu_\alpha \partial_y^\beta \partial_y \psi(u,v)$, being $\mu_\alpha = \rho v_\alpha$. With the fractional Blasius coefficient $B_\beta = (\alpha/\beta)(v_a/v_2)^{\beta-1}\partial_y^{\beta-1}(g''(\xi))\big|_{\xi=0}$ and the IR $R_{x\beta} = ux^\beta/v_\alpha$, the stress and the friction factor can be written as in (4):

$$\tau_{xy} = (\beta/\alpha)(v_2/v_\alpha)B_\beta(\rho U^2)(1/R_{x\beta})^{\frac{\beta}{1+\beta}}(x^{-(1-\beta)}), \quad f_\beta \approx 8B_\beta(1/R_{l\beta})^\theta \tag{4}$$

The Hadamard functionals are expressions defined as $H_b(x) = x^{b-1}/\Gamma(b)$, $b > 0$ function, where Γ is the gamma function, which acts on locally integral functions and have the semi group property allowing them to maintain their shapes under the convolution operation while orders are added. The friction force per bi-normal, or cross-horizontal, unit length is calculated as $F_f = 2\int_0^l \tau_{xy}dx$. In the friction force the stress expression is reintroduced as in Eq. (4), where the Hadamard's functionals are recognized, which allows their representation as fractional derivatives. Being $\gamma = \beta^2/(1+\beta)$, and $C(\beta) = \gamma B_\beta \Gamma(-\gamma)C_1(\beta,\sigma)$, and denoting by $*_l$ the convolution product, being l the limit integration, the friction force per unit bi-normal width, is a fractional derivative as shows in (5):

$$F_f = 2\left(\rho U^2\right)C(\beta)\frac{x^{-\gamma-1}}{\Gamma(-\gamma)} *_l \left(\frac{1}{R_{x\beta}}\right)^{\sigma} = 2\left(\rho U^2\right)C(\beta)D_x^{\gamma}\left(\frac{1}{R_{x\beta}}\right)^{\sigma} \qquad (5)$$

Where, after derivations, the result should be evaluated at $x = l$. Then, with Eq. (4), is obtained so that σ and the friction factor satisfy the relationship in (6). While (5) is described as a transformation through fractional derivation of a multi-fractal of dimension $\beta\sigma$ to another analogue of dimensions θ. It is possible to show that the exponent $\beta\sigma$ depends on longitudinal pressure gradient.

$$f_\beta = 8C(\beta)D_x^{\gamma}\left(1/R_{x\beta}\right)^{\sigma}, \quad \theta = \beta(\sigma + 1/(1+\beta)) \qquad (6)$$

3 Applications

Our approach can be applied in the structural analysis of the turbulent boundary layer (TBL). For smooth or regular surface boundaries, the wall zone contains the viscous sub layer that extends to about 5 units or characteristic lengths from the wall; and the mix sub layer to about 30 characteristic lengths. Then the inertial layer, with a sub layer of the fully developed turbulence, from about 30 up to 400 units from the wall. And, the rest corresponds to the wake. Meanwhile, for rough surfaces the characteristic length decreases, because there is a downward displacement due to increased losses of momentum produced by the roughness (Clifford et al. 1993; Levi 1989).

For low and intermediate Reynolds numbers, the multifractal is described by $8B_\beta/f_\beta = (R_e)^{\theta}$, with resolution $1/R_e$, dimension θ, and local singularity β, (Mercado et al. 2009; Landau and Lifshitz 1987; Riedi and Scheuring 1997). For the dimensionless velocities the following representation is proposed $U/(U_*/\sqrt{B_\beta}) = (R_e)^{\theta/2}$. The height y is dimensionless expressed with the characteristic length v_2/U_*, obtaining $(U_*y)/v_2$, as the Reynolds number associated with the shear velocity, which may be taken as the resolution of the multifractal dimension $\theta/2$, which is written,

$$U/U_* = U_+ = \left(1/\sqrt{B_\beta}\right)(y_+)^{\theta/2} = \left(1/\sqrt{B_\beta}\right)((U_*y)/v_2)^{\theta/2} \qquad (7)$$

This type of profile is shown in Barenblatt et al. (1997). It is observed that if $\theta \to 2$ it recovers the approximation of the wall law $U_+ \approx y_+$. Later the slope is obtained observing that the profile is increasing since $\theta \geq 0$. The curvature shows a primary convex sub-phase if $\theta \geq 2$ even for $\theta = 4$, where a parabolic behavior could be achieved; and later a second concave sub-phase for $\theta \leq 2$; in this way a Blasius type law $U = y^{1/7}$ is included. Then it is possible to describe the mix layer from the inflection point up to the fully turbulent layer, and later the intermittent wake trail. For the totally turbulent sub-layer, the Nikuradse formula must be employed, so that $U/U_* = \sqrt{8/f_\beta} = \sqrt{2/C_f} = \sqrt{2}A \ln R_e$, and the logarithmic profile results $U_+ = \sqrt{2}A\ln(y_+)$, which in addition to being linear with the logarithm, is increasing and concave (Rouse 1946).

The roughness height is matched with the boundary condition of the logarithmic profile of the inertial layer. According to the Nikuradse's results, for regular or smooth surfaces, the thickness of the viscous sub-layers is around 100 times the roughness height; meanwhile for rough surfaces the thickness of the sub-layer is a bit higher (Rouse 1946; Levi 1989).

To parameterize the roughness height, the property of invariance in the form of the hydraulic gradient is used J. The same gradient is made dimensionless by means of the hydraulic radius R_h, and of the shear velocity; and the length U_*^2/g is obtained with g as the gravity. The condition of form invariance is guaranteed if the roughness is $y_r = CU_*^2/g$, which corresponds with the structure of the Charnock roughness (1995) for the ocean surface. Moreover, for smooth surfaces the constant C depends on the Kármán and Nikuradse constants. In the sand transport caused by turbulent winds the boundary roughness is described as in the above paragraph but the parameter C has a range of values in the interval $[0.02, 0.05]$, for the flow of sand in wind tunnels and up to 0.18, for field experiments (Clifford et al. 1993). In turn, the ripples of the sea surface caused by wind, generates a rough surface, allowing to approximate the friction factor by the formula of Nikuradse, with $U_+ = \left(1/\sqrt{B_\beta}\right)((U_*y)/v_2)^{\theta/2}$, as a profile, while the roughness height controls vertical variation of wind speed (Powell et al. 2003).

The interaction between the flow and the river bed gives place to coherent structures in the wall which present quantitative characteristics as the roughness height. Also the same profile is applied to the previous, so that if $\theta \rightarrow 2$, then $U_+ \approx (U_*y)/v_2 = y_+$. As characteristic length is taken $v_2/U_* = v_2\sqrt{\rho/\tau}$. With the data cited in (Clifford et al. 1993), according to A. Kirkbride, Chap. 7, its numerical value is $1.31 \times 10^{-6}\sqrt{(1000)/5} = 1.8526 \times 10^{-6}$. So the roughness length is of the Charnock type, with constant values in the interval $[0.02, 0.05]$ and even with 0.18 as another value. So that, the result of the viscous layer for 0.05 is of the order of $10^7 \times (1/9.8)\left(\sqrt{5/1000}\right)^2 \times (0.05) = 2.7296 \times 10^{-3}$. However, the reported value of the thickness of the viscous sub-layer is 2.1×10^{-7}m. Nevertheless, the known inferior limiting value for gravel is 2.0×10^{-3}m which is comparable to the coefficient in $C = 0.05$. In turn, the comparison between the roughness length, and the thickness of the viscous layer, for $C = 0.02$, is $y_r/\delta' = 1.024 X 10^{-5}/10^{-7} = 102.4$ and falls into the expected order. While for $C = 0.18$, is $y_r/\delta' \approx 9.1836 \times 10^{-5}/10^{-3} \approx 1/100$; and in this case, it is concluded that the viscous sub-layer cannot develop owing to the fact that roughness greatly exceeds the length of the thickness of this sub-layer.

4 Conclusions

- The interaction of the flow with a flat surface shows that the Blasius friction factor is characterized by an inverse power law of the indexed Reynolds number, and it is represented as a fractional derivative. It is possible to do a similar analysis for other friction factor formulas, such as those of Nikuradse, and Prandtl-Kármán.
- An estimate for the Blasius exponent which presents a multifractal structure is established, and it is reduced to the well-known value of ½ for the viscous sublayer when the spatial occupancy index (SOI) comes close to the unit.
- Combining the multi-fractal expression with the non-dimensional velocity, and introducing the dimensionless roughness height, the four sub-layers of the turbulent boundary layer can be described.

References

Barenblatt GI, Chorin AJ, Prostokishin VM (1997) Scaling laws for fully developed turbulent flow in pipes: discussion of experimental data. Proc Natl Acad Sci USA Appl Math 94:773–776

Clifford NJ, French JR, Hardisty J (1993) Turbulence, perspectives on flow and sediment transport. Sand transport response to fluctuating wind velocity, pp 304–334

Landau LD, Lifshitz EM (1987) Fluid mechanics. Pergamon Press, Oxford, p 539

Levi E (1989) El Agua según la Ciencia, Conacyt Ed. Castell Mexicana, México

Mercado JR, (2010) Ecuación Blasius fraccional. Proceedings: Congreso Latinoamericano de Hidráulica, Punta del Este, Uruguay. Ref. Number 237

Mercado JR, Ramírez J, Perea H, Íñiguez M (2009) La ecuación Navier-Stokes fraccional en canales de riego. Revista de Matemática: Teoría y Aplicaciones, (International Journal on Mathematics: Theory and Applications); RMTA-082-2009. (Submitted)

Mercado JR, Olvera E, Perea H, Íñiguez M (2010) La ecuación Saint-Venant fraccional. Revista de Matemática: Teoría y Aplicaciones, (International Journal on Mathematics: Theory and Applications); RMTA-030-2010. (Submitted)

Powell MD, Vickery PJ, Reinhold TA (2003) Reduced drag coefficient for high wind speeds in tropical cyclones. Nature 422:279–280

Riedi RH, Scheuring I (1997) Conditional and relative multifractal spectra. Fractals 5(1):153–168

Rouse H (1946) Elementary mechanics of fluids. Dover Publications, New York, p 376

Zhang X (2011) Stochastic lagrangian particle approach to fractal Navier-Stokes equations, ArXiv: 1103.0131v1 [Math.PR]

Relativistic Hydrodynamics and Dynamics of Accretion Disks Around Black Holes

Juan Carlos Degollado and Claudia Moreno

Abstract We give a brief overview of a formulation of the equations of general relativistic hydrodynamics, and one method for their numerical solution. The system of equations can be cast as first-order, hyperbolic system of conservation laws, following a explicit choice of an Eulerian observer and suitable vector of variables. We also present a brief overview of the numerical techniques used to solve this equation, providing an example of their applicability in one scenario of relativistic astrophysics namely, the quasi periodic oscillations of a thick accretion disk.

1 Introduction

Relativistic hydrodynamics is a branch of physics related to the study of both: flows in which the velocities attained by the fluid as a whole approaching the speed of light, or those for which the strength of the gravitational is very strong and it is mandatory a description in terms of Einsteins theory of gravity. Scenarios

J. C. Degollado (✉) · C. Moreno
Departamento de Astrofísica Teórica, Instituto de Astronomía,
Universidad Nacional Autónoma de México, Circuito Exterior Ciudad Universitaria,
Apdo 70-264, CU, 04510 Mexico, D.F., Mexico
e-mail: jcdegollado@ciencias.unam.mx

C. Moreno
e-mail: claudia.moreno@cucei.udg.mx

J. C. Degollado · C. Moreno
Departamento de Matemáticas y Departamento de Física,
Centro Universitario de Ciencias Exactas e Ingeniería, Universidad de Guadalajara,
Revolución 1500 Colonia Olímpica, 44430 Guadalajara, Jalisco, Mexico

J. Klapp et al. (eds.), *Fluid Dynamics in Physics, Engineering and Environmental Applications*, Environmental Science and Engineering,
DOI: 10.1007/978-3-642-27723-8_45, © Springer-Verlag Berlin Heidelberg 2013

involving compact objects such as supernova explosions leading to neutron stars, micro quasars, active galactic nuclei, and coalescing neutron stars, all contain flows at relativistic speeds and, in many cases, strong shock waves. The correct description of the dynamics and evolution of such astrophysical systems strongly relies on the use of accurate large-scale numerical simulations. Our claim is to present a brief overview of a 3 + 1 formulation of general relativistic hydrodynamics, applied to a simple example, the dynamics of an accretion disk around a stationary black hole.

The simplest model of accretion disks in General Relativity is the one proposed by Abramowicz et al. (1978). This model is based on the assumption that the disk is formed by a barotropic fluid that has reached a stationary configuration in the background of a black hole space time. We chose this configuration as initial data and after adding a radial perturbation, we study the response of the disk that is reflected as a quasi periodic oscillation that resembles the epicyclic frequencies of point particles orbiting a black hole.

2 General Relativistic Hydrodynamics

The General Relativistic Hydrodynamics (GRHD) equations are crucial in situations involving strong gravitational fields or flows approaching the speed of light. Generally these set of equations are described by local conservation laws as in the Newtonian description case but are now encoded in the stress-energy tensor conservation law $\nabla_\mu T^{\mu\nu} = 0$, and in the matter current density conservation law $\nabla_\mu J^\mu = 0$, where ∇ denotes the covariant derivative compatible with the metric of the space-time.

The previous system of equations is closed once an equation of state (EOS) is chosen. Nowadays the EOS has become very sophisticated in order to take into account processes such as molecular interactions, relativistic effects or nuclear physics, see for example Font (2007) for a review. However the most widely employed EOS in numerical simulations in astrophysics are the ideal fluid EOS, $p = (\Gamma - 1)\rho\varepsilon$ where Γ is the adiabatic index, and the polytropic EOS given by $p = \kappa\rho^\Gamma$. These EOS are very useful because their simplicity and because they capture enough information of the system to give us its thermal properties.

3 HRSC-Methods

The main theoretical aspects to construct a High Resolution Shock Capturing Method (HRSC) in 3 + 1 general relativity can be found in Banyuls et al. (1997), Font (2007). In these references the GRHD equations were written as a first-order, flux-conservative, hyperbolic system which is very convenient to work numerically with. Among the most popular methods to solve the GRHD equations are the approximate Riemann solvers. These, being computationally much cheaper than the exact solvers yield equally accurate results. These methods are based in the solution of Riemann problems corresponding to a new system of equations obtained by a suitable linearization of the original one. The spectral decomposition of the Jacobian matrices is the basis of such solvers and these methods extend the ideas used for linear hyperbolic systems. This is the approach followed by an important subset of shock-capturing schemes, the so-called Godunov-type methods.

The Godunov's theorem, Leveque (1992), Leveque (1998) states that monotone linear systems for solving partial differential equations are at most first order accurate. This is an important caveat when one is trying to describe a flux with high precision. HRSC are used to avoid spurious oscillations that appear when a simple spatial discretization is used, due to the discontinuities of the solution or due to very high gradients. In order to avoid spurious oscillations, schemes with a total variation diminishing property were introduced. However even when one use a high order scheme to solve the equations in the whole numeric domain the method will decrease its order near the discontinuities Font (2007).

In order to get a higher degree of accuracy numerically and avoid spurious oscillations, one could use functions of average states in the cells. For each cell, the right and left states can be obtained by a slope limiter, these values are used to solve the Riemann problem in the boundaries of each cell and the degree of accuracy of the method relies in the order of the approximation used to reconstruct the fluxes in each boundary cell.

4 Stationary Accretion Disks

A stationary disk can be modeled as a perfect fluid which follows circular orbits around the black hole. Keeping the space time as fixed, one only has to worry about the dynamics of the fluid in a given metric. This is known as a test fluid aproximation. The stress energy tensor for a perfect fluid is given by $T^{\mu\nu} = \rho h u^\mu u^\nu + p g^{\mu\nu}$, where ρ is the rest mass density, h is the specific enthalpy $h = 1 + p/\rho + \epsilon$, ϵ is the internal energy density and p is the pressure of the fluid. The four velocity of the fluid is assumed as $u^\mu = (u^t, 0, 0, u^\phi)$ such that the motion has only an angular component (but radial dependence). The angular velocity of the fluid Ω and the angular momentum per unit of mass l are defined as $\Omega = u^\phi/u^t$, $l = -u_\phi/u_t$. Which implies the following conditions:

$$(u_t)^{-2} = -\frac{g_{\phi\phi} + 2l\,g_{t\phi} + l^2 g_{tt}}{g_{t\phi} - g_{tt}g_{\phi\phi}}, \quad \Omega = -\frac{g_{tt}l + g_{t\phi}}{g_{t\phi}l + g_{\phi\phi}}. \tag{1}$$

The equation of motion for the fluid are obtained from the projection of the conservation of the stress energy tensor $h_i^\mu \nabla_\nu T^\nu{}_\mu = 0$. For the stationary case we obtain a Bernoulli's type equation

$$\frac{\nabla_i p}{\rho h} = -\nabla_i ln(u_t) + \frac{\Omega \nabla_i l}{1 - \Omega l}, \tag{2}$$

where h_i^k is the projector to the normal surfaces to u^μ. For a barotropic fluid, the surfaces of constant pressure coincide with the equipotentials $\nabla_i W = 0$ and are given by the Boyer's condition:

$$\int_0^p \frac{dp}{\rho h} = W_{in} - W = -\ln \frac{u_t}{u_{t_{in}}} + F(l), \quad F(l) = \int_{l_{in}}^l \frac{\Omega dl}{1 - \Omega l}.$$

The sub-index *in* refers to the inner edge of the disk. The simplest model to build an accretion disk around a Schwarzschild black hole is based in imposing a constant angular momentum distribution within the disk $l = l_0$ and a a polytropic EOS $p = \kappa \rho^\Gamma$. Under these suppositions u_t can be calculated in a closed form.

$$-u_t = r \sin \theta \sqrt{\frac{r - 2M}{r^3 \sin^2 \theta - (r - 2M) l_0^2}}, \tag{3}$$

where we have used the usual for the Schwarzschild metric. Since l is assumed to be constant then $F(l) = 0$ and substituting this into (3). With this, it is possible to get the potential W. The explicit form of W is not very important but what it is the fact that the equipotentials can be closed $W < 0$, or open $W > 0$ and the geometry of the equipotentials is fixed by the value of l_0. In particular the presence of the *cusp* in the equatorial plane is necessary to allow the accretion without viscous stresses. The position of the cusp (r_{cusp}) and the center of the disk (r_c) are very easy to find because in these points the pressure gradients compensate the centrifugal force $\nabla_i W = 0 = \nabla_i p$ an the particles move freely following geodesics.

5 Evolution Equations

If one defines a system of coordinates adapted to Eulerian observers resembling the $3 + 1$ decomposition of the space time, the quantities they measure are the rest mass density D, the density of momentum along the i direction S_i and the density of total energy E. It is convenient, for numerical purposes and in order to recover the Newtonian limit, to define the total energy minus the energy density $\tau = E - D$. With these set of variables, it is found that the stress energy conservation and the conservation of momentum imply a system of equations with a conservative structure

$$\frac{\partial}{\partial t}(\sqrt{\gamma}\,\mathbf{q}) + \frac{\partial}{\partial x^i}(\sqrt{-g}\,\mathbf{f}^i) = \mathbf{S}, \tag{4}$$

where the state vector and the fluxes are given by:

$$\mathbf{q} := (D, S_j, \tau),$$

$$\mathbf{f}^i := \left(D\left(v^i - \frac{\beta^i}{\alpha}\right), S_j\left(v^i - \frac{\beta^i}{\alpha}\right) + p\delta^i_j, \tau\left(v^i - \frac{\beta^i}{\alpha}\right) + pv^i\right). \tag{5}$$

Finally the sources $\mathbf{S}$ are given by:

$$\mathbf{S} = \alpha\sqrt{\gamma}\left(0, T^{\mu\nu}\left(\partial_\mu g_{\nu j} - \Gamma^\lambda_{\nu\mu}g_{\lambda j}\right), \alpha\left(T^{\mu 0}(\partial_\mu ln\alpha) - T^{\mu\nu}\Gamma^0_{\nu\mu}\right)\right). \tag{6}$$

Where γ_{ij} is the metric induced in the spatial surfaces of the $3+1$ decomposition, γ is its determinant, α is the lapse function, β^i is the shift vector, v^i corresponds to the 3-velocity defined as $v^i = \frac{u^i}{\alpha u^0} + \frac{\beta^i}{\alpha}$. Written in this form it is possible to use advanced numerical techniques to handle in an appropriate way the discontinuities inherent of the nonlinearity of the equations such as shocks. The integral form of the system (4) can be written as:

$$\int_\Omega \frac{\partial}{\partial t}(\sqrt{\gamma}\,\mathbf{q})d\Omega + \int_\Omega \frac{\partial}{\partial x^i}(\sqrt{-g}\,\mathbf{f}^i)d\Omega = \int_\Omega \mathbf{S}d\Omega. \tag{7}$$

The idea behind HRSC methods is to solve this integral form by defining the numerical fluxes, which are recognized as approximations to the time-averaged fluxes across the cell interfaces. These fluxes depend on the solution at those interfaces, $\mathbf{q}(x^j + x^j/2, t)$, during a timestep: $\mathbf{f}_{j+\frac{1}{2}} \approx \frac{1}{\Delta t}\int_{t^n}^{t^{n+1}} \mathbf{f}(\mathbf{q}(x^{j+1/2}, t))$.

These numerical fluxes can be obtained by solving a collection of local Riemann problems, and the evolution of the state vector can be done in a direct way by means of a method of lines.

6 An Example: Quasi Periodic Oscillations of Accretion Disks

One of the motivations to study thick accretion disks where the observational evidence of quasi periodic variations of luminosity in some X-ray sources. These systems contains a neutron star or a black hole surrounded by an accretion disk. Many of these objects are binary systems of a few solar masses that emit in the X-ray spectrum and their periods are from fractions of hour until dozen of days. These variations in luminosity are associated to instabilities or oscillations in the accretion disks around these objects. The Quasi Periodic Oscillations (QPO's) in X-ray sources could reflect some of the intrinsic properties of the central object such as its mass or angular momentum. Numerical simulations of an accretion disk perturbed in radial direction have shown that the disk has radial movements that are close to the epicyclic frequencies. As an example, in system SAX J1808.4-3658 the variations in luminosity could be produced by the forced movement of the accretion disk that induce resonances at that frequencies, see Lee (2005).

In order to model such system, one could start with the stationary solution described above and then induce a radial perturbation and study the dynamics of the disk. If the perturbation is small enough, the disk oscillates and only a small part of it falls into the hole. The oscillation of the system could be understood in terms of a non linear response of the disk when is perturbed by an external source. Previews results, Rubio-Herrera and Lee (2005), Zanotti et al. (2005) showed that the axisymmetric oscillations correspond to p-modes and consequently describe the QPO's. These results showed that the oscillations behave as sound waves globally trapped within the disk an that the proper frequencies appear in a 3:2 ratio for a distribution of constant angular momentum for the disk.

In our case the system of equations was solved using a 2d code where a high resolution shock-capturing method was implemented. We used an approximate Riemann solver that explodes the characteristic decomposition of the jacobian matrix. Second order accuracy in space and time was achieved by means of a piecewise linear cell reconstruction (minmod) and a second order total variation diminishing Runge Kutta scheme. The numerical grid in this case consists of 500×100 zones in radial and angular directions. We use a constant angular momentum distribution $l = 3.6$ that lies between the value for the marginally stable and the marginally bound orbits in order to get a bound configuration. The polytropic index was $\Gamma = 4/3$ and $\kappa = 0.9 \times 10^9 \, N \, m^2 \, (Kg)^{-4/3}$. The radial extension of the disk was of $35.27 \, Km$ while the center of the disk, in which the maximum of the density is located, was initially at $r_c = 33.10 \, Km$, the frequency was obtained by means of a Fourier transform of the maximum of density as a function of time. For this case, we found that the disk oscillates at $230 \, Hz$ while the corresponding epicyclic frequency for single particle located at the position of the center is about $283.42 \, Hz$. We also found that the besides the main frequency there are another resonance at $\sim 345 \, Hz$, however the signal is very weak in this case, this is due to the perturbation of the disk is a global one ie. we impose a radial velocity in the whole disk and the effective perturbation is no strong enough to capture the whole spectrum of resonances.

7 Conclusions

We have presented a brief overview of the equations of General Relativistic Hydrodynamics and the main ideas of a numerical method frequently used to solve them. Also we showed how to build a thick disk with a constant angular momentum distribution. We solved numerically the equations of GRHD and reproduce some results about the QPO's of thick accretion disks. These results are consistent with the previous studies both numerical an analytical in which it is showed that the accretion disk oscillates with frequencies that follow a regular pattern. Also we found that the frequencies of relativistic disks tend to be lower than the ones on Newtonian disks, this is somehow expected since the gravitational description close the horizon of the black hole needs a general relativistic treatment.

Acknowledgments This research was supported in part by DGAPA UNAM and FOMIX-CONACYT grant 149481. CM acknowledge PROSNI 2011 support.

References

Abramowicz M, Jaroszynski M, Sikora M, (1978) Relativistic, accreting disks. Astron Astrophys 63:221–224

Font J (2007) Numerical hydrodynamics and magnetohydrodynamics in general relativity. Living Rev Rel 11:7

Banyuls F, Font J, Marti J, Miralles J (1997) Numerical 3+1 general relativistic hydrodynamics: a local characteristic approach. Astrophys J 476:221–231

Leveque R J (1992) Numerical methods for conservation laws. Birkhauser, Basel.

Leveque R J (1998) Computational methods for astrophysical fluid flow. Springer, Berlin, p 27

Lee WH (2005) Forced oscillations in fluid tori and quasi-periodic oscillations. Astronomische Nachrichten 326:838–844

Rubio-Herrera E, Lee W (2005) Oscillations of thick accretion discs around black holes. Mon Not Roy Astron Soc 357:L31–L34

Zanotti O, Font J, Rezolla L, Montero P (2005) Dynamics of oscillating relativistic tori around Kerr black holes. Mon Not Roy Astron Soc 356:1371–1382

Low-Re μUAV Rotor Design

Oscar Rubio, Fidel Gutiérrez, Juan Carlos Zuñiga
and Marcelo Funes-Gallanzi

Abstract This paper presents a novel approach to the design, optimization and testing of rotors for small helicopters using Bezier curves as airfoils. Blade design with low Reynolds numbers (5,000–60,000) is seriously affected by air viscosity; therefore our approach to this scenario is to analyze the propellers through the use of 3D BEM/VPM (Blade Element Momentum/ Vortex panel method) theory, further refined by estimates of blade tip and blade root losses. In addition, the code used throughout this report creates a CAD file that allows the propeller to be visualized in 3D. This report also contains the rotor build for a micro-helicopter with its results compared to the performance of current propellers; the rotor features a 15 cm blade under a 10,000 Reynolds number at cruising speed.

Keywords Low-Re rotors · MAV · μUAV · Bezier airfoils · BEM/VPM

O. Rubio (✉)
Mechatronics Student, ITESM Campus, Guadalajara, Mexico
e-mail: O.Rubio.Rivera@gmail.com

F. Gutiérrez
AVNTK S.C, Av. Chapalita, 1143, Guadalajara, Jalisco, Mexico
e-mail: fidel@avntk.com

J. C. Zuñiga
Department of Mathematics, Universidad de Guadalajara,
Av. Revolución 150, Guadalajara, Jalisco, Mexico
e-mail: juan.zuniga@red.cucei.udg.mx

M. Funes-Gallanzi
Ardita Aeronáutica SA de CV San José, 3016 Guadalajara, Jalisco, Mexico
e-mail: mfg@ardita-aeronautica.com

J. Klapp et al. (eds.), *Fluid Dynamics in Physics, Engineering and Environmental Applications*, Environmental Science and Engineering,
DOI: 10.1007/978-3-642-27723-8_46, © Springer-Verlag Berlin Heidelberg 2013

1 Introduction

Recently, micro-air vehicles (MAVs) have gained a lot of interest because they have many promising civil and military applications including inspection of difficult-to-access structures, search and rescue, deliver of micro payloads, communications relay and remote/distributed sensing. An unmanned system represents an emerging sector of the aerospace market. Rotary-wing MAVs have had military-focused application by armed forces around the world, however, their used in the civilian sector have increased due to the advantages of these kinds of aerial machines will bring to their business.

Despite the growth in the micro-helicopter industry, one of the disadvantages in the design of MAVs is that in order to move with agility, their rotors consume vast amounts of energy and as a consequence this decreases their flight time.

In order to improve flying time and payload capacity, a set of simulations were created to research the design of low-Reynolds number rotors, by giving the user the freedom to design the airfoil, followed by its analysis with the capability of tip/root corrections to be made, and in addition, allows the optimization and 3D assembly of the propeller for manufacture. The objective of our research is challenging because high-accuracy implies a 3D viscous flow modelling using experimental data correlations to account for the fact that viscous effects cannot be scaled. The design of low-Reynolds number rotors is complex, which is why only few dare to undertake in their research and development, since viscous effects are a major influence in the flow-fields of these types of propellers. Based on a large body of previous work (Bohorquez 2003; Bohorquez 2010; Harrington 1951), we have implemented one of the most advanced algorithms that exist in this area of aeronautics.

At first stage, we created a 3D blade and sliced into a set of airfoils using Bezier curves, as this type of arcs work better on an airfoil at low-Reynolds numbers (Bohorquez 2010). Then, we performed a viscous analysis of the profile, in order to obtain its aerodynamic coefficients and this information was then filtered to pick the best airfoils. The finest profiles were evaluated with a combination of BEM/VPM, taking into account the root and tip losses. After these simulations were done, we created the coordinates of the entire blade to allow it to be exported into CAD format.

The objective is to create a complete rotor design, from the airfoil to the propeller, with all the engineering simulations needed to cover the required criteria. This rotor design would increase MAV flight time allowing them to get more work done with the same amount of energy or increasing the maximum payload allowed.

2 Blade Design Program

The blade design program is a series of computational codes that are used to design and refine low-Reynolds number propellers. It enables the user to create airfoils using Bezier curves so that the new blade can be modified to the specific purpose of its application. The mathematical methods used within these codes are Blade Element

Airfoil 0.1 mm thick

Fig. 1 It shows a simple airfoil built-up with Bezier curves. A Bezier curve was chosen due to its easy manufacture, geometry and high performance benefits in small scale-vehicle

Momentum theory (BEM) and Vortex Panel Method (VPM), combined with Bezier approximations and MIT's *xfoil* subsonic airfoil development system (Bohorquez 2003; Bohorquez 2010; Harrington 1951; Drela and Youngren 2001). The only requirement for the user is to input a three-dimensional design through the parameters needed to describe the blade, such as: camber, reflex, thickness, number of blade slices, radius and the operating low-Reynolds number.

The simulations consist of several steps, such as:

1. Airfoil construction.
2. Airfoil aerodynamic analysis.
3. Airfoil Data Validation.
4. Blade analysis.
5. 3D Coordinates export.

3 Airfoil Construction

The airfoil construction is done in this step, where the user introduces their parameters, the profiles are built with Bezier curves. The Bezier curve satisfies the restriction specified by the designer, this arc-shape conform a single unit, and this will give the form of the airfoil. Figure 1 shows an example of such arc-shape.

The user needs to type the parameters for camber, reflex and thickness of the profile; it can introduce different values for each factor. The simulations were created by combining each set of parameters to create different shapes, the objective in this stage is to design large batches files of wings with a certain range of restrictions and evaluate all of them at once and take the best choice. Finally, we created a file with 2D coordinates for every single profile designed.

4 Airfoil Aerodynamic Analysis

In order to analyze every single airfoil created, we performed several *xfoil*-analysis. This is done by using the *xfoil* code written by Drela and Youngren (2001). The *xfoil* algorithm contains a series of routines and functions for the viscous analysis of the airfoil plus many other features.

Since, it was created many profiles until this stage; it needs to automate the reading and execution of the code. We have created MATALB interface in which the user just needs to set up the range of parameters such as the *angle of attack*, the

Fig. 2 It shows the data acquire from the aerodynamic analysis to the airfoil. Among all the results, it shows both lift and drag coefficients (CL & CD respectively) of the control surface. Also, its shows the angle of attack considered

```
--------------------------------------------------------------------
|        XFOIL           Version 6.94                              |
| Calculated polar for: BEZ062017010                              |
|                                                                  |
| 1 1 Reynolds number fixed              Mach number fixed        |
| xtrf =   1.000 (top)          1.000 (bottom)                    |
| Mach =   0.000      Re =      0.010 e 6       Ncrit =    7.000   |
|  alpha     CL        CD         CDp        CM     Top_Xtr Bot_Xtr|
| -------  -------   --------   --------   --------  ------- -------|
| -20.000  -1.0365   0.27181    0.26088    0.1497   1.0000  0.1316 |
| -19.800  -1.0287   0.26822    0.25732    0.1484   1.0000  0.1349 |
| -19.600  -1.0242   0.26762    0.25677    0.1467   1.0000  0.1371 |
| -19.400  -1.0182   0.26559    0.25481    0.1451   1.0000  0.1391 |
|    .        .         .          .          .        .       .   |
|    .        .         .          .          .        .       .   |
|    .        .         .          .          .        .       .   |
|  19.400   1.0210   0.26283    0.24729   -0.0911   0.3292  1.0000 |
|  19.600   1.0257   0.26416    0.24864   -0.0927   0.3285  1.0000 |
|  19.800   1.0306   0.26560    0.25009   -0.0943   0.3276  1.0000 |
|  20.000   1.0359   0.26726    0.25180   -0.0959   0.3264  1.0000 |
--------------------------------------------------------------------
```

height blade as well as the operating the Reynolds number. The MATLAB interface searches both angle of attack and Reynolds number desired and it using *xfoil* to evaluate each airfoil, it creates a new data-base. After, it examines the profile, it is written a file for each profile, where it specifies both the lift and drag generated in each angle of attack. An example can be shown in Fig. 2.

At this point we have the aerodynamic analysis of each profile, we need to know if the desired characteristics has a real blade geometry.

5 Airfoil Data Validation

In order to validate the data, it needs to check for congruence of every single profile. That is why we have created a code that allows the user to verify the output files generated in the 2D analysis, the examination is done using different criteria based on experimental results.

The purpose of this MATALB interface is to make easy for the user to look for the right set of profiles, because there are many data profiles, it takes a long time to go on each file and check the lift and drag coefficients they have at each angle of attack. That is why we decided to make a batch program where the user just puts in the parameters he wants and the computer will read all the files and outputs a file with the names of the best airfoils, subjected to the experimental data.

This *"chosen airfoils"* file will help us in the next step, since we know which profiles are the best and that fit the restriction of the project we can do a 3D viscous analysis, just on these airfoils.

6 Blade Analysis

We now have everything we need to construct the whole blade in 3D and analyze it. In this part of the algorithm, we are going to obtain the final results of performance of our design and now the user can choose which profile is the best and fits his needs.

The designer needs to introduce some variables, so we can create the whole blade, these parameters are the "chosen airfoils" file with the names of the best airfoils, number of blades, blade radius, type of taper, root and tip measurements, blade cut out, collective pitch and twist type.

We can then launch the program to initiate the simulation of blade performance. The simulation uses BEM/VPN theory, BEM give us a basic insight of the blade performance but this useful results cannot be used as a standalone code to design propellers, that why the idea arose to combine it with VPM. The VPM theory calculates the pressure distribution around the arbitrary airfoil based upon an assembly of vortices of appropriate strength between coordinates specified in the airfoil coordinate file.

After we get the results from the calculations, we also include the losses generated at the tip and the root, giving our program 15% more efficiency. The amalgamation of this entire test will give you an estimate of blade performance which is the closest result to reality. After the mathematical iterations have finished we are going to see the performance graphs of each blade, these diagrams are:

- Trust vs RPM
- Torque vs RPM
- Power vs RPM
- FOM vs RPM

We can now compare each blade and select the best one for the job, or evaluate where the designs are lacking some characteristics we need and change some variables to adjust it as we desire. This is when aerodynamic evaluation ends and if arrived to an acceptable design, we can take it to the next stage, which is to get the CAD design so we can go and manufacture it.

7 3D Coordinate Creator

We are at the final phase, this module generates the X, Y, Z coordinate of the profile that we designed. The user just needs to input the name of the airfoil that is going to be formed and the blade attributes.

The code will perform lineal interpolation to build the blade from the root to the tip, and at the end the output there is a single text file containing the propeller coordinates in millimeters and a rough graph of the design in MATLAB as is shown in Fig. 3.

With the coordinates of the file, we can import it into a mechanical computer aided design, this will draw the curves that constitute the blade, as shown in the Fig. 4, so we just need to unite them as a solid, and then finally we have our blade design ready to be manufactured.

Fig. 3 It shows the wired-design seen in 3D. It shows the length of the blade that we designed and the measurement of the root and tip chord. The program assembles the wing by generating 50 profiles and stacking them up each other. The graph generated by MATLAB

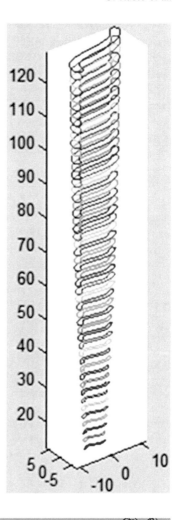

Fig. 4 The airfoils is shown when was exported to a CAD environment. Note that those profiles can form a solid body

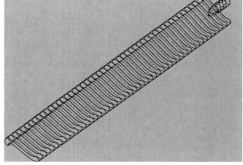

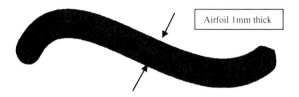

Fig. 5 The profile shown is the one we designed, as you see the airfoil is thicker than in Fig. 1, where the airfoil was 0.1 mm and now it is 1 mm wide. This change was due to manufacturing processes and load capacity of the MAV. It also has increase in camber and presents a more profuse reflex on the tail

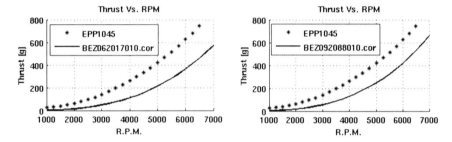

Fig. 6 Thrust graphs. (*Left*) current blades; (*Right*) designed blades. We observe that the new blades outrun the current ones by almost a 100 g. The increase of camber has improved our design. NACA EPP1045 ideal

8 New Rotor Design

As stated before, we are using the codes to design a new rotor for micro-helicopters with the purpose of enhancing the flying time of the aircraft.
The blade design we require has the following restrictions:

- Low-Reynolds number (10,000).
- 15 cm radius rotor.
- Above 1 mm thickness.
- Low energy consumption.
- High load capacity.

So we are restricted by these parameters, we need to fulfill all of them so the performance can be improved and fit in the micro-helicopter. The 1 millimeter and above thickness arises due to the resolution of the machines available for manu-facture. We designed several airfoils for this blade but the only airfoil that exceeded all the rest is the one shown in Fig. 5, that has a high camber and reflex, this is for compensating the losses generated by the width.

The performance of the designed airfoil improves on a previous one by 50%, making this rotor suitable for manufacturing. The first improvement we can see is the thrust where the new design has over 17% more as shown in Fig. 6.

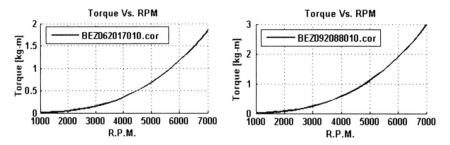

Fig. 7 Torque graphs. (*Left*) current blades; (*Right*) designed blades. As the graphs are showing our new design has amplified to 3 [kg-m] @ 7 R.P.M. from 1.8 [kg-m] @ 7 R.P.M. This means that we can carry more weight without demanding an extra effort to the motors. NACA EPP1045 ideal

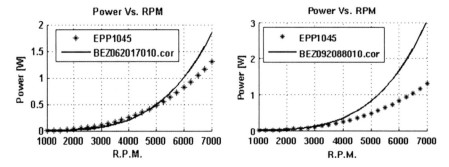

Fig. 8 Power graphs. (*Left*) current blades; (*Right*) designed blades. Due to the modifications to the width where we increase ten times the original size. We can see and increment of the power in 1.2 W. NACA EPP1045 ideal

The next comparison highlights the difference between the blades as shown in Fig. 7, the new blade has a maximum torque of 3.0 kg/m and the current one has a 1.85 kg/m, this means we have 63% more torque.

The power requirement of the new design needs more watts, but that is because it is thicker, but even with this loss our design works better because our curve is more linear and the thrust and torque are greater as shown in Fig. 8.

We can see in the FOM (Figure-of-Merit) graph that the designed blade is better than current blade (Fig. 9), since it has surpassed the performance of the previous rotor, we decided to manufacture it.

We introduced the resulting curves from the 3D generator coordinates program, and the propeller blade we got is the one shown in the Fig. 10.

In order to eliminate more losses or instability in the tip we have added wing-lets, this will help the performance of the blade steadiness and reduce vibrations on the rotor, the lateral wing-lets are shown in Fig. 11.

In order to attach the blade to the motor we have designed a fitting which not only will support the blades and connect them to the motor, but that can also change the angle of attack, with a simple mechanism, it can be adjusted to 0, +10, −10, +5, −5 and +15 degrees (Fig. 12). Adjustment of the angle of attack will

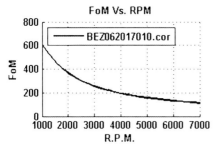
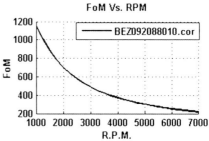

Fig. 9 FOM graphs. (*Left*) current blades; (*Right*) designed blades. It can observes than the new blades overcome the existing ones, even after the modification in thickness to the latest propeller, it is heavier and with a bigger area it generates more drag; but regardless of this it is superior

Fig. 10 It shows the witness the solid body of the blade in CAD format. We are able to see the blade and evaluate the angle of twist, length, and chord measurements and perform structural test with the desire material to manufacture

Fig. 11 The blade with the lateral fin added. Compare with Fig. 10, we can observe a more smooth finish to prevent air bubbles at the end of the wing, and to give stability to the rotors by minimizing vibrations

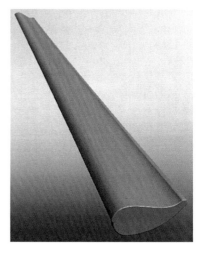

Fig. 12 Rotor concector fitting is an element that we have incorporated to the system, so the user can customize the angle of attack without having to manufactured a whole different blade. The propellers have a connection rod attach to their body with the holes properly arrange in diverse angles. Modifying the angle of attack is going to alter the performance of the rotor to enhance some properties but of course decreasing other ones

Fig. 13 It shows the final CAD model of the rotor, this is a great way to see the whole rotor; we can evaluate the measures of it and perceive how it looks like before we send it to manufacture. The drawing is going to be used in the rapid prototyping machine for its construction. Also notice the connection rotor-blade we mentioned in Fig. 12

allow testing/modification of the rotor behavior without designing a new one or including servos in to the system.

The (Fig. 13) shows the complete rotor, ready to be manufactured..

9 Conclusions

The main results in this research, the development and optimization of software able to create, analyze and build micro helicopter blades, and a specific propeller micro-helicopter design, manufactured at the available facilities at ITESM.

Furthermore, Bezier curves make perfect airfoils in this low-Reynolds number range, because of the following characteristics:

1. Higher maximum lift.
2. Lower minimum drag.
3. Low dependency of lift and drag coefficients with low-Reynolds number.

4. Minimal hysteresis loops when cycling Reynolds numbers or angle of attack.
5. Low sensitivity to ambient turbulence.

The airfoil analysis in 2D gives us a good basic knowledge of the aerodynamic behavior of the profile, but the more informative results are obtained in the 3D viscous analysis, at which point we can see the performance of the blade, and with the incorporation of root and tip losses we obtain 15% more accuracy.

The combination of BEM/VPM allows a complete examination of the propeller, the BEM gives us a rough view of interaction between the fluid and the blade, but with VPM we add a fine touch at the end, evaluating the conduct of the airfoils and air passing around them. With the implementation of these two theories we acquire a closer approximation.

With the help of the coordinate generator code, it is easier to see and build a rotor; this program helps the user to construct the blade instead of just making the calculations. With CAD design the users can get a glimpse of the blade and how it will look like if they machine it. The machines to manufacture nowadays need the CAD drawing to make the design real, so with this program you will save some time.

The new blade that we designed shows great performance overcoming the old rotors the helicopter has installed, we can see and improve in almost all the features, thrust 17%, torque 63% and the FoM is 400 units higher at low RPM, all of these variables will make the helicopter consume less energy with greater capability.

The software developed, not only gives the user freedom to design complex low-Re 3D blades, but it also outputs a manufacturing-ready file, helps to take theory to practice, extending Mexican manufacturing, engineering, design and industrial capabilities.

Finally, the program herein described is a complete compound tool for propeller engineering design. It takes you from the designing of the airfoil, to the evaluation of each profile, filter the best designs, analyze them with 3D viscous theories and build the blade on CAD software.

References

Abbot D (1959) Theory of wing sections. Dover, Canada

Asher DP (2010) A low order model for vertical axis wind turbines. Am Inst Aeronaut Astronaut 28th AIAA Applied Aerodynamics Conference, 2010–4401, 1–9

Bohorquez SS (2003) Design, analysis and hover performance of a rotary wing micro air vehicle. J Am helicopter Soc 80–87

Bohorquez PS (2010) Small rotor optimization using blade element momentum theory and hover tests. J Aircr 47, 268–283

Castillo LD (2005) Modelling and control of mini-flying machines. Springer, London

Conslisk (1997) MODERN HELICOPTER. Helicopter Aerodyn 29, 515–567

Drela M, Youngren H (2001) XFOIL 6.9. United States

Gupta L (2005) Comparison of momentum and vortex method for the aerodynamic analysis of wind turbines. Am Inst Aeronaut Astronaut 43rd AIAA Aerospace Sciences meeting and Exhibit, 1–24

Harrington R (1951) Full scale tunnel investigation of the static-thrust performance of a coaxial helicopter rotor, NACA

Hazra J (2007) Aerodynamic Shape Optimization of Airfoils in. Aerospace Computing Lab (ACL) report 2007-4, Stanford university, USA

Little J, Moler C (1994) Matlab. Natick, Massachussets, United States: The MathWorks, Inc

Kunz (2003) Aerodynamics and design. Dissertation, 1–204

Morris H (2000) Design of micro air vehicles and flight test validation. pp 1–22

Pestov S (1998) jEdit

Reid M (2006) Thin/cambered/reflexed airfoil development. A Thesis Submitted in the Requirement for Master of Science in Mechanical Engineering, Rochester Institute of Technology, 1–184

Sunada YY (2002) Comparison of wing characteristics. J Aircr 39(2), 331–338

Taamallah DR (2005) Development of a rotorcraft mini-uav system demonstrator. IEEE 2:11.A.2-1–11.A.2-15

Oscillation Characteristics of a Vertical Soft Pipe Conveying Air Flow

Héctor Manuel De La Rosa Zambrano and Anne Cros

Abstract This study takes place in the "fluid–structure interaction" area. The experiment consists of a large vertical thin fabrics tube whose lower extreme is fixed to an air blower exit. For a sufficient intensity of the air flow, the tube stands up. Whereas these kinds of systems have different characteristics than previous studies (Païdoussis 1998) we find that at the instability threshold, the tube oscillates with a frequency of the same order of magnitude as in these previous works.

1 Introduction

A garden-hose undulates if the water which flows through is strong enough. The different stability regimes which appear in a flexible tube conveying a flow have been studied for more than half a century. The books written by Païdoussis (1998, 2004) relate with great detail and explanations the different solutions of problems found by many authors. Recent works (Doaré and de Langre 2002) still deal with the "garden-hose" instability.

The tube undulations are generated by a complex interaction between the flow and the elastic wall of the tube. The theoretical equation involves partial, spatial and temporal, derivatives of the transversal displacement $w(x,t)$ of the tube, where x is the longitudinal coordinate and t is time. The equation can be written as:

H. M. De La Rosa Zambrano · A. Cros (✉)
Departamento de Física, Universidad de Guadalajara,
Av. Revolución 1500, Col. Olímpica, 44430 Guadalajara,
Jalisco, Mexico
e-mail: anne@astro.iam.udg.mx

J. Klapp et al. (eds.), *Fluid Dynamics in Physics, Engineering and Environmental Applications*, Environmental Science and Engineering,
DOI: 10.1007/978-3-642-27723-8_47, © Springer-Verlag Berlin Heidelberg 2013

$$El\frac{\partial^4 w}{\partial x^4} + m_f v^2 \frac{\partial^2 w}{\partial x^2} + 2m_f v \frac{\partial^2 w}{\partial x \partial t} + (m_f + m_t)\frac{\partial^2 w}{\partial t^2} = 0, \tag{1}$$

where EI (Nm^2) is the tube flexural rigidity (E being the Young's modulus of the material and I the second moment of inertia of the tube section), m_f (kg/m) is the mass of fluid contained per unit length of pipe, v is the flow velocity and m_t (kg/m) the tube mass per unit length. This equation is valid only for a "snake-like" spatial shape of the tube. In our experiment, the tube exhibits "break points" where the first spatial derivative is undefined. Nevertheless during the transient dynamics presented in this work, this equation may apply.

From Eq. (1) four nondimensional parameters can be constructed. The first is the nondimensional velocity

$$u = \sqrt{\frac{m_f}{EI}} Lv. \tag{2}$$

The ratio between the fluid mass and the total mass of the system is

$$\beta = \frac{m_f}{m_f + m_t}. \tag{3}$$

The gravity parameter represents the influence of gravity on the fluid-pipe system:

$$\gamma = \frac{(m_f + m_t)L^3}{EI} g. \tag{4}$$

Conventionally, $\gamma < 0$ for standing pipes, $\gamma = 0$ for horizontal pipes and $\gamma > 0$ for hanging pipes.

The last parameter is the nondimensional frequency:

$$\Omega = \sqrt{\frac{m_f + m_t}{EI}} \omega L^2, \tag{5}$$

where ω (rad/s) is the pipe oscillation frequency.

2 Experimental Set-Up

The experimental set-up is very simple and is composed of low-cost elements.

The pipes are made in our laboratory by heat-sealing a plastic membrane. This material has a Young's modulus $E = 4.5$ MPa, a mass per surface unit $\rho_s = 0.083$ kg/m^2 and a thickness $e = 0.1$ mm. We constructed tubes with two different diameters: 2.3 and 3 cm. The tubes were initially 140 cm-long and were cut to have smaller lengths.

The set-up scheme is shown in Fig. 1. The air pump is connected to a potentiometer to regulate the intensity of the air flow, so that v can be varied from 0 to

Fig. 1 Experimental set-up.
See text for details

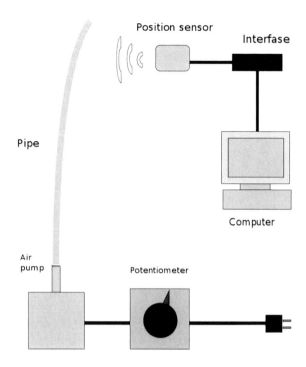

Position sensor

Interfase

Pipe

Computer

Air
pump   Potentiometer

25 m/s. The pipe is fixed to the pump exit and the position sensor is fixed at the tube free end level. The position sensor is PASCO CI-6742A: it has a spatial resolution of 1 mm and records data each millisecond.

The air velocity is maintained so that the tube stands up without movement. Then v is slightly increased and the position sensor is started. The data $w(x_0,t)$ are saved and we do the same procedure after the tube was cut 2 cm shorter.

3 Results

An example of the signal $w(x_0,t)$ is shown in Fig. 2. It can be seen that the oscillation amplitude of the tube increases until the tube "breaks" and falls down, that is to say that a break point appears (generally near the bottom of the tube, see Cros et al. 2011). The Fourier spectrum (Fig. 2, right-hand side) shows that the signal is essentially periodic. By using Eq. (5), the nondimensional frequencies Ω are plotted against the nondimensional "length" γ of the tube. The graphic is shown in Fig. 3.

Different points in Fig. 3 which have the same value of γ correspond to different values of the air speed v. As v is greater, Ω is greater too. Moreover, we can observe that as $|\gamma|$ increases, Ω decreases. These characteristics are coherent with the experimental and theoretical analysis of Pa (1998) for standing-up tubes. Moreover, the previous theory and experiments show that Ω should vary between

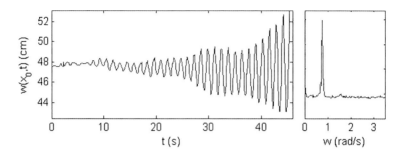

Fig. 2 Example of the temporal evolution $\omega(x_0,t)$ of the tube extreme oscillation. *Right* Fourier spectrum

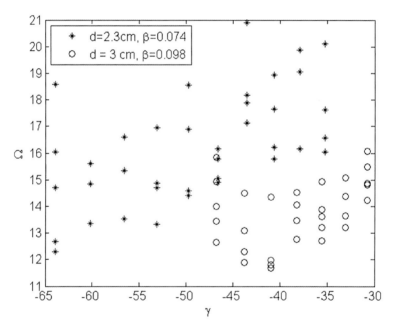

Fig. 3 Nondimensional frequencies Ω as a function of γ, for tubes of two different diameters

14 and 16 for $\beta < 0.3$. The order of magnitude of our experimental values agrees with these observations.

Finally, the Ω values are slightly lower for $\beta = 0.098$ than those for $\beta = 0.074$. The theory indeed predicts that Ω always decreases when β increases when $\beta < 0.3$.

4 Discussion

Whereas the sky dancer does not exhibit a smoothly undulating spatial shape when it is "dancing", it appears to behave like a thicker-walled pipe during transient dynamics. The transient dynamics occur at the instability threshold, when the tube cannot stand up anymore at higher air speed. Initially in a straight vertical position, the tube begins oscillating with amplitude increasing with time. When the oscillation reaches a certain amplitude, the tube "breaks" and a chaotic regime develops (Castillo Flores and Cros 2009).

During the transient dynamics, the oscillation frequencies of the tube fit well with previous theories (Païdoussis 1998). Those previous studies describe undulations of thicker-walled tubes as due to a complex competition between the stabilizing tube rigidity and the destabilizing centrifugal forces. The centrifugal forces are exerted by air flow in *curved* portions of the tube (Cros et al. 2011). In the thicker-walled pipes, no "break point" is observed as in our system, the tube only curves in on itself.

In the transient regime studied here, the tube does not break. So Eq. (1) can apply to this study. Nevertheless, after the tube "breaks", this equation does not describe our system anymore. Then we should wonder what is the difference between the thicker-walled tubes and our "soft" pipe? Our tube has a very low rigidity indeed. So, is the tube rigidity sufficient to play the role of the stabilizing force? The thin wall of our soft tube appears to tense because of a positive pressure inside the tube. The first pressure measurements show that these pressure forces are two orders of magnitude greater than the rigidity effects. So whereas thick-walled tubes theory is valid in the oscillating transient regime, it does not apply anymore in the chaotic regime. We think that pressure could play the role of the stabilizing effect in our system.

5 Conclusion

This work deals with a soft vertical tube conveying an ascending air flow. The air velocity is initially fixed in order to make the tube straight and immobile. When the air speed is gradually increased until it reaches a threshold, the tube oscillates with an increasing amplitude. It finally falls down. After this undulating transient dynamics, the tube "dances" with broken spatial shapes. We measured the oscillation frequencies of the transient regime. We found that the order of magnitude and the evolution of the nondimensional frequencies when the tube length varies fit well with previous theories. These previous works deal with thicker-walled tubes that do not exhibit "break points". We conclude that thicker-walled tubes theory applies to our transient regime. Nevertheless we think that a stabilizing effect other than the (very low) tube rigidity acts in our system, that could be a tensioning effect generated by the positive pressure inside the tube.

References

Castillo Flores F, Cros A (2009) Transition to chaos of a vertical collapsible tube conveying air flow. J Phys Conf Ser 166:012017

Cros A, Rodriguez Romero JA, Castillo Flores F (2011) Sky dancer. In: Klapp J et al (eds) A complex fluid–structure interaction. Experimental and theoretical advances in fluid dynamics. Springer, p 517

Doaré O, de Langre E (2002) The flow-induced instability of long hanging pipes. Eur J Mech A Solids 21:857–867

Païdoussis MP (1998) Fluid–structure interactions, slender structures and axial flow, vol 1. Academic, London, p 572

Païdoussis MP (2004) Fluid–structure interactions, slender structures and axial flow, vol 2. Academic, London, p 1585

Capillary Rise in a Convergent Hele-Shaw Cell

C. A. Vargas, A. Medina and F. Aragón

Abstract The capillary rise of a viscous liquid into a convergent Hele-Shaw cell whose plates make a wedge-like channel along the downwards coordinate z has been analyzed theoretically by using the lubrication theory. The solution to this problem allows to determine the existence of a continuous flow which is atypical in problems of capillary rise.

1 Introduction

The situation where a viscous liquid is capillary forced to flow within gaps formed by two very close-together irregularly spaced plates is commonly encountered in the exploitation of naturally fractured oil reservoirs where gaps conduits water, gas and oil (Barenblatt et al. 1990; Dietrich et al. 2005), among others. Flows like the aforementioned are also of theoretical and academic interest (see, for instance Park and Homsy 1984; Eastathopoulos et al. 1999; Sánchez et al. 2004). In this ambit the configuration of two close-together vertical parallel plates of uniform separation has been known as the Hele-Shaw configuration meanwhile a configuration

C. A. Vargas (✉)
Laboratorio de Sistemas Complejos, Departamento de Ciencias Básicas,
UAM Azcapotzalco, Av. San Pablo 180, Azcapotzalco, 02200 Mexico, DF, Mexico
e-mail: cvargas@correo.azc.uam.mx

A. Medina · F. Aragón
Instituto Politécnico Nacional, SEPI ESIME Azcapotzalco, Av. de las Granjas 682,
Col. Santa Catarina, Azcapotzalco, 02250 Mexico, DF, Mexico
e-mail: amedinao@ipn.mx

F. Aragón
e-mail: micme2003@yahoo.com.mx

J. Klapp et al. (eds.), *Fluid Dynamics in Physics, Engineering and Environmental Applications*, Environmental Science and Engineering,
DOI: 10.1007/978-3-642-27723-8_48, © Springer-Verlag Berlin Heidelberg 2013

where such separation decreases in the flow direction will be named here the convergent Hele-Shaw cell.

Moreover, locally, many systems can be understood, in a first approximation, as a convergent cell, for instance, when two spherical grains are close-together the space can be visualized as two very close planes. Consequently, understanding the flow behavior in such a domain is of considerable interest to engineering.

The flow condition that will be studied here is corresponding to a converging flow. This is made by admitting the fluid through an inlet located near the base of the cell, and draining it through an area located near the vertex. The reverse produces a diverging flow that now is not of interest.

It is well known that there are analytical exact solutions for the creeping flow in convergent and divergent wedges (Jeffrey 1915; Bond 1925) but, due to the capillary flow is a film flow, in this work we used the lubrication theory in order to get a simpler and direct treatment.

The aim of the present paper is to provide solutions for the converging capillary flow in narrow enclosures formed by two inclined planes.

2 Theoretical Model

We assume the existence of a flow whose origin is due to the imbalance between the capillary and the hydrostatic pressures just at the upper free surface. The flow is continuously maintained due to the Hele-Shaw cell was submerged into the liquid reservoir. Thus, the largest separation among the plates will be smaller than the capillary length, $l_c = (\sigma/\rho g)$, where ρ is the liquid's density, g is the gravity acceleration and σ is the surface tension. For water, this length is of around 2 mm.

Therefore, it is valid to maintain the assumption of a one-dimensional film flow and it can be treated with the lubrication theory. The plates in the cell make a small angle α, and they can or not be touched at the vertex $z = 0$ (Fig. 1).

In this case the equations of the lubrication can be given using polar coordinates where the z-coordinate point downwards.

Thus the Reynolds equation that expresses the mass conservation is

$$\frac{\partial}{\partial z}\left(\frac{h^3}{12\mu}\frac{\partial P}{\partial z}\right) = 0, \tag{1}$$

where P is the pressure and $h = \alpha z$ is the non uniform separation among the plates. Similarly, the equation of momentum conservation is in this case

$$q_z = -\frac{h^3}{12\mu}\frac{\partial P}{\partial z}, \tag{2}$$

where q_z is the flow, per unit length, along z. Equations (1) and (2) will be solved obeying the following relations

Fig. 1 Scheme of a convergent Hele-Shaw cell

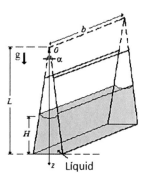

$$\text{at } z = L, \quad P = 0, \tag{3}$$

they indicate that the pressure at the liquid entry, is null, and

$$\text{at } z = L - H, \quad P = -\frac{2\sigma \cos(\theta_c + \alpha)}{h} + \rho g H. \tag{4}$$

This later relation assures that at the front of the liquid, measured from the vertex of the cone, $z = 0$, are acting the capillary and hydrostatic pressures. Moreover, in Eq. (4) H is the liquid actual position (position of the free surface) where the pressure has such a form, if $(L–H) \gg h$, remember that θ_c characterizes the contact angle. At $z = (L–H)$ the kinematic condition is the assertion that the fluid does not cross the free surface, which is expressed as

$$h \frac{dH}{dt} = -q_z, \ z = L - H. \tag{5}$$

Equations (1), (2) and (5) and their respective boundary conditions allow the solution of the problem. Thus, Eq. (1) immediately yields that

$$\frac{h^3}{12\mu} \frac{\partial P}{\partial z} = A(t), \tag{6}$$

where $A(t)$ is a time function: The integration of the later equation formally gives

$$\int_0^{-\frac{2\sigma \cos(\theta_c + \alpha)}{\alpha(L-H)} + \rho g H} dP = \frac{12\mu A}{\alpha^3} \int_L^{(L-H)} \frac{dz}{z^3} \tag{7}$$

and the integration of Eq. (7) allows us to determine A

$$A = -\frac{\sigma \alpha^2 \cos(\theta_c + \alpha)}{3\mu} \frac{L^2(L - H)}{H(H - 2L)} + \frac{\rho g \alpha^3}{6\mu} \frac{L^2(L - H)^2}{(H - 2L)}. \tag{8}$$

The use of Eqs. (2), (6) and (8) in the kinematic condition (5) lets us find the differential equation for the capillary rise

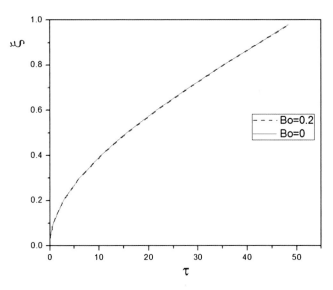

Fig. 2 Dimensionless evolution of the front of the liquid as a function of dimensionless time. Plots were made for two Bond numbers $Bo = 0$ (*dashed line*) and $Bo = 0.2$ (*continuous red line*)

$$\frac{dH}{dt} = \frac{L}{(H - 2L)}\left[-\frac{\sigma\alpha\cos(\theta_c + \alpha)}{3\mu}\frac{L}{H} + \frac{\rho g\alpha^2}{6\mu}L(L - H)\right]. \tag{9}$$

The equilibrium heights will be attained if $dH/dt = 0$. The use of this condition in Eq. (9) yields the heigths

$$H_{eq} = \frac{1}{2}L \pm \frac{1}{2}\sqrt{L^2 + 4l^2}. \tag{10}$$

where $l^2 = 2\sigma\cos(\theta_c + \alpha)/\rho g\alpha$ is an effective capillary length. Notice that $\sqrt{L^2 + 4l^2} > L$ if $l > 0$.

This result assures that the liquid will drip if the physical condition in the vertex allows it.

The dimensional form of the previous equations is obtained by introducing the dimensional variables $\xi = H/L$ and $\tau = t/t_c$ where $t_c = 3\mu L/\sigma$. Consequently, Eq. (9) transforms into

$$\frac{d\xi}{d\tau} = \frac{1}{(\xi - 2)}\left[Bo\frac{\alpha^2}{2}(1 - \xi) - \frac{\alpha\cos(\theta_c + \alpha)}{\xi}\right], \tag{11}$$

where $Bo = \rho g L^2/\sigma$ is the Bond number. Whenever $d\xi/d\tau = 0$, the dimensionless height is now

$$\xi_{eq} = \frac{1}{2} + \frac{1}{2}\sqrt{1 + \frac{4\cos(\theta_c + \alpha)}{\alpha Bo}}. \tag{12}$$

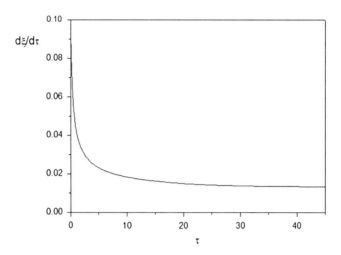

Fig. 3 Plot of the dimensionless speed of the front of the liquid as a function of the dimensionless time. It is also evident that the flow is not arrested in the cell

The plus sign allows to find that $\xi_{eq} > 1$ if $4\cos(\theta_c + \alpha)/\alpha Bo > 0$.

An asymptotic solution of (11) is obtained if $Bo = 0$, because of the resulting differential

$$(2 - \xi)\xi\frac{d\xi}{d\tau} = \alpha\cos(\theta_c + \alpha), \tag{13}$$

has the solution

$$3\xi^2 - \xi^3 = 3\alpha\cos(\theta_c + \alpha)\tau. \tag{14}$$

For $\xi \ll 1$, it is found that

$$\xi^2 = \alpha\cos(\theta_c + \alpha)\tau, \tag{15}$$

which is the Washburn law.

3 Numerical Solution

If $Bo \neq 0$ Eq. (11) should be solved numerically. We supposed that the water fills the cell ($\rho = 998$ kg/m^3, $\sigma = 0.072$ N/m, $\mu = 1.002$ cP) of $L = 0.1$ m height, $\alpha = 30°$ and perfect wetting, *i.e.*, $\theta_c = 0°$. Figure 2 shows the dimensionless plot of the front of the liquid, ξ, as a function of time, τ. It is worth notice that the effect of gravity is not substantial

In Fig. 3 we show the plot of the speed of the front of the liquid. Because the speed does not vanish a continuous flow in the cell occurs.

4 Conclusions

This work analyzed the problem of the capillary rise of a viscous liquid in a convergent Hele-Shaw cell. The model was based on the lubrication theory and theoretical results allow to find the existence of a continuous flow in such a system. More studies are necessary in order to understand when the capillary flow will be arrested.

References

Barenblatt GI, Entov VM, Ryzhik VM (1990) Theory of Fluid Flows through Natural Rocks. Kluwer, Dordrecht

Bond WN (1925) Viscous flow through wide angle cones. Phil Mag 50:1052–1066

Dietrich P, Helmig R, Sauter M, Kongeter HJ, Teustch G (2005) Flow and transport in fractured porous media. Springer, Berlin

Eastathopoulos DN Nicholas MG, Devret B (1999) Wettability at high temperatures 1st edn. Vol 3 of Pergamon materials series. Pergamon, New York

Jeffrey B (1915) Steady motion of a viscous fluid. Phil Mag Ser 6(29):455–465

Park C-W, Homsy GM (1984) Two-phase displacement in Hele Shaw cells: theory. J Fluid Mech 139:291–308

Sanchez M, Sanchez F, Pérez-Rosales C, Medina A, Treviño C (2004) Imbibition in a Hele-Shaw cell under a temperature gradient. Phys Lett A 324:14–21

Manufacturing of Polymeric Micro-Lenses by Drip Injection

Miguel Ortega, Abel López-Villa, Guadalupe Juliana Gutiérrez and Carlos A. Vargas

Abstract In this work we study an alternative and innovative method for the manufacture of polymeric micro-lenses and how the characterization of its physical, mechanical and optical parameters is relevant for their application. We study the equilibrium shapes of drops that emerge slowly from vertical thick-walled tubes, as a result, it is found that the sizes and shapes of drops depend on the Bond number, the injection pressure and the value of the contact angle, so that finally with this technique we are able to make lenses. The manufacture will be made though drip injection since it is easier to control the shape of the lens.

1 Introduction

The optics is one of the most important technologies used in both the industry and life. Many sectors use polymeric materials with advanced properties in applications like electronic devices (CDs and DVDs, bar code readers), illumination (LEDs encapsulation), (Wijshoff 2006) medical applications (lenses for sensors and actuators), automation (video cameras), security and biometric analysis (fingerprints readers).The tendency of this type of market is the micro optical, the main reason is to

M. Ortega · A. López-Villa (✉) · G. J. Gutiérrez
Instituto Politécnico Nacional SEPI ESIME Azcapotzalco, Av. de las Granjas 682,
Col. Santa Catarina, Azcapotzalco, 02250 Mexico, DF, Mexico
e-mail: abelvilla77@hotmail.com

C. A. Vargas
Departamento de Ciencias Básicas, Universidad Autónoma Metropolitana Unidad
Azcapotzalco, Av. San Pablo 180, Azcapotzalco, 02200 Mexico, DF, Mexico
e-mail: cvargas@correo.azc.uam.mx

J. Klapp et al. (eds.), *Fluid Dynamics in Physics, Engineering and Environmental Applications*, Environmental Science and Engineering,
DOI: 10.1007/978-3-642-27723-8_49, © Springer-Verlag Berlin Heidelberg 2013

Fig. 1 Schematic of the drop shape emerging in a thick-walled tube: **a** drop with good wetting (contact angle $<\pi/2$) and **b** drop with poor wetting (contact angle $>\pi/2$)

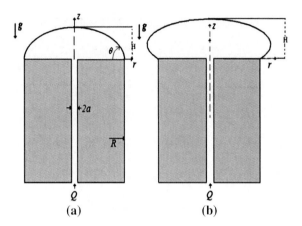

(a)   (b)

reduce the size and weight of the pieces, maintain the same properties or fulfill the requirements of new applications.

The formation of drops is a common phenomenon and it is a great interesting study for diverse processes from the natural to the industrial ones (Lopez-Villa 2010; Ortiz 2009; Schwarz et al. 2008; Pailler-Mattei et al. 2006; Oliver et al. 1977). With a controlled and analytical study this phenomenon of drops formation has enormous practical consequences. One of them is that the characterization of the form of a drop can be very useful to quantify important properties like the surface tension or the wetting of a liquid on any substrate. In a general case, the liquid may have a good wetting when $\theta < \pi/2$ and it has a poor wetting when $\theta < \pi/2$. At this work we will analyze the drop shape on both cases according to our purposes.

2 Formulation of the Problem

We consider that a drop emerges, on the top or the bottom of a vertical thick-walled tube. The drop is generated at a very low flow rate so that, the drop shape can be described by an equation of balance between the hydrostatic, the injection and the capillary pressure, *i.e.*, in such conditions the drop remains in equilibrium and with the same shape. In this case the description of the drop shape can be made by using the Young–Laplace equation (Longuet-Higgins et al. 1991; Landau 1969; Liñan et al. 2003; de Genes 2004; Middleman 1995).

A scheme of the problem of slow injection (quasi-static case) of a liquid in a thick-walled cylinder of inner radius a and outer radius $R = a + d$ is shown in Fig. 1. On the drop surface the capillary pressure, p_c, must be equal to the sum of the liquid injection pressure, p_0, hydrostatic pressure, $\rho g z$, and the atmospheric pressure, p_a (1), *i.e.*,

$$p_c = p_0 - \rho g z - p_a. \tag{1}$$

Note that the injection pressure must be greater than the hydrostatic pressure, i.e., $p_0 > \rho g z$ when the drop emerges at the top of the tube, if it emerges at the bottom of the tube these pressures are added. The use of this condition allows the drop growth.

Since equation $p_c = \sigma \nabla \cdot \mathbf{n}$ (Longuet-Higgins et al. 1991; Landau et al. 1969; Liñan et al. 2003; de Genes 2004, Middleman 1995) where $\mathbf{n}$ is a normal unit vector to the outer drop surface, and with $P = p_0 - p_a$, if

$$\nabla \cdot \mathbf{n} = \left(\frac{1}{R_1} + \frac{1}{R_2} \right), \tag{2}$$

where R_1 and R_2 are the radius of curvature.

We have that the differential dimensionless equation for the free surface in cylindrical coordinates (3), is

$$-\frac{\frac{d^2 \xi}{d\zeta^2}}{Bo\left[\left(\frac{H}{R}\right)^2 + \left(\frac{d\xi}{d\zeta}\right)^2 \right]^{3/2}} + \frac{1}{Bo\,\xi\left[\left(\frac{H}{R}\right)^2 + \left(\frac{d\xi}{d\zeta}\right)^2 \right]^{1/2}} = p - \zeta, \tag{3}$$

where $\xi = r/R$, $\zeta = z/H$ and $p = P/\rho g H$. R is the outer radius and H is the maximum drop height and the Bond number is $Bo = \rho g R^2/\sigma$. The dimensionless boundary conditions are

$$\text{at } \zeta = 1 \;\; : \;\; \xi = 0 \text{ and } \frac{d\xi}{d\zeta} \to \infty,$$
$$\text{at } \zeta = 0 \;\; : \;\; \xi = 1 \text{ and } \frac{d\xi}{d\zeta} = \tan\theta, \tag{4}$$

and the dimensionless volume takes the form

$$\int_0^1 \xi^2(\zeta) d\zeta = \frac{V}{\pi R^3} \tag{5}$$

It should be noted that the static contact angle values are in the range $[\theta, \pi/2 + \theta]$.

3 Experiments

The work material with which the experimentation will be done is SYLGARD® 184 SILICONE ELASTOMER of the Dow Corning mark, since it possesses the necessary characteristics for the experimentation and also fulfills the properties a material must have for optical applications. A micro pump is used (Fig. 2) to inject

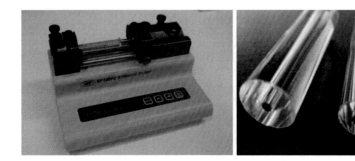

Fig. 2 Micropump and capillaries of diverse sizes

polymer, with the help of a syringe, the polymer is deposited in the capillary of diverse inner diameter by means of a tube that connects the syringe and the thick-walled capillary (Shih et al. 2006).

The slow growth of the drops was hardly obtained by injecting the liquid of test to the tube with a hose connected to it and a level of liquid superior to the level of exit of liquid in the tube. This method was suitable for dripping upwards as well as downwards

The idea is to observe what happens on different contact angles and different number of Bond and in that way to be able to validate the theoretical results. Figure 3 shows silicone droplets emerging from capillaries at different numbers of Bond.

In some cases, the material presents bubbles due to the mixture of the resin with the catalyst, in Fig. 4 we can observe the bubbles which the capillaries present, to eliminate these, it is necessary to introduce the mixture in a vacuum device to eliminate them, for that reason, a device was designed to fulfill the function.

Thanks to the adjustments done in the experimental part it was possible to obtain a bubbles free lens.

The used values of Bond number give us spheres shapes as a result, where the volumes are determined by the contact angle of the materials, in this case $\theta = 130°$ in the poor wetting and $\theta = 30°$ in the good wetting.

4 Results and Discussion

With the aid of the design software, RHINOCEROS, modeling and using a complement software called RHINOSHOE, we can determine areas, volumes and centers of gravity from the images acquired with the fast camera, the achievement of these data supports the calculated analytical data and gives the validity to the work done. The procedure is developed by the software support, the outlines of the lens are modeled; as long as the image to analyze is black and the bottom white, see Fig. 5.

Fig. 3 Experiments with variation of the number of bond

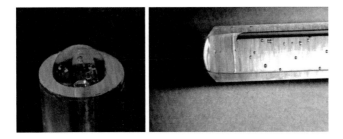

Fig. 4 Lens that presents bubbles in its interior

Fig. 5 Generated bidimensional profiles from photos obtained experimentally from drops which emerge downwards and upwards. They were used to find centers of mass, volume and area

The outline is made in 1:1 scale model with respect to the image applying the command "Raster to Vector", which will create a contour of the image; as a result, the 2D element is obtained, whom center of mass and area can be acquired.

In order to obtain the volume of the image, first 3D is needed to turn the flat piece into a body (solid), then a revolution axis is generated and a 360° turn is done, therefore generating the solid in 3D. As a consequence, we get the volume, see Fig. 6.

A drop growing at a low flow rate is considered in equilibrium and the only way for it to grow is a change in the injection pressure. When the pressure changes the drop equilibrium conditions also change. In our case, the change in the drop volume is due to a change in the contact angle, which is possible to be measured.

Different values of the Bond number and contact angle give us different shapes of lens, see Fig. 7.

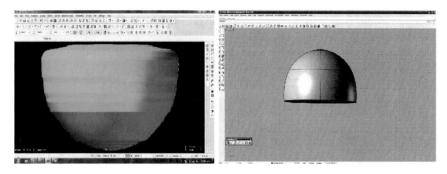

Fig. 6 We observe how a 3D solid is generated

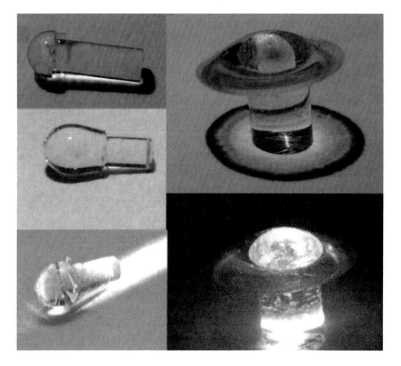

Fig. 7 Finished lens

5 Conclusions

In this paper we studied theoretically the drop shapes that emerge slowly from thick-walled tubes and through various experiments we established that the drop shapes are heavily dependent on injection pressure, the contact angle and the Bond number. The results are very similar to those obtained theoretically, the forms of the lenses which have been obtained approximate to the shapes that were predicted.

References

deGennes P G, Brochard-Wyart F, Quéré D (2004) Capillarity and wetting phenomena: drops, bubbles, pearls, waves. Springer, Berlin

Landau LD, Liftshitz EM (1969) Mecánica de Fluidos, Ed. Reverté

Liñan Martínez A, Rodríguez Fernández M, Higuera FJ (2003) Mecánica de Fluidos, Lecciones1–22, third course, Madrid

Longuet-Higgins Kerman MB, Lunde K (1991) The release of air bubbles from an underwater nozzle. J. Fluid Mech 230:365–390

López-Villa A (2010) Crecimiento y desarrollo de burbujas en líquidos viscosos en geometrías confinadas, Ph. D. Thesis Mexican Petroleum Institute

Middleman S (1995) Modeling axisymmetric flows dynamics of films, jets, and drops. Academic, San Diego, pp 257–287

Oliver JF, Huh C, Mason SG (1977) Resistance to Spreading of Liquids by Sharp Edges. J Colloid Sci Interface 59:568–581

Ortiz A, López-Villa A, Medina A, Higuera FJ. (2009) Formación de burbujas en líquidos viscosos contenidos en conos y cilindros, Revista Mexicana de Física, vol 55, Num 3, p 166

Pailler-Mattei CS, Vargiolu R, Zahouani H (2006) Analysis of adhesion contact of human skin in vivo, Contact Angle, wettability and adhesion. In: KL Mittal (ed) vol 4. pp 501–514

Schwarz B, Eisenmenger-Sittner C, Steiner H (2008) Construction of a high-temperature sessile drop device. Surf Eng Surf Instrum Vac Technol 82:186–188

Shih T, Chen C (2006) Fabrication of PDMS (polydimethylsiloxane) microlens and diffuser using replica molding. Elsevier, Taiwan

Wijshoff H (2006) Drop formation mechanisms in piezo-accoustic inkjet, Technologies B.V

Part VI
Gallery of Fluids

Visualization of Flow Inside a Ranque-Hilsch Tube

David Porta Zepeda, Carlos Echeverría Arjonilla, Catalina Elizabeth Stern Forgach and Marcos Ley Koo

A Ranque-Hilsch tube is a device that separates—in absence of mobile parts—a flow into two fractions; one hot and the other one cold.

The tube operates by introducing air at high pressure (5 atm) into a tube that is opened at both ends. The air is injected by a nozzle connected laterally, so that it introduces a flow tangential to the surface.

First Place of the Gallery of Fluids, XVII National Congress of the Fluid Dynamics Division of the Mexican Physical Society, Guadalajara, Jalisco, Mexico, November 8–11, 2011.

D. P. Zepeda (✉) · C. E. Arjonilla · C. E. S. Forgach · M. L. Koo
Facultad de Ciencias, Universidad Nacional Autónoma de México,
Ciudad Universitaria 3000, Col. Copilco Universidad,
Del. Coyoacán, 04360 Mexico, D.F., Mexico
e-mail: alviond@gmail.com

C. E. Arjonilla
e-mail: carlosea1982@gmail.com

C. E. S. Forgach
e-mail: catalina@ciencias.unam.mx

M. L. Koo
e-mail: mlk@ciencias.unam.mx

J. Klapp et al. (eds.), *Fluid Dynamics in Physics, Engineering and Environmental Applications*, Environmental Science and Engineering,
DOI: 10.1007/978-3-642-27723-8_50, © Springer-Verlag Berlin Heidelberg 2013

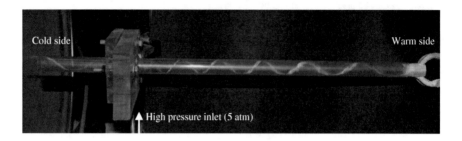

A Ranque-Hilsch tube was built in acrylic, based on the article by C. L. Stong, The "Hilsch" Vortex Tube, in The Amateur Scientist, Scientific American.

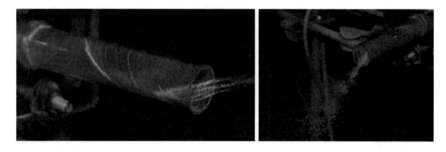

The flow was visualized by introducing tracer particles (baby powder) at 5 atm. On the left side of the device, air comes out at 14 °C and, on the right side at 25 °C. The photographs were taken at a room temperature of 20.5 °C.

It is possible to observe that air flows in a helical trajectory along the tube.

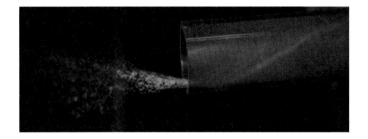

In this case the temperature difference achieved was of 11 °C between the ends of the Ranque-Hilsch tube.

Acknowledgments The authors acknowledge support from DGAPA UNAM through project PAPIIT IN119509.

For information: Carlos Echeverría Arjonilla, carlosea1982@gmail.com, David Porta Zepeda, alviond@gmail.com.

Splashing of Solid Spheres Impinging in Various Fluids

Sergio Valente Gutierrez Quijada, Martha Yadira Salazar Romero and Catalina Elizabeth Stern Forgach

A solid object with spherical symmetry splashes when it impinges on a liquid surface. A cavity is formed and structures with fancy shapes appear surrounding it. As the sphere sinks, it is surrounded by small bubbles, and once it reaches the bottom a jet is ejected as a consequence of the conservation of energy.

Second Place of the Gallery of Fluids, XVII National Congress of the Fluid Dynamics Division of the Mexican Physical Society, Guadalajara, Jalisco, Mexico, November 8–11, 2011.

S. V. G. Quijada (✉) · M. Y. S. Romero · C. E. S. Forgach
Facultad de Ciencias, Universidad Nacional Autónoma de México,
Ciudad Universitaria 3000, Col. Copilco Universidad,
Del. Coyoacán, 04360 Mexico, D.F., Mexico
e-mail: valentegq@hotmail.com

M. Y. S. Romero
e-mail: yayafisica@gmail.com

C. E. S. Forgach
e-mail: catalina@ciencias.unam.mx

J. Klapp et al. (eds.), *Fluid Dynamics in Physics, Engineering and Environmental Applications*, Environmental Science and Engineering,
DOI: 10.1007/978-3-642-27723-8_51, © Springer-Verlag Berlin Heidelberg 2013

On the left, as the sphere enters a container with water, a cavity in the shape of a cup is formed. The photograph in the middle shows that the cup breaks and a layer of bubbles surrounds the spheres. The image on the right shows a very energetical jet ejected from the bottom.

The structures are quite different when the sphere impinges in silicon oil. The image on the left shows a bigger cavity, both wider and taller. In the middle image, the sphere has separated from the top cavity. Finally, on the right, and inverse crown is formed and a jet, that seems less energetical than in the previous case, is ejected.

Acknowledgments The authors acknowledge support from DGAPA UNAM through project PAPIME PE104907.

Bubbles in Isotropic Homogeneous Turbulence

Ernesto Mancilla, Roberto Zenit, Gabriel Ascanio and Enrique Soto

An experimental study of the bubble deformation in an isotropic homogeneous turbulent flow field was carried out. It is of a fundamental importance to understand the rate of coalescence and breakup in two phase disperse flows. The size of bubbles or droplets has a significant impact on the heat and mass transfer rates in such flows (Risso and Fabre 1998). Several mechanisms has been identified: capillary instabilities, turbulence, to name a few. The turbulent flow field was characterized by Particle Image Velocimeter (PIV) measurements and the bubble dynamics was recorded with a high speed camera both measurements were taken simultaneously. The deformation process was characterized by quantifying bubble aspect ratio at each instant of the process and by measuring the local turbulent Reynolds number at a distance of the diameter size around the bubble. The deformation process was governed by the gradient of turbulent stresses that accelerate some parts of the bubbles surfaces, which induces a redistribution of internal flow and causes a pressure difference inside the bubble which lead to its eventual breakup when this try to recover its original shape.

Third Place of the Gallery of Fluids, XVII National Congress of the Fluid Dynamics Division of the Mexican Physical Society, Guadalajara, Jalisco, Mexico, November 8–11, 2011.

E. Mancilla · R. Zenit (✉)
Instituto de Investigaciones en Materiales, Circuito Exterior, Cd. Universitaria, Coyoacán, 04510 Mexico, D.F., Mexico
e-mail: zenit@unam.mx; faermara@hotmail.com

G. Ascanio · E. Soto
Centro de Ciencias Aplicadas y Desarrollo Tecnológico, Circuito Exterior, Cd. Universitaria S/N, 04510 Mexico, D.F., Mexico
e-mail: gabriel.ascanio@ccadet.unam.mx

E. Soto
e-mail: enrique.soto@ccadet.unam.mx

J. Klapp et al. (eds.), *Fluid Dynamics in Physics, Engineering and Environmental Applications*, Environmental Science and Engineering,
DOI: 10.1007/978-3-642-27723-8_52, © Springer-Verlag Berlin Heidelberg 2013

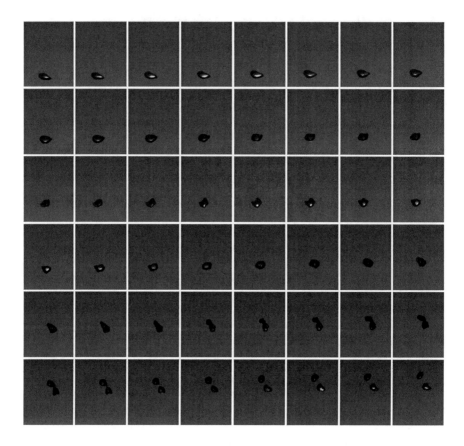

The sequence of images shows a typical bubble undergoing a break up process. The time step between frames is 5 μs; the initial diameter of the bubble is 3 mm. The flow is characterized by $Re_\lambda = 0.623$, $We_t = 0.04$. It can be observed that the bubble undergoes significant deformation and oscillation prior to the actual breakup. We have identified the critical conditions for this process to occur.

Reference

Risso F, Fabre J (1998) Oscillation and breakup of a bubble immersed in a turbulent field. J Fluid Mech 372:323–355

CDF on Graphic Cards

C. Malaga, J. Becerra, C. Echeverría and F. Mandujano

Abstract Examples of 2 and 3D simulations using graphi cards. The equations of motion were solved using a Lattice Boltzmann Method and a semi-lagrangian method.

Examples of 3D flows simulated with Lattice-Boltzmann Methods (LBM). Top left: flow past an airfoil shaped obstacle (3145728 nodes) with streamlines showing the wing tip vortex formation. Top right: oscillatory flow at the exit of a tube at Womersley's number 14.35 (4194304 nodes), streamlines show a vortex ring. Bottom: flow past a sphere at Reynold's number 240 (1048576 nodes) when the wake becomes unsteady, streamlines show vortical structures shed from the sphere.

Thermal-LBM simulation of the convective flow on a cylinder of radius four times its height (1199744 nodes). The cylinder is heated from the bottom and cooled from the top, Rayleigh number is 2×10^4. Top: isothermal surfaces, a sector has been removed to show the internal structure. Bottom: the velocity vector field, stream lines and a surface of constant velocity magnitude are shown in different sectors of the cylinder.

Non-linear shallow water simulation of a wave tank (65536 nodes). The free surface is shown. A perturbation of the bottom surface at the middle of the channel produces the wave dispersion. A dissipative (viscous) term is included at the end of the channel to avoid reflections.

C. Malaga (✉) · J. Becerra · C. Echeverría · F. Mandujano
Facultad de Ciencias, Ciudad Universitaria 3000, Col. Copilco Universidad,
04360 Del. Coyoacán, D.F., Mexico

J. Klapp et al. (eds.), *Fluid Dynamics in Physics, Engineering and Environmental Applications*, Environmental Science and Engineering,
DOI: 10.1007/978-3-642-27723-8_53, © Springer-Verlag Berlin Heidelberg 2013

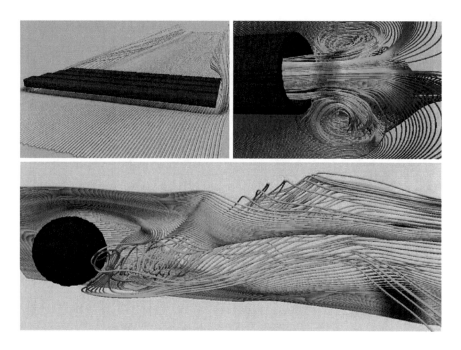

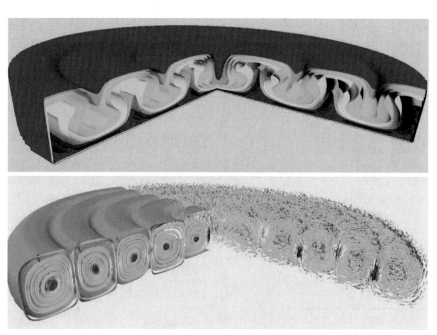

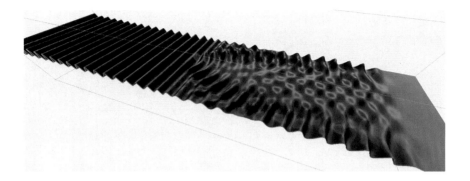

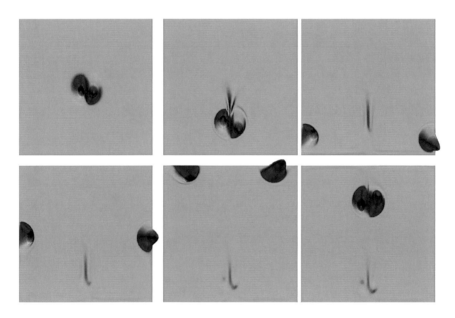

An inviscid 2D vortex dipole moving in a square box. A sequence of different times of the vorticity surface evolution is shown.

Printed by Publishers' Graphics LLC
CAMZ130908.20.05.103